Wave Propagation in Random Media (Scintillation)

Invited papers of a conference held
3–7 August 1992
University of Washington
Seattle, Washington

Support

Support for this meeting has been provided by the

U.S. Army Research Office

National Science Foundation

International Commission for Optics

Participating Organizations

Acoustical Society of America (ASA)

European Optical Society

Institute of Physics (IOP)

International Commission for Optics (ICO)

International Union of Radio Sciences (URSI)

Optical Society of America (OSA)

SPIE—The International Society for Optical Engineering

Cover: Figure 1 from "Astronomical speckle masking, speckle spectroscopy, and long-baseline interferometry," K.-H. Hofmann, G. Weigelt, p. 188 this publication.

Wave Propagation in Random Media (Scintillation)

V. I. Tatarskii
University of Colorado, CIRES/WPL, NOAA

A. Ishimaru
University of Washington

V. U. Zavorotny
University of Colorado, CIRES/WPL, NOAA

Editors

SPIE PRESS

Copublished by

SPIE—The International Society for Optical Engineering
Bellingham, Washington USA

Institute of Physics Publishing
Bristol and Philadelphia

Library of Congress Cataloging-in-Publication Data

Wave propagation in random media (scintillation) / V. I. Tatarskii, A. Ishimaru, V. U. Zavorotny, editors.
p. cm.
"Invited papers of a conference held 3-7 August 1992, Seattle, Washington."
Includes bibliographical references and index.
ISBN 0-8194-1062-4
1. Radio waves—Scintillation—Congresses. I. Tatarskii, V. I. (Valer 'ian Il 'ich) II. Ishimaru, Akira, 1928-
III. Zavorotny, V. U.
QC676.7.S35W39 1993
531' . 1133--dc20 93-14591
CIP

SPIE—The International Society for Optical Engineering
ISBN 0-8194-1062-4

Institute of Physics Publishing
ISBN 0 7503 0292 5

Copublished by
SPIE—The International Society for Optical Engineering
P.O. Box 10, Bellingham, Washington 98227-0010 USA
Telephone 206/676-3290 (Pacific Time) Fax 206/647-1445

Institute of Physics Publishing
Techno House
Redcliffe Way, Bristol, BS1 6NX, England

U.S. Editorial Office:
Institute of Physics Publishing, The Public Ledger Building, Suite 1035,
Independence Square, Philadelphia, PA 19106 USA

Printed in the United States of America.

Table of Contents

SECTION 3 Reconstruction of Source and Medium Parameters Using Scintillation

SECTION 4 Theory of Wave Scintillation and Its Applications

SECTION 5 New Theoretical Approaches in Wave Propogation in Random Media

Organizing Committee

Chairmen

A. Ishimaru
University of Washington, USA

V. I. Tatarskii
University of Colorado, CIRES/WPL, NOAA, USA

V. A. Banakh
Institute of Atmospheric Optics, Tomsk, Russia

M. J. Beran
Tel Aviv University, Israel

B. Beland
Hanscom Air Force Base, USA

L. Bissonnette
Defence Research Establishment, Valcartier, Canada

G. Bourgois
Observatoire de Paris, Meudon, France

P. Chavel
Institut d'Optique/CNRS, Orsay Cedex, France

M. I. Charnotskii
University of Colorado, CIRES/WPL, NOAA, USA

S. F. Clifford
Wave Propagation Lab, NOAA, USA

J. Codona
AT&T Bell Labs, USA

W. A. Coles
University of California, San Diego, USA

A. Consortini
University of Florence, Italy

T. J. Cornwell
National Radio Astronomy Observatory, USA

J. C. Dainty
Imperial College of Science and Technology, London, UK

R. Dashen
University of California, San Diego, USA

L. Durvasula
DARPA/DSO, USA

L. M. Erukhimov
Scientific Research Radiophysics Institute, N. Novgorod, Russia

R. G. Frehlich
University of Colorado, USA

S. M. Flatté
University of California, Santa Cruz, USA

Organizing Committee

W. Flood
Army Research Office, USA

S. Hanson
Riso National Laboratory, Denmark

R. J. Hill
Wave Propagation Laboratory, NOAA, USA

E. Jakeman
Royal Signals and Radar Establishment, UK

D. Lynch
Aerospace Corporation, USA

R. Manning
NASA, USA

W. B. Miller
U.S. Army Atmospheric Sciences Laboratory, USA

V. E. Ostashev
Acoustic Institute, Russian Academy of Sciences, Russia

B. J. Rickett
University of California, San Diego, USA

J. C. Ricklin
U.S. Army Atmospheric Sciences Laboratory, USA

C. L. Rino
Vista Research, Inc., USA

F. Roddier
University of Hawaii, USA

G. C. Rousset
ONERA, Chatillon Cedex, France

V. I. Shishov
Lebedev Physical Institute, Moscow, Russia

R. Spindel
Applied Physics Lab, University of Washington, USA

L. Tsang
University of Washington, USA

V. Thiermann
Forschungsinstitut für Optik, Tubingen, FRG

D. Winebrenner
Applied Physics Lab, University of Washington, USA

R. Woo
Jet Propulsion Laboratory, USA

K. C. Yeh
University of Illinois, USA

V. U. Zavorotny
University of Colorado, CIRES/WPL, NOAA, USA

Preface

In the past several years, there has been a surging interest in wave propagation in random media and an increasing awareness that waves in random media encompass many diverse fields such as atmospheric optics, ocean acoustics, radio physics, astronomy, plasma physics, and condensed-matter physics. In recognition of this interest and recent developments, the first International Meeting on Wave Propagation in Random Media was held in Seattle in August 1992, cochaired by Valerian I. Tatarskii and Akira Ishimaru, with Rod Frehlich as Executive Secretary. The purpose of the meeting was to foster communication and exchange of ideas relating to problems of wave propagation in continuous random media and to promote a multidisciplinary approach to research in scintillation by bringing together the optics, radio, and acoustics communities in this endeavor. The conference attracted two hundred scientists from thirteen countries, including thirty from the former Soviet Union.

All the invited review papers were presented orally, and most of these papers are included in this book. In addition, 130 papers presenting new results were exhibited as posters, and these will be published in *Waves in Random Media,* Institute of Physics Publishing, United Kingdom.

The invited review papers were organized into sessions on scintillation phenomena, statistical description of scintillation, imaging through turbulence, theory, and remote sensing using scintillation. The coverage of the topics is diverse, including optical scintillation in the atmosphere; acoustic fluctuations in the ocean; and radio waves through the ionosphere, interstellar plasmas, planetary atmosphere, and solar wind. Theoretical methods include extended Huygens-Fresnel principle, asymptotic analysis for solutions of moment equations, path-integral techniques, two-scale expansion, multiscreen analysis, and extensive numerical simulations. Also discussed are imaging through turbulence, speckle interferometry, adaptive optics, optical remote sensing, and random gravitational lenses. Throughout the sessions and the panel discussions, it became clear that great progress has been made in wave propagation in random media in many diverse fields. However, it also became evident that further investigations are needed in several areas. Numerical simulations have been recognized as an important benchmark. Intermittency, power law and other spectra, pulse propagation, beam waves, double passage and enhanced backscattering, higher moments, correlations in space and time, spatial and temporal frequencies, and anisotropic random media are among those topics that require further intensive study. Also noted is the need to relate theories to the real atmosphere and ocean for useful applications.

All in all, it was a successful international meeting where scientists and engineers from different disciplines came together to discuss the common theme of wave propagation in random media. This is also an indication that the field of waves in random media has become an important scientific discipline, and it is hoped that the scientific interactions exhibited in this conference will continue to contribute to the advancement of the field. It is also hoped that future conferences will include topics on waves in discrete scatterers and rough surfaces.

In this book, we depart from the original arrangement of the papers for the meeting and present the material in an order more convenient for reading and understanding.

The first section provides an introduction to scintillation phenomena, including theoretical ideas, historical reviews, and classical experimental results for various types of natural media such as the atmosphere, ionosphere, interplanetary plasma, and oceans. Section 2 gives a more detailed review of scintillation measurements of various media and their interpretations based on recent results. Section 3 contains papers whose main goal is to obtain information about the spatial structure of sources or medium characteristics from the scintillating signals. The first three papers are devoted to the problem of image restoration using three principally different approaches, and the other three deal with the remote sensing of the turbulent parameters of such different media as a terrestrial atmosphere or interplanetary plasma.

The next two sections address the theory of wave propagation in random media. Section 4 starts with a review of classical theoretical results concerning the solution of the fourth moment equation, using the perturbation theory and asymptotic methods. The following papers give an idea of how this theory and other similar techniques can be applied to more specific problems such as pulse and beam wave propagation, double passage of a beam through an extended medium, and even propagation in space, taking into account the gravitational effects. Section 5 demonstrates the attempts to improve conventional theory. The authors of this section emphasize the development of more powerful analytical and computational methods that will allow expansion of the applicability of theoretical methods for the description of the diverse and still puzzling scintillation phenomena.

The financial support given by the U.S. Army Research Office, the National Science Foundation, and the International Commission for Optics is gratefully acknowledged. We would also like to thank the following participating organizations for their support: ASA, European Optical Society, ICO, IOP, OSA, SPIE, URSI, U.S. Army Atmospheric Science Laboratory, and NOAA Wave Propagation Laboratory. We would like to express our deep gratitude, as well, to Rod Frehlich, Executive Secretary of the Scintillation Meeting, for his invaluable support during the editing of the book.

V. I. Tatarskii
A. Ishimaru
V. U. Zavorotny

Section 1

Scintillation Phenomena

REVIEW OF SCINTILLATION PHENOMENA

V.I. Tatarskii

Cooperative Institute for Research in Environmental Sciences,
University of Colorado / NOAA / Wave Propagation Laboratory,
Boulder, Colorado 80309-0216

1.Introduction

Many different types of waves exist: acoustic waves, electromagnetic (visible and radio) waves, mechanical waves in solid bodies (including seismic waves), magnetohydrodynamical waves in plasma, probability waves in quantum physics, and so on. For these different types of waves we can observe many similar phenomena such as interference, diffraction and refraction, scattering, and phase and intensity fluctuations. As a striking example of similar phenomena in quite different branches of physics, we can mention the backscattering enhancement of waves in random media and Anderson weak localization in solid state physics. Because the similar phenomena have similar physical treatments, we can use the simplest really visual consideration: the optics of inhomogeneous media.

We can usually distinguish between regular and random inhomogeneities. Examples of regular inhomogeneities are the mean pressure and temperature profiles in the atmosphere, the salinity profile in the ocean, and the mean electron density profile in the ionosphere. All these profiles result in corresponding profiles of refractive index for electromagnetic waves, sound, plasma waves, and so on. In this review we consider only the influence of random inhomogeneities on wave propagation, although sometimes the mean profile has an essential influence on fluctuation phenomena (in the case of waveguide, for instance).

Fluctuations of the parameters of a medium practically always occur. They can be caused, for example, by thermal fluctuations, by turbulence, and by different instabilities in plasma. For the problem of wave propagation in the atmosphere, oceans, ionosphere, and space plasma, the most essential are usually fluctuations of refractive index, caused by turbulent fluctuations of the following:

Temperature and humidity (for electromagnetic waves in the troposphere).
Velocity and temperature (for sound waves in the atmosphere).
Magnetic field and electron density (for electromagnetic waves in the ionosphere and space).
Temperature, velocity,and salinity (for sound waves in the ocean).

Different practical applications of the theory of wave fluctuations are essential:
(1) Coordinate and velocity measurement accuracy by radars, lidars, and sonars.
(2) Intensity fluctuations of laser beams, stars, radio sources, and sound sources.

(3) Phase fluctuations and related phenomena: spatial coherence, antenna gain, and angular resolution.
(4) Spectrum transformation in a turbulent medium.
(5) Propagation of radio waves beyond the horizon.
(6) Information loss caused by turbulent fluctuations.
(7) Thermal blooming in a random medium.
(8) Remote sensing of medium and source parameters (in the atmosphere, ocean, earth, and space).

2. Turbulent refractive index fluctuations

The refractive index $n(\mathbf{r})$ in a turbulent medium is a random function depending on the position $\mathbf{r}$ and time. We can describe the spatial structure of the refractive index by the correlation function $B_n(\mathbf{r},\mathbf{r}') = <\tilde{n}(\mathbf{r})\tilde{n}(\mathbf{r}')>$. The subscript n denotes that this correlation function describes the refractive index fluctuations. Here, $\tilde{n} = n - <n>$ is the deviation of the mean value. In turbulence theory the *structure* function $D_n(\mathbf{r},\mathbf{r}')$ is usually used:

$$D_n(\mathbf{r},\mathbf{r}') = <(\tilde{n}-\tilde{n}')^2> .$$

These two functions are related by the formula

$$D_n(\mathbf{r},\mathbf{r}') = B_n(\mathbf{r},\mathbf{r}) + B_n(\mathbf{r}',\mathbf{r}') - 2B_n(\mathbf{r},\mathbf{r}').$$

In case of statistical *homogeneity* all the statistical characteristics do not depend on position, but only on the relative position. In this case,

$$D_n(\mathbf{r}-\mathbf{r}') = 2[B_n(0) - B_n(\mathbf{r}-\mathbf{r}')],$$

where

$$D_n(\mathbf{r}-\mathbf{r}') = D_n(\mathbf{r},\mathbf{r}'), \qquad B_n(\mathbf{r}-\mathbf{r}') = B_n(\mathbf{r},\mathbf{r}').$$

The typical picture of these functions is shown in Fig. 1.
We can usually consider the function $D_n(\mathbf{r}-\mathbf{r}')$ as *isotropic* for small enough values of the vector $\mathbf{r}-\mathbf{r}'$. This means that D_n does not depend on the direction of $\mathbf{r}-\mathbf{r}'$, but only on its value $\rho = |\mathbf{r}-\mathbf{r}'|$.

The function $D_n(\rho)$ is quadratic for $\rho << l_0$. The *inner scale* l_0 shows the scale of smallest inhomogeneities. For $\rho << l_0$ $n(\mathbf{r})$ is a random linear function. For very large Reynolds numbers the structure function $D_n(\rho)$ has, in the range $l_0 << \rho << L_0$, the form

$$D_n(\rho) = C_n^2\rho^{2/3}, \qquad l_0 << \rho << L_0.$$

L_0 is the *outer scale* of turbulence, and it describes the scale of the largest inhomogeneities. For $\rho >> L_0$ the correlation function $B_n(\rho) \cong 0$ and the structure function $D_n(\rho) \cong 2\sigma_n^2 = \text{Const}$. Here, $\sigma_n^2 = <\tilde{n}^2>$ is the variance of the refractive index. The inhomogeneities with the scales $\rho >> L_0$ are statistically independent.

In stratified media (atmosphere, ocean) L_0 may depend on direction. We can match the function $D_n(\rho)$ in the points $\rho = l_0$ and $\rho = L_0$. If $D_n(\rho) = M\rho^2$ for $\rho < l_0$ and $D_n(\rho) = C_n^2\rho^{2/3}$ for $\rho > l_0$, matching in the point $\rho = l_0$ leads to the formula $M = C_n^2 l_0^{-4/3}$. For the point $\rho = L_0$ we obtain in the same way $C_n^2 L_n^{2/3} = 2\sigma_n^2$. Thus, the representation of $D_n(\rho)$ in different ranges has the form:

$$D_n(\rho) = \begin{cases} C_n^2 l_0^{-4/3} \rho^2 & \text{for } \rho << l_0, \\ C_n^2 \rho^{2/3} & \text{for } l_0 << \rho << L_0, \\ C_n^2 L_0^{2/3} & \text{for } \rho >> L_0. \end{cases}$$

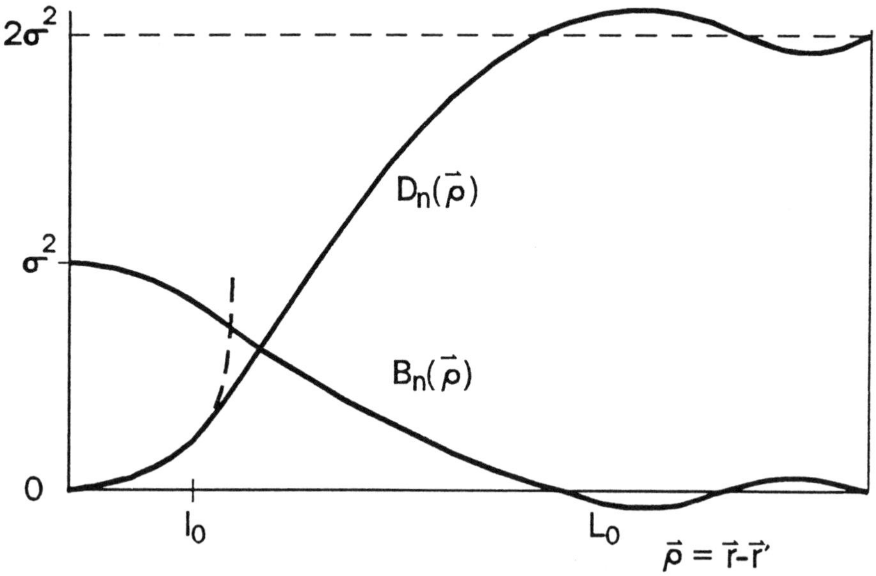

Fig. 1. Typical behavior of the correlation $B_n(\rho)$ and the structure $D_n(\rho)$ functions of the refractive index in a turbulent medium

The typical value of l_0 in the atmosphere is about 1 cm. The value of L_0 in the atmosphere depends on elevation. For the boundary layer of the atmosphere $L_0 = 0.4z$, z being the height. For the free atmosphere L_0 is about 100 m in the vertical direction and several kilometers in the horizontal direction. In the ionosphere l_0 is of the order of several tens of centimeters. In the ocean the Reynolds numbers are less than in the atmosphere and stratification is more essential.

The three-dimensional spectrum, corresponding to $D_n(\rho) \propto \rho^{2/3}$, has the form $\Phi(\kappa) = AC_n^2 \kappa^{-11/3}$, where $A \cong 0.033$ and κ is a wavenumber.

3. Scattering

In considering the scattering of waves, it is convenient to represent the distribution of refractive index as a sum of sinusoidal components (Fourier theorem). We consider first the scattering of waves by one of these components with distance d between the corresponding planes. From Fig. 2 we obtain the following relations:

$$\alpha = \frac{\pi}{2} - 2\varphi; \qquad l_1 = \frac{d}{\cos\varphi}; \qquad l_2 = l_1 \sin\alpha = l_1 \cos 2\varphi;$$

$$s = l_1 + l_2 = l_1(1 + \cos 2\varphi) = \frac{d}{\cos\varphi} \times 2\cos^2\varphi \ = \ 2d\cos\varphi.$$

If $s = 2d\cos\varphi = \lambda$ (or 2λ, $3\lambda, \ldots$), the scattering in the corresponding direction is essential, because for this direction all the scattering waves have the same phase and enhance each other.

If $\theta = \pi - 2\varphi$ is the scattering angle, then $\cos\varphi = \sin(\theta/2)$ and the last relation takes the form of the Bragg condition:

$$2d\sin(\theta/2) = \lambda.$$

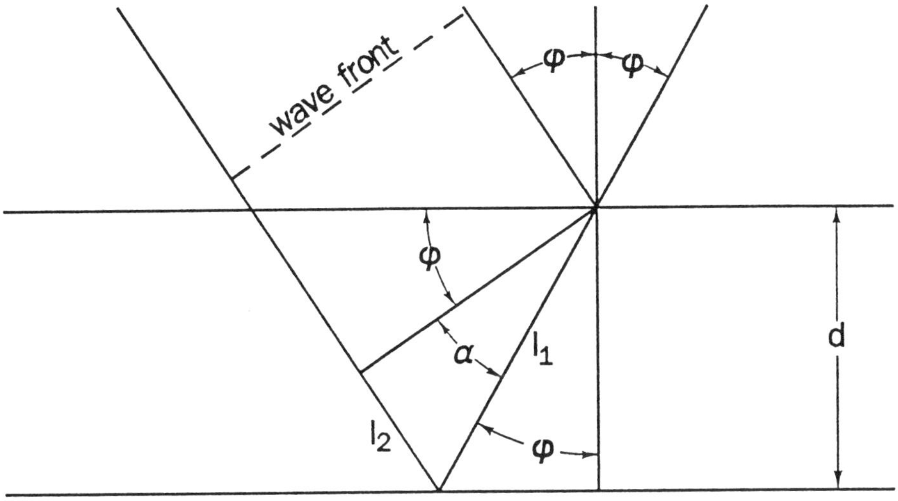

Fig. 2. Scattering geometry

We can express the same condition in terms of spatial frequency $K = 2\pi/d$ and wave number $k = 2\pi/\lambda$:

$$K = 2k\sin(\theta/2).$$

If the scattering from a *single* plane is *very small*, we can obtain a significant result only under this condition. This means that if we represent the inhomogeneities as a sum of sinusoidal gratings, only one of them, satisfying the Bragg condition, can be the origin of the scattering for given λ, θ.

In the statistical case this means that the scattering cross section σ is determined by the spatial spectrum of inhomogeneities $\Phi(\kappa)$ with the argument K (see Fig. 3):

$$\sigma \propto \Phi(2k\sin(\theta/2)).$$

This relation is the basis of the theory of wave scattering and its applications. For example, if we measure $\sigma = \sigma(\theta)$ we can estimate the spectrum Φ.

4. Phase fluctuations

4.1 Total phase shift fluctuations

The largest inhomogeneities, with the scale L_0 characterized by the largest values of refractive index fluctuations and their contributions to the phase shift, are the most essential. Because of this we can consider the contribution of only these largest inhomogeneities to the whole phase shift. For small-angle fluctuations the contribution of the single i- th inhomogeneity to the phase shift is equal to $\delta S_i = k\delta n_i L_0$. Because the mean value $< \delta n_i > = 0$, the same is valid for δS fluctuations: $< \delta S_i > = 0$. For the mean square of δS_i we obtain

$$< (\delta S_i))^2 >= k^2 L_0^2 < (\delta n_0)^2 >= k^2 L_0^2 \sigma_n^2.$$

The total phase shift ΔS is equal to sum of all contributions:

$$\Delta S = \sum \delta S_i; \qquad < \Delta S >= 0.$$

The mean square of ΔS is given by

$$< \Delta S^2 >= \sum_i \sum_j < \delta S_i \delta S_j > .$$

But the inhomogeneities with the scale L_0 are statistically independent for $i \neq j$. Thus only the terms with $i = j$ are nonzero:

$$< \Delta S^2 >= \sum < (\delta S_i)^2 > .$$

The number N of terms in the sum is given by the ratio $N = L/L_0$. Thus, we obtain the simple result

$$< \Delta S^2 >= N < (\delta S_i)^2 > \propto k^2 \sigma_n^2 L_0 L.$$

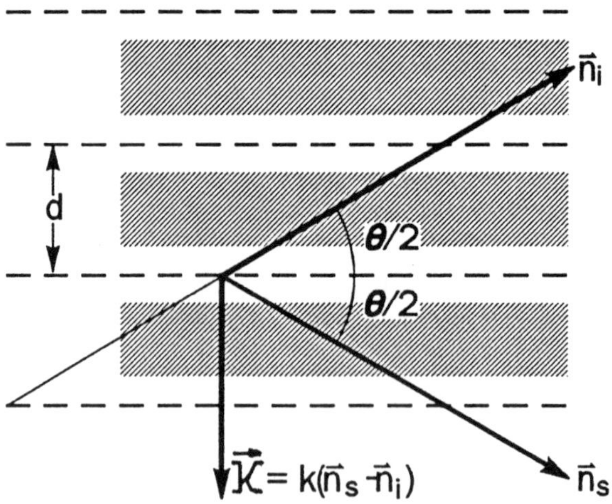

Fig. 3. Definition of scattering vector.

Because we obtained this formula using a qualitative approach, there is no numerical coefficient in it. We assert that the mean square of the total phase shift fluctuations is proportional to the distance, to the outer scale of inhomogeneities, and to the variance of refractive index.

4.2 Transverse phase difference.

To solve the problem of transverse phase differences, we discriminate the three types of inhomogeneities according to their scales l (see Fig. 4):

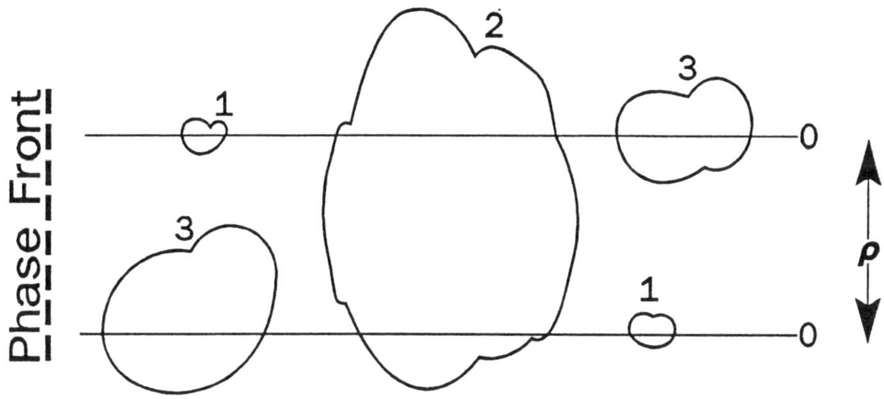

Fig. 4. Three types of inhomogeneities with respect to separation of observation points.

(1) $l << \rho$. Usually these inhomogeneities are not essential, because their contribution, $< \tilde{n}^2(l) >$ is small for small l.
(2) $l >> \rho$. These irregularities are not essential because they contribute the same phase shift to both rays and produce no phase difference.
(3) $l \simeq \rho$. These irregularities are most important and we will consider only their influence. The single i-th inhomogeneity with the scale ρ creates the phase difference δS_i:

$$\delta S_i = k\rho(n' - n'').$$

The mean value and the variance of δS_i are given by

$$< \delta S_i >= 0, \quad < \delta S_i^2 >= k^2\rho^2 < (n' - n'')^2 > .$$

The number N of inhomogeneities of the scale ρ is given by $N = L/\rho$, and the variance of the total phase difference $\Delta S = \sum \delta S_i$ is equal to

$$< \Delta S^2 >= N < (\delta S_i)^2 >,$$

or

$$< \Delta S^2 > = D_S(\rho) = \text{Const } k^2 L\rho D_n(\rho),$$

where $D_n(\rho) =< (n' - n'')^2 >$ is the *structure function* of refractive index. Because $< \Delta S^2 >$ is really the phase structure function, we use the notation $D_S(\rho)$ for this quantity.

For turbulence, $D_n(\rho) = C_n^2 \rho^{5/3}$ in the region $l_0 << \rho << L_0$ and

$$D_S(\rho) = <\Delta S^2> = \text{Const } C_n^2 k^2 L \rho^{5/3}.$$

The case $\rho << l_0$ requires special consideration. For this case the most essential inhomogeneities have the scale l_0, according to Fig. 5.

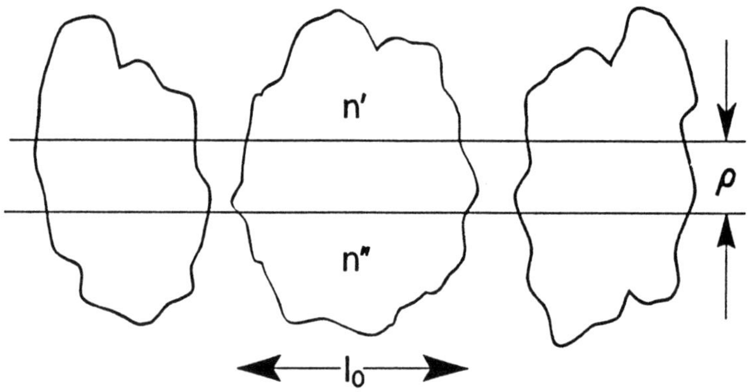

Fig. 5. Typical picture of inhomogeneities in the case of small separation with respect to inner scale of turbulence.

We can calculate $D_S(\rho)$ in a similar way: $\Delta S_i = k(n' - n'')l_0$, $N = L/l_0$, and $D_S(\rho) \propto k^2 L l_0 D_n(\rho)$ for $\rho << l_0$.
Taking into account the form of $D_n(\rho)$ for $\rho << l_0$, we obtain for $\rho << l_0$:

$$D_n(\rho) = \text{Const } C_n^2 l_0^{-1/3} L k^2 \rho^2.$$

4.3 Some phenomena related to the phase fluctuations

4.3.1 Angle of arrival

Angle of arrival α measured by an interferometer with base ρ is related to phase difference ΔS and to optical path difference Δl by these relations (see Fig. 6):

$$k\Delta l = \Delta S; \qquad \alpha = \Delta l/\rho = \Delta S/(k\rho);$$
$$<\alpha^2> = <\Delta S^2>/(k\rho)^2 = D_S(\rho)/(k\rho)^2.$$

Thus we obtain

$$<\alpha^2> \propto \begin{cases} C_n^2 L l_0^{-1/3}, & \text{for } \rho << l_0, \\ C_n^2 L \rho^{-1/3}, & \text{for } \rho >> l_0. \end{cases}$$

Correct angle-of-arrival fluctuations correspond to $\rho << l_0$. But if the aperture diameter ρ, used for the α measurement, is bigger than l_0, the resulting fluctuations of α will be less, as a result of averaging fluctuations over the aperture.

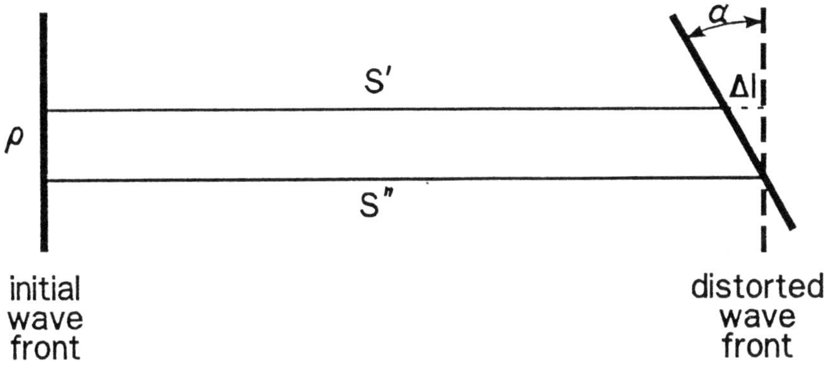

Fig. 6. Relation between phase difference ΔS and angle of arrival.

4.3.2 Mutual coherence function

In the case considered, the path length $L >> \rho$, thus the phase difference consists of the large number of statistically independent terms. That is why ΔS is the Gaussian random variable. Let us consider the field $U = Ae^{iS}$. Its coherence function is determined by the relation

$$\Gamma(\mathbf{r}', \mathbf{r}'') = < U'U''^{*} > .$$

In many cases we can neglect the amplitude fluctuations in comparison with the phase fluctuation. In this case

$$\Gamma(\mathbf{r}', \mathbf{r}'') = A^2 < \exp[i(S' - S'')] > .$$

The value $\xi = S' - S''$ has a Gaussian probability distribution. Thus,

$$< e^{i\xi} > = \frac{1}{\sqrt{2\pi}\sigma} \int_{-\infty}^{\infty} \exp\left\{ - \frac{(\xi - < \xi >)^2}{2\sigma^2} + i\xi \right\} d\xi .$$

Here, $< \xi > = 0$, and

$$\sigma^2 = < [\xi - < \xi >]^2 > = < [S' - S'']^2 > = D_S(\mathbf{r}' - \mathbf{r}'').$$

After calculating this well-known integral, we obtain the result

$$< \exp(i\xi) > = \exp(-\sigma^2/2).$$

Using this result, we obtain for the mutual coherence function

$$\Gamma(\mathbf{r}', \mathbf{r}'') = A^2 \exp\{-(1/2) D_S(\mathbf{r}' - \mathbf{r}'')\}.$$

Let us consider the physical meaning of Γ. If the distance $\rho = |\mathbf{r}' - \mathbf{r}''|$ is small enough, so that $D_S(\rho) << 1$, then we can consider the wave front as regular. But if $D_S(\rho) \simeq 1$, than $\Gamma(\rho) \simeq 0.6\Gamma(0)$ and the field fluctuations become statistically independent. Thus, the value ρ_c, which is determined by the relation $D_S(\rho_c) = 1$, has the physical sense of a coherence radius of the field. If the aperture diameter D is small in comparison with ρ_c, it is possible to realize its diffractional limit. In the opposite case it is impossible.

If the coherence radius is big in comparison with the inner scale of turbulence, it determines by the equation

$$C_n^2 k^2 L \rho_c^{5/3} = 1.$$

We obtain from this equation

$$\rho_c \propto (C_n^2 \, k^2 \, L)^{-3/5}.$$

From this expression it follows that coherence radius decreases if the distance L increases. Note that in homogeneous media, coherence radius increases together with distance. The typical value of ρ_c in the atmosphere for visible light is about 10 cm.

5. Intensity fluctuations

First we consider this phenomenon on the basis of ray optics. We can represent the inhomogeneities as a collection of random positive and negative lenses. Let us consider the effect of a single lens (see Fig. 7).

We assume that the plane of observation is located far from the focal point F: $L << F$. Let us calculate the changing of the beam radius. According to Fig. 7, $\alpha = R/F$, $\Delta R = \alpha L = RL/F$, $F = R/\Delta n(R)$. If we substitute F into ΔR, then

$$\Delta R = \frac{RL}{R/\Delta n(R)} = L\Delta n(R).$$

The energy conservation law connects intensity I with the beam's crosssection πR^2:

$$IR^2 = \text{Const}.$$

If R changes by ΔR, then I changes by ΔI and $R^2 \Delta I + 2\, IR\, \Delta R = 0$. Thus,

$$\frac{\Delta I}{I} = -2\frac{\Delta R}{R}.$$

Let us apply this relation to a single random lens. We will not account for the numerical coefficients. Substituting ΔR, we obtain:

$$\frac{\Delta I_i}{I} \propto \frac{\Delta R_i}{R_i} \propto L\frac{\Delta n(R_i)}{R}.$$

The mean value of this expression is zero, because $< \Delta n > = 0$. The mean square is

$$\frac{< (\Delta I_i)^2 >}{I^2} \propto L^2 \frac{< \Delta n^2(R_i) >}{R_i^2}.$$

We know that $< \Delta n^2(R) > \propto R^2$ for $R << l_0$. Thus, in this region the ratio considered does not change if R changes. But $< \Delta n^2(R_i) > \propto R_i^{2/3}$ for $R >> l_0$, and the ratio considered decreases if R_i increases. This means that the smallest inhomogeneities are the most essential. Thus, to calculate the intensity fluctuations we can take into account only the inhomogeneities of the scale l_0.

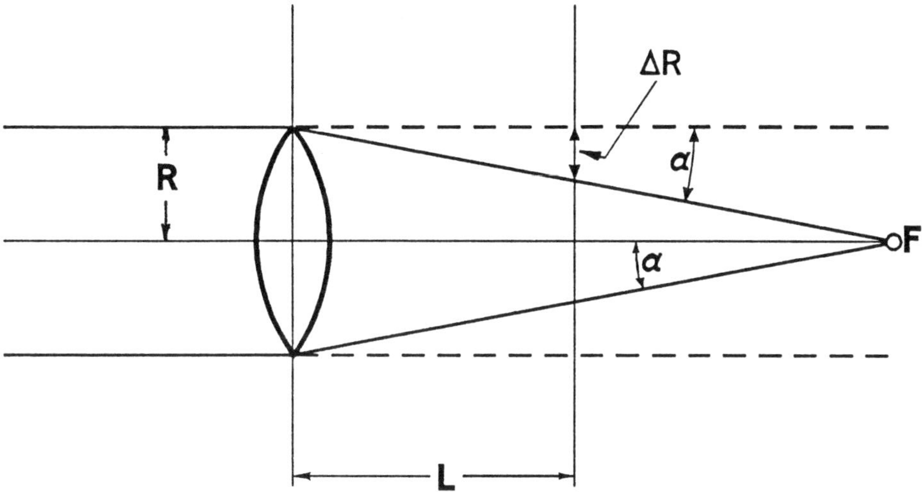

Fig. 7. Geometry of beam focusing.

We obtain the result for the i-th lens:

$$\frac{(\Delta I_i)^2}{I^2} \propto L^2 \frac{D_n(l_0)}{l_0^2}.$$

To obtain the total effect of all random lenses, we have to multiply this expression by the total number of inhomogeneities, $N = L/l_0$. The result is

$$\beta^2 = \frac{< (\Delta I)^2 >}{I^2} \propto D_n(l_0)\ (L/l_0)^3.$$

If we substitute into this the general formula $D_n(l_0) = C_n^2 l_0^{2/3}$, we obtain

$$\beta^2 \propto C_n^2\ l_0^{-7/3}\ L^3.$$

This formula is valid only if geometrical optics is applicable to the problem. We consider now what that means. The well-known condition $\lambda << l_0$ must hold for geometrical optics to be valid. But there exists one more essential condition, $\sqrt{\lambda L} << l_0$. If we consider the shadow of some body at the screen, we can obtain its diffractional spreading, which has the scale $\sqrt{\lambda L}$. The shadow will be similar to the object only if this spreading is small in comparison with the scale of the body.

If $R << \sqrt{\lambda L}$, the diffractional spreading becomes more essential than focusing or defocusing of lenses. For both cases the result is almost the same:

diffractional spreading. This means that lenses with a diameter less than $\sqrt{\lambda L}$ cannot focus or defocus the beam. Thus the smallest lenses that can do this are lenses with a scale $\sqrt{\lambda L}$. To obtain the expression for the mean square of intensity fluctuations, we can simply replace l_0 by $\sqrt{\lambda L}$ in the formula for β^2. The result is

$$\beta^2 \propto D_n(\sqrt{\lambda L})\,(L/\sqrt{\lambda L})^3.$$

For case of turbulence, $D_n(\rho) = C_n^2 \rho^{2/3}$ and the last formula takes the form

$$\beta^2 \propto C_n^2\, k^{7/6}\, L^{11/6}.$$

This formula takes into account diffractional effects. *But it corresponds to the geometrical optics description of large-scale inhomogeneities* ($R > \sqrt{\lambda L}$).

The values

$$\beta^2 = \frac{<(\Delta I)^2>}{I^2} \quad \text{and} \quad \beta_0^2 = 0.31\, C_n^2\, k^{7/6}\, L^{11/6}$$

can be obtained from direct independent measurements of I fluctuations and C_n^2. Values of 2β and $2\beta_0$ are compared in Fig. 8.

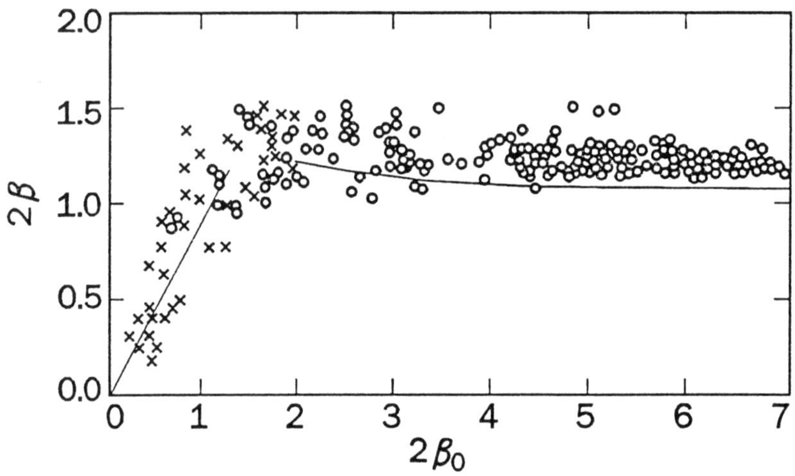

Fig. 8. Comparison between measured values of r.m.s. intensity fluctuations (β) and their estimation on the basis of perturbation theory (β_0).

We can distinguish between two regions in Fig. 8: the region of weak fluctuations, corresponding to $\beta_0 < 0.8$, and the region of strong (saturated) fluctuations, $\beta_0 > 0.8$. In the first region the theoretical values show a good fit with the experimental values. The reason for disagreement between results in the region of strong fluctuations is the supposition $L << F$ used in the theory. The theoretical description of the strong fluctuations is rather complicated. We present here a qualitative explanation of this phenomenon.

We can introduce the distance L_s for which $\beta_0^2 = 1 : C_n^2\, k^{7/6}\, L_s^{11/6} = 1$. This distance separates the regions of weak and strong fluctuations.

Because $C_n^2\, k^{7/6}\, L_s^{11/6} = 1$ and $D_S(\rho_c) = C_n^2\, k^2\, L\, [\rho_c(L)]^{5/3} = 1$, we obtain for the ratio of these two expressions:

$$k^{5/6}\, L\, [\rho_c(L)]^{5/3} / L_s^{11/6} = 1.$$

From this relation we find the expression for coherence radius at a distance L in terms of L_s:

$$\rho_c(L) = (L_s/L)^{3/5}\sqrt{\lambda L_s}$$

and

$$\frac{\rho_c(L)}{\sqrt{\lambda L}} = (L_s/L)^{11/10}.$$

It is clear that $\rho_c > \sqrt{\lambda L}$ in the region of weak fluctuations and $\rho_c < \sqrt{\lambda L}$ in the region of strong fluctuations. From a physical point of view, this means that there exist many independent random secondary sources of a field inside the Fresnel zone. Thus, the Fresnel zone plays no more role in the process of field propagation. The general behavior of $\sqrt{\lambda L}$ and ρ_c as a function of L is presented at the Fig. 9.

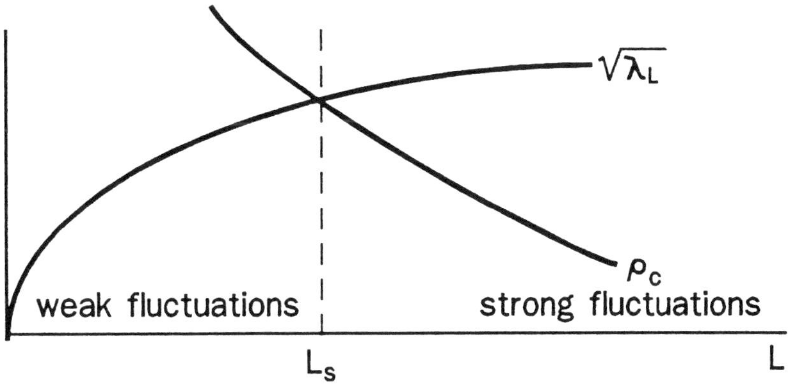

Fig. 9. Dependence of coherence radius ρ_c and the first Fresnel zone radius $\sqrt{\lambda L}$ on the distance L.

Intensity I is defined as $I = uu^*$ where u is an optical field; its mean value is $< I > = < uu^* >$. For a Gaussian distribution it is easy to obtain the second statistical moment of I :

$$< I^2 > = < uuu^*u^* > = 2 < uu^* >^2; \quad < I^2 > - < I >^2 = < uu^* >^2 = < I >^2 .$$

Thus, for $L >> L_s$,

$$\beta^2 = \frac{< I^2 > - < I >^2}{< I >^2} = 1.$$

Statistical properties of the field in the region of strong (saturated) fluctuations are similar to the statistics of thermal radiation (but with a different temporal spectrum), because both of them are Gaussian. The origin of this similarity is the very large number of statistically independent sources in both cases.

6. Mathematical methods in the theory of wave propagation in random media

6.1 Geometrical optics

This is the simplest approach to the quantitative description of our problem. The equations of this method are

$$(\nabla\theta)^2 = n^2(\mathbf{r}), \qquad \mathrm{div}\{[A(\mathbf{r})]^2\,\mathrm{grad}\theta(\mathbf{r})\} = 0.$$

Here, $\theta = S/k$ is the eikonal, and A is the amplitude. First we have to solve the eikonal equation, and then we can determine A. It is impossible to solve the eikonal equation in the general form for arbitrary $n(\mathbf{r})$. Perturbation theory is used in different forms to solve this equation. This approach leads to satisfactory results for phase fluctuations and explains well the different phenomena, mainly related to phase. Geometrical optics approximation is restricted by the following three conditions:

$$(1)\ \lambda << l_0; \quad (2)\ \sqrt{\lambda L} << l_0; \quad (3)\ \beta_0^2 << 1.$$

The two first conditions allow us to use the eikonal equation. The third condition is the consequence of using perturbation theory.

6.2 Perturbation theory

This approach is based on the expansion of the solution of the wave equation

$$\Delta u + k^2 n^2(\mathbf{r}) u = 0$$

in powers of the small quantity $\epsilon(\mathbf{r}) = n^2 - 1$. This approximation can be applied to various situations; only the condition $|\ \delta u\ | << |\ u\ |$ must hold. It is a universal method to describe different *small* effects. Because the perturbation method is also used in geometrical optics, it is often more convenient to apply it directly to the wave equation, rather then to the eikonal equation.

6.3 Parabolic approximation

This approximation takes into account only scattering in the forward direction. The field u is represented in the form

$$u(x,y,z) = \exp(ikz)v(x,y,z),$$

and v is supposed to change slowly in the z-direction. The parabolic equation

$$2ik\frac{\partial v}{\partial z} + \frac{\partial^2 v}{\partial x^2} + \frac{\partial^2 v}{\partial y^2} + k^2[n^2 - 1]v = 0$$

for the smooth amplitude $v(x,y,z)$ corresponds to the Fresnel approximation of the Green function. This equation is also usually solved by the perturbation technique.

6.4 Rytov's method

This is a perturbation method applied to the function $\psi = \log[v(x,y,z)]$. The real and imaginary parts of ψ,

$$\chi = \mathrm{Re}[\psi] = \log[A(x,y,z)], \; S = \mathrm{Im}[\psi],$$

are the logarithm of amplitude and the phase of a wave. This method extends the geometric optics, and makes it possible to consider the diffraction influence on the amplitude and on the phase of a wave. As a perturbation method, it is also restricted by the condition

$$< \chi^2 > << 1.$$

This method is very useful for many applications.

6.5 Markovian approximation

This method *is not a perturbation* method. It uses another small *statistical* parameter: the ratio of the longitudinal correlation length of refractivity fluctuations to the longitudinal correlation length of field fluctuations. It is possible to apply this expansion to the parabolic equation, which neglects the backscattering. This method describes well the region of strong (saturated) fluctuations. But it is applicable only in the case when it is possible to neglect the backscattering.

6.6 Diagram technique

This is a very general method for analyzing the structure of perturbation series. It allows us to summarize the infinite subsequences of perturbation series for the statistical moments of a field. The most essential result of this method is the derivation of the radiative transfer equation on the basis of the wave equation.

6.7 Functional integral method

This method uses a special representation of the exact solution of the wave equation in the form of infinite-fold integral. It is a source of different approximations.

OPTICAL SCINTILLATION OBSERVATION IN RUSSIA

A. S. Gurvich

Institute of Atmospheric Physics, Russian Academy of Sciences

Pyzhevskii, 3, Moscow, 109017, Russia

Abstract

A brief history of Russian experimental studies of atmospheric scintillations during the period 1956-1979 is given. Only observations that have significant meaning for understanding of the phenomenon are discussed.

The observation of scintillations has a long history. Astronomers were the first scientists who saw stellar scintillations and twinkling of star images through telescopes. Sir Isaac Newton was probably the first to give a scientific explanation of these phenomena. He wrote in his famous book "Opticks":

"If the Theory of making Telescopes could at length be fully brought into Practice, yet there would be certain Bounds beyond which Telescopes could not perform. For the Air through which we look upon the Stars, is in a perpetual Tremor; as may be seen by the tremulous Motion of Shadows cast from high Towers, and by the twinkling of the fix'd Stars. But these Stars do not twinkle when viewed through Telescopes which have large apertures. For the Rays of Light which pass through divers parts of the aperture, tremble each of them apart, and by means of their various and sometimes contrary Tremors, fall at one and the same time upon different points in the bottom of the Eye, and their trembling Motions are too quick and confused to be perceived severally. And all these illuminated Points constitute one broad lucid Point, composed of those many trembling points confusedly and insensibly mixed with one another by very short and swift Tremors, and thereby cause the Star to appear broader than it is, and without trembling of the whole. Long Telescopes may cause Objects to appear brighter and larger then short ones can do, but they cannot be so formed as to take away that confusion of the Rays which arises from the Tremors of the Atmosphere. The only Remedy is a most serene and quiet Air, such as may perhaps be found on the tops of the highest Mountains above the grosser Clouds".

It is notable that his recommendations for reducing the influence of turbulence on the image quality in large telescopes have turned out to be fundamental today.

Scientists in Russia began to study atmospheric optical scintillations later than Sir Isaac Newton. As a modern scientific problem it appears in A. M. Obukhov's paper "Effect of weak inhomogeneities in the atmosphere on sound and light propagation." [1] In this paper, Obukhov introduced diffraction in the calculations of light intensity fluctuations using the smooth perturbation method developed by S. M. Rytov[2] to solve the stochastic wave equation. Obukhov also formulated the problem of recovering turbulence characteristics using scintillation measurements.

V. I. Tatarskii developed Obukhov's theory for turbulence with a power law spectrum.[3] Measurements of light scintillations began in 1956. By that time, measurements of turbulent parameters for the atmospheric boundary layer had been developed at the Institute of Atmospheric Physics.[4,5] Light beams were propagated over a uniform flat surface that was free of obstacles. The main features of these measurements was independent determination of C_n^2 from micrometeorological observations and simultaneous measurements of scintillation statistics. Comparisons the theory with the observations were possible for various propagation paths and different meteorological conditions. The results showed that the variance of scintillations depends on the propagation distance L as $L^{11/6}$, which was predicted by Tatarskii's theory, in contrast to the geometrical optics approximation that predicts L^3. The results were published in the Astronomical Journal[6] because at that time, astronomers were interested in light scintillations caused by atmospheric turbulence.

The next experiments, made in 1957, were prepared more carefully. Electronic equipment was capable of the real-time estimation of variance, frequency spectrum, and the probability density function of scintillations.[7] It is necessary to remember that there were no personal computers, and even analogue correlators and spectra analyzers were not commonly used to measure atmospheric turbulence. The experiment consisted of two-point measurements of light intensity fluctuations. The results of these measurements[8] provided good experimental confirmation of all the predictions of Tatarskii's theory. One of the results, which was surprising for many scientists, was that the typical scale of scintillation is defined by diffraction. This unexpected result invalidated different models of the atmospheric inhomogeneities structure that were based on observations of stellar scintillations.

In these measurements, a common filament lamp was used as the light source. The brightness of such a source is small, therefore it was possible to make observations of scintillations only at night (to exclude the daylight background) and at distances not exceeding 2 km. The turbulence at night is relatively weak hence it was possible to measure weak fluctuation of light intensity. At the same time it was clear that the theory of scintillations, based on some methods of perturbation, had a restricted range of applicability. However, at that time there were no reliable theoretical or experimental estimations of this regime. The behavior of scintillations at large values of the parameter $\sigma_T^2 = 1.23 C_n^2 k^{7/6} L^{11/6}$ was still unknown. Measurements at large values of σ_T^2 were very interesting for the experimenters because there were no theoretical predictions of scintillation statistics.

In 1963 use of a lamp with high-pressure mercury-vapor discharge as the light source in 1963 made it possible to start measurements during the day when turbulence was higher by ten times or more than at night. The first results[9] showed that the measured scintillation variance is a function of the parameter σ_T^2, which is obtained from the theory of smooth perturbations. However, quantitative predictions of this theory are correct only for $\sigma_T^2 < 1$. These measurements showed that there exists a knee at $\sigma_T^2 \simeq 1$ and then there is the region of strong fluctuation of light intensity in which the variance changes very slowly. This paper had many defects, but the experimental results showed clearly the existence of two different regimes of scintillations: weak scintillations, which are described by the smooth perturbation methods, and strong fluctuations.

The scintillation measurements were repeated in 1967[10] with improved equipment, which made it possible to correct some defects in the previous measurements and more carefully study strong scintillation. These measurements showed that there exists a region of random focusing located between the regions of weak and strong scintillations. The variance of scintillations has a maximum in this region. However, the finite angular size of the light source prevented a more detailed study of strong scintillations.

More reliable results were obtained using a laser as the light source. To study the principal phenomena of the scintillation in the plane wave limit, a high-quality collimator with a diameter of $50cm$ was used and scintillations were measured in the central region of a wide laser beam propagated over a uniform surface. The direct measurements of the variance and the mean of intensity logarithm[11,12] revealed their dependence on the parameter $\beta_o = 1.23C_n^2 k^{7/6} L^{11/6}$. The coherence and high brightness of the laser source made it possible to use small detectors of diameter $(\lambda L)^{1/2}/10$. These results showed, however, that the limit of the normal distribution of the field in strong scintillation is not approved as was assumed in some theoretical works.

In Russia, the theoretical study of scintillations quickly developed at approximately the same time. Two excellent reviews[13,14] were published in Soviet Physics - Uspekhi, and provided a complete picture of the theory of scintillations. The work of the theoreticians motivated a new interest in the study of the statistical moments of field fluctuations.

Measurements of the second moment of the field began a new period of observations. Solution of the equation for the second moment of the field (the coherence function) has a very simple form for a plane wave. Therefore, it was easy to compare the measurements of the spectra of the coherence function with the theoretical predictions. For such a comparison the model of the turbulence spectra was used, which included the inner scale of turbulence. The results of these measurements showed good agreement between the experimental data and the theory based on the Markovian approximation.[15] The results also showed that measurements of coherence, or observations of phenomena that depend on coherence, make a good foundation for remote sensing of the turbulence. Such methods are independent of the magnitude of scintillations.

Measurements of scintillations in a wide collimated laser beam were began in 1970 using very small detectors, with a diameter of about $(\lambda L)^{1/2}/70$. These measurements showed the saturation of scintillation variance.[16] The measurements of probability distributions showed that normal field approximation was not observed at large values of the parameter β_o^2. The small aperture made it possible to measure the scintillation spectra.[17] Results of the measurements showed that there are two spatial scales at large values of the parameter β_o that are in good agreement with the prediction based on the Markovian approximation.[18]

The high degree of temporal coherence of a laser beam made it possible to measure a four-point coherence function of a complex field.[19] These results are in agreement with predictions based on the Markovian approximation.[20]

At the same time numerical methods of solutions for the fourth moment equation were developed. A comparison of experimental results and numerical solutions for the variance of intensity and spectra of scintillations[21] showed agreement between numerical solutions and experimental data. Thus an era of experimental and theoretical studies was completed.

However, small discrepancies between the measured and calculated values were significant and could not be explained by experimental errors. It is necessary to remember that the modern theory of scintillations is a true physical theory, while scintillation measurements in the atmosphere are geophysical observations. It is almost impossible to control atmospheric turbulence. The interpretation of experimental data is based on some assumptions about spatial turbulence spectra, on statistical homogeneity of the optical paths and on the hypothesis of frozen turbulence. Such assumptions are very difficult to verify and should be studied separately. For these reasons it is very difficult to expect better agreement between the experimental data and the theoretical predictions for the infinite plane wave propagation in turbulent media.

All experimenters are now in strong competition in the study of the basic features of scintillation. This an idea probably comes to every experimenter while reading the paper by Martin and Flatté[22] and seeing the pictures of computer simulation of scintillations. Such computer experiments will probably make it possible to obtain reliable data on the probability distribution or spatial structure of scintillations. Computer experiments have their own defects but these defects are exactly known. Therefore, one can study the effect of these defects.

However, scintillation measurements in the atmosphere, ocean, and interplanetary space constitute a direct contact with reality. Nature may give scientists such riddles which are very pleasant and interesting to solve.

In this paper, only the principal Russian experiments important for understanding the physics of scintillation are mentioned. It is impossible to describe in a short paper all Russian experimental studies of scintillations for different waves in different turbulent media. For example, the applied studies of optical scintillations, carried out basically at the Institute of Atmospheric Optics in Tomsk under the direction of V. E. Zuev, are the saubject of a special monograph.[23] At present, experimental studies of scintillations are continueing in Russia.

Acknowledgment

The author is grateful to Dr. R. Frehlich of the University of Colorado CIRES for suggesting the subject of this paper and for his useful comments.

References

1. A. M. Obukhov, "Effect of weak inhomogeneities in the atmosphere on sound and light propagation," *Izv. Akad. Nauk SSSR*, Ser. Geofiz., No.2, 155-165 (1953) (in Russian).

2. S. M. Rytov, "Diffraction of light by ultrasonic waves," *Izv. Akad. Nauk SSSR*, Ser. Fiz., No.2, 223-259 (1937) (in Russian).

3. V. I. Tatarskii, *Wave Propagation in a Turbulent Medium*, McGraw-Hill, New York (1961).

4. A. S. Monin and A. M. Obukhov, "Basic turbulent mixing laws in the atmospheric surface layer," *Trudy Geofiz. Inst. AN SSSR*, No.24(151), 163-187 (1954) (in Russian).

5. V. I. Tatarskii, "Microstructure of the temperature field in the atmospheric surface layer," *Izv. Akad. Nauk SSSR*, Ser. Geofiz., No.6, 689-699 (1956) (in Russian).

6. V. I. Tatarskii, A. S. Gurvich, M. A. Kallistratova, and L. V. Terent'eva, "The effect of meteorological conditions on scintillation intensity in the surface layer of the atmosphere," *Astron. Zh.*, **35**(4), 623 (1958) (in Russian).

7. V. M. Bovsheverov, A. S. Gurvich, V. I. Tatarskii, and L. R. Tsvang, "Instruments for the statistical analysis of turbulence," *Proc. Conf. on Stellar Scintillation*, (AN SSSR Press, Moscow), 26-32 (1959) (in Russian).

8. A. S. Gurvich, V. I. Tatarskii, and L. R. Tsvang, "Scintillation of terrestrial light sources," *Proc. Conf. on Stellar Scintillation*, AN SSSR Press, Moscow, 33 (1959) (in Russian).

9. M. E. Gracheva and A. S. Gurvich, "Strong fluctuations in the intensity of light propagated through the atmosphere close to the earth," *Radiophys. and Quantum Electron.*, **8**(4), 511-515 (1965).

10. M. E. Gracheva, "Investigation of the statistical properties of strong fluctuations in the intensity of light propagated through the atmosphere near the earth," *Radiophys. and Quantum Electron.*, **10**(6), 424-433 (1967).

11. M. E. Gracheva, A. S. Gurvich, and M. A. Kallistratova, "Dispersion of strong atmospheric fluctuations in the intensity of laser radiation," *Radiophys. and Quantum Electron.*, **13**(1), 40-42 (1970).

12. M. E. Gracheva, A. S. Gurvich, and M. A. Kallistratova, "Measurements of the average amplitude of a light wave propagating in turbulent atmosphere," *Radiophys. and Quantum Electron.*, **13**(1), 36-39 (1970).

13. Yu. N. Barabenkov, Yu. A. Kravtsov, S. M. Rytov, and V. I. Tatarskii, "Status of the theory of propagation of waves in a randomly inhomogeneous medium," *Sov. Phys.-Usp.*, **13**(5), 551-680 (1971).

14. A. M. Prokhorov, F. V. Bunkin, K. S. Gochelashvili, and V. I. Shishov, "Propagation of laser radiation in randomly inhomogeneous media," *Sov. Phys.-Usp.* **17**(6), 826-847 (1975). [See also A. M. Prokhorov, F. V. Bunkin, K. S. Gochelashvili, and V. I. Shishov, "Laser irradiance propagation in turbulent media," *Proc. IEEE* **63**(5), 790-811 (1975)].

15. A. V. Artem'ev and A. S. Gurvich, "Experimental study of coherence-function spectra," *Radiophys. and Quantum Electron.*, **14**(5), 580-583 (1971).

16. M. E. Gracheva, A. S. Gurvich, S. O. Lomadze, Vl. V. Pokasov, and A. S. Khrupin, "Probability distribution of strong fluctuations of light intensity in the atmosphere," *Radiophys. and Quantum Electron.*, **17**(1), 83-87 (1974).

17. A. S. Gurvich and Vl. V. Pokasov, "The frequency spectra of strong fluctuations of laser radiation in a turbulent atmosphere," *Radiophys. and Quantum Electron.*, **16**(6), 701-704 (1973).

18. K. S. Gochelashvili and V. I. Shishov, "Saturated fluctuations in the laser radiation intensity in a turbulent medium," *Sov. Phys. JETP*, **39**(4), 605-609 (1974).

19. A. S. Gurvich and V. Kan, "Measurements of the four-point coherence function of a laser radiation field in a turbulent atmosphere," *Radiophys. and Quantum Electron.*, **22**(2), 131-135 (1979).

20. V. U. Zavorotnyi, "Four-point function for field-coherence in a turbulent medium in a region of strong fluctuations of wave intensity," *Radiophys. and Quantum Electron.*, **22**(8), 677-683 (1979).

21. A. S. Gurvich, B. S. Yelepov, Vl. V. Pokasov, K. K. Sabel'feld and V. I. Tatarskii, "Spatial structure of strong fluctuations of light intensity in a turbulent medium," *Radiophys. and Quantum Electron.*, **22**(2), 135-141 (1979).

22. J. M. Martin and S. M. Flatté, "Intensity images and statistics from numerical simulation of wave propagation in $3-D$ random media," *Appl. Opt.*, **27**(11), 2111-2126 (1988).

23. V. E. Zuev, V. A. Banakh, and V. V. Pokasov, *Optics of the Turbulent Atmosphere*, Gidrometeoizdat, Leningrad (1988) (in Russian).

ATMOSPHERIC PROPAGATION

R. J. Hill
National Oceanic and Atmospheric Administration
Wave Propagation Laboratory
Boulder, Colorado 80303 USA

1. Introduction

I will not discuss here many aspects of atmospheric scintillation, such as measurements in Russia, optical remote sensing, or vertical propagation and tropospheric turbulence layers, because other conference participants will describe those. I will touch very lightly on theoretical aspects because there are many theoretical reports in this conference proceedings. My discussion is mainly phenomenological and pedagogical, and limited to horizontal propagation a few meters above flat ground. I will relate observed scintillation to the behavior of atmospheric turbulence.

2. Strong scintillation

The conference presentation included a motion picture of scintillating irradiance. The film shows a laser beam after it had passed through 600 m of atmosphere at night, demonstrating laser beam breakup and the motion of the irradiance patches in the direction of the cross-path wind. Next, the series of film segments from very high-speed photography shows a 22-cm-diameter portion of the irradiance of a diverged wave after it had passed through 1 km of atmosphere 1 m above the ground; the refractive structure parameter C_n^2 was from 10^{-13} to 10^{-12} $m^{-2/3}$, indicating that the scintillation was strong. The cross-path wind component was only from 0.3 to 1.5 m/s; film frame rates were 4000/s. The smallest irradiance patches do not move across the image; rather, they form and vanish after moving roughly the distance of their diameter. This behavior was previously deduced from statistics of irradiance for finite apertures.[1] The irradiance patches the size of the scattering disk do move as in frozen turbulence, as evidenced by the operation of saturation-resistant large-aperture cross-wind scintillometers.[2] Of course, for a diverged wave, the random medium would have to rotate like a solid body around the transmitter to cause the irradiance pattern to move in a frozen manner. That the small patches of irradiance in the film have a boiling appearance indicates that in strong scintillation the irradiance pattern is very sensitive to relative motion of phase screens. I believe that even if we had a widely collimated wave, simulating a plane wave, one would see a similar boiling pattern for irradiance in strong scintillation caused by air moving at somewhat different cross-path speeds at different path positions. This may be why there is only qualitative agreement between the experimental temporal spectra of irradiance obtained during strong scintillation and theoretical spectra based on the frozen turbulence hypothesis.[3]

Coles and Frehlich[4] showed 1 s of irradiance data from a 1063-m path. Their Fig. 9 has many intensity spikes similar to those in our motion picture; many spikes are greater than 50 times the mean intensity. Their figure also shows clumps of spikes that represent the large scale of the intensity covariance in strong scintillation. Very evident is the small-scale behavior on the size of the coherence length and the larger-scale clumps of spikes; the latter have a size equal to the wavelength times path length divided by coherence length.

Intensity covariance observed in that strong turbulence is shown in Fig. 11 of Coles and Frehlich.[4] The long tail of covariance is apparent, as is the very sharp rise at the coherence length size. The Fresnel scale is about 1 cm in their experiment. During that same experiment, for a quite small refractive structure parameter, they measured an intensity covariance in moderately weak scintillation. It does have a scale characteristic of the Fresnel scale, as seen in their Fig. 10. This covariance compares well with that calculated on the basis of weak scattering theory.

3. Atmospheric boundary-layer turbulence

Acoustic and radar sounding of the atmosphere reveal the diurnal behavior of the boundary layer. The important thing that I would like to point out is that at night one sees many layers of turbulence close to the ground. As the sun rises, convection begins: hot, rising turbulent air has a large refractive structure parameter, but the down-moving, cool air has a very small refractive structure parameter. A capping inversion having a relatively large refractive structure parameter forms aloft. As the sun goes up, this capping layer lifts. At night, one can see wave activity close to the ground, but in the daytime, such wave activity is mostly in the capping layer. The horizontal spatial scale of these rising, hot thermal plumes increases with the depth of that layer, as does the scale of cool, descending interplume air, which has very little scattering strength in it. This creates an intermittent turbulent structure.

Wilczak and Tillman[5] showed temperature from thermometers at heights of 14, 4, 3, and 1 m, as well as from thermometers separated horizontally by as much as 100 m. By looking at the temperature at 14 m above the ground, and using frozen flow to convert time into spatial separation, one can see that thermal plumes are very extensive in size. Looking at temperature from another thermometer 1 m above the ground and directly below the first thermometer, one can see much more randomized temperature turbulence filling the volume much more completely with almost equal scattering strength. At 1-m height compared with 14-m height, the fraction of time occupied by temperature minima is very much diminished. However, the imprint of the large thermal plumes still exists even at 1 m above the ground. Wilczak and Tillman[5] showed isotherms measured in this convective turbulence structure at a height of 4 m. These isotherms were measured to a distance of 50 m transverse to the wind direction and to a distance of 1000 m in the wind direction. The regions of hot, rising plumes and cooling, descending plume air are evident in their graphs. One can imagine that a 50-m propagation path could be entirely immersed in the strong scattering plume at one time and later immersed in air that was essentially devoid of scattering strength. On the other hand, if the propagation path were very long, covering 1000 m, there would be a sufficient number of these scattering regions to fill the entire path at various positions. That is, for a long propagation path, there would be a number of different phase screens.

4. Intermittency and weak scintillation

On a 50-m path in such convective turbulence, one can observe very weak scintillation of intensity. The probability distribution of irradiance is very different from log normal, as shown by Churnside and Frehlich.[6] This is because of the extreme intermittency of the convective turbulence. If one does a conditional averaging, choosing short segments of data having equal half-second variances, then the irradiance probability density has a very good log-normal fit.[6] However, for long averaging times, one does not expect log-normal irradiance probability density in the case of very intermittent turbulence because, as we saw from the data by Wilczak and Tillman,[5] there are many turbulence structures that are quite significantly larger than the propagation path length. On the other hand, consider the data of Hill et al.[7] for which the propagation path was at a height of 4 m on a long path (1374 m), propagating at a long wavelength of 1.7 mm such that the irradiance variance was quite small (0.02), despite the rather large value of $C_n^2 = 6 \times 10^{-12}$ m$^{-2/3}$. In this case, the thermal plume structure was even more pronounced than for that 50-m-path data of Churnside and Frehlich.[6] Yet, Hill et al.[7] obtained a very close fit to the log-normal probability density. The log-normal fit was clearly better than a Gaussian fit. Gaussian, of course, is the limit of the log normal for zero variance. Other effects of intermittency have been discussed by Tatarskii and Zavorotny.[8]

5. Absorption fluctuations

Another phenomenon that occurs in the atmosphere is absorption fluctuations. These affect irradiance statistics but have negligible effect on phase statistics. Absorption fluctuations cause irradiance fluctuations that are most easily described as whole-path-integrated attenuation. The largest atmospheric structures contribute to the absorption fluctuations. Thus the resulting irradiance fluctuations are as statistically nonstationary as is the atmosphere's path-integrated attenuation. Measurement time series of increasing duration simply increase the measured irradiance variance caused by absorption fluctuations. This is the same as for the effect of refraction fluctuations on the total phase of a wave. Irradiance structure functions and phase structure functions obey local stationarity because of the spatial filtering effect of spaced receivers. Hill et al.[9] showed a dramatic case of absorption fluctuations wherein, during a 40-min data run, the humidity suddenly dropped by 1 g m^{-3} over the entire propagation path. When a millimeter wave of 2-mm wavelength propagates, the obvious result is that the irradiance of that wave will suddenly increase as the absorption suddenly decreases. The temporal power spectrum of irradiance obtained before and after the humidity increase has the classic shape predicted using frozen refractive turbulence and weak scintillation.[9] However, this spectrum, obtained over the whole 40-min run including the increase in absorption, has an enormous low-frequency enhancement, caused by the absorption fluctuations, that contributes 90% of the total irradiance variance. The probability density of irradiance becomes bimodal because it was a log-normal distribution, having one mean irradiance value for part of the data run, and a log-normal distribution, having a different mean for another part of the data run. Consequently, the effects of absorption fluctuations also cause deviations from log-normal irradiance statistics even in weak scattering.

6. Outer-scale turbulence structure and its influence on propagation

For describing the largest-scale atmospheric fluctuations, propagation theorists ubiquitously use the von Kármán spectrum of refractive index fluctuations. The von Kármán spectrum is given by 0.033 $C_n^2\,(K^2 + L_0^{-2})^{-11/6}$, where K is spatial wave number and L_0 is an outer scale; this formula becomes a constant for $K \ll 1/L_0$ and becomes proportional to $K^{-11/3}$ for $K \gg 1/L_0$. This von Kármán spectrum is appropriate for turbulence in a pipe because a pipe has only one spatial scale: its diameter. The atmosphere has many scales. In the atmospheric boundary layer, one should consider at least the height above ground, the Obukhov length, and the depth of the boundary layer; even larger scales or scales of surface inhomogeneity can be relevant to propagation problems.

Although the von Kármán spectrum is used by nearly everyone making calculations, the atmosphere is really quite different. Comparison of the one-dimensional von Kármán spectrum with the temporal and spatial spectra measured by Kaimal et al.[10,11] shows that the von Kármán spectrum is utterly inadequate. The variance-containing range in the atmosphere is very much broader, more than several decades broader than that in the von Kármán spectrum. Comparing what theorists use for making predictions with what the atmosphere presents to us, we see very little correspondence in the large-scale range. For no value of outer scale do we have a good connection between what we use theoretically and what the atmosphere presents us to use. Phase structure functions are sensitive to the variance-containing range of the refractive-index spectrum; optical phase data by Lukin and Pokasov[12] shows disagreement with the theory based on the von Kármán spectrum, and with theory based on other single-scale model spectra as well.

Hill et al.[7] used much longer wavelengths, e.g., 1.7-mm wavelength and 1.34-km path length and a height above ground of 3.7 m; the Fresnel scale is 0.6 m. One can measure the structure function of phase at much larger receiver separations in this millimeter wave case than in the optical case. (Hill et al.[7] used separations as large as 10 m.) Hill et al.[7] do not observe the five-thirds prediction on the basis of the inertial range. They observe a definite outer-scale behavior that is not quantitatively predicted by the von Kármán spectrum. Certainly, the anisotropy of the turbulence must enter into this calculation, and that has not been done. The message is that the large-scale behavior of the atmosphere has an important effect on many measured statistics.

7. Inner-scale turbulence structure and its influence on propagation

In this section, I address the smallest scale in atmospheric turbulence, the inner scale l_0. Figure 1 is a simple graph showing on the abscissa the typical inner scales that we see several meters above the ground and on the ordinate the typical propagation path lengths that are used, from a few tens of meters to a few kilometers. The regime of strong scintillation is separated from that of weak scintillation by the horizontal line corresponding to a value of 0.3 for the Rytov variance. The vertical position of this line corresponds to $\lambda = 0.63$ µm and $C_n^2 = 10^{-13}$ m$^{-2/3}$. The line $0.6\sqrt{Z/k} = l_0$ (where Z is propagation path length and k is optical wavenumber) separates the geometrical optics region on the right from diffractional optics on the left. The line $\rho_0 = l_0$ also separates these regimes for strong scattering. On the left side of the lines $\rho_0 = l_0$ and

$0.6\sqrt{Z/k} = l_0$, the inertial range is an important description for the propagation. On the right side of these lines, the dissipation range is important, so the inner scale is an important parameter.

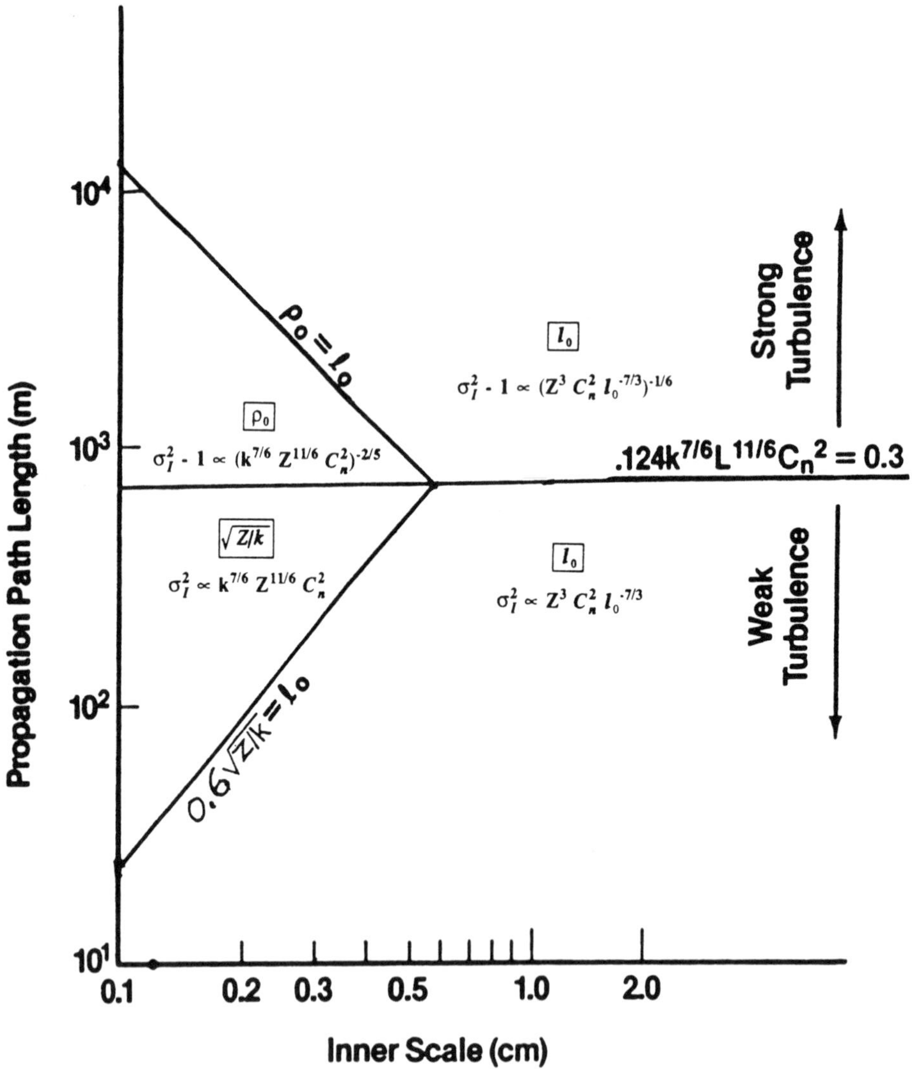

Fig. 1. Parameter regimes in which inner-scale effects are important for atmospheric propagation of spherical waves.

The four asymptotic formulas for irradiance variance are shown in the four regimes in Fig. 1. The approach to unity of the irradiance variance for strong scintillation is shown for those two cases. In each box is the spatial scale of refractive fluctuations that influences propagation in each region; counter-clockwise from lower right, these are the inner scale, inner scale, coherence length, and Fresnel scale. The region where the inner scale is dominant is to the right of the two sloped lines, $\rho_0 = l_0$ and $0.6\sqrt{Z/k} = l_0$. For a millimeter wavelength, these sloped lines would move off the graph toward the right, and outer-scale L effects would begin to occur.

The lesson of Fig. 1 is that the inner scale is a very important parameter in describing optical scintillation for most propagation path lengths most of the time. The traditional spectrum used to describe inner-scale fluctuations has a Gaussian rolloff. The dissipation range having Gaussian rolloff is often referred to as the "Tatarskii spectrum," but "traditional spectrum" is a more appropriate name. It has never been supported by data. In fact, the real atmosphere displays a different behavior in the dissipation range. The true dissipation-range spectrum for the atmosphere is given in Refs. 13 and 14. Quantitative agreement of theory with weak scattering data requires use of this spectrum; use of the ad hoc traditional spectrum gives disagreement with data.[15,16,17] There is a spectral bump at the high wavenumber end of the inertial range (the beginning of the dissipation range). This bump gives variances that are greater than variances from predictions based on the traditional spectrum.[14,17,18] From Fig. 1, we see that use of the true spectrum is necessary for quantitative theoretical prediction in the case of strong atmospheric scintillation as well. We performed an experiment in which irradiance variance, inner-scale, and refractive structure parameter were all measured simultaneously.[19] The results clearly show the effect of the inner scale in strong scintillation. During this conference, Martin[20] compared the results with numerical simulation. Agreement was good because the correct atmospheric refractive-index spectrum was used in the simulation.

The inner scale also affects the large scale of the intensity covariance for strong scintillation. One can see that clearly in aperture-averaged scintillation measurements.[21] Use of an inertial range spectrum in the asymptotic theory gives poor agreement with the aperture-averaged data. An experiment in strong scintillation in which aperture-averaged scintillation was measured simultaneously with inner scale and refractive structure parameter showed that the asymptotic theory[21] combined with the true atmospheric spectrum[13] produces quantitative agreement with the data.

8. Probability distribution of irradiance in strong scintillation

Reference 22 shows normalized irradiance moments measured in strong scintillation. They are compared with moments of a K-distribution and a log-normal distribution. The formulas for these probability distributions and their moments are given in Refs. 23, 24, 28. The irradiance moment data are closer to the K-distribution moments, but above them, which means that the data cannot possibly be K-distributed. The fact that the moments of the log-normal distribution are above the measured moments, even though orders of magnitude above, means that the log-normal is a possibly accurate distribution for the data. Severe underestimation of measured moments is caused by too few samples[24-27] and inadequate receiver dynamic range, the latter usually resulting from amplifier saturation.[28,29] In fact, comparing the probability densities instead of moments, one sees that the data are closer to the log-normal distribution than to the K-distribution.[24] The data are even closer to the log-normally modulated exponential distribution.[24] However, these probability distributions are all rather ad hoc models; they are not derived from first principles.

Aperture averaging returns the probability distribution to log normal in strong scattering.[24] Even using a 3-mm aperture when the coherence length is 1.4 mm produces a histogram that is very close to log normal.[1] Thus, even an aperture diameter slightly larger than the coherence length produces log-normal intensity statistics in strong scintillation. Effects of finite apertures and bandwidths on the probability distributions have been investigated.[30,31]

Churnside and Hill[24] showed that for an irradiance variance of 3.5, which is not particularly large for a diverged wave in atmospheric turbulence, fourth-order and higher-order moments are very poorly estimated even if the probability density is measured to the 10^{-7} level, which requires 10 million independent samples and requires an amplifier saturation level set very high so that we have occurrences of spikes 100 times the mean value. One cannot get more than, say, 20 million samples in the atmosphere because the sun goes down. Consequently, we are at the limit at which we can perhaps measure the second and third moments of irradiance in strong scintillation in the atmosphere, but we cannot measure fourth-, fifth-, or higher-order moments.

The interest in high-order irradiance moments arose partly because theory that is valid into the strong-scintillation regime is formulated in terms of field moments, from which irradiance moments are obtainable. Since fourth and higher irradiance moments are not measurable in the atmosphere, it follows that theoretical field moments of order greater than six need not be considered. Description of atmospheric scintillation by field moments of all orders is inappropriate. Practical applications draw attention to quantities that are not described by field moments. Examples are the log-intensity variance, for remote sensing applications, and the joint probability distribution of observing irradiance exceeding a given value for a given duration, for laser eye safety.[1] Numerical simulation can help with such statistics.

9. Summary

I conclude by contrasting the important aspects of millimeter-wave and visible-light propagation in Table 1 and by presenting the following summary of my major points:

- The atmospheric refractive-index spectrum at high wave numbers (inertial and dissipation ranges) is known and required for quantitative comparison with measurements of visible-light scintillation.

- The atmospheric refractive-index spectrum at low wave numbers (variance-containing range) is more variable and lesser known, and needed for comparison with measurements, especially for millimeter-wave phase difference.

- Propagationists usually use refractive-index spectra for both variance-containing and dissipation ranges that are not correct for the atmosphere.

- Absorption fluctuations can have a marked effect on intensity statistics for millimeter and infrared waves, but not on phase statistics.

- For propagation purposes, turbulence is or is not characterized as intermittent depending on whether or not the propagation path length exceeds the spatial scale of atmospheric structures that are effective in modulating the scattering strength and whether or not the observation duration exceeds the time for passage of these atmospheric structures.

- Log-normal intensity statistics are not expected or observed for weak scintillation from intermittent turbulence, not expected because some turbulence structures exceed the propagation path length.

- A propagation path much longer than the intermittent structures produces log-normal intensity statistics for weak scintillation.

- When the large intensity spikes in strong scintillation form, they move a distance about equal to their diameter before they vanish. The large intensity patterns move as for frozen turbulence.

- The large intensity spikes have size proportional to the wave coherence length, independent of their peak value of intensity.

- For strong scintillation, aperture averaging using an aperture greater than or equal to twice the coherence length results in a nearly log-normal probability density of intensity.

- The intensity probability density is not radically different from log normal in atmospheric strong scintillation, contrary to suppositions of extreme departure from log normal on the basis of measurements of moments of intensity.

- Because of the limitations of amplifiers and the number of samples attainable, it is not possible to measure intensity moments of order four and greater for atmospheric strong scintillation.

- The asymptotic theory gives quantitative agreement with aperture-averaged measurements of scintillation for aperture diameters significantly larger than the Fresnel length. The inner scale has a significant effect even for apertures much larger than the inner scale. Hence the large-scale intensity covariance in strong scintillation is sensitive to inner scale.

Table 1. Millimeter-wave scintillation relative to visible-light scintillation

	Millimeter-Waves	Visible Light
Importance of saturation of scintillation	None; any statistic is formulable using perturbation theory; extended Huygens-Fresnel principle applies	Often; only field moments are formulable for strong scintillation
Importance of effects of inner scale	None	Very; especially on intensity statistics
Importance of effects of outer-scale structure	Very; especially for phase difference at large spacings, and on long-time-scale intensity	Little; mostly on large-separation phase-difference structure function
Importance of absorption fluctuations	Very; especially for on intensity fluctuations, atmospheric non-stationarity is important	None

10. References

1. R. J. Hill, J. H. Churnside, and D. H. Sliney, "Measured statistics of laser beam scintillation in strong refractive turbulence relevant to eye safety," *Health Phys.*, **53**, 639-647 (1987).

2. T. Wang, G. R. Ochs, and S. F. Clifford, "Wind measurements by the temporal cross-correlation of the optical scintillations," *Appl. Opt.*, **20**, 4073-4081 (1981).

3. M. E. Gracheva, A. S. Gurvich, S. S. Kashkarov, and Vl. V. Pokasov, "Similarity relations and their experimental verification for strong intensity fluctuations of laser radiation," in *Laser Beam Propagation in the Atmosphere*, J. W. Strohbehn, ed., Springer-Verlag, Berlin (1978).

4. W. A. Coles and R. G. Frehlich, "Simultaneous measurements of angular scattering and intensity scintillation in the atmosphere," *J. Opt. Soc. Am.*, **72**, 1042-1048 (1982).

5. J. M. Wilczak and J. E. Tillman, "The three-dimensional structure of convection in the atmospheric surface layer," *J. Atmos. Sci.*, **37**, 2424-2443 (1980).

6. J. H. Churnside and R. G. Frehlich, "Experimental evaluation of log-normally modulated Rician and IK models of optical scintillation in the atmosphere," *J. Opt. Soc. Am. A*, **6**, 1760-1766 (1989).

7. R. J. Hill, R. A. Bohlander, S. F. Clifford, R. W. McMillan, J. T. Priestley, and W. P. Schoenfeld, "Turbulence-induced millimeter-wave scintillation compared with micrometeorological measurements," *IEEE Trans. Geosci. Remote Sensing*, **26**, 330-342 (1988).

8. V. I. Tatarskii and V. U. Zavorotny, "Wave propagation in random media with fluctuating turbulent parameters," *J. Opt. Soc. Am. A*, **2**, 2069-2076 (1985).

9. R. J. Hill, S. F. Clifford, R. J. Lataitis, and A. D. Sarma, "Scintillation of millimeter-wave intensity and phase caused by turbulence and precipitation," in *Atmospheric Propagation in the UV, Visible, IR, and MM-Wave Regions and Related Systems Aspects*, Conf. Proc. No. 454, AGARD, Neuilly sur Seine, France.

10. J. C. Kaimal, J. C. Wyngaard, Y. Izumi, and O. R. Coté, "Spectral characteristics of surface-layer turbulence," *Quart. J. Roy. Meteorol. Soc.*, **98**, 563-589 (1972).

11. J. C. Kaimal, R. A. Eversole, D. H. Lenschow, B. B. Stankov, P. H. Kahn, and J. A. Businger, "Spectral characteristics of the convective boundary layer over uneven terrain," *J. Atmos. Sci.*, **39**, 1098-1114 (1982).

12. V. P. Lukin and V. V. Pokasov, "Optical wave phase fluctuations," *Appl. Opt.*, **20**, 121-135 (1981).

13. R. J. Hill, "Models of the scalar spectrum for turbulent advection," *J. Fluid Mech.*, **88**, 541-562 (1978).

14. R. J. Hill and S. F. Clifford, "Modified spectrum of atmospheric temperature fluctuations and its application to optical propagation," *J. Opt. Soc. Am.*, **68**, 892-899 (1978).

15. E. Azoulay, V. Thiermann, A. Jetter, A. Kohnle, and Z. Azar, "Optical measurement of the inner scale of turbulence," *J. Phys. D: Appl. Phys.*, **21**, 541-544 (1988).

16. G. R. Ochs and R. J. Hill, "Optical-scintillation method of measuring turbulence inner scale," *Appl. Opt.*, **24**, 2430-2432 (1985).

17. R. J. Hill and G. R. Ochs, "Inner-scale dependence of scintillation variances measured in weak scintillation," *J. Opt. Soc. Am.*, **A9**, 1406-1411 (1992).

18. R. J. Hill and G. R. Ochs, "Fine calibration of large-aperture optical scintillometers and an optical estimate of inner scale of turbulence," *Appl. Opt.*, **17**, 3608-3612 (1978).

19. A. Consortini, F. Cochetti, J. H. Churnside, and R. J. Hill, "Inner-scale effect on intensity variance measured for weak, intermediate, and strong atmospheric scintillation," *J. Opt. Soc. Am.* (submitted).

20. J. Martin, Article in this proceedings.

21. R. G. Frehlich and G. R. Ochs, "Effects of saturation on the optical scintillometer," *Appl. Opt.*, **29**, 548-553 (1990).

22. G. Parry, "Measurement of atmospheric turbulence-induced intensity fluctuations in a laser beam," *Opt. Acta*, **28**, 715-728 (1981).

23. G. Parry and P. N. Pusey, "K distributions in atmospheric propagation of laser light," *J. Opt. Soc. Am.*, **69**, 796-798 (1979).

24. J. H. Churnside and R. J. Hill, "Probability density of irradiance scintillations for strong path-integrated refractive turbulence," *J. Opt. Soc. Am. A*, **4**, 727-733 (1987).

25. N. Ben-Yosef and E. Goldner, "Sample-size influence on optical scintillation analysis. 1: Analytical treatment of the higher-order irradiance moments," *Appl. Opt.*, **27**, 2167-2171 (1988).

26. E. Goldner and N. Ben-Yosef, "Sample-size influence on optical scintillation analysis. 2: Simulation approach," *Appl. Opt.*, **27**, 2172-2177 (1988).

27. R. G. Frehlich and J. H. Churnside, "Statistical properties of estimates of the moments of laser scintillation," *J. Mod. Opt.*, **36**, 1645-1659 (1989).

28. A. Consortini and G. Conforti, "Detector saturation effect on higher-order moments of intensity fluctuations in atmospheric laser propagation measurements," *J. Opt. Soc. Am. A*, **1**, 1075-1077 (1984).

29. A. Consortini and R. J. Hill, "Reduction of the moments of intensity fluctuations caused by amplifier saturation for both the K and the log-normally modulated exponential probability densities," *Opt. Lett.*, **12**, 304-306 (1987).

30. R. J. Hill and J. H. Churnside, "Observational challenges of strong scintillations of irradiance," *J. Opt. Soc. Am. A*, **5**, 445-447 (1988).

31. J. H. Churnside, R. J. Hill, G. Conforti, and A. Consortini, "Aperture size and bandwidth requirements for measuring strong scintillation in the atmosphere," *Appl. Opt.*, **28**, 4126-4132 (1989).

On Ionospheric Scintillations

K. C. Yeh
Department of Electrical and Computer Engineering
University of Illinois at Urbana-Champaign
Urbana, IL 61801-2991
U. S. A.

A. W. Wernik
Space Research Centre
Polish Academy of Sciences
Warsaw, Poland

Abstract

Ionospheric scintillation studies are concerned with investigating the behavior of radio waves after propagating through the turbulent ionosphere. The ionosphere is a dispersive medium. In the high frequency limit, its fractional perturbation in the dielectric permittivity caused by electron density fluctuations is inversely proportional to the square of radio frequency. Therefore, multi-frequency scintillation measurements have been used to check for internal consistencies in the modeling effort and the theory itself. Furthermore, the ionosphere is accessible to probing by multi-technique which can help determine the environment through which radio waves propagate. In this review we discuss several theories that have been used to interpret the scintillation data. Also discussed are effects caused by nonfrozen flows and the backscatter enhancement in a double passage problem.

1. Introduction

The fading of the ionospherically reflected signal has been known for a long time, ever since shortly after the discovery of the ionosphere (Mimno, 1937). However, the discovery of transionospheric scintillations occurred later and can be traced to 1946 (Hey et al., 1946) when short-period irregular fluctuations at 64 MHz were sometimes observed in radiations coming from Cygnus. Subsequent experiments identified the ionospheric origin for the cause of fluctuations. Theories were then developed to explain the observed phenomenon (Hewish, 1952; see also early reviews by Ratcliffe, 1956; Booker, 1958). The sources used were cosmic and hence were incoherent and sometimes extended. After the launch of Sputnik in 1957 the sources transmitted by the satellite as radio beacons have become coherent and point-like (for an early review, see Yeh and Swenson, 1964). Since then there have been quite a few satellites dedicated to radio beacon studies (e.g., Davies et al., 1975; Fremouw et al., 1978) as well as many so-called satellites of opportunities.

It is now well accepted that ionospheric scintillations of radio signals originate from random electron density irregularities acting as scatterers. In almost all experiments the temporal variations of the wave parameters (such as amplitude, phase, angle of arrival, polarization, etc.) are measured. The temporal variations are caused by (a) the convection of irregularities, (b) the evolution of irregularities as they are carried by the drift. The temporal statistics (such as the first few moments) can be related to the spatial statistics when the ergodicity condition holds (Marple, 1987). This is frequently assumed in analyzing experimental data, although there may exist cases for which a strict ergodicity is not expected.

To relate the statistics of the measured wave parameters to the statistics of the electron density fluctuations requires an understanding of wave behavior propagating in a random medium. The ionosphere as a propagation medium differs from other random media in several aspects which should be clarified. After several decades of experimentation we have learned the existence of ionospheric irregularities with scales extending over an eight-decade range. A composite spectrum was therefore constructed by Booker (1979) showing irregularities with scales of several thousand kilometers (close to Earth radius) down to scales as small as centimeters. With irregularities spanning an eight-decade range, the potential for scattering electromagnetic waves propagating through them must necessarily span a likewise eight-decade frequency range. However, since the intensity of electron density fluctuations falls off quickly with wave number and the ionosphere as a propagation medium is dispersive, radio frequencies much above 3 GHz (corresponding to a wavelength of 10 cm) rarely experience the scintillation phenomenon except in the equatorial anomaly region during solar maximum. The dispersive ionosphere has a dielectric permittivity given by

$$\varepsilon(\mathbf{r}, t) = \langle\varepsilon\rangle \, [1 + \varepsilon_1(\mathbf{r}, t)] \tag{1}$$

where the background dielectric permittivity is

$$\langle\varepsilon\rangle = \varepsilon_o(1 - f_{po}^2/f^2) \tag{2}$$

The presence of free electrons modifies the free space dielectric permittivity ε_o by making it radio frequency f dependent. The background plasma frequency f_{po} depends on electron density and its F region peak varies globally with season and time of day from several MHz to 18 MHz. The fractional fluctuation ε_1 in the dielectric permittivity from the mean given in (1) is

$$\varepsilon_1(\mathbf{r}, t) = -\frac{(f_{po}/f)^2}{1 - (f_{po}/f)^2} \frac{N(\mathbf{r}, t)}{N_o} \tag{3}$$

where N_o is the background electron density (with a corresponding plasma frequency f_{po}) and N is the electron density fluctuations. In most transionospheric scintillation experiments the frequency is much higher than the plasma frequency, i.e., $f \gg f_{po}$. Hence from (2), $\langle\varepsilon\rangle \approx \varepsilon_o$ and from (3) $|\varepsilon_1| \ll 1$. Even though $|\varepsilon_1|$ is small, in the high frequency limit ε_1 is approximately inversely proportional to frequency squared. This frequency dependence is frequently exploited in multi-frequency scintillation experiments as will be discussed later.

Also present in the ionosphere is the Earth's magnetic field which makes the propagation medium anisotropic. The first order effect in the high frequency limit is the introduction of polarization fading (Yeh and Liu, 1967) which are sometimes observed under extreme scintillation conditions at the equatorial anomaly regions (Lee et al.,

1982). Also to be noted is the inhomogeneous nature of the ionospheric background. Several studies (e.g., Liu, 1968) have been made and they show that most of the scintillation effects are imparted to the wave near the peak of F region. In this review both the anisotropic and inhomogeneous nature of the ionosphere will be ignored. Instead the ionospheric profile will be replaced by a slab of homogeneous ionosphere containing electron density irregularities. The wave will experience scattering and phase perturbations in the slab. On exiting from the ionosphere the wave further experiences diffraction. This is known as the slab model which will be used as the propagation geometry in this review.

2. Scintillation theories and observations

In an isotropic plasma with $f >> f_{po}$, the dielectric fluctuations are very small compared to unity. If the temporal variations in N are much slower than the period of the wave and the irregularities are much larger than the wavelength, the vector wave equation reduces to the scalar wave equation

$$\nabla^2 E + k^2[1 + \varepsilon_1(\mathbf{r}, t)] E = 0 \tag{4}$$

where E is a component of the electric field and $k^2 = k_o^2<\varepsilon>$ with k_o equal to the free space wave number. Under forward scatter conditions it is convenient to work with the complex amplitude $u(\mathbf{r}, t)$ defined by

$$E = u(\mathbf{r}, t)\, e^{ikz} \tag{5}$$

where the wave propagates in the z direction. The harmonic time dependence $\exp(-i\omega t)$ is understood and suppressed. Substituting (5) into (4) and making the assumption $2k\, |\partial u/\partial z| >> |\partial^2 u/\partial z^2|$, valid when the wavelength is small compared with irregularities, the complex amplitude is found to satisfy the parabolic equation

$$2ik\, \partial u/\partial z + \nabla_\perp^2 u = -k^2\, \varepsilon_1(\mathbf{r}, t)\, u \tag{6}$$

Here $\nabla_\perp^2 = \partial^2/\partial x^2 + \partial^2/\partial y^2$ is the transverse Laplacian. In the slab model of thickness L, (6) is valid for $0 \le z \le L$. Within the slab the wave suffers diffraction (as represented by the transverse Laplacian term in (6)) and random phase shifts (as represented by the right-hand side of (6)). The solution to (6) that takes both effects into account can be expressed as a path integral (Dashen, 1979; Lee et al., 1981). Computationally it has been found convenient to take into account diffraction and phase shift sequentially in a split-step scheme (DiNapoli and Deavenport, 1979). These are the topics of other reviews and will not be amplified further here. Suffice it to say that there exist two approximate solutions that have been popularly adopted in the literature: One is the phase screen approximation valid when $k/L >> 1/r_o^2$ where r_o is the outer scale of the irregularities and the other is the Rytov approximation valid under the weak scattering conditions. These approximate solutions are discussed separately in the following. A more extensive review of the theory can be found elsewhere (Yeh and Liu, 1982; Yeh et al., 1985).

An equivalent statement to the approximation $k/L >> 1/r_o^2$ is that the first Fresnel zone $\sqrt{\lambda L}$ is much smaller than the outer scale r_o. Under this approximation the diffraction term in (6) can be ignored altogether, yielding

$$u(\boldsymbol{\rho}\ z, t) = A_o \exp[i\phi(\boldsymbol{\rho}, z, t)] \tag{7}$$

where the phase is given by

$$\phi(\boldsymbol{\rho}, z, t) = \frac{k}{2}\int_o^z \varepsilon_1(\boldsymbol{\rho}, z', t)\, dz' = -\lambda r_e N_T(\boldsymbol{\rho}, z, t) \tag{8}$$

The approximation leading to (7) and (8) is commonly known as the phase screen approximation or model. In (8), λ is the wavelength, $r_e = e^2\mu_o/4\pi m$ is the classical electron radius and $N_T(\rho, z, t)$ is the fluctuation of the integrated electron density (or electron content) from 0 to z. For the slab model the complex amplitude u at the bottom (i.e., z = L) of the slab is interested. This can be obtained from (7) and (8) by letting z = L. The complex amplitude $u(\rho, L, t)$ becomes the "initial" condition for the diffraction of the field below the ionospheric slab. Diffraction below the slab is described by the Fresnel diffraction formula (e.g., Born and Wolf, 1964). If ϕ has a Gaussian probability distribution, the mean complex amplitude $<u>$ is found to be

$$<u> = A_o\, e^{-\sigma_\phi^2/2}\, e^{ikz} \tag{9}$$

where σ_ϕ is the variance of the phase $\phi(\boldsymbol{\rho}, L, t)$ at the phase screen. The second moment of u can be computed by making the further approximation $|\phi| << 1$, known as the shallow screen approximation. For a shallow screen, the spectrum of the logarithmic amplitude χ and the phase departure S have the respective expressions

$$F_\chi(\boldsymbol{\kappa}, z) = F_\phi(\boldsymbol{\kappa}) \sin^2(\kappa^2 z/2k) \tag{10}$$

$$F_S(\boldsymbol{\kappa}, z) = F_\phi(\boldsymbol{\kappa}) \cos^2(\kappa^2 z/2k) \tag{11}$$

where $\boldsymbol{\kappa}$ is the transverse spectral wave vector and F_ϕ is the spectrum of the phase at the phase screen. These expressions show that the diffraction process can be viewed as spatial filtering on the phase spectrum. The filtering functions given by the trigonometric functions in (10) and (11) are sometimes called the Fresnel filtering functions. Experimentally a shallow phase screen is approached in daytime when the scintillation is weak. An example showing clearly the effects produced by the interlaced Fresnel filtering nulls between the amplitude and the phase is depicted in Figure 1. An extensive discussion of the ionospheric scintillation index S_4 in a power law environment has been carried out in a series of papers by Rino and his coworkers. The scintillation index S_4 is defined as the variance of intensity, viz.,

$$S_4^2 = \frac{<I^2> - <I>^2}{<I>^2} = \frac{<(uu^*)^2> - <uu^*>^2}{<uu^*>^2} \tag{12}$$

and it is therefore a fourth moment quantity. For weak scintillations, Rino (1979) has shown, for $2 < p < 6$

$$S_4^2 = 4\sqrt{\pi}\, r_e^2\, \lambda^2 (L \sec\theta)\, \sigma_N^2\, \kappa_o^{p-3}\, \Gamma\left(\frac{p/2+1}{2}\right) \Gamma\left(\frac{3-p/2}{2}\right) Z^{p/2-1}\, g/(p/2-1) \quad (13)$$

In (13) θ is the incident angle of the wave on the slab, σ_N is the variance of the fluctuating electron density, $Z = \lambda z_R \sec\theta/2\pi$ with $z_R = zz_s/(z + z_s)$, z_s is the distance of the slab to the source, g is a geometric factor dependent on the nature of anisotropy ($g = 1$ for isotropic irregularities), p is the spectral index and κ_o is equal to $2\pi/r_o$. From (13) the scintillation index S_4 is seen to follow the frequency scaling law f^{-n} where $n = (p + 2)/4$. Such a frequency dependence is valid only for weak scintillations. For strong scintillations using a deep one-dimensional power law phase screen κ^{-p_1}, numerical calculations (Rino, 1980; Rino and Owen, 1984) show that S_4 does not exceed unity unless $p_1 \geq 3$ and the frequency dependence is weak. For $p_1 \geq 3$, S_4 can exceed unity but eventually saturates at unity (Rayleigh limit). In the intermediate regime where $S_4 \geq 1$, the scintillation is said to be in the "focusing" region.

The second approximation to (6) is the Rytov approximation. This approximation is valid for weak scintillations only. The two-dimensional spectra for the logarithmic amplitude χ and the phase departure S can be found (Wernik and Liu, 1975), respectively as

$$F_\chi(\boldsymbol{\kappa}, z) = (\pi k^2 L/4)\left[1 - \frac{2k}{\kappa^2 L} \sin\frac{\kappa^2 L}{2k} \cos\frac{\kappa^2(z - L/2)}{k}\right] \Phi_\varepsilon(\boldsymbol{\kappa}, 0) \quad (14)$$

$$F_S(\boldsymbol{\kappa}, z) = (\pi k^2 L/4)\left[1 + \frac{2k}{\kappa^2 L} \sin\frac{\kappa^2 L}{2k} \cos\frac{\kappa^2(z - L/2)}{k}\right] \Phi_\varepsilon(\boldsymbol{\kappa}, 0) \quad (15)$$

where $\Phi_\varepsilon(\boldsymbol{\kappa}, 0)$ is the three-dimensional spectrum of ε_1 at $\kappa_z = 0$. The respective spatial filtering is altered somewhat but reduces to the phase screen results given by (10) and (11) in the limit $L \to 0$. As L increases, the Fresnel oscillations tend to be smoothed out, especially when the slab approaches the ionospheric scale height (Liu and Yeh, 1977). Spectra computed at a number of frequencies using these formulas have agreed very well with those computed from the multi-frequency scintillation data (e.g., Umeki et al., 1977).

As the irregularity slab thickness increases and/or σ_N increases, the scintillation strength such as S_4 will increase as well. In this case the scattering may be strong and the weak scatter approximation is no longer valid. To take the multiple scattering into account one can either approach it by using the path integral (Dashen, 1979) or by finding the equations satisfied by the moments of the wave function (Lee, 1974; Tatarskii, 1971). In addition to the forward scatter approximation, the Markov approximation and the assumption that ε_1 is a Gaussian process are generally made. The multiple scattering characteristics will emerge phenomenologically in the scintillation data in several ways. The S_4 scintillation index saturates to unity indicating that a Rayleigh limit has been reached. This has been shown by using multiple frequency scintillation data (Rino and Liu, 1982). As a consequence of this saturation effect, the frequency scaling law f^{-n} with $n = (p + 2)/4$, valid under weak scatter conditions, is not expected to be valid in strong scatter regime. This has been shown previously (e.g., Yeh and Liu, 1982), but probably the most extensive data is that used by Bhattacharyya et al. (1990) and reproduced in Figure 2. This plot is based on the amplitude scintillation data taken at two frequencies (40 MHz and 140 MHz) transmitted by ATS-6 and recorded at Ootacamund, India. The ratio of S_4 at these two frequencies is computed, from which the frequency exponent n can be calculated. For weak scintillations, n approaches 1.42, giving a spectral index $p = 3.68$. As the scintillation index increases, the frequency exponent decreases, showing the effect of multiple scattering. As S_4 at 140 MHz reaches unity (Rayleigh limit), S_4 at 40 MHz must certainly have reached this limit as well, yielding the result $n = 0$. This anchors the extreme right point of the curve at $n = 0$ when S_4 at 140 MHz approaches unity. The intermediate behavior between these two extremes is given by the experimental data as indicated by the solid line in Figure 2. In addition to the violation of frequency scaling law, the multiple scattering has the effect of reducing the correlation distance or correlation time. In the weak scatter regime the correlation distance is determined by the Fresnel distance which is proportional to $f^{-1/2}$. Thus as the frequency decreases, the correlation distance increases. However, in the ionospheric plasma, a lower frequency experiences a stronger scatter which tends to decrease the correlation distance. The net result is that with decreasing frequency, the correlation distance first increases showing domination by Fresnel filtering as one would expect in the weak scatter limit. This increase slows and reaches a peak when Fresnel filtering is balanced by the multiple scattering. Further decrease in frequency beyond this point will mean decrease in correlation distance since effects are now dominated by the multiple scattering (Yeh and Liu, 1982). Such a behavior is plotted in Figure 3. The de-correlation under multiple scattering conditions is alternately revealed as spectral broadening. A numerical calculation by Franke which attempts to model this spectral broadening effect in a multi-frequency scintillation experiment shows very good comparison (Franke and Liu, 1983).

In the foregoing discussions, the temporal dependence of the turbulent medium is implicitly ignored. If the flow is frozen in, i.e., $\varepsilon_1(\mathbf{r}, t) = \varepsilon_1(\mathbf{r} - \mathbf{v}t)$, the spatial characteristics can be translated into temporal characteristics. In this case the scintillation data recorded at two stations must show perfect correlation except for a time delay. However, the scintillation data recorded at two stations spaced one hundred meters apart, for example, do show a strong correlation but not a perfect correlation. The perfect frozen-in model must thus be generalized. Several models have been proposed by Wernik et al. (1983); they all have to do with manners that $\mathbf{v}$ varies in the random ionosphere. The radar data seem to show that $\mathbf{v}$ varies randomly from a mean value (Kudeki and Franke, 1986), which leads to the well-known locally frozen-in model (Shkarofsky, 1968). An example showing the decrease of maximum cross-correlation with the increasing spacing between two stations is depicted in Figure 4. Experimental studies to investigate the time-space structures of scintillation data by using spaced-station observations have been carried out by several groups (Rino and Livingston, 1982; Costa et al., 1988; Vacchione et al., 1987 and Bhattacharyya et al., 1989). Such studies have been concerned with the anisotropic nature of the irregularities, the measurement of the mean drift v_o and its rms velocity fluctuations σ_v.

3. The double passage problem

The scattering of electromagnetic waves from a velocity fluctuating dielectric turbulence has been investigated several decades ago (Booker and Gordon, 1950). The theory shows the scattering cross-section σ_{so} is proportional to the irregularity power spectrum at the Bragg vector. As pointed out by de Wolf (1971), with increasing turbulence strength, the wave may undergo cumulative forward scattering and single backscattering. Furthermore, the fact that the backscattered wave passes through the same medium as the incident wave, the correlation so gained will enhance the cross-section calculated in the absence of such a correlation (Vinogradov et al., 1973; Gochelashvili and Shishov, 1981). The calculation of this double passage problem can be carried out by using the propagator $K(\mathbf{R}_1, \mathbf{R}_2)$ where $z_2 \geq z_1$. The propagator K is just the solution to the parabolic equation (6) under the condition that the beam is initially (i.e., $z_2 = z_1$) infinitesimally narrow (Lee et al., 1981). It can be formally represented as the path integral but its moments simplify to analytic expressions only under either Rytov limit or Rayleigh limit. For volume scattering, the scattered field $E_s(\mathbf{R}_R)$ at the receiver becomes

$$E_s(\mathbf{R}_R) = -(E_o/4) \int_{V_s} K(\mathbf{R}_T, \mathbf{R})\, \varepsilon_1(\mathbf{R})\, K(\mathbf{R}, \mathbf{R}_R)\, e^{-i2kz}\, d^3R \tag{16}$$

where $z_R \geq z \geq z_T$ and $z_R - z_T \approx z$. Using this expression the properties of the mutual coherence function (Yeh, 1983) and the backscatter cross-section (Yang and Yeh, 1985; Lure et al., 1989) have been investigated. By including both the multiple scattering effect and the Fresnel effect, the scattering cross-section σ_s is found to be

$$\sigma_R(\mathbf{R}) = (k^4/16\pi^2)\, \mathrm{Re} \int A_\varepsilon(\mathbf{r}')\, F(\mathbf{R}, \mathbf{r}')\, M(\mathbf{R}, \mathbf{r}')\, d^2r' \tag{17}$$

where A_ε is the one-dimensional Fourier spectral content of the correlation function B_ε at 2k, $F(\mathbf{R}, \mathbf{r}')$ is a Fresnel factor, and M is a multiple scattering factor. Define an enhancement factor η by

$$\eta = \sigma_s/\sigma_{so} \tag{18}$$

The enhancement factor η in the Rayleigh limit is found to decrease quickly as the receiver separates from the transmitter (Lure et al., 1989). In particular for backscattering from a scatterer imbedded in the random medium the enhancement factor simplifies to

$$\eta = 1 + S_4^2 \tag{19}$$

In the Rayleigh limit, the one-way scintillation index is $S_4 = 1$ and the backscattering cross-section becomes twice that in the absence of scintillation. This effect has been dramatically demonstrated experimentally by Knepp and Houpis (1991). For this experiment, the ALTAIR radar at Kwajalein is programmed to track the Soviet calibration satellite Cosmos 1427 which has a 2-meter diameter. For a passage of this satellite, the mean radar cross-section, the two-way S_4 scintillation index and the mean signal-to-noise ratio are plotted in Figure 5 after acquiring the satellite. An enhancement in the mean cross-section occurs simultaneously with the increased scintillation. This enhancement is real and is not caused instrumentally as the signal is strong in comparison with noise. In addition to using the scintillation theory they have also related the enhancement factor to the Nakagami scintillation statistics (Knepp and Houpis, 1991), obtaining (19). This result has been further amplified by Fremouw and Ishimaru (1992).

4. Conclusion

The discovery of ionospheric scintillation has had a history of several decades, first for reflected waves and later for transionospheric waves. A steady progress has been made especially after the availability of a coherent source on an orbiting satellite. Since the ionospheric medium is dispersive, the frequency dependence of fluctuations in the dielectric permittivity can be exploited to our advantage. To interpret the scintillation data in remote sensing applications one needs to know the behavior of waves while propagating through the irregular ionosphere. Here the theoretical development of wave propagation in random media has helped.

For transionospheric applications the wave frequency is usually much higher than the plasma frequency. In this case the radio ray follows mainly a straight line and the scattering from fluctuations in electron density is mainly in the forward direction. The space-time structure of the field expressed in terms of spectrum, scintillation index, drift velocity, rms velocity, and the scattering cross-section in a double passage problem has been extensively investigated and is reviewed here. For applications to ionospheric reflections the wave frequency may be comparable or even smaller than the ionospheric plasma frequency. The radio ray in this case is usually curved and the scattering cone expands to a much larger angle than the transionospheric case. The theory becomes more difficult (e.g., Wagen and Yeh, 1989; Tinin et al., 1992) and is not reviewed here.

The study of ionospheric scintillation derives its interests from its application to satellite communication, global mapping of ionospheric irregularities and plasma instabilities for the creation of these irregularities. Both experimental and theoretical studies in these applications have been engaged in actively by many investigators and are outside the scope of this review.

Acknowledgment This work was supported by grants from the National Science Foundation under ATM 91-21660 and INT 92-00765.

References

Bhattacharyya, A., K. C. Yeh and S. J. Franke (1992), Deducing turbulence parameters from transionospheric scintillation measurements, *Space Science Review*, to appear.

Bhattacharyya, A., R. G. Rastogi and K. C. Yeh (1990), Signal frequency dependence of ionospheric amplitude scintillations, *Radio Sci.*, **25**, 289-297.

Bhattacharyya, A., S. J. Franke and K. C. Yeh (1989), Characteristic velocity of equatorial F region irregularities determined from spaced receiver scintillation data, *J. Geophys. Res.*, **94**, 11959-11969.

Booker, H. G. (1958), The use of radio stars to study irregular reflection of radio waves in the ionosphere, *Proc. IRE*, **46**, 298-314.

Booker, H. G. (1979), The role of the acoustic gravity waves in the generation of spread F and ionospheric scintillation, *J. Atmos. Terr. Phys.*, **41**, 501-515.

Booker, H., and W. Gordon (1950), A theory of radio scattering in the troposphere, *Proc. IEEE*, **38**, 401-412.

Born, M., and E. Wolf (1964), *Principles of Optics*, Macmillan, New York.

Costa, E., P. F. Fougere and S. Basu (1988), Cross-correlation analysis and interpretation of spaced receiver measurements, *Radio Sci.*, **23**, 141-162.

Dashen, R. (1979), Path integral for waves in random media, *J. Math. Phys.*, **20**, 894-920.

Davies, K., R. B. Fritz, R. N. Grubb and J. E. Jones (1975), Some early results from the ATS-6 radio beacon experiments, *Radio Sci.*, **10**, 785-799.

de Wolf, D. A. (1971), Electromagnetic reflection from an extended turbulent medium: Cumulative forward-scatter single-backscatter approximation, *IEEE Trans. Antennas Propag.*, **AP-19**, 254-262.

DiNapoli, F. R., and R. L. Deavenport (1979), Numerical models of underwater acoustic propagation, in *Topics in Current Physics, Ocean Acoustics*, edited by J. A. DeSanto, pp. 79-185, Springer-Verlag, New York.

Franke, S. J., and C. H. Liu (1983), Observations and modeling of multi-frequency VHF and GHz scintillations in the equatorial region, *J. Geophs. Res.*, **88**, 7075-7085.

Fremouw, E. J., and A. Ishimaru (1992), Intensity scintillation index and mean apparent radar cross section on monostatic and bistatic paths, *Radio Sci.*, **27**, 539-543.

Fremouw, E. J., R. L. Leadbrand, R. C. Livingston, M. D. Cousins, C. L. Rino, B. C. Fair and R. A. Long (1978), Early results from the DNA wideband satellite experiments: Complex signal scintillation, *Radio Sci.*, **13**, 167-187.

Gochelashvili, K. S., and V. I. Shishov (1981), Propagation of reflected radiation in a randomly inhomogeneous medium, *Sov. J. Quantum Electron., Engl. Transl.*, **11**, 1182-1184.

Hewish, A. (1952), The diffraction of galactic radio waves as a method of investigating the irregular structure of the ionosphere, *Proc. Roy. Soc.*, **A214**, 492-514.

Hey, J. S., S. J. Parsons and J. W. Phillips (1946), Fluctuations in cosmic radiation at radio frequencies, *Nature* (London), **158**, 247.

Knepp, D. L., and H. L. F. Houpis (1991), Altair VHF/UHF observations of multipath and backscatter enhancement, *IEEE Trans. Antennas Propag.*, **AP-39**, 528-534.

Kudeki, E., and S. J. Franke (1986), Radar interferometer estimate of zonal drift variability during spread F, *Geophys. Res. Lett.*, **13**, 1117-1120.

Lee, L. C. (1974), Wave propagation in random medium: A complete set of moment equations with different wave numbers, *J. Math. Phys.*, **15**, 1413-1435.

Lee, M. C., A. Das Gupta, J. A. Klobuchar, S. Basu and S. Basu (1982), Depolarization of VHF geostationary satellite signals near the equatorial anomaly crests, *Radio Sci.*, **17**, 399-409.

Lee, S. Y., C. H. Liu and K. C. Yeh (1981), Waves in random continuum described by path integrals, in *Research Topics in EM Wave Theory*, edited by J. A. Kong, Chapter 2, Wiley, New York.

Liu, C. H. (1968), Wave propagation in a random medium with stratified background, *Radio Sci.*, **3**, 551-559.

Liu, C. H., and K. C. Yeh (1977), Model computations of power spectra for ionospheric scintillation at GHz frequencies, *J. Atmos. Terr. Phys.*, **39**, 149-156.

Lure, Y. M., C. C. Yang and K. C. Yeh (1989), Enhanced backscattering phenomenon in a random continuum, *Radio Sci.*, **24**, 147-159.

Marple, S. L., Jr. (1987), *Digital Spectral Analysis with Applications*, Prentice Hall, Inc., Englewood Cliffs, NJ.

Mimno, H. R. (1937), The physics of the ionosphere, *Rev. Modern Physics*, **9**, 1-43.

Ratcliffe, J. A. (1956), Some aspects of diffraction theory and their application to the ionosphere, *Rep. Progr. Phys.*, **19**, 188-267.

Rino, C. L. (1979), A power law phase screen model for ionospheric scintillation 2. Strong scatter, *Radio Sci.*, **14**, 1147-1155.

Rino, C. L. (1980), Numerical computations for a one-dimensional power law phase screen, *Radio Sci.*, **15**, 41-47.

Rino, C. L., and C. H. Liu (1982), On the parameterization of intensity scintillation of transionospheric radio signal, *Radio Sci.*, **17**, 279-284.

Rino, C. L., and J. Owen (1984), Numerical simulations of intensity scintillation using the power law phase screen model, *Radio Sci.*, **19**, 891-908.

Rino, C. L., and R. C. Livingston (1982), On the analysis and interpretation of spaced-receiver measurements of transionospheric radio waves, *Radio Sci.*, **17**, 845-854.

Shkarofsky, I. P. (1968), Generalized turbulence space-correlation and wave number spectrum-function pairs, *Can. J. Phys.*, **46**, 2133-2153.

Tatarskii, V. I. (1971), The effects of the turbulent atmosphere on wave propagation, *Natl. Tech. Inform. Serv.*, Springfield, VA.

Tinin, M. V., N. T. Afanasyev, S. M. Mikheev, A. P. Pobedina and O. V. Fridman (1992), On some problems of the theory of radio wave propagation in a randomly inhomogeneous ionosphere, *Radio Sci.*, **27**, 245-255.

Umeki, R., C. H. Liu and K. C. Yeh (1977), Multifrequency spectra of ionospheric amplitude scintillation, *J. Geophys. Res.*, **82**, 2752-2760.

Vacchione, J. D., S. J. Franke and K. C. Yeh (1987), A new analysis technique for estimating zonal irregularity drifts and variability in the equatorial F region using spaced receiver scintillation data, *Radio Sci.*, **22**, 745-756.

Vinogradov, A. G., Yu A. Kravtov and V. I. Tatarskii (1973), The effect of intensification on backscattering by bodies that are situated in a medium having random inhomogeneities, *Radiofizika*, **16**, 1064-1070.

Wagen, J.-F., and K. C. Yeh (1989), Simulation of HF propagation and angle of arrival in a turbulent ionosphere, *Radio Sci.*, **24**, 196-208.

Wernik, A. W., and C. H. Liu (1975), Application of the scintillation theory to ionospheric irregularities studies, *Artificial Satellites*, **10**, 37-58.

Wernik, A. W., C. H. Liu and K. C. Yeh (1983), Modeling of spaced-receiver scintillation measurements, *Radio Sci.*, **18**, 743-764.

Yang, C. C., and K. C. Yeh (1985), The behavior of the backscattered power from an intensely turbulent ionosphere, *Radio Sci.*, **20**, 319-324.

Yeh, K. C. (1983), Mutual coherence functions and intensities of backscattered signals in a turbulent medium, *Radio Sci.*, **18**, 159-165.

Yeh, K. C., and C. H. Liu (1967), Wave propagation in a random medium with anisotropic background, *IEEE Trans. Antennas Propagat.*, **AP-15**, 539-542.

Yeh, K. C., and C. H. Liu (1982), Radio wave scintillations in the ionosphere, *Proc. IEEE*, **70**, 324-360.

Yeh, K. C., and G. W. Swenson, Jr. (1964), F-region irregularities studied by scintillation of signals from satellites, *J. Res. Nat. Bur. Stand. (Radio Sci.)*, **68D**, 881-894.

Yeh, K. C., C. H. Liu and M. Y. Youakim (1975), A theoretical study of the ionospheric scintillation behavior caused by multiple scattering, *Radio Sci.*, **10**, 97-106.

Yeh, K. C., C. H. Liu and S. J. Franke (1985), Manifestations of multiple scattering characteristics in transionospheric radio signals, in *Multiple Scattering of Waves in Random Media and Random Rough Surfaces*, edited by V. V. Varadan and V. K. Varadan, The Pennsylvania State University, 291-310.

Figure Captions

Figure 1 Phase and amplitude of 140 MHz signals transmitted by the geostationary satellite ATS-6 and received at Ootacamund, India. The trend and detrended phase and amplitude are shown on the top two panels, respectively. The computed power spectra for the phase (solid line) and the amplitude (dashed line) are plotted on the bottom panel. The interlaced Fresnel nulls are indicated by vertical arrows. (After Bhattacharyya et al., 1992)

Figure 2 Illustrating frequency exponent n as a function of S_4 at 140 MHz. For each data bin the number of data points is indicated by a number, its average is plotted as a circle and its standard deviation is indicated by the vertical line. The formula used to compute the frequency exponent n is shown in the middle of the figure as insert. (After Bhattacharyya et al., 1990)

Figure 3 The correlation distance d_c as a function of frequency f for an ionosphere with f_{po} = 10 MHz, L = 100 or 400 km, p = 4, r_o = 300 m, z = 300 km and σ_N = 5%. (After Yeh et al., 1975).

Figure 4 The cross-correlation of the amplitude scintillation data spaced at 0, 122, 244 and 366 m apart. A locally frozen-in model is used with a drift velocity of 80 m/s and an rms velocity fluctuation of 20 m/s. (After Wernik et al., 1983)

Figure 5 Average radar cross-section in dB above m^2, the two-way scintillation index S_4 and the mean signal-to-noise ratio of ALTAIR radar returns at 422 MHz from Cosmos 1427. (After Knepp and Houpis, 1991)

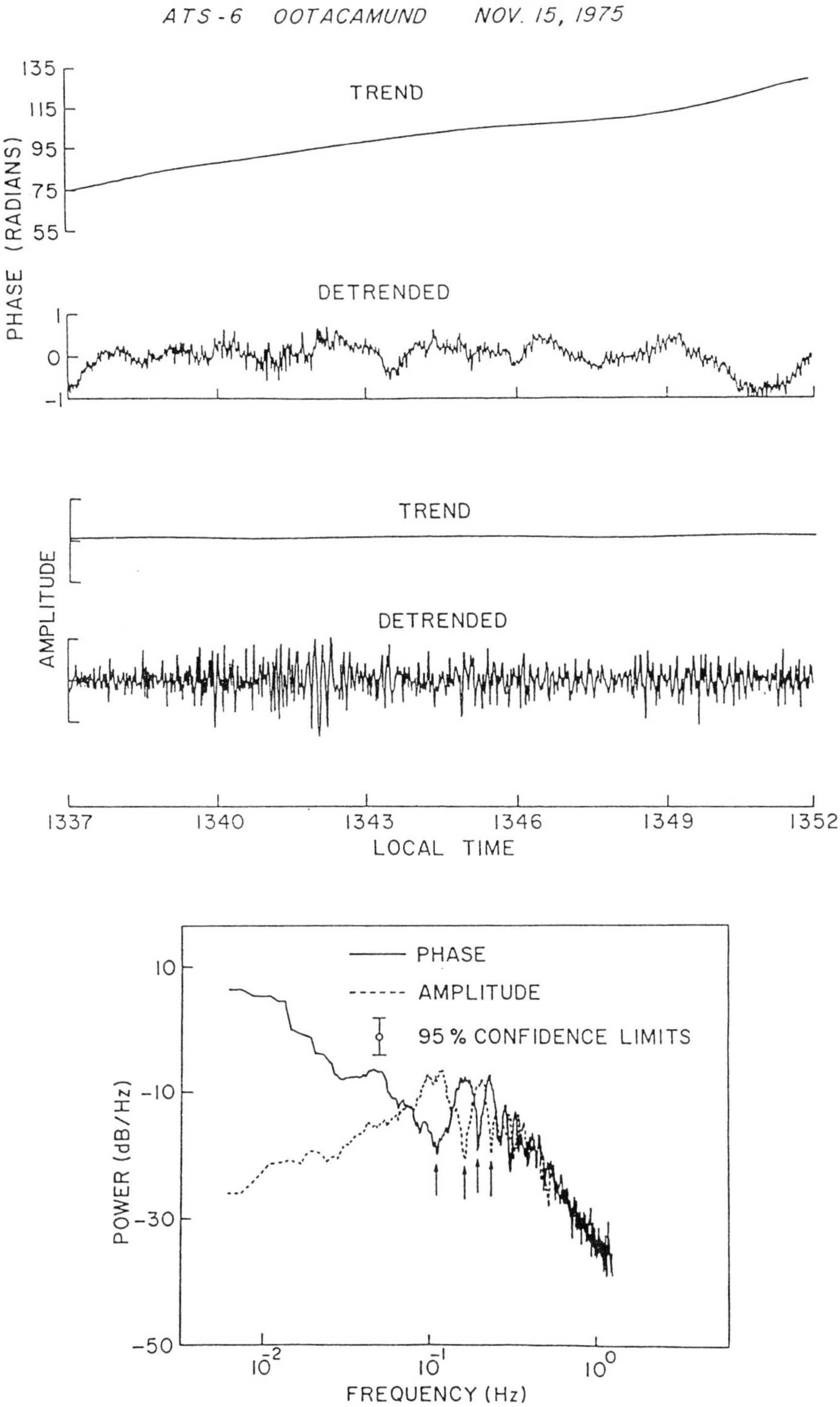
ATS-6 OOTACAMUND NOV. 15, 1975
PHASE (RADIANS)
135
115
95
75
55
TREND
DETRENDED
1
0
-1
AMPLITUDE
TREND
DETRENDED
1337
1340
1343
1346
1349
1352
LOCAL TIME
PHASE
AMPLITUDE
95 % CONFIDENCE LIMITS
POWER (dB/Hz)
10
-10
-30
-50
10^{-2}
10^{-1}
10^{0}
FREQUENCY (Hz)

Figure 1

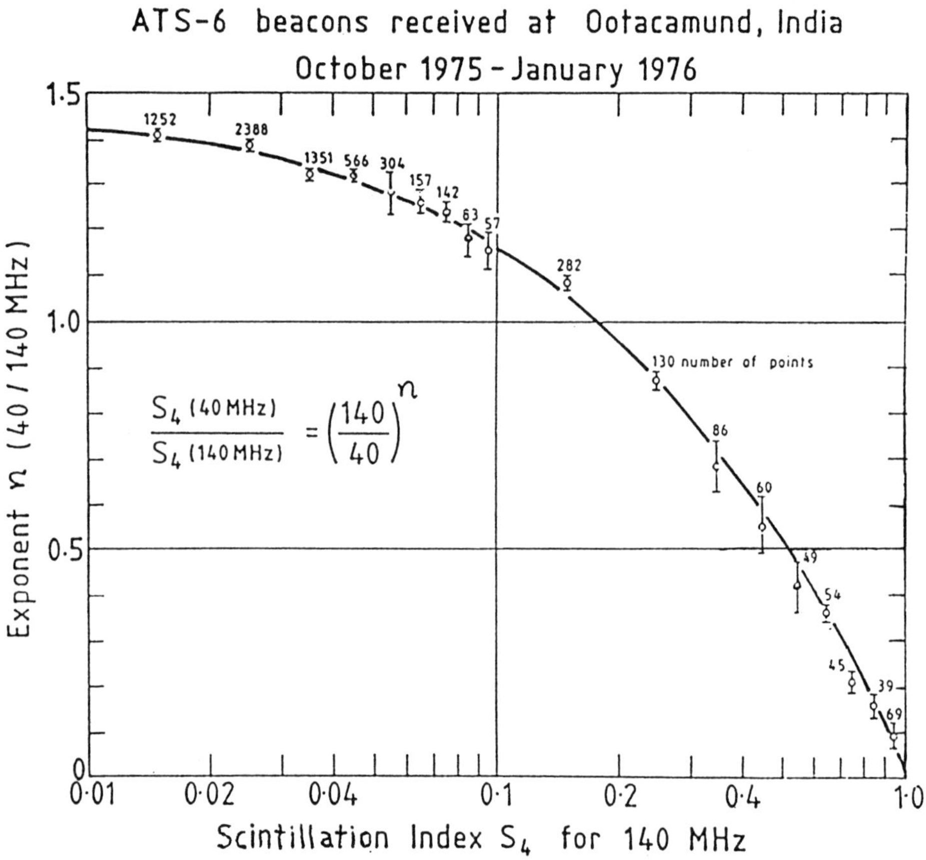
ATS-6 beacons received at Ootacamund, India
October 1975 - January 1976
Exponent n (40 / 140 MHz)
Scintillation Index S4 for 140 MHz
1.5
1.0
0.5
0
0·01
0·02
0·04
0·1
0·2
0·4
1·0
1252
2388
1351
566
304
157
142
63
57
282
130 number of points
86
60
49
54
45
39
69
S4 (40 MHz) / S4 (140 MHz) = (140/40)^n

Figure 2

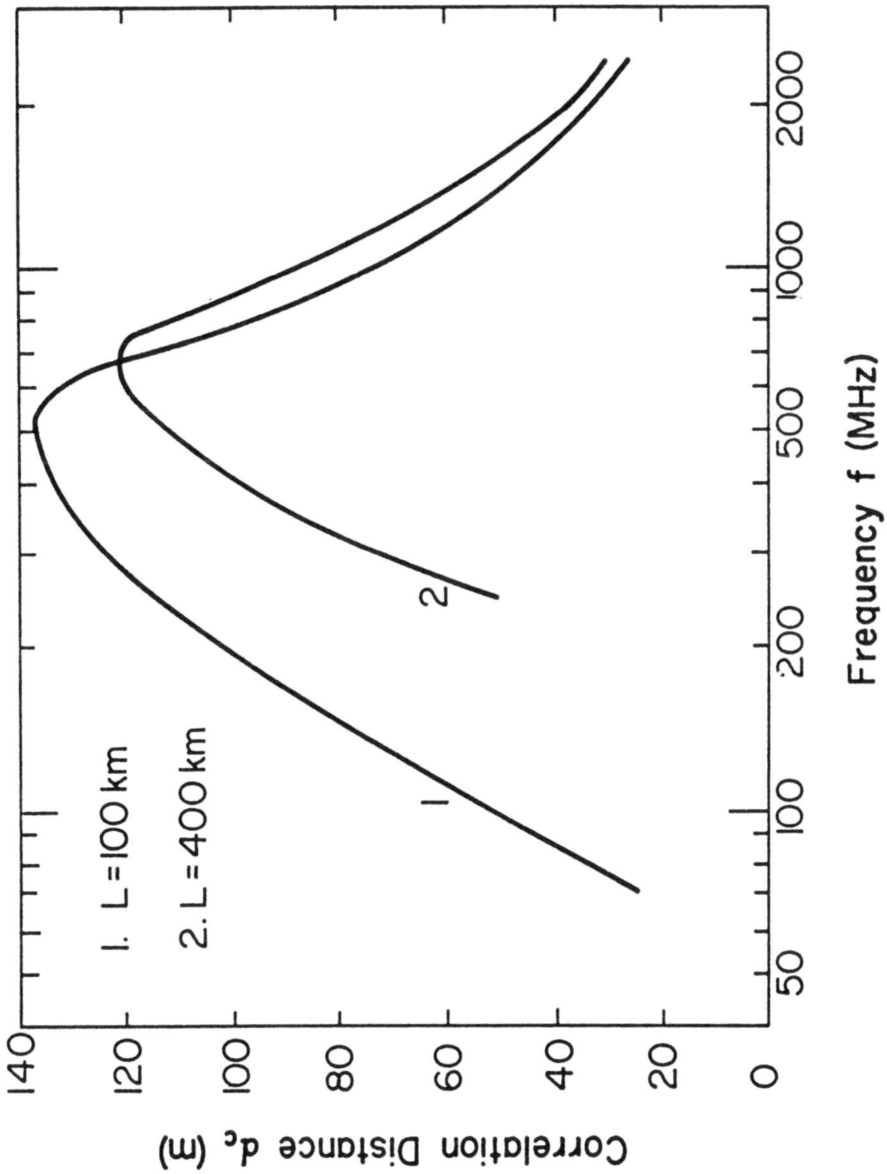

Correlation Distance d_c (m)
0
20
40
60
80
100
120
140
Frequency f (MHz)
50
100
200
500
1000
2000
1. L = 100 km
2. L = 400 km
1
2

Figure 3

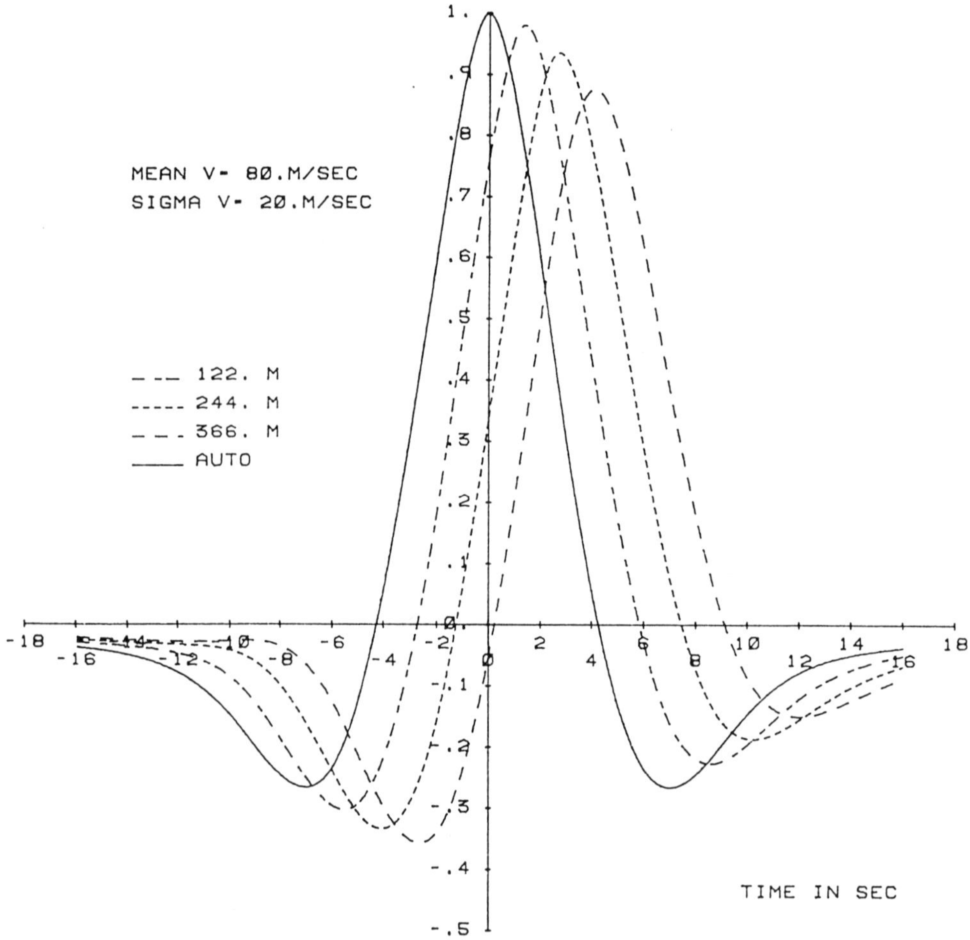

Figure 4

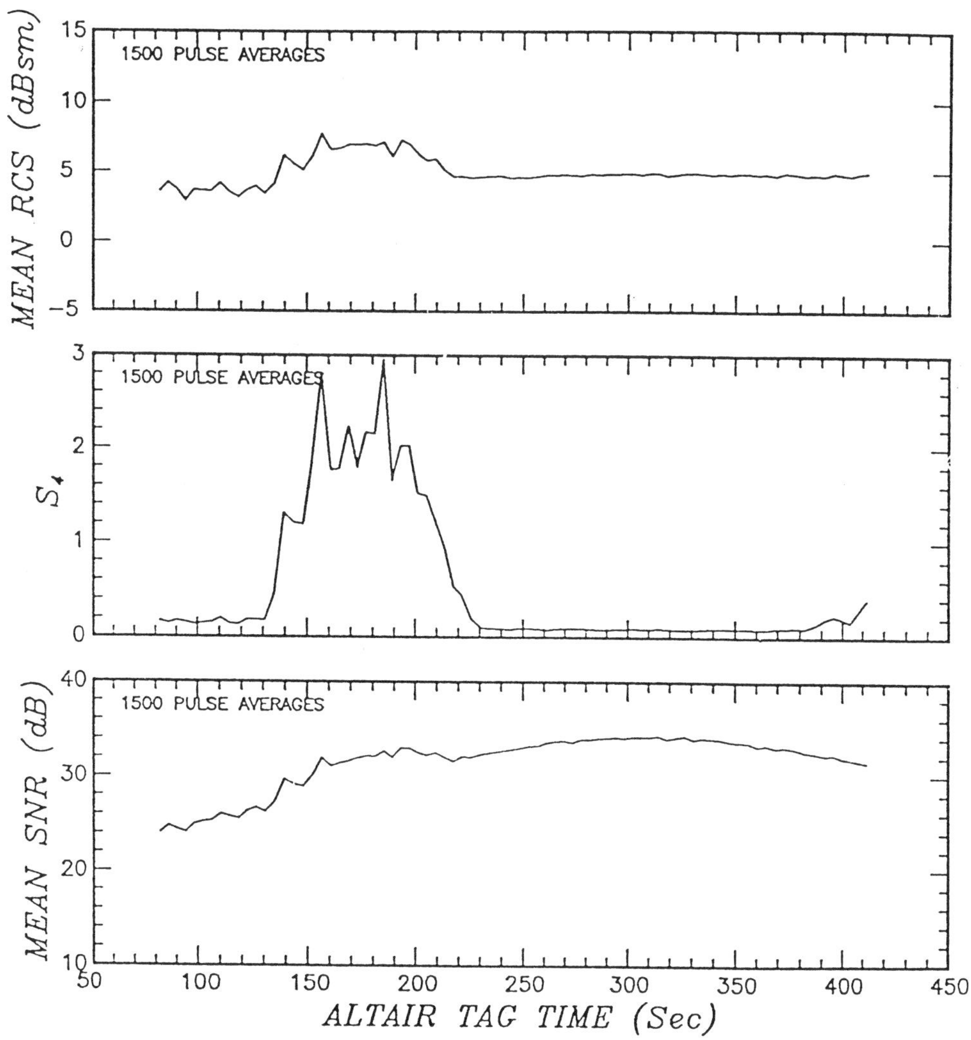

1500 PULSE AVERAGES
MEAN RCS (dBsm)
15
10
5
0
−5
1500 PULSE AVERAGES
S_4
3
2
1
0
1500 PULSE AVERAGES
MEAN SNR (dB)
40
30
20
10
50
100
150
200
250
300
350
400
450
ALTAIR TAG TIME (Sec)

Figure 5

SPACECRAFT RADIO SCINTILLATION AND SOLAR SYSTEM EXPLORATION

Richard Woo
Jet Propulsion Laboratory
California Institute of Technology
Pasadena, CA 91109, USA

1. Introduction

When a wave propagates through a turbulent medium, scattering by the random refractive index inhomogeneities can lead to a wide variety of phenomena that have been the subject of extensive study.[1–3] The observed scattering effects include amplitude or intensity scintillation, phase scintillation, angular broadening, and spectral broadening, among others. In this paper, I will refer to these scattering effects collectively as scintillation. Although the most familiar example is probably the twinkling of stars (light wave intensity scintillation by turbulence in the Earth's atmosphere), scintillation has been encountered and investigated in such diverse fields as ionospheric physics,[4–7] oceanography,[8] radio astronomy,[9–11] and radio and optical communications.[12,13]

Ever since planetary spacecraft began exploring the solar system, scintillation has appeared during the propagation of spacecraft radio signals through planetary atmospheres, planetary ionospheres, and the solar wind. Early studies of these phenomena were motivated by the potential adverse effects on communications and navigation,[14–17] and on experiments that use the radio link to conduct scientific investigations. Examples of the latter are radio occultation measurements (described below) of planetary atmospheres to deduce temperature profiles,[18,19] and the search for gravitational waves.[20] However, these concerns soon gave way to the emergence of spacecraft radio scintillation as a new scientific tool for exploring small-scale dynamics in planetary atmospheres and structure in the solar wind, complementing *in situ* and other remote sensing spacecraft measurements, as well as scintillation measurements using natural (celestial) radio sources.

The purpose of this paper is to briefly describe and review the solar system spacecraft radio scintillation observations, to summarize the salient features of wave propagation analyses employed in interpreting them, to underscore the unique remote sensing capabilities and scientific relevance of the scintillation measurements, and to highlight some of the scientific results obtained to date. Special emphasis is placed on comparing the remote sensing features of planetary and terrestrial scintillation measurements, and on contrasting spacecraft and natural radio source scintillation measurements. I will first discuss planetary atmospheres and ionospheres, and then the solar wind.

2. Planetary atmospheres

When a spacecraft flies behind a planet as seen from Earth, its radio signal penetrates the atmosphere of the planet (see Fig. 1). Because the spacecraft is occulted by the

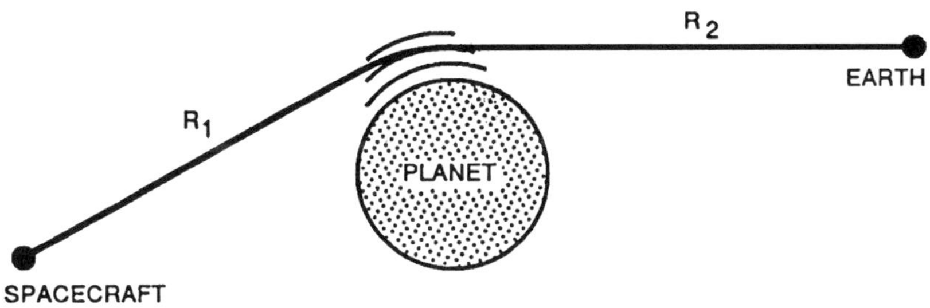

Fig. 1. Geometry of planetary radio occultation measurements. R_1 and R_2 are the limb-to-spacecraft and limb-to-Earth distances, respectively.

atmosphere, measurements made in this fashion are referred to as radio occultation measurements. Since first conducted during the Mariner IV flyby of Mars,[21] the radio occultation method has served as a workhorse of planetary exploration by providing vertical profiles of total density in the neutral atmosphere and plasma density in the ionosphere. As in propagation in the terrestrial atmosphere, radio scattering by both electron density irregularities in the ionosphere and temperature inhomogeneities in the neutral atmosphere can give rise to scintillation phenomena in radio occultation measurements.

Unlike with the terrestrial ionosphere and the solar wind, intensity scintillation measurements of planetary atmospheres using celestial radio sources were not possible before the advent of space probes. The Fresnel size R_F (radius of the first Fresnel zone) for occultation measurements is given by $(\lambda R)^{1/2}$, where $R = R_1R_2/(R_1+R_2)$, and R_1 and R_2 are the spacecraft-to-limb and limb-to-Earth distances, respectively (see Fig. 1). In the case of spacecraft occultations, the spacecraft is much closer to the planet than Earth, so that $R_2 \gg R_1$ and $R_F \approx (\lambda R_1)^{1/2}$. Typically, $R_1 \sim 10^4 - 10^5$ km, so that for an observing frequency at S-band (2.3 GHz or 13-cm wavelength), R_F is several kilometers, and smaller than the scale height of the atmosphere. For a celestial radio source, $R_1 \gg R_2$, and $R_F \approx (\lambda R_2)^{1/2}$. Even if the planet were only 1 AU from Earth, R_F at S-band would be 140 km, greatly exceeding the scale height of the atmosphere. Intensity scintillation measurements of planetary atmospheres at centimeter wavelengths using natural radio sources are, therefore, not feasible. At optical wavelengths, the Fresnel size $(\lambda R_2)^{1/2}$ is indeed small enough that stellar occultations have been used to probe planetary atmospheres.[22] However, centimeter wavelength spacecraft measurements are complementary, as they probe the much deeper atmospheric regions that are not accessible to Earth-based optical measurements. Moreover, spacecraft radio scintillation measurements so far represent the only means by which we can measure and study small-scale structures in these deeper regions.

Entry probes have also explored the atmospheres of the planets, and their radio links can be used for observing scintillation. In these missions, R_F is given by $(\lambda L)^{1/2}$, where L is the propagation distance through the atmosphere. When compared with the spacecraft-to-limb distance for a flyby or orbiting spacecraft, the shorter propagation distance L results in a significantly smaller Fresnel size, and hence much weaker amplitude scintillation for entry probe missions.

When comparing scintillation observed during radio occultation with scintillation measurements of the terrestrial atmosphere and ionsophere, it is important to remember that, owing to the larger Fresnel size, radio occultation measurements probe scale sizes that are significantly larger. Also, during occultation measurements, the radio path sweeps across the atmosphere, so that the frozen-in hypothesis used in translating temporal to spatial scales during wave propagation analyses is well satisfied. Although this sweeping motion precludes measuring winds in the atmosphere, it does provide useful information on the variation of scintillation level with altitude.

2.1 Weak scattering theory

The development of theoretical analyses for interpreting scintillation during propagation in turbulent planetary atmospheres has closely followed the evolution of spacecraft radio measurements. Fortunately, in many instances, the observed scintillation has been strong enough for clear detection, yet sufficiently weak for weak scattering theory to apply. Radio scintillation by a planetary atmosphere was first investigated in the case of Venus using both entry probe[23–25] and occultation measurements,[26] and based on the variance of amplitude scintillation. Unfortunately, the early studies of Venus were confused by substantial differences between the levels of turbulence inferred from flyby and entry probe missions.[26–29] These inconsistencies led to wave propagation analyses for the variance[30–31] and frequency spectra[32] of both amplitude and phase scintillation based on a turbulent atmosphere whose extent and outer scale of turbulence L_0 were both smaller than R_F, and whose inhomogeneities were isotropic and described by the Kolmogorov spectrum. These theoretical results were used to model-fit the Mariner 5 amplitude scintillation measurements, and obtain information on the characteristics of the turbulence in the atmosphere of Venus.[33]

The discrepancy between radio occultation and entry probe measurements was resolved with two events. First, the addition of a second frequency at X-band (8.4 GHz, 3.6-cm wavelength) on the flyby mission Mariner 10 made it possible to determine the frequency dependence of the amplitude scintillation, and consequently conclude that L_0 was not smaller, but in fact larger, than R_F.[34] Second, analysis of the Pioneer Venus entry probe scintillation measurements showed that the observed amplitude fluctuations can be dominated by the effects of probe motion combined with the nonuniform probe antenna power pattern, rather than atmospheric propagation.[35]

An important effect not included in the interpretation of the Mariner 10 scintillation measurements was that of defocusing due to differential refraction, which reduces the Fresnel size in the vertical direction[36–40] and causes attenuation in the signal.[41] When defocusing was considered, the computed power spectra did not fit the data as well as that which excluded defocusing.[37,42] For the computed spectra that accounted for defocusing to match the measurements, the turbulence model had to undergo one more iteration, a change from isotropic to anisotropic turbulence elongated in the horizontal direction.[43] Stellar occultation measurements conducted with an Earth-orbiting satellite have found similar large-scale horizontally elongated inhomogeneities in the upper terrestrial atmosphere.[44,45]

Shown in Fig. 2 are computed examples of the power spectra of log-amplitude scintillation during radio occultation.[43] Fluctuation frequency ω is normalized to the free space Fresnel frequency $\omega_F = 2\pi v(\pi R_F)^{-1/2}$. The magnitude v of the transverse ray path

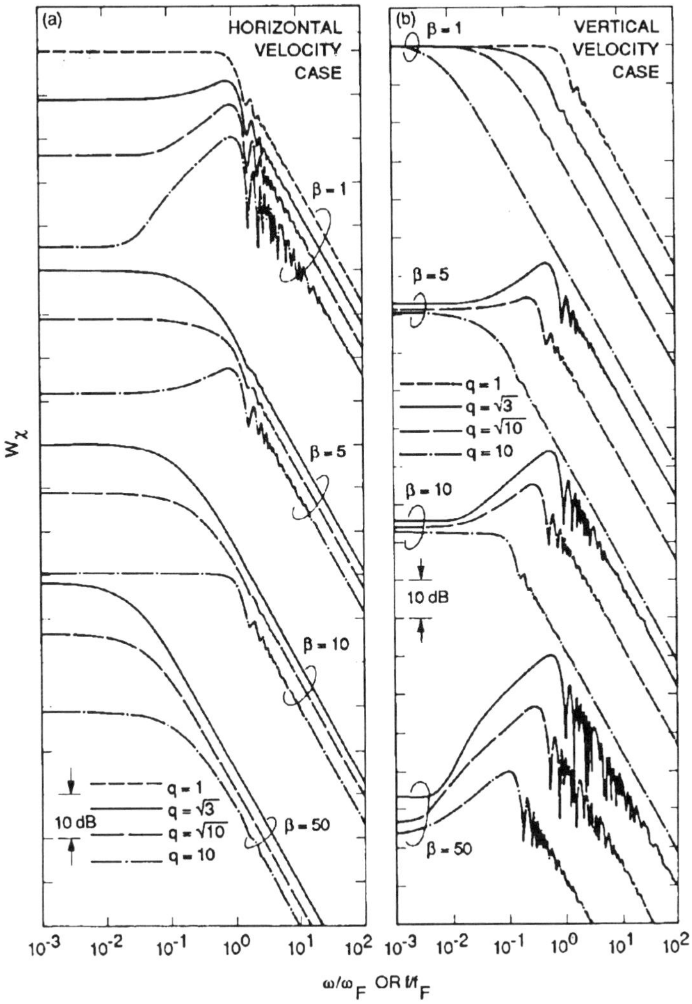

Fig. 2. Sample frequency spectra of log-amplitude scintillation during radio occultation by a planetary atmosphere, illustrating the dependence on inhomogeneity axial ratio β, defocussing attenuation q^2, and transverse radio path velocity **V**.[43] The results are for the Kolmogorov spectrum (p=11/3). The left and right panels correspond to the horizontal ($v_v = 0$) and vertical velocity ($v_h = 0$) cases, respectively.

velocity **V** is given by $v^2 = v_h^2 + v_v^2$, where v_h and v_v are the horizontal and vertical components of **V**, respectively. The spectra in Fig. 2 correspond to the vertical ($v_h = 0$) and horizontal ($v_v = 0$) velocity cases. The effect of defocusing was accounted for by using geometrical optics to describe the unperturbed fields in the first Rytov solution. The parameter q^2 is the power reduction factor due to defocusing. The anisotropic refractive index inhomogeneities were assumed to be elongated in the horizontal direction with axial ratio β, and their spatial wavenumber spectrum described by a power-law with three-dimensional spectral index p (where p = 11/3 is the Kolmogorov case) and outer scale size

L_o. Similar results have been obtained for phase scintillation, and the coherences of amplitude and phase scintillation operating at one or two frequencies.[43]

As seen from Fig. 2, the amplitude scintillation spectra depend strongly on four parameters: q, β, p and V. However, only two of these are free parameters, because V is known from the spacecraft trajectory and occultation geometry, and q from the observed radio signal attenuation. Moreover, p and β are essentially independently determined, since β shifts the spectrum in frequency, while the spectral index p changes the slope of the high-frequency asymptote of the spectrum.[43] The spectra are independent of L_o because $L_o \gg R_F$. It is interesting to note that when $\beta = q$, the shape of the spectrum is identical to that in which the irregularities are isotropic and defocusing is absent. In other words, from the standpoint of the shape of the spectrum, defocusing effects can be offset by the anisotropy of the irregularities. It is for this reason that reasonable agreement was obtained when the Mariner 5 and 10 spectra were originally fit with theoretical spectra that did not account for defocusing but assumed isotropic irregularities.[34]

2.2 Strong scattering theory

Before discussing application of the weak scattering results, it is important to point out that strong scattering has been observed in propagation through the tropospheres of the outer planets and Titan,[46] and analysis of these data require theoretical results that account for strong scattering. The strong scattering case for planetary radio occultation measurements has been examined[47] and useful numerical calculations of the spectra carried out.[46] The computed spectra follow and extend earlier studies[48–50] to include defocusing, anisotropic irregularities and a range of values for spectral index p. One difference between weak and strong scattering spectra is that there are more free parameters in the strong scattering case, making it more difficult to meaningfully fit the data. To my knowledge, no analyses of strong amplitude scintillation observations in the presence of defocusing have been published.

2.3 Venus

Since radio occultation measurements by Mariner 5 first revealed regions of enhanced amplitude scintillation located around 45 and 60 km,[33] extensive radio occultation measurements[34,51–55] have shown that they are a permanent feature of the atmosphere of Venus. Radio occultation measurements of Venus are limited by critical refraction to altitudes higher than about 35 km.[56] Subsequent *in situ* measurements by the Pioneer Venus probes[57] showed that the regions of enhanced scintillation coincided with those of high static stability. It is interesting that the VEGA balloon measurements detected considerable wind fluctuation near 55 km,[58] where scintillation was minimal. These results are consistent with each other, since the atmosphere at this altitude is neutral, and small temperature fluctuations would be expected.[59]

The orbiting spacecraft Pioneer Venus was equipped with both S- and X-band radio links. Shown in Fig. 3 is a typical time history of the Pioneer Venus S-band signal amplitude during an entrance occultation;[52] the time history for the X-band signal is similar. Region A corresponds to the upper region of enhanced scintillation located in the vicinity of the

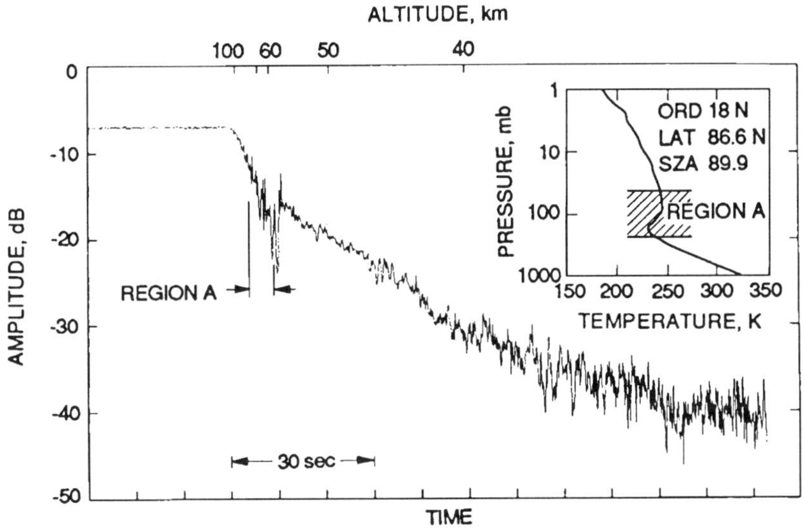

Fig. 3. Time history of Pioneer Venus S-band signal amplitude during radio occultation by the atmosphere of Venus.[52] Ray path altitudes are indicated along the top of the figure. Region A corresponds to the upper region of enhanced scintillation in the vicinity of 60-km altitude. The inset pressure-temperature profile deduced from the radio occultation measurements[56] shows that region A is located around the tropopause of Venus.

tropopause (see inserted temperature profile deduced from Doppler measurements of the atmosphere[56]). The observed attenuation in this region is due primarily to differential refraction or defocusing rather than to microwave absorption.[60] Comparison of measurements from adjacent occultations indicated that the scintillation was more representative of random turbulence than permanent layers. After detrending the S- and X-band signals by fitting quadratic time functions, the log-amplitude frequency spectra and the log-amplitude coherence coh_χ were computed and are displayed in Fig. 4. The observed spectra and coh_χ were fit with the theoretical spectra described in the previous section, and the results superimposed. The theoretical curves correspond to Kolmogorov turbulence ($p = 11/3$) and anisotropic irregularities with axial ratio $\beta = 20$ (q is estimated to be $10^{1/2}$ from Fig. 3). The agreement between the observed and theoretical results is remarkable, especially in view of the constraints imposed by the relationship between S- and X-band scintillation and the paucity of free parameters. Similar fits with large axial ratio β have been obtained for the lower region of turbulence, located around 45 km.[55] The structure constants of refractive index and temperature fluctuations for the upper region of enhanced scintillation[55] are at least an order of magnitude higher than those found near the terrestrial tropopause.[61–63] However, the temperature inhomogeneities in the terrestrial measurements are considerably smaller than the kilometer scale sizes probed in the radio occultation measurements.

To summarize the characteristics of the temperature inhomogeneities, good agreement is found between the observed and theoretical spectra when the inhomogeneities are modeled by a spatial wavenumber spectrum that is power-law and approximately Kolmogorov for scale sizes smaller than R_F (~1 km), and when the inhomogeneities are elongated in the horizontal direction with axial ratio β (ratio of horizontal to vertical scales) greater than 10.

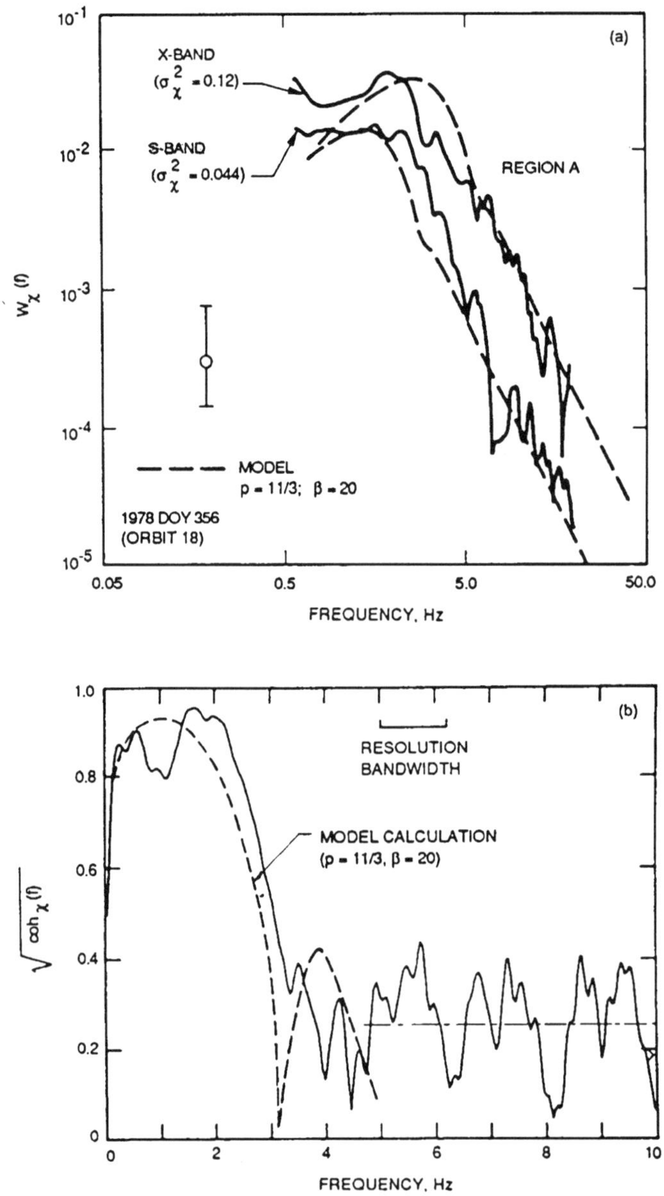

Fig. 4. Comparison of theoretical model (dashed) and observed (solid) frequency spectra and coherence of Pioneer Venus S- and X-band log-amplitude scintillation observed in region A.[52]

While these inferred characteristics do not make any assumptions about the small-scale dynamical processes in the atmosphere, they provide clues.

The anisotropy of the inhomogeneities and the coincidence of the regions of enhanced scintillations with regions of peak static stability are consistent with the scintillation being a manifestation of trapped gravity waves.[52,64] On the other hand, the Kolmogorov spectrum suggests atmospheric turbulence. The upper atmosphere of Venus is stably stratified and similar to that of Earth, where waves and turbulence have both been observed. Identification of the dynamical processes in this region is hampered by the possibility that gravity waves and turbulence may coexist, and the difficulty of reliably distinguishing gravity waves and turbulence without the benefit of high-resolution temperature and wind measurements.[65] For this reason, atmospheric turbulence has also been considered by assuming that it exists in thin layers, and order-of-magnitude estimates of the eddy diffusion coefficient K and the energy dissipation rate ε obtained.[59]

2.4 Titan

Both weak and strong amplitude scintillations have been observed by Voyager 1 during S- and X-band radio occultation measurements of the atmosphere of Titan,[66] a satellite of Saturn. As with Venus, good agreement was found when theoretical weak scattering intensity scintillation spectra similar to those described above were compared with the observed spectra around the altitude of 44 km. Refractivity index inhomogeneities in the scale size range of a few hundred meters to a few kilometers were found to be horizontally elongated and to exhibit a power-law spectrum. In the entry occultation case, $\beta = 4$ and $p = 3.1$, while in the exit case, $\beta = 6$ and $p = 3.7$.

The altitude profile of scintillation index m was estimated based on refractive index or density fluctuations caused by a vertically propagating gravity wave and using Voyager 1's radio occultation measurements of the background thermal structure of Titan's atmosphere.[67] The agreement between the estimated and observed altitude profiles of m provides support for a hypothesis of small-scale stratospheric gravity waves in the atmosphere of Titan, possibly driven by a cloud-free convective region in the lowest few kilometers of the atmosphere.[66] Estimates of net upward flux of energy and horizontal momentum due to the gravity waves were also obtained.

2.5 Uranus

Phase and amplitude scintillation observed by Voyager 2 during S- and X-band radio occultation measurements of Uranus have been used to identify and characterize an equatorial wave in the stratosphere of Uranus.[68] Similar to the case of Titan, but using a more rigorous theory for atmospheric waves, the observed quasi-periodic vertical variations in atmospheric density were shown to be consistent with either an eastward or westward (relative to the mean flow) inertia-gravity wave mode. Comparison between entry and exit occultation measurements provided constraints on the meridional and zonal structure of the wave. All Kelvin, Yanai and Rossby wave modes were inconsistent with the observations. Estimates of eddy diffusion coefficient K were obtained based on expectations of wave breaking. A spectral analysis of the observed phase and amplitude scintillation was not performed in this study.

3. Planetary ionospheres

Planetary ionospheric scintillation was first detected and investigated during Pioneer 10 and 11 S-band radio occultation measurements of Jupiter.[69,70] The observed attenuation due to differential refraction was very small because the ionospheric scale height is large and the angle of refraction small. When analyzing ionospheric scintillation measurements, the effect of defocusing can, therefore, be neglected. The electron density irregularities were observed to be anisotropic and aligned in a preferred direction throughout the ionosphere. If the irregularities are assumed to be aligned along the magnetic field, as is most likely the case, ionospheric scintillation provides a method for observing the magnetic field direction in the ionosphere of magnetic planets for which *in situ* measurements have so far not been available. This information can be useful in refining the planetary magnetic field models, which are based on *in situ* magnetic field measurements made by spacecraft that are farther away from the planet (to avoid radiation effects and ring hazards) and, therefore, less sensitive to the higher order terms of the spherical harmonic expansion of the field.[71–73]

3.1 Weak scattering theory

Theoretical power spectra of log-amplitude and phase scintillation based on the first Rytov solution have been obtained for planetary ionospheric scintillation;[70] typical results are shown in Fig. 5. The anisotropic electron density irregularities with axial ratio β are described by a power-law spectrum with three-dimensional spectral index p and outer scale size L_o, and are assumed to be aligned along the magnetic field direction (ϕ, ψ), where ϕ is the angle that the magnetic field in the plane of the sky makes with the transverse velocity of the line-of-sight path (or **V** as defined in Section 2.1), and ψ the angle the magnetic field makes with the line-of-sight. The spectra do not depend on L_o because $L_o \gg R_F$. Since the planetary magnetic field is nearly in the plane of the sky ($\psi = 90°$), and the spectra do not vary in the range of $\psi = 70$–$90°$, the spectra are essentially independent of ψ.[70] As can be seen from Fig. 5, the spectra are very weakly dependent on β, but strongly dependent on the angle of most interest ϕ, the magnetic field orientation in the plane of the sky. Since the spectra are essentially dependent on just ϕ, which has a broad range of values, ϕ is well determined except for two ambiguities: one stems from two possible mirror image solutions of ϕ (since the spectra depend only on the absolute value of ϕ); the other from the fact that ϕ is not affected by a reversal in magnetic field direction.

3.2 Strong scattering measurements

Monochromatic spacecraft radio signals can undergo spectral broadening as they propagate through planetary ionospheres during radio occultation. Measurements of this scattering phenomenon, first observed during measurements of the solar corona,[74] complement the measurements of amplitude and phase scintillation, and are especially useful for studying the electron density spectrum under strong scattering conditions. This is due to the fact that interpretation of spectral broadening measurements depends on the second moment theory, which is better understood under strong scattering conditions than the fourth moment required for amplitude scintillation.[75] The use of spectral broadening observations for this purpose has been demonstrated using Pioneer 10 and 11 measurements of the Jovian ionosphere under weak scattering conditions,[76] and in measurements of the solar wind under strong scattering conditions.[75] Similar observations have been conducted of the terrestrial ionosphere using satellite radio signals.[77]

3.3 Jupiter

Shown in Fig. 6 are the model and observed frequency spectra of the S-band log-amplitude scintillation during Pioneer 11 entry occultation measurements of the Jovian ionosphere.[70] The corresponding regions in the ionosphere are indicated on the electron

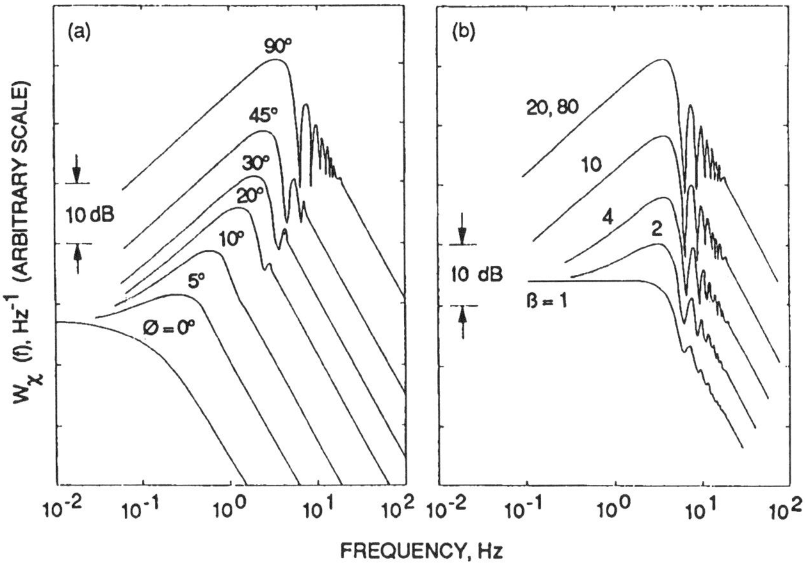

Fig. 5. Sample frequency spectra of S-band log-amplitude scintillation during radio occultation by the ionosphere of a magnetic planet, illustrating the strong dependence on magnetic field direction ϕ in the plane of the sky with respect to the transverse velocity V of the ray path (left panel: $\psi = 90°$), and weak dependence on axial ratio β of the electron density irregularities for large β (right panel: $\phi = 0°$).[70]

density profile deduced from the radio occultation measurements.[78] These results illustrate that, as the radio path descends through the ionosphere, amplitude scintillation increases, and eventually rises above the system noise level.

The Pioneer 10 and 11 results show that, while the electron density irregularities are isotropic in the collision-dominated lower ionosphere, they are anisotropic with large axial ratio ($\beta > 5$) in the upper ionosphere. The irregularities are described by a power-law spectrum that is close to Kolmogorov for scale sizes smaller than the Fresnel size of about 6 km. Comparison of magnetic field orientation ϕ deduced from the Pioneer measurements with magnetic field models show differences of up to 10° with magnetic field models.[71]

Further results have been obtained from Voyager 1 and 2 radio occultation measurements, which were carried out at both S- and X-bands.[79] The Fresnel size in the case of the Voyager measurements was larger on account of the longer fly-by distances of the Voyager spacecraft. More important, the addition of X-band made it possible to observe phase scintillation, and hence obtain information on scale sizes exceeding the Fresnel size. The Voyager measurements showed that over the approximate scale size range of 1–200 km, the spatial wavenumber spectrum was power-law with spectral index p in the range of 3.0–3.7, and axial ratio β in the range of 2–10. No evidence for either an inner or outer scale was found. Occultation deduced magnetic field orientation and spacecraft-based magnetic field models[71–73] differed by as much as 13° near the equator. The differences in magnetic field orientation observed by Pioneer and Voyager imply that a series of occultation measure-

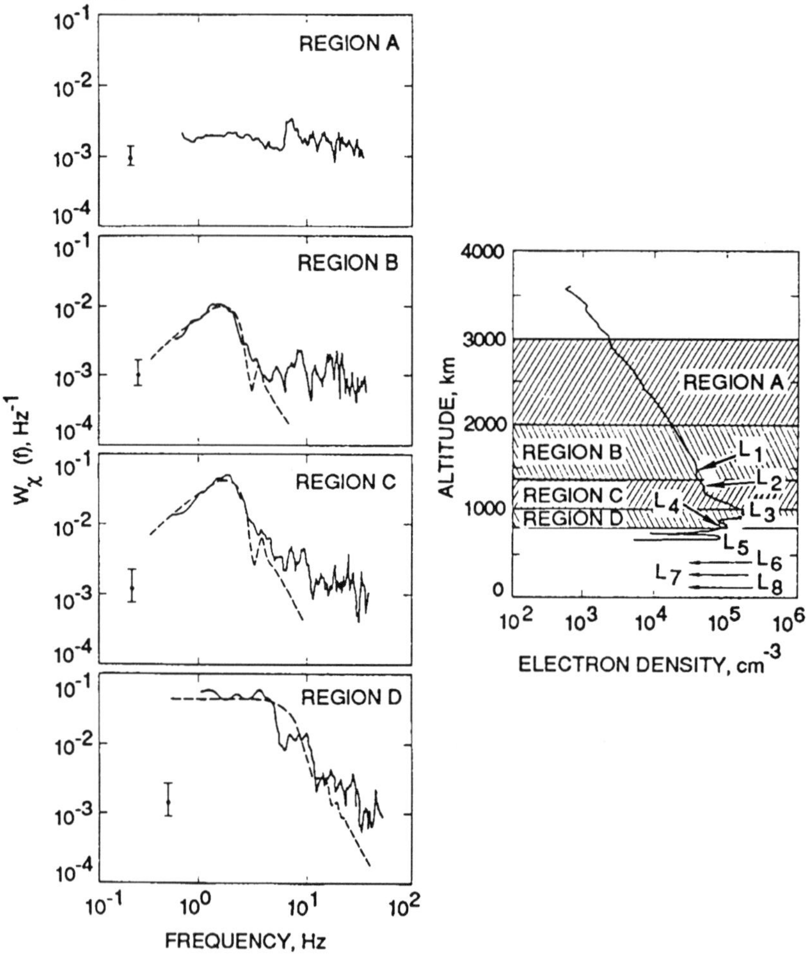

Fig. 6. Comparison of theoretical model (solid) and observed (dashed) frequency spectra of Pioneer 11 S-band amplitude scintillation as the radio path descended through the Jovian ionosphere.[70] For the model spectra, $\beta = 40$ (except in region D for which $\beta = 1$) and $\phi = 15°$. The corresponding altitude ranges over which the spectra were observed are indicated on the electron density profile deduced from the Pioneer 11 radio occultation measurement.[78]

ments at a number of locations in the Jovian ionosphere, as is planned during the Galileo encounter of Jupiter,[80] might contribute significantly to the refinement of the Jovian magnetic field structure.

3.4 Saturn

Although Pioneer 11 was the first spacecraft to explore Saturn (within the instrumental limits of its radio system) neither intensity scintillation nor spectral broadening was detected in the ionosphere.[80] Amplitude and phase scintillation at both S- and X-band frequencies were, however, observed during radio occultations by Voyager 1 and 2.[81] As in the case of Jupiter, good agreement was obtained when fitting the observed amplitude and phase spectra with computed spectra based on weak scattering theory. In the case of Saturn, the ionospheric irregularities in the approximate scale size range of 2-60 km followed a power-law spectrum with spectral index near 3.5 and axial ratio $\beta > 4$. The differences between occultation measurements and magnetic field models[82-83] in the case of Saturn appear smaller than those in the case of Jupiter, but also suggest a small adjustment of model parameters that might improve the field representation.[81]

3.5 Venus

Despite extensive radio occultation measurements of the non-magnetic planet Venus following the Mariner 5 observations in 1967, only recently has S-band scintillation due to the ionosphere been detected.[84] This stems from the fact that S-band ionospheric scintillation is apparently confined to the region near the subsolar region. It was first observed in the radio link of the orbiting spacecraft Pioneer Venus when Pioneer Venus traversed the upper ionosphere of Venus. High-resolution electron density measurements by the on-board Langmuir probe,[85] as well as propagation calculations, confirmed that the scintillation was caused by electron density irregularities located in the topside ionosphere but below the ionopause.

Simultaneous Pioneer Venus plasma and magnetic field measurements revealed that these topside plasma irregularities were a consequence of the plasma's reaction to maintain pressure balance during the penetration of large-scale magnetic fields in the ionosphere, a situation that occurs most frequently in the subsolar region when the solar wind dynamic plasma pressure exceeds the ionospheric plasma pressure.[86,87] As with large-scale magnetic fields, the disturbed plasma and resulting scintillation are a manifestation of high-dynamic solar wind interaction with the ionosphere. Remote sensing radio occultation measurements are, therefore, useful for investigating magnetization of the ionosphere due to high-dynamic solar wind interaction when *in situ* magnetic field measurements are not available, such as during solar minimum conditions.[88]

The global morphology of ionospheric scintillation at Venus is quite different from that of the terrestrial ionosphere, where scintillation is observed in both polar and equatorial regions, with peaks occurring during nighttime, and scintillation activity higher during solar maximum than solar minimum.[7] In the non-magnetic planet of Venus, scintillation occurs in the subsolar region and falls off quickly with increasing solar zenith angle. It is more pervasive and extends to slightly higher solar zenith angles during solar minimum, when the Venus ionospheric pressure is lower than at solar maximum. This decrease in pressure is caused not only by a drop in solar ionizing radiation flux,[89] but also by changes in the underlying thermosphere.[90]

3.6 Mars

Our understanding of the solar wind interaction with the ionosphere of Mars is poor because of the lack of *in situ* magnetic field measurements of the Martian ionosphere. Since the solar wind interaction may be expected to be similar to that of Venus,[91] scintillation in the subsolar region may also be expected. Unfortunately, radio occultation measurements of Mars are necessarily limited to solar zenith angles greater than about 46° by the Mars-Earth-Sun geometry, and, within instrumental limits, these have not revealed scintillation.[88]

4. Solar wind

Since *in situ* plasma measurements by Mariner 2 confirmed its existence in 1962,[92] many spacecraft have probed the solar wind directly and our knowledge has grown enormously. *In situ* measurements, however, have been confined to the ecliptic plane and to heliocentric distances greater than approximately 0.3 AU. One exception in the near future is Ulysses, which is on its way to probing the polar regions of the solar wind, albeit at distances of 1.7–2.9 AU.[93] Because of the absence of near-Sun observations, many outstanding questions remain.[94] Some of these can only be answered when a mission such as Solar Probe,[94] which will send a spacecraft to within a few solar radii of the sun, is finally flown. In the meantime, scintillation measurements serve as an invaluable means for observing these unexplored regions of the solar wind.

Unlike the case of planetary atmospheres, radio scintillation and scattering measurements of the solar wind or interplanetary medium have not only been possible with celestial radio sources, they have been conducted extensively. The earliest observations were of angular broadening,[95–97] but these were soon joined by intensity scintillation (often called IPS, for interplanetary scintillation) using compact radio sources.[98] Intensity scintillation measurements of the solar wind are possible, because, although the Fresnel size is comparable to that of the planetary case (~140 km at S-band), it is dwarfed by the vast expanse of the solar wind. Intensity scintillation is a path-integrated measurement of electron density fluctuation. However, due to the rapid fall-off of electron density fluctuation with heliocentric distance, intensity scintillation essentially probes the region of solar wind in the vicinity of the closest approach point.[9]

For over two decades, extensive IPS measurements have contributed much to our knowledge of the properties of the solar wind in regions inaccessible to direct measurement. For instance, IPS measurements have shown the acceleration of the near-Sun solar wind,[99–101] the existence of a turbulent envelope surrounding the Sun,[99,102] the high anisotropy of electron density irregularities near the Sun,[100] the latitudinal and solar cycle variation of the solar wind beyond 0.5 AU,[103,104] and the three-dimensional extent of interplanetary disturbances beyond 0.5 AU.[105–107]

4.1 Spacecraft radio scintillation observations

Intensity scintillation measurements can also be carried out with spacecraft radio signals during occultation by the solar wind. At centimeter wavelengths, spacecraft radio signals are ideally suited for monitoring and probing the near-Sun solar wind. But, since spacecraft generally venture no closer than within about 0.5 AU of the Sun, the Fresnel size for both spacecraft and celestial radio sources is comparable. One difference between planetary and solar wind measurements is that, since the motion of the spacecraft is slower than the solar wind speed, spacecraft radio scintillation measurements, like natural radio source measurements, can provide information on the wind.

At first glance, spacecraft signals may not appear to offer substantial advantages over natural radio sources. Intensity scintillation using point-source spacecraft radio signals are of course easier to interpret, since intrinsic source-size effects obviously need not be considered. The principal advantages of spacecraft radio signals, however, lie in the fact that they are coherent, so that spectral broadening[75] and phase[108,109] (or correspondingly Doppler, since Doppler can be obtained from phase by differentiating the phase time series[110]) scintillation are additional scattering measurements that can be made. These observations complement the extensive IPS measurements by providing new abilities for remote sensing of both small and large-scale phenomena in the solar wind.[111–112] Let me illustrate.

Fig. 7 shows the normalized rms of intensity scintillation, scintillation index m, that has been observed with Viking and Helios radio signals at S-band (13-cm wavelength) as a function of heliocentric distance in solar radii.[113] Similar results have been obtained with other spacecraft and observing frequencies.[114–116] The $R^{-1.55}$ radial dependence of m under weak scattering conditions implies a fall-off of $R^{-2.05}$ for electron density fluctuation, a result consistent with that obtained from natural radio source measurements.[117,118] The results in Fig. 7 show the limited heliocentric distance range over which intensity scintillation can be observed. Closer than about 10 R_o, the intensity scintillation is strong and m saturates near unity, while beyond about 40 R_o, poor signal-to-noise ratio precludes the detection of intensity scintillation. Moving the heliocentric distance range of intensity scintillation measurements closer or farther from the Sun necessitates raising or lowering the observing frequency, respectively.

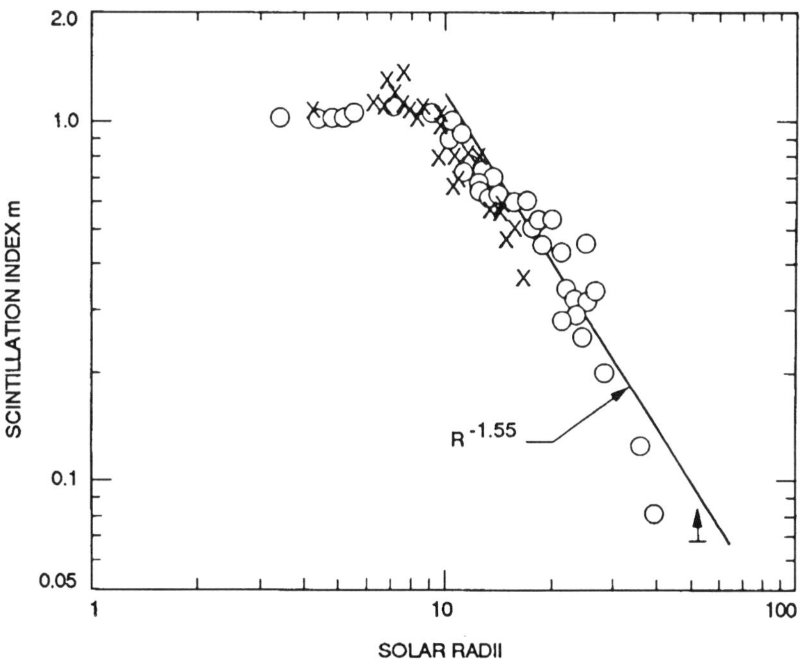

Fig. 7. Variation of S-band scintillation index m with heliocentric distance in solar radii showing saturation at around $10R_o$.[113] Spacecraft radio source is about 2 AU from Earth.

As with scintillation index m in the case of intensity scintillation, rms Doppler scintillation σ_D is useful for characterizing the strength of Doppler scintillation. Shown in Fig. 8 are plots of the radial variation of 1-hr averages of σ_D for three sets of superior conjunction data;[119] σ_D is estimated based on 10 sec S-band Doppler scintillation measurements and computed over 3-min periods, and hence sensitive to scale sizes of less than about 5×10^4 km. Since Doppler or phase scintillation is not limited by Fresnel filtering to responding to scale sizes smaller than R_F (~140 km), and electron density fluctuation is considerably larger for sizes greater than R_F,[109] σ_D is observed all the way out to near 1 AU. Fig. 8 also shows that, unlike intensity scintillation near the Sun, Doppler scintillation does not saturate. Thus, a major feature of Doppler/phase scintillation is that it can be observed over the entire heliocentric distance range, from the vicinity of the Sun to near the orbit of Earth.[110,115,120] An additional difference between m and σ_D is that, whereas m responds to electron density fluctuation, σ_D reflects both electron density fluctuation and solar wind speed.[110]

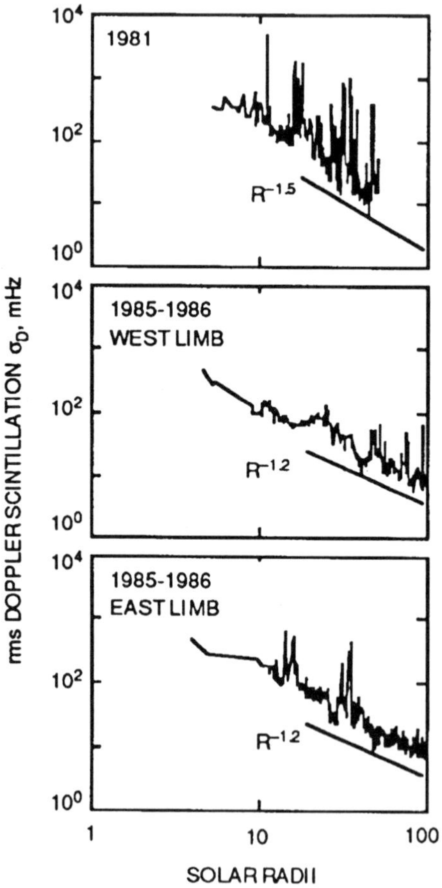

Fig. 8. Variation of S-band rms Doppler scintillation σ_D with heliocentric distance in solar radii during Pioneer Venus superior conjunctions in 1981 (solar maximum conditions) and 1985–1986 (solar minimum conditions).[119] The two sets of data in 1985–1986 pertain to entry (west limb) and exit (east limb) occultations.

As mentioned earlier, scattering from electron density irregularities in the solar wind also causes the monochromatic spacecraft radio signals to undergo spectral broadening, an effect that was first observed by Mariner 4 and Pioneer 9.[74,121] The degree of broadening can be characterized by bandwidth B, the full width between frequencies enclosing half the area under the radio signal spectrum. As with σ_D, spectral broadening bandwidth B does not saturate, and reflects electron density fluctuation and solar wind velocity.[75,110] The inherent line width of the spacecraft signal limits spectral broadening measurements to heliocentric distances closer than about 20 R_o at S-band,[110] and about 30 R_o with 32-cm wavelength radio signals.[115] Turning to highlights of the solar wind results obtained from spacecraft radio scintillation measurements, I will next discuss contributions to solar wind turbulence, large-scale phenomena, and interplanetary disturbances.

4.1 Solar wind turbulence

Phase scintillation and spectral broadening measurements are especially useful for investigating the characteristics of the turbulent electron density irregularities in the solar wind. The penalty imposed by the Fresnel filtering limitation on the remote sensing abilities of intensity scintillation is more severe for the solar wind than for planetary atmospheres, because both the extent and range of scale sizes of the solar wind exceed R_F by a large amount. Thus, the availability of phase scintillation measurements has been an important factor in clarifying earlier questions about the density spectrum,[10,122,123] and in establishing that the density spectrum follows a Kolmogorov power-law at large scales over a wide range of heliocentric distances in the inner heliosphere not directly observed by spacecraft measurements.[116,120,124,125] In analyzing the phase scintillation measurements, interpretation under strong scattering conditions has been examined,[126] and simulations have been carried out.[125] Given that the solar wind is a collisionless plasma, the degree to which the Kolmogorov spectrum describes the density fluctuations has been surprising, and the subject of theoretical investigation.[127–130]

Like intensity scintillation, spectral broadening senses the smaller scales of electron density irregularities. However, as mentioned earlier, when intensity scintillation is strong, as is the case close to the Sun, deduction of the electron density spectrum from spectral broadening measurements is more straightforward, because it is based on the second rather than the fourth moment.[75] Also, the presence of velocity fluctuation and irregularity anisotropy does not increase the uncertainty in spectral index estimates as it does for intensity scintillation.[102,131,132] Spectral broadening measurements have shown that there is a flattening of the small-scale density spectrum with respect to the Kolmogorov spectrum close to the Sun that appears to be related to the turbulent envelope surrounding the Sun.[114–116,120,133] I will discuss scintillation transients representing interplanetary disturbances later, but it is interesting to note that the level of the power-law density spectrum rises, and the spectral index steepens approximately to the Kolmogorov case during the passage of transients.[125,134,135]

Occasionally, spacecraft missions are equipped with linearly polarized radio signals so that Faraday rotation, which provides information on magnetic field as well as electron density, can be observed. In an analysis of solar wind Faraday rotation fluctuations observed with Helios S-band radio signals, the variance of magnetic field fluctuations relative to the mean field was found to be considerably larger than that of electron density fluctuations.[136] Furthermore, the magnetic field fluctuations were consistent with the hypothesis that the Sun radiates Alfvén waves with sufficient energies to heat and accelerate high-speed solar wind streams. Two-station observations of Faraday rotation fluctuations have been conducted, and a cross correlation analysis performed to obtain time delays interpreted as the speeds of the Alfvén waves.[137]

4.2 Large-scale solar wind phenomena

The Pioneer Venus results shown in Fig. 8 correspond to solar minimum (1985–1986) and solar maximum (1981) phases of the solar cycle. Superimposed on the background solar wind, which decreases with heliocentric distance, is solar wind structure including transients representing interplanetary disturbances. Scintillation transients will be discussed in the next section. A line has been drawn on each plot representing a power-law fall-off with radial distance for the background σ_D (lower envelope of the data) in the range of 20-100 R_o. For electron density fluctuations described by a Kolmogorov spatial wavenumber spectrum, $\sigma_D \propto \sigma_{ne} v^{5/6}$, where σ_{ne} is the rms electron density fluctuation and v is the solar wind speed transverse to the radio path.[110] With the assumption that $\sigma_{ne} \propto n_e$ and an approximation that $v^{5/6} \approx v$, where n_e is electron density, σ_D characterizes mass flux density. Furthermore, the radial variation of mass flux density is given by multiplying the radial variation of σ_D by $R^{-1/2}$, where R is heliocentric distance.[110] Thus, the results in Fig. 8 imply that mass flux density varies as R^{-2} and $R^{-1.7}$ during solar maximum and solar minimum, respectively. These Doppler scintillation measurements suggest that the solar wind is approximately spherically symmetric during solar maximum, but slightly converging in the ecliptic plane during solar minimum. This latter result is consistent with 1975 measurements during the previous solar minimum,[110] and may be related to the formation of polar coronal holes during solar minimum.[138]

The results in Fig. 8 show that beyond 20 R_o the background Doppler scintillation is stronger during solar minimum than during solar maximum. If the assumptions discussed above are independent of solar cycle, this implies that mass flux density in the ecliptic plane is higher during solar minimum than during solar maximum. This result is consistent with the solar cycle variation of mass flux density observed by *in situ* Helios and IMP 7/8 plasma measurements near 1 AU.[139,140]

4.3 Transients representing interplanetary disturbances

In meter wavelength IPS observations, scintillation index transients caused by increases in compressed turbulence have been used to detect and study large dynamic solar wind structure beyond 0.5 AU, e.g., propagating interplanetary disturbances and corotating features.[105–107,141] Although enhancements in spectral broadening[74,134] bandwidth B and rms Doppler scintillation[139] σ_D can be used for similar studies near the Sun, they offer considerable advantages.

First, since neither spectral broadening nor Doppler/phase scintillation saturates when strong, both have a dynamic range that is much greater than that of intensity scintillation. This is especially important near the Sun where very substantial enhancements are often observed.[134,143] Second, the sensitivity of spectral broadening and Doppler/phase scintillation to changes in dynamic structure in the solar wind is higher than that of intensity scintillation, since as mentioned before, the former respond to solar wind speed in addition to electron density fluctuation. Third, the ability to use Doppler scintillation to observe the solar wind from the vicinity of the Sun all the way out to near 1 AU permits not only study of the evolution of solar wind structure with heliocentric distance, but also direct comparison of Doppler scintillation with *in situ* spacecraft plasma measurements near the Earth's orbit.[144] Finally, because continuous tracking of spacecraft signals is possible, the temporal resolution of both B and σ_D is much higher than that of IPS observations conducted with natural radio sources. The latter are carried out at meridian transit, so that each individual radio source can only be tracked for at most a couple of hours a day. On the other hand, because the number of spacecraft available for scintillation measurements is limited, IPS observations using multiple natural radio sources are better suited for mapping and studying the evolution of large-scale structures in the plane of the sky.[107] These features, together with the fact that centimeter-wavelength measurements observe the region close to the Sun and bridge the gap between solar and direct measurements farther out, enable spacecraft scintillation observations to address several outstanding questions about interplanetary disturbances near the Sun. Examples of these include the structure and evolution of shocks and corotating streams near the sun, and the interplanetary manifestation of solar events and coronal mass ejections.[145–147]

The features of spacecraft radio scintillation measurements are readily apparent in observations of a fast-moving flare-associated shock wave conducted on August 18, 1979 at 13.1 R_0.[134–142] Shown in Fig. 9 are time histories of (1) 3-min rms Doppler scintillation σ_D at S-band, (2) spectral broadening bandwidth B at both S- and X-bands integrated over 1-min, and (3) 1-min averages of log-amplitude scintillation σ_χ^2 at X-band (S-band amplitude scintillation is not shown because it is saturated). Since B and σ_D respond to both electron density fluctuation and solar wind velocity, profiles of velocity can only be deduced from S-band measurements of B or σ_D if an independent measure of electron density fluctuation, as from simultaneous weak-scattering X-band intensity scintillation measurements, is available. This has happened in the case of only one other shock, that observed on April 23, 1981.[148] Included in Fig. 9 are the deduced time histories of the structure constants of refractive index and electron density fluctuation, and solar wind flow speed for the August 18, 1979 shock. Results obtained from these scintillation measurements revealed not only a surprisingly high shock speed close to the sun, but also rapid deceleration as the shock propagated away from the Sun.[142,149] Although not available in this instance, another approach to extracting velocity from spactral broadening measurements is to simultaneously conduct angular broadening observations to independently determine density fluctuation.[110,131]

The August 18, 1979 shock observations demonstrate the significant advantages of spectral broadening and Doppler scintillation measurements. If S-band log-amplitude variance or scintillation index alone were monitored, this shock would probably have escaped detection, because the scintillation was saturated. Yet, in the case of spectral broadening and Doppler scintillation, a dramatic increase by a factor of around 60 was observed during passage of the shock. The deduced profiles in Fig. 9 show that this enhancement is due to a factor of about 13 increase in solar wind flow and a factor of about 4 increase in rms electron density fluctuation.

The timing of shock crossing near the Sun observed in Doppler scintillation measurements can be combined with shock detection times by *in situ* plasma measurements farther

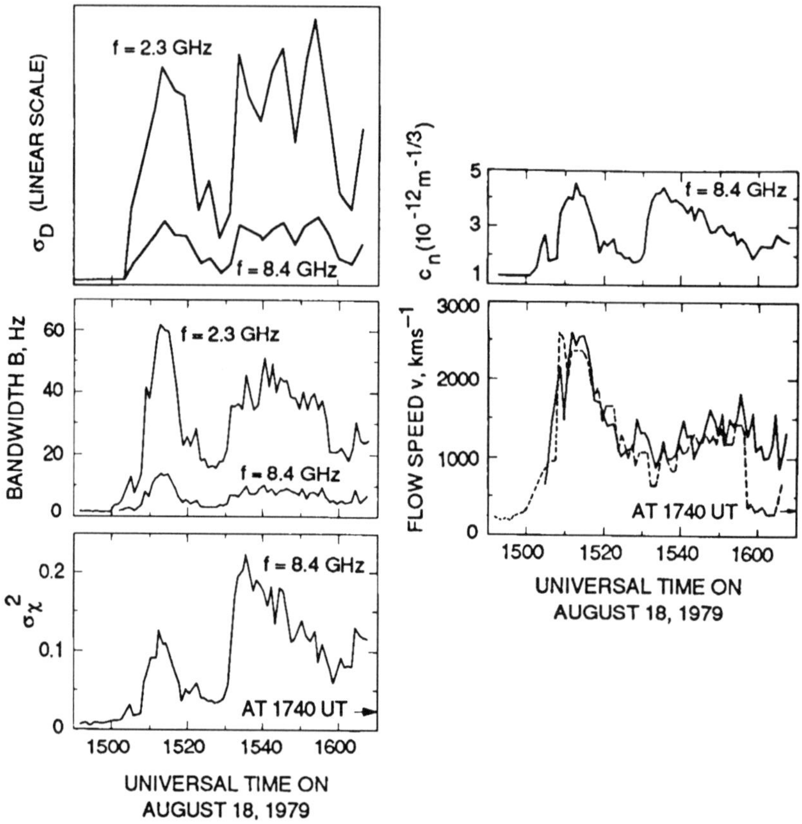

Fig. 9. Scintillation observations of a flare-associated shock wave observed on August 19, 1979 by Voyager 1 when its radio path was probing at a radial distance of 13.1 R_o.[134] The radio measurements shown on the left side include S- and X-band rms Doppler scintillation σ_D, S- and X-band spectral broadening bandwidth B, and X-band variance σ_χ^2 of log-amplitude scintillation (S-band log-amplitude scintillation is saturated). The deduced solar wind parameters shown on the right side include structure constant cn of refractive index fluctuations, and solar wind flow speed v. Note the substantial enhancement in observed radio and deduced solar wind parameters during passage of the interplanetary shock.

away from the sun and white-light coronagraph measurements of coronal mass ejections to study the evolution of shock propagation.[142,144] Multiple spacecraft Doppler scintillation can also provide information on shock propagation speeds.[143] Results of these studies, combined with the spectral broadening measurements of the August 18, 1979 and April 23, 1981 shocks,[113,148] reinforce the significance of observing interplanetary shocks close to the Sun with scintillation measurements. There is a substantially wider range of shock speeds in the vicinity of the sun, with the fastest moving shocks decelerating the most while propagating away from the sun. The fast and rapidly decelerating shocks appear to be associated with regions of very energetic and active solar flares, while the shocks with relatively low speeds, which may decelerate or move uniformly through the interplanetary medium, are associated with events characterized by the eruption and ejection of cool filamentary material from coronal quiescent regions.[148,150] Time-dependent MHD models have been used in a parametric study of these interplanetary shocks.[151]

Doppler scintillation transients such as those just described represent disturbances in the solar wind. A synoptic study of these transients has shown that, around solar maximum, they occur at a rate of about 0.29 transients per day, with the rate decreasing markedly during solar minimum.[152] Some of the scintillation transients during solar minimum probably correspond to corotating high speed streams,[119] and these represent the first measurements of such streams close to the sun. When combined with theoretical models,[153] these measurements promise to improve our understanding of the formation and evolution of high-speed streams near the sun. The observed solar cycle dependence of the frequency of transients is consistent with that of the occurrence of shocks,[154] coronal mass ejections observed by white-light coronagraphs,[155] and counterstreaming solar wind halo electron events.[156]

Comparison of near-continuous Doppler scintillation and *in situ* plasma measurements of the same solar wind over a period of 3 months under solar maximum conditions showed that there was a near one-to-one correspondence between Doppler scintillation transients and interplanetary shocks.[144] Displayed in Fig. 10 are simultaneous time histories of Helios 1 *in situ* measurements of proton temperature T_p, bulk solar wind speed v_p, proton density n_p, proton mass flux density $n_p v_p$, and Pioneer Venus rms Doppler scintillation σ_D for July 19–28, 1982, when the Pioneer Venus radio path was probing nearly the same solar region around 100 R_O as observed *in situ* by Helios 1. Although σ_D approximates mass flux density as discussed in Section 4.2, the similarity through active and quiet phases of the solar wind is striking. These results demonstrate the efficacy of using Doppler scintillation measurements as a proxy for spacecraft plasma measurements when studying a wide range of solar wind structure in the vicinity of the Sun.

The ability of Doppler scintillation or spectral broadening transient measurements to detect and locate interplanetary shocks near the Sun with a relatively high degree of confidence[144] is useful for locating shocks in other solar and interplanetary observations of coronal mass ejections.[149] Comparisons of simultaneous spectral broadening and white-light coronagraph measurements of the same coronal mass ejections suggest that the shock front precedes the white-light front.[157,158]

4.4 Solar wind velocity

Ever since it was first shown[159] that multistation IPS observations of compact radio sources could be used to estimate solar wind velocity, extensive measurements have yielded invaluable information on this key solar wind parameter.[103,104] Although spacecraft radio scintillation measurements supplement the small but growing pool of velocity estimates close to the Sun, contrary to the other areas discussed above, they do not appear to offer exceptional advantages over IPS measurements with natural radio sources. Velocity estimates have been deduced from spaced receiver IPS measurements using spacecraft signals,[113] but the number and choice of receiver baselines with existing spacecraft tracking stations is limited.

Solar wind velocity estimates have also been obtained with spaced receiver phase scintillation[20,160–162] measurements. The baseline options are also very limited, but since

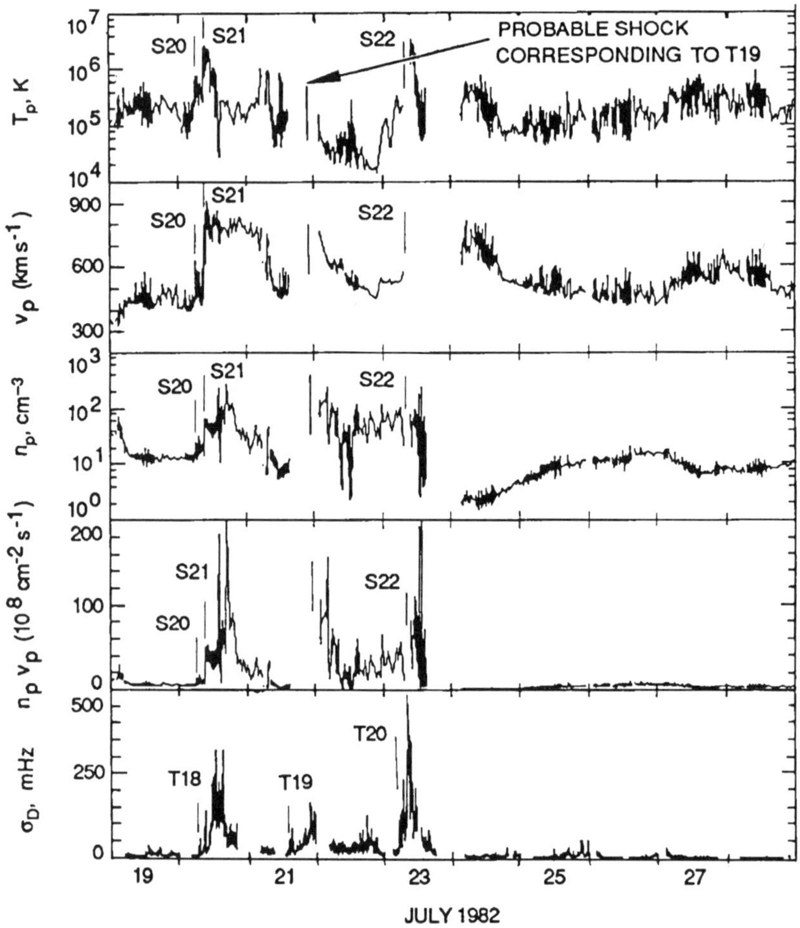

Fig. 10. Comparison of Pioneer Venus Doppler scintillation and Helios 1 *in situ* plasma measurements of the same solar wind region located at about 100 R_o on July 19-28, 1982.[141] The *in situ* measurements are of proton temperature T_p, bulk solar wind speed v_p, proton density n_p, and proton mass flux density $n_p v_p$. Note the marked similarity between $n_p v_p$ and σ_D through both quiet and active phases of the solar wind.

observations can also be made of the solar wind region near the Earth's orbit, they can be compared with *in situ* measurements to validate the method. Although source size is not a factor in interpreting single-station intensity scintillation measurements conducted with spacecraft radio signals,[108,163,164] anisotropy, velocity fluctuation, and strong scintillation, all of which exist near the Sun,[100,101] can contribute to increased uncertainties in the velocity estimates.[102,132] Beyond 40 R_o, where anisotropy and wind fluctuation are reduced, velocity estimates from measurements using natural radio sources appear to be well determined.[165] Finally, the ability to use multiple natural radio sources is a clear advantage for IPS measurements using natural radio sources, because simultaneous measurements in radial alignment along streams are possible.[101]

5. Summary

Radio scintillation effects, initially regarded as noise and a nuisance to the communications link and navigation of deep space missions, have provided opportunities for validating new wave propagation analyses and establishing spacecraft radio scintillation as a tool for exploring small-scale dynamics in the solar system. Noteworthy in the wave propagation analyses have been the inclusion and verification of the effect of defocusing due to differential refraction on weak scattering amplitude scintillation, and the validation of second moment mutual coherence function theory by spectral broadening observations under both weak and strong scattering conditions. A factor that has facilitated these studies is the ubiquitous power-law spatial wavenumber spectrum, often close to the Kolmogorov case, that seems to universally describe the diverse turbulent media.

In remote sensing of planetary atmospheres and ionospheres, amplitude scintillation plays a key role because Fresnel filtering produces features in the amplitude scintillation spectrum that allow the anisotropy of the inhomogeneities and its orientation to be determined accurately. Phase scintillation complements amplitude scintillation measurements by providing information on scales larger than R_F, and dual-frequency measurements help establish that $L_0 >> R_F$. The situation is quite different for the solar wind. Here, amplitude scintillation measurements with natural radio sources have not only been available, they have been used extensively to explore inaccessible regions of the solar wind. Although spectral broadening and phase scintillation, made possible by coherent spacecraft signals, add to the capabilities of remote sensing with amplitude scintillation in both planetary atmospheres and the solar wind, the benefit is farther reaching in the case of the solar wind. First, since the range of scale sizes in the solar wind is extensive, and R_F by comparison very small, the ability to observe scales larger than R_F becomes very important. Second, the range of scintillation levels in the solar wind due to spatial and temporal variations exceeds that in atmospheres and ionospheres by a substantial amount, so that scattering measurements that do not saturate when strong make a profound difference in the investigation of solar wind structure, particularly near the Sun.

Scintillation phenomena, which have been observed on virtually every planetary mission since exploration of the solar system began with space probes over two decades ago, yield measurements not available from either direct or other remote sensing instruments. Owing to the larger Fresnel size, radio occultation measurements probe scale sizes that are larger than those in terrestrial scintillation measurements. To date, investigations based on scintillation measurements have already been carried out for the atmospheres of Venus, Titan and Uranus, the ionospheres of Jupiter, Saturn and Venus, and the solar wind. These have resulted in new information on temperature inhomogeneities, turbulence, and waves in planetary atmospheres, electron density irregularities and magnetic field in planetary ionospheres, solar wind interaction with planetary ionospheres, and structure and dynamics in the near-Sun region of the solar wind. Continued observations of the near-Sun solar wind during the superior conjunctions of current and future deep space probes, as well as future observations planned for Jupiter[80] and Mars,[166] promise to add to the growing body of scientific knowledge already obtained by exploiting the unique features of the radio links of spacecraft used in solar system exploration.

Acknowledgment

It is a pleasure to acknowledge the many people within the NASA Flight Projects and the Deep Space Network organizations whose efforts contributed to the success of many of the results summarized in this paper. I am especially grateful to A. Ishimaru for introducing me to the subject of scintillation, and who together with my colleagues, J. Armstrong, R. Goldstein, W. Kendall, J. Luhmann, A. Kliore, W. Sjogren, and F.C. Yang, have made the study of noisy signals both exciting and fun. This paper describes research carried out at the Jet Propulsion Laboratory, California Institute of Technology, under a contract with the National Aeronautics and Space Administration.

References

1. V. Tatarskii, *The Effects of the Turbulent Atmosphere on Wave Propagation*, National Technical Information Service, Springfield, Va. (1971).

2. A. Ishimaru, *Wave Propagation and Scattering in Random Media*, Vol. 2, Academic Press, New York (1978).

3. B.J. Uscinski, *The Elements of Wave Propagation in Random Media*, McGraw-Hill, New York (1977).

4. S. Basu and S. Basu, "Equatorial scintillations — a review," *J. Atmos. Terr. Phys.*, 43 (5/6), 473–489 (1981).

5. B.G. Fejer and M.C. Kelley, "Ionospheric irregularities," *Rev. Geophys. Space Sci.*, 18, 401–454 (1980).

6. K.C. Yeh and C.H. Liu, "Radio wave scintillations in the ionosphere," *Proc. IEEE*, 70, 324–360 (1982).

7. J. Aarons, "Global morphology of ionospheric scintillations," *Proc. IEEE*, 70, 360–378 (1982).

8. S. M. Flatté (Ed.), *Sound Transmission Through a Fluctuating Ocean*, Cambridge University Press, New York (1979).

9. A. Hewish, "Observations of the solar plasma using radio scattering and scintillation methods," *Solar Wind*, NASA Spec. Publ. SP-308, 477–493 (1972).

10. J.R. Jokipii, "Turbulence and scintillations in the interplanetary plasma," *Ann. Rev. Astron. Astrophys.*, 11, 1–28 (1973).

11. J.M. Cordes, B.J. Rickett, and D.C. Backer (Eds.), *Radio Wave Scattering in the Interstellar Medium*, AIP Conference Proceedings 174, American Institute of Physics, New York (1988).

12. R.S. Lawrence and J.W. Strohbehn, "A survey of clear-air propagation effects relevant to optical communications," *Proc. IEEE*, 58, 1523–1545 (1970).

13. E. Brookner, "Atmosphere propagation and communciation channel model for laser wavelengths," *IEEE Trans. Comm. Tech.*, COM-18, 396–416 (1970).

14. R. Woo, W. Kendall, A. Ishimaru, and R. Berwin, "Effects of turbulence in the atmosphere of Venus on Pioneer Venus radio," *JPL Technical Memorandum 33-644*, Jet Propulsion Laboratory, Pasadena, Calif., June 30, 1973.

15. T.A. Croft, V.R. Eshleman, E.A. Marouf, and G.L. Tyler, "Preliminary review and analysis of effects of the atmosphere of Venus on radio telemetry and tracking of entry probes," *Final report SEL Project 3239*, Stanford Electronics Laboratory, Stanford, Calif., October 1972.

16. J.H. Yuen, R.Woo, and T. Grant, "Effects of the Jovian ionosphere on entry probe communications," *Proceedings of the Effect of the Ionosphere on Space and Terrestrial Systems Symposium*, Washington DC, 586-590, Jaunary 24–26, 1978.

17. J. W. Armstrong, R. Woo, and M. Koerner, "Starprobe — Coronal plasma turbulence effects on tracking and telemetry," *AIAA paper 82-0044*, Issue 6, 820–827 (1982).

18. V.R. Eshleman and B.S. Haugstad, "Effects of turbulence on average refraction angles in occultations by planetary atmospheres," *Icarus*, 34, 396–405 (1978).

19. W.B. Hubbard, "On the atmosphere with exponential turbulence," *Icarus*, 39, 473–478 (1979).

20. J.W. Armstrong, R. Woo, and F.B. Estabrook, "Interplanetary phase scintillation and the search for very low frequency gravitational radiation," *Astrophys. J.*, 230, 570–574 (1979).

21. A.J. Kliore, D. Cain, G.S. Levy, V.R. Eshleman, G.Fjeldbo, and F.D. Drake, "Occultation experiment: Results of the first direct measurement of Mars' atmosphere and ionosphere," *Science*, 149, 1243–1248 (1965).

22. D.M. Hunten and J. Veverka, "Stellar and spacecraft occultations by Jupiter: A critical review of derived temperature profiles," in *Jupiter: The Giant Planet*, edited by T. Gehrels, University of Arizona Press, Tucson, 247–283 (1976).

23. M.A. Kolosov, O.I. Yakovlev, and A.I. Efimov, "Study of the propagation of decimeter radiowaves in the atmosphere of Venus with the aid of AIS Venera 4," *Dokl. Akad. Nauk SSSR, Geofiz.*, 182, 93–94 (1968).

24. M.A. Kolosov, O.I. Yakovlev, and A.I. Efimov, "Propagation of radiowaves in the atmosphere of Venus from the data of Venera-4, Venera-5 and Venera-6," *Radio. i Electron.*, 15, 2473–2478 (1970).

25. O.I. Yakovlev, A.I. Efimov, T.S. Timofeeva, and K.M. Shvachdin, "Power flow and decimeter radiowave fluctuation at the descent of probes Venera-7 and Venera-8 to the planet surface," *Kosm. Issle.*, 12, 100–103 (1974).

26. A.S. Gurvich, "An estimate of characteristics of small-scale turbulence in the Venus atmosphere obtained by radio-signal fluctuations of interplanetary stations Venera-4 and Mariner 5," *Iz. Atmos. Oceanic Phys.*, 5, 1172–1178 (1969).

27. G.S. Golitsyn and A.S. Gurvich, "An estimate of refractive index fluctuations in the Venus atmosphere from Mariner 5 data," *J. Atmos. Sci.*, 28, 138–140 (1971).

28. D.A. De Wolf, "Evidence for severe microwave signal fading in the atmosphere of Venus due to turbulence," *J. Geophys. Res.*, 75, 1202–1208 (1970).

29. D.A. De Wolf, "Atmospheric turbulence on Venus: Venera 4,5,6, and Mariner 5 estimates," *J. Geophys. Res.*, 76, 3154–3158 (1971).

30. A. Ishimaru, A new approach to the problem of wave fluctuations in localized smoothly varying turbulence, *IEEE Trans. Ant. Prop.*, AP-21, 47–53 (1973).

31. R. Woo and A. Ishimaru, "Remote sensing of the turbulence characteristics of a planetary atmosphere by radio occultation of a space probe," *Radio Sci.*, 8, 103–108 (1973).

32. R. Woo and A. Ishimaru, "Effects of turbulence in a planetary atmosphere on radio occultation," *IEEE Trans. Ant. Prop.*, AP-22, 566–573 (1974).

33. R. Woo, A. Ishimaru, and W.B. Kendall, "Observations of small-scale turbulence in the atmosphere of Venus by Mariner 5," *J. Atmos. Sci.*, 31, 1698–1706 (1974).

34. R. Woo, "Observations of turbulence in the atmosphere of Venus using Mariner 10 radio occultation measurements," *J. Atmos. Sci.*, 32, 1084–1090 (1975).

35. R. Woo, J.W. Armstrong, and W.B. Kendall, "Measurements of turbulence in the Venus atmosphere deduced from Pioneer Venus multiprobe radio scintillations," *Science*, 205, 87–89 (1979).

36. G. Fjeldbo and V.R. Eshleman, "Atmosphere of Venus as studied with the Mariner 5 dual radio-frequency occultation experiment," *Radio Sci.*, 4, 879–897 (1969).

37. R. Woo, F.C. Yang, and A. Ishimaru, "Wave propagation in a random medium with inhomogeneous background," paper presented at the *1975 USNC/URSI Meeting*, Boulder, Colo., Oct. 20–23, 1975.

38. A.T. Young, "Scintillations during occultations by planets, I. An approximate theory," *Icarus*, 27, 325–357 (1976).

39. B.S. Haugstad, "Effects of inhomogeneous background on radiation propagating through turbulent planetary atmospheres," *Radio Sci.*, 13, 435–440 (1978).

40. W.B. Hubbard, J.R. Jokipii, and B.A. Wilking, "Stellar occultations by turbulent planetary stmospheres: A wave optical theory including a finite scale height," *Icarus*, 34, 375–395 (1978).

41. A.J. Kliore, "Radar and radio occultation of the planets," *Physics of the Solar System*, NASA Spec. Publ. 300, 295–349 (1972).

42. B.S. Haugstad, "Turbulence in planetary occultations, IV. Power spectra of phase and intensity fluctuations," *Icarus*, 37, 322–335(1979).

43. R. Woo, A. Ishimaru, and F.C. Yang, "Radio scintillations during occultations by turbulent planetary atmospheres," *Radio Sci.*, 15, 695–703 (1980).

44. A.S. Gurvich, "Scintillation spectra during observations of star obscurations by the earth's atmosphere," *Atm. Optics*, 2, 188–193 (1989).

45. G.M. Grechko, A.S. Gurvich, V. Kan, S.V. Kireev, and S.A. Savchenko, "Anisotropy of spatial structures in the middle atmosphere," *Adv. Space Res.*, 12(10), 169–175 (1992).

46. D.P. Hinson, "Strong scintillations during atmospheric occulations: Theoretical intensity spectra," *Radio Sci.*, 21, 257–270 (1986).

47. B.S. Haugstad, "Turbulence in planetary occultations: A strong scattering formulation including an inhomogeneous background," *Radio Sci.*, 17, 565–573 (1982).

48. V.H. Rumsey, "Scintillations due to a concentrated layer with a power-law turbulence spectrum," *Radio Sci.*, 10, 107–114 (1975).

49. M. Marians, "Computed scintillation spectra for strong turbulence," *Radio Sci.*, 10, 115–119 (1975).

50. C.L. Rino, "Numerical computations for a one-dimensional power law phase screen," *Radio Sci.*, 15, 41–47 (1980).

51. T.S. Timofeeva, O.I. Yakovlev, and A.I. Efimov, "Radio wave fluctuation and turbulence of the nightime Venusian atmosphere from radioscopy data of the space probe Venera 9," *Cosmic Res.*, 16, 185-293 (1978).

52. R. Woo, J.W. Armstrong, and A. Ishimaru, "Radio occultation measurements of turbulence in the Venus atmosphere by Pioneer Venus," *J. Geophys. Res.*, 85, 8031–8038 (1980).

53. T.S. Timofeeva, A.I. Efimov, and O.I. Yakovlev, "Turbulence in the atmosphere of Venus from Venera 9 and 10 radio fluctuations," *Cosmic Res.*, 18, 775–782 (1980).

54. M.A. Kolosov, O.I. Yakovlev, A.I. Efimov, S.S. Matyugov, T.S. Timofeeva, E.V. Chub, A.G. Pavelyev, A.I. Kucheryavenkov, I.E. Kalashnikov, and O.E. Milekhin, *Acta Astronautica*, 7, 219–234 (1980).

55. R. Woo, J.W. Armstrong, and A.J. Kliore, "Small-scale turbulence in the atmosphere of Venus," *Icarus*, 52, 335–345 (1982).

56. A.J. Kliore and I.R. Patel, "Vertical structure of the atmosphere of Venus from Pioneer Venus orbiter radio occultations," *J. Geophys. Res.*, 85, 7957–7962 (1980).

57. A. Seiff, D.B. Kirk, R.E. Young, R.C. Blanchard, J.T. Findlay, G.M. Kelly, and S.C. Sommer, "Measurements of thermal structure and thermal contrasts in the atmosphere of Venus and related dynamical observations: Results from the four Pioneer Venus probes," *J. Geophys. Res.*, 85, 7903–7933 (1980).

58. R.Z. Sagdeev, V.M. Linkin, V.V. Kerzhanovich, A.N. Lipatov, A.A. Shurupov, J.E. Blamont, D. Crisp, A.P. Ingersoll, L.S. Elson, R.A. Preston, C.E. Hildebrand, B. Ragent, A. Seiff, R.E. Young, G. Petit, L. Boloh, Yu. N. Alexandrov, N.A. Armand, R.V. Bakitko, and A.S. Selivanov, "Overview of VEGA Venus balloon *in situ* meteorological measurements," *Science*, 231, 1411–1414 (1986).

59. R. Woo and A. Ishimaru, "Eddy diffusion coefficient for the atmosphere of Venus from radio scintillation measurements," *Nature*, 289, 383–384 (1981).

60. J.B. Cimino, C. Elachi, A.J. Kliore, D.J. McCleese, and I.R. Patel, "Polar cloud structure as derived from the Pioneer Venus orbiter," *J. Geophys. Res.*, 85, 8082–8088 (1980).

61. R.K. Crane, "A review of radar observations of turbulence in the lower stratosphere," *Radio Sci.*, 15, 177–193 (1980).

62. K.S. Gage, J.L. Green, and T.E. Van Zandt, "Use of Doppler radar for the measurement of atmospheric turbulence parameters from the intensity of clear-air echoes," *Radio Sci.*, 15, 407–416 (1980).

63. B. Balsley and V.L. Peterson, "Doppler-radar measurements of clear air turbulence at 1290 MHz," *J. Appl. Meteorol.*, 20, 266–274 (1981).

64. S.S. Leroy and A.P. Ingersoll, "Small-scale gravity waves in Venus's middle atmosphere," *Bull. Am. Astron. Soc.*, 24, 997–998 (1992).

65. E.M. Dewan, "On the nature of atmospheric waves and turbulence," *Radio Sci.*, 20, 1301–1307 (1985).

66. D.P. Hinson and G.L. Tyler, "Internal gravity waves in Titan's atmosphere observed by Voyager radio occultation," *Icarus*, 54, 337–352 (1983).

67. G.F. Lindal, G.E. Wood, H.B. Hotz, D.N. Sweetnam, V.R. Eshleman, and G.L. Tyler, "The atmosphere of Titan: An analysis of the Voyager 1 radio occultation measurements," *Icarus*, 53, 348–363 (1983).

68. D.P. Hinson and J.A. Magalhães, "Equatorial waves in the stratosphere of Uranus," *Icarus*, 94, 64–91 (1991).

69. R. Woo and F.C. Yang, "Measurements of electon density irregularities in the ionosphere of Jupiter by Pioneer 10," *J. Geophys. Res.*, 81, 3417–3422 (1976).

70. R. Woo and F.C. Yang, "Measurements of the magnetic field orientation in the Jovian ionosphere deduced from Pioneer 10 and 11 scintillation observations," *J. Geophys. Res.*, 83, 5245–5255 (1978).

71. E.J. Smith, L. Davis, Jr. and D.E. Jones, "Jupiter's magnetic field and magnetosphere," in *Jupiter: The Giant Planet*, edited by T. Gehrels, University of Arizona Press, Tucson, 778–829 (1976).

72. M.H. Acuña and N.F. Ness, "The main magnetic field of Jupiter," *J. Geophys. Res.*, 81, 2917–2922 (1976).

73. J. Connerney, "Magnetic field of Jupiter: A generalized inverse approach," *NASA Tech. Memo. 82146*, Goddard Space Flight Center, Greenbelt, Md. (1981).

74. R.M. Goldstein, "Superior conjunction of Pioneer 6," *Science*, 166, 598–600 (1969).

75. R. Woo, F.C. Yang, and A. Ishimaru, "Structure of density fluctuations near the sun deduced from Pioneer-6 spectral broadening measurements," *Astrophys. J.*, 210, 593–602 (1976).

76. R. Woo and J.W. Armstrong, "Spectral broadening measurements of the ionospheres of Jupiter and Saturn," *Nature*, 287, 309–311 (1980).

77. C.L. Rino and J. Owen, "The mutual coherence function for transionospheric radiowaves," *Radio Sci.*, 17, 675–683 (1982).

78. G. Fjeldbo, A. Kliore, B. Seidel, D. Sweetnam, and D. Cain, "The Pioneer 10 radio occultation measurements of the ionosphere of Jupiter," *Astron. Astrophys.*, 39, 91–96 (1975).

79. D.P. Hinson and G.L. Tyler, "Spatial irregularities in Jupiter's upper ionosphere observed by Voyager radio occultations," *J. Geophys. Res.*, 87, 5275–5289 (1982).

80. H.T. Howard, V.R. Eshleman, D.P. Hinson, A.J. Kliore, G.F. Lindal, R. Woo, M.K. Bird, H. Volland, P. Edenhofer, M. Pätzold, and H. Porsche, "Galileo radio science investigations, *Space Sci. Rev.*, 60, 565–590 (1992).

81. D.P. Hinson, "Magnetic field orientations in Saturn's upper ionosphere inferred from Voyager radio occultations," *J. Geophys. Res.*, 89, 65–73 (1984).

82. D.L. Chenette and L. Davis, Jr., "An analysis of the structure of Saturn's magnetic field using charged particle absorption signatures," *J. Geophys. Res.*, 87, 5267–5274 (1982).

83. J. Connerney, N.F. Ness, and M.H. Acuña, "Zonal harmonic model of Saturn's magnetic field from Voyager 1 and 2 observations," *Nature*, 298, 44–46 (1982).

84. R. Woo, W.L. Sjogren, J.G. Luhmann, A.J. Kliore, and L.H. Brace, "Solar wind interaction with the ionosphere of Venus inferred from radio scintillation," *J. Geophys. Res.*, 94, 1473–1478 (1989).

85. L.H. Brace, R.F. Theis, J.P. Krehbiel, A.F. Nagy, T.M. Donahue, M.B. McElroy, and A. Pedersen, "Electron temperatures and densities in the Venus ionosphere: Pioneer Venus orbiter electron temperature probe results," *Science*, 203, 763–765 (1979).

86. J.G. Luhmann, R.C. Elphic, C.T. Russell, J.D. Mihalov, and J.H. Wolfe, "Observations of large-scale steady magnetic fields in the dayside Venus ionosphere," *Geophys. Res. Lett.*, 7, 917–920 (1980).

87. J.L. Phillips, J.G. Luhmann, and C.T. Russell, "Growth and maintenance of large-scale magnetic fields in the dayside Venus ionosphere," *J. Geophys. Res.*, 89, 10676–10684 (1984).

88. R. Woo and A.J. Kliore, "Magnetization of the ionospheres of Venus and Mars: Results from radio occultation measurements," *J. Geophys. Res.*, 94, 11073–11081 (1991).

89. J.G. Luhmann, "The solar wind interaction with Venus," *Space Sci. Rev.*, 44, 241–306 (1986).

90. G.M. Keating and S. Bougher, "Neutral upper atmospheres of Venus and Mars," *Adv. Space Res.*, 7(12), 57–71 (1987).

91. J.G. Luhmann, C.T. Russell, F.L. Scarf, L.H. Brace, and W.C. Knudsen, "Characteristics of the Marslike limit of the Venus-solar wind interaction," *J. Geophys. Res.*, 92, 8545–8557 (1987).

92. M. Neugebauer and C.W. Synder, "Mariner 2 observations of the solar wind," *J. Geophys. Res.*, 71, 4469–4484 (1966).

93. E.J. Smith, K.-P. Wentzel, and D.E. Page, "Ulysses at Jupiter: An overview of the encounter," *Science*, 257, 1503–1507 (1992).

94. W. Feldman et al., "Solar probe scientific rationale and mission concept," *JPL Rpt. D-6797*, Jet Propulsion Laboratory, Pasadena, Calif., Nov. 1989.

95. A. Hewish, "The scattering of radio waves in the solar corona," *Mon. Not. R. Astro. Soc.*, 118, 534–546, 1958.

96. V.V. Vitkevich, "Radio astronomical observations of moving plasma clouds in the solar supercorona," *Sov. Astron.*, 4, 897–903 (1961).

97. W.C. Erickson, "The radio-wave scattering properties of the solar corona," *Astrophys. J.*, 139, 1290–1311 (1964).

98. A. Hewish, P.F. Scott, and D. Wills, "Interplanetary scintillation of small diameter radio sources," *Nature*, 203, 1214–1217 (1964).

99. R.D. Ekers and L.T. Little, "The motion of the solar wind close to the sun," *Astron. Astrophys.*, 10, 310–316 (1971).

100. J.W. Armstrong, W.A. Coles, M. Kojima, and B.J. Rickett, "Solar wind observations near the sun," in *The Sun and the Heliosphere in Three Dimensions*, edited by R.G. Marsden and D. Reidel, 59–64 (1986).

101. W.A. Coles, R. Esser, U.-P. Løvhaug, and J. Markkanen, "Comparison of solar wind velocity measurements with a theoretical acceleration model," *J. Geophys. Res.*, 96, 13849–13859 (1991).

102. S.L. Scott, W.A. Coles, and G. Bourgois, "Solar wind observations near the sun using interplanetary scintillation," *Astron. Astrophys.*, 123, 207–215 (1983).

103. M. Kojima and T. Kakinuma, "Solar cycle evolution of solar wind speed structure between 1973 and 1985 observed with the interplanetary scintillation method," *J. Geophys. Res.*, 92, 7269–7279 (1987).

104. B.J. Rickett and W.A. Coles, "Evolution of the solar wind structure over a solar cycle: Interplanetary scintillation velocity measurements compared with coronal observations," *J. Geophys. Res.*, 96, 1717–1736 (1991).

105. B.J. Rickett, "Disturbances in the solar wind from IPS measurements in August 1972," *Sol. Phys.*, 43, 237–247 (1975).

106. S.J. Tappin, A. Hewish, and G.R. Gapper, "Tracking a major interplanetary disturbance," *Planet. Space Sci.*, 31, 1171–1176 (1983).

107. T. Watanabe and R Schwenn, "Large-scale propagation properties of interplanetary disturbances revealed from IPS and spacecraft observations," *Space Sci. Rev.*, 51, 147–173 (1989).

108. R. Woo, "Multifrequency techniques for studying interplanetary scintillation," *Astrophys. J.*, 201, 238–248 (1975).

109. R. Woo, F.C. Yang, K.W. Yip, and W.B. Kendall, "Measurements of large-scale density fluctuations in the solar wind using dual-frequency phase scintillations," *Astrophys. J.*, 210, 568–574 (1976).

110. R. Woo, "Radial dependence of solar wind properties deduced from Helios 1/2 and Pioneer 10/11 radio scattering observations," *Astrophys. J.*, 219, 727–739 (1978).

111. R. Woo, "Measurements of the solar wind using spacecraft radio scattering observations," in *Study of Travelling Interplanetary Phenomena 1977*, ed. by M. Shea, D. Smart and S. Wu, D. Reidel, Dordrecht, Holland, 81–100 (1977).

112. M.K. Bird and P. Edenhofer, "Remote sensing observations of the solar corona," in *Physics of the Inner Heliosphere I*, eds. R. Schwenn and E. Marsch, Springer-Verlag, Berlin, 13–97 (1990).

113. J.W. Armstrong and R. Woo, "Solar wind motion within 30 R_O: Spacecraft radio scintillation observations," *Astron. Astrophys.*, 103, 415–421 (1981).

114. J.M. Martin, "Voyager Microwave Scintillation Measurements of solar wind plasma parameters," *Sci. Rpt., D208-1985-1*, Center for Radar Astronomy, Stanford Univ., Stanford Calif., August 1985.

115. M.A. Kolosov, O.I. Yakovlev, A.I. Efimov, V.I. Rogal'sky, V.M. Razmanov, and V.K. Shtrykov, "Decimeter radio wave propagation in the turbulent plasma near the sun, using Venera 10 spacecraft," *Radio Sci.*, 17, 664–674 (1982).

116. N.A. Armand, A.I. Efimov, and O.I. Yakovlev, "A model of the solar wind turbulence from radio occultation experiments," *Astron. Astrophys.*, 183, 135–141 (1987).

117. G. Bourgois, "Scintillations interplanétaire des radiosources à 2695 MHz," *Astron. Astrophys.*, 2, 209–217 (1969).

118. J.W. Armstrong and W.A. Coles, "Interplanetary scintillations of PSR 0531+21 at 74 MHz," *Astrophys. J.*, 220, 346–352 (1978).

119. R. Woo and J.W. Armstrong, "Observations of large-scale structure in the inner heliosphere with Doppler scintillation measurements," *Solar Wind Seven*, COSPAR Colloquia Series, 3, 319–322 (1992).

120. R. Woo and J.W. Armstrong, "Spacecraft radio scattering observations of the power spectrum of electron density fluctuations in the solar wind," *J. Geophys. Res.*, 84, 7288–7296 (1979).

121. J.V. Hollweg and J.V. Harrington, "Properties of solar wind turbulence deduced by radio astronomical measurements," *J. Geophys. Res.*, 73, 7221–7230 (1968).

122. W.M. Cronyn, "Density fluctuations in the interplanetary plasma; Agreement between spaceprobe and radio scattering observations," *Astrophys. J.*, 171, L101–L105 (1972).

123. N.A. Lotova, "Current ideas concerning the spectrum of the irregularities in the interplanetary plasma," *Sov. Phys.-Usp.*, 18, 292–300 (1975).

124. P.S. Callahan, "Columnar content measurements of the solar-wind turbulence near the sun," *Astrophys. J.*, 199, 227–236 (1975).

125. W.A. Coles, W. Liu, J.K. Harmon, and C.L. Martin, "The solar wind density spectrum near the sun: Results from Voyager radio measurements," *J. Geophys. Res.*, 96, 1745–1755 (1991).

126. A. Ishimaru, "Phase fluctuations in a turbulent medium," *Appl. Opt.*, 16, 3190–3192 (1977).

127. J.C. Higdon, "Density fluctuations in the interstellar medium — Evidence for anisotropic magnetogasdynamic turbulence (I): Model and astrophysical sites," *Astrophys. J.*, 285, 109–123 (1984).

128. J.C. Higdon, "Density fluctuations in the interstellar medium — Evidence for anisotropic magnetogasdynamic turbulence (II): Stationary structures," *Astrophys. J.*, 309, 342–361 (1986).

129. D. Montgomery, M.R. Brown, and W.H. Matthaeus, "Density fluctuation spectra in magnetohydrodynamic turbulence," *J. Geophys. Res.*, 92, 282–284 (1987).

130. G. Zank and W.H. Matthaeus, "Nearly incompressible fluid dynamics," *Solar Wind Seven*, COSPAR Colloquia Series, 3, 587–594 (1992).

131. R. Woo, F.C. Yang, and A. Ishimaru, "Probing the solar wind with radio measurements of the second moment field," *Astrophys. J.*, 218, 557–568 (1977).

132. S.L. Scott, *Density Spectrum and Velocity of the Solar Wind Inferred Scintillation Observations*, Ph.D. thesis, Univ. of Calif. Sand Diego, La Jolla, Calif. (1978).

133. J.K. Harmon and W.A. Coles, "Spectral broadening of planetary radar signals by the solar wind," *Astrophys. J.*, 270, 748–757 (1983).

134. R. Woo and J.W. Armstrong, "Measurements of a solar flare-generated shock wave at 13.1 R_o," *Nature*, 292, 608–610 (1981).

135. R. Woo and J.W. Armstrong, "Observations of the electron density spectrum near the sun during spectral broadening transients," *Solar Wind Seven*, COSPAR Colloquia Series, 3, 583–586 (1992).

136. J.V. Hollweg, M.K. Bird, H. Volland, P. Edenhofer, C.T. Stelzried, and B.L. Seidel, "Possible evidence for coronal Alfvén waves," *J. Geophys. Res.*, 87, 1–8 (1982).

137. M.K. Bird, H. Volland, A.I. Efimov, G.S. Levy, B.L. Seidel, and C.T. Stelzried, "Coronal Alfvén waves detected by radio sounding during the solar occultations of the Helios spacecraft," *Solar Wind Seven*, COSPAR Colloquia Series, 3, 147–150 (1992).

138. J.B. Zirker (ed)., *Coronal Holes and High Speed Wind Streams*, Colorado Associated University Press, Boulder (1977).

139. R. Schwenn, "Large-scale structure in the interplanetary medium," in *Physics of the Inner Heliosphere I*, eds. R. Schwenn and E. Marsch, Springer-Verlag, Berlin, 159–161 (1990).

140. J.M. Ajello, A.I. Stewart, G.E. Thomas, and A. Graps, "Solar cycle study of interplanetary Lyman-Alpha variations: Pioneer Venus Orbiter sky background results," *Astrophys. J.*, 317, 964–986 (1987).

141. Z. Houminer, "Scintillation observations of the interplanetary plasma," in *Study of Travelling Interplanetary Phenomena 1977*, ed. by M. Shea, D. Smart and S. Wu, D. Reidel, Dordrecht, Holland, 119–141 (1977).

142. R. Woo, J.W. Armstrong, N.R. Sheeley, Jr., R.A. Howard, M.J. Koomen, and D.J. Michels, "Doppler scintillation observations of interplanetary shocks within 0.3 AU," *J. Geophys. Res.*, 90, 154–162 (1985).

143. R. Woo, "A synoptic study of Doppler scintillation transients in the solar wind," *J. Geophys. Res.*, 93, 3919–3926 (1988).

144. R. Woo and R. Schwenn, "Comparison of Doppler scintillation and *in situ* spacecraft plasma measurements of interplanetary disturbances," *J. Geophys. Res.*, 96, 21227–21244 (1991).

145. A.J. Hundhausen et al., "Coronal transients and their interplanetary effects," *NASA Rep. Ref. Publ. 1120*, June 1984.

146. R. Schwenn, "Relationship of coronal transients to interplanetary shocks: 3D aspects," *Space Sci. Rev.*, 44, 139–168 (1986).

147. S.W. Kahler, "Solar flares and coronal mass ejections," *Ann. Rev. Astron. Astrophys.*, 30, 113–141 (1992).

148. R. Woo, J.W. Armstrong, R. Schwenn, and F. Tang, "Radio scintillation measurements of an interplanetary shock at 15.8 R_O," *EOS Trans. AGU*, 71, 590 (1990).

149. H.V. Cane, R.G. Stone, and R. Woo, "Velocity of the shock generated by a large east limb flare on August 18, 1979," *Geophys. Res. Letts.*, 9, 897–900 (1982).

150. H.V. Cane, S.W. Kahler, and N.R. Sheeley, Jr., "Interplanetary shocks preceded by solar filament eruptions," *J. Geophys. Res.*, 91, 13321–13329 (1986).

151. Z. Smith and M. Dryer, "MHD study of temporal and spatial evolution of simulated interplanetary shocks in the ecliptic plane within 1 AU," *Solar Phys.*, 129, 387–405 (1990).

152. J.W. Armstrong, R. Woo, and C. Chang, "Solar cycle dependence of interplanetary disturbances observed as Doppler scintillation transients by Pioneer Venus within 0.3 AU," *EOS Trans. AGU*, 73, 237 (1992).

153. V. Pizzo, "The evolution of corotating stream fronts near the ecliptic plane in the inner solar system 2. Three dimensional tilted-dipole fronts," *J. Geophys. Res.*, 96, 5405–5420 (1991).

154. P.M. Volkmer and F.M. Neubauer, "Statistical properties of fast magnetoacoustic shock waves in the solar wind between 0.3 and 1 AU: Helios-1 and -2 observations," *Ann. Geophys.*, 3, 1–12 (1985).

155. D.F. Webb, "The solar cycle variation of the rates of CMEs and related activity," *Adv. Space Res.*, 11(1), 37–40 (1991).

156. J.T. Gosling, D.J. McComas, J.L. Phillips, and S.J. Bame, *J. Geophys. Res.*, 97, 6531–6535 (1992).

157. R. Woo, J.W. Armstrong, N.R. Sheeley, Jr., R.A. Howard, D.J. Michels, and M.J. Koomen, "Simultaneous radio scattering and white-light observations of a coronal transient," *Nature*, 300, 157–159 (1982).

158. M.K. Bird, H. Volland, R.A. Howard, M.J. Koomen, D.J. Michels, N.R. Sheeley, Jr., J.W. Armstrong, B.L. Seidel, C.T. Stelzried, and R. Woo, "White-light and radio sounding observations of coronal transients," *Solar Phys.*, 98, 341–368 (1985).

159. P.A. Dennison and A. Hewish, "The solar wind outside the plane of the ecliptic," *Nature*, 213, 343–346 (1967).

160. R. Woo, "Measuring solar wind velocity with spacecraft phase scintillations," *Nature*, 266, 514 (1977).

161. A.I. Efimov, O.I. Yakovlev, V.H. Shtrykov, V.I. Rogal'sky, and V.F. Tikhonov, "Observations from different points of fluctuations in frequency and phase of radio waves that are scattered by plasma around the sun," *Radio. i Electron.*, 26, 311–318 (1981).

162. O.I. Yakovlev, A.I. Efimov, V.P. Yakubov, O.M. Korsak, A.S. Kartonov, A.L. Erofeev, and S.N. Rubtsov, "Occultation of the plasma near the sun and the speed of the solar wind: Fluctuations in the frequency and phase of radio waves at two separated points," *Izv. Vyssh. Uchebn. Zaved. Radiofiz.*, 32, 531–537 (1989).

163. A.T. Young, "Interpretation of interplanetary scintillation," *Astrophys. J.*, 168, 543–562 (1971).

164. G.L. Tyler, J.F. Vesecky, M.A. Plume, H.T. Howard, and A. Barnes, "Radio wave scattering observations of the solar corona: First-order measurements of expansion velocity and turbulence spectrum using Viking and Mariner 10 spacecraft," *Astrophys. J.*, 249, 318–332 (1981).

165. P.K. Manoharan and S. Ananthakrishnan, "Determination of solar-wind velocities using single-station measurements of interplanetary scintillation," *Mon. Not. R. Astr. Soc.*, 244, 691–695 (1990).

166. G.L. Tyler, G. Balmino, D.P. Hinson, W.L. Sjogren, D.E. Smith, R. Woo, S.W. Asmar, M.J. Connally, C.L. Hamilton, and R.A. Simpson, "Radio science investigations with Mars Observer," *J. Geophys. Res.*, 97, 7759–7779 (1992).

Random-Media Effects in Ocean Acoustics: An Introduction

Stanley M. Flatté
Physics Department and Institutes of Tectonics and Marine Sciences
University of California at Santa Cruz
Santa Cruz, CA 95064 U.S.A.
Telephone: 408/459-2090
FAX: 408/459-2127
email: smf@pacific.ucsc.edu

An Invited Talk at the International Meeting for
Wave Propagation in Random Media, Seattle, WA, August 3-7, 1992

Abstract. As a problem in waves in random media, ocean acoustics has a number of unusual features. The propagation takes place in a deterministic waveguide called the sound channel. The sound-speed fluctuation spectrum in the ocean is inhomogeneous, anisotropic, and internally time-varying (not advected). The waveguide called the ocean sound channel is described. Observations of phase and amplitude fluctuations in space and time on acoustic waves transmitted in the ocean are categorized. The ocean processes that cause the dominant part of observed acoustic fluctuations are discussed, with emphasis on internal waves. The use of ocean acoustics in tomographic probes of ocean processes is discussed.

I. Introduction

The theory of waves in random media was first quantitatively developed for a realistic medium by workers interested in optical propagation through the atmosphere. The model for the medium was taken from Kolmogorov's pioneering work in homogeneous, isotropic, turbulence with a power-law spectrum. The success of this work, which was summarized in the influential book by Tatarskii,[1] consisted of describing phase and amplitude fluctuations in the weak-fluctuation regime.

By the mid 1970's, ocean acousticians had begun to realize that the concept of homogeneous, isotropic turbulence was not relevant to most of the phenomena in their field. The long acoustic wavelengths and long ranges between sources and receivers imply that the ocean processes being probed are considerably larger than one meter. But temperature variability in the ocean (which is the dominant cause of sound-speed variation) at scales larger than one meter is dominated by ocean *stratification* in depth: that is, the tendency for colder, denser, heavier water to underly warmer, less dense water. The stratification is continuous, so that the ocean is filled at all depths with a field of waves. Each wave is a generalization of the more familiar surface waves, which depend on the density discontinuity between air and water.

The spatial spectrum of internal waves is anisotropic; variations in temperature tend to be pancake-like, being isotropic in the horizontal plane but with a smaller correlation length in the vertical. The spatial spectrum is also inhomogeneous, having larger strength near the surface of the ocean and decreasing roughly exponentially with a characteristic vertical scale of several hundred meters. Figure 1 shows a simulated field of internal waves.

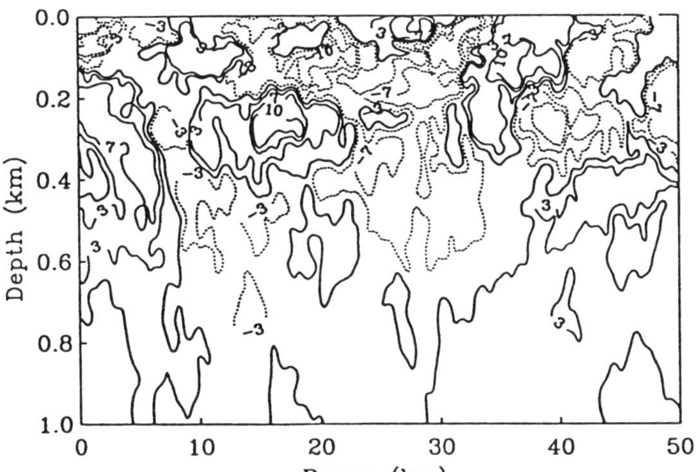

Figure 1. Contour plot of sound-speed fluctuation due to a simulated field of internal waves. The numbers are sound-speed fluctuation in units of 0.1 m/s. Note the increased variability near the ocean surface and the increase in correlation length with depth. Note also the anisotropy of the field. (The horizontal scale is compressed by a factor of 33.)

The relationship between the spatial and temporal variations due to internal waves is different from the oft-used Taylor hypothesis of spatially frozen variations advected by a mean wind. The dispersion relation of internal waves follows from the behavior of density with depth. The physical principle to keep in mind is that of a pendulum, whose frequency is independent of the size of the spatial excursions. In the same way, large-scale temperature variations in the ocean experience larger restoring forces, resulting in larger velocities, and hence the frequency of motion is roughly independent of spatial scale. The empirical ocean internal-wave spectrum factorizes into two pieces: a function of vertical wavenumber and a function of frequency. This has profound effects on predictions of acoustic fluctuations.

Finally, the propagation of acoustic signals ***in the absence of internal-wave fluctuations*** occurs in the ocean sound channel, which causes rays to be curved trajectories rather than the simple straight lines often used in other wave propagation analyses. See Figure 2.

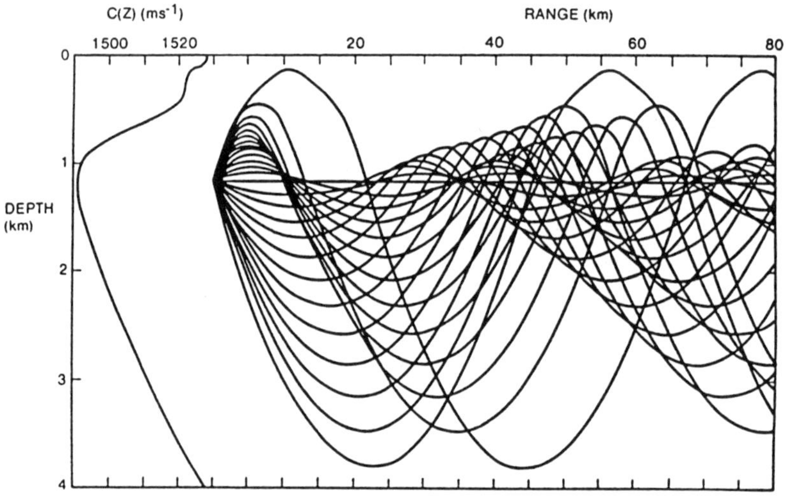

Figure 2. An example sound-speed profile with depth, and the resulting ray paths from a source on the sound-channel axis. Note the caustics and shadow zones formed due to the sound channel.

The confluence of all these factors — anisotropy, inhomogeneity, space-time correlation, non-Kolmogorov spectrum, background sound channel, and interest in strong fluctuations — required a new approach to wave propagation theory. A new approach was found in the form of the path-integral technique, which takes advantage of the geometrical optics approximation for propagation in the absence of medium fluctuations (but with the background sound channel waveguide), and accounts for physical-optics effects in adding the medium fluctuations (internal waves in the ocean case). The successes of this approach are described in detail in a 1979 book[2] and a 1983 review article.[3] A brief summary of that work is given in Section II of this paper.

Both the weak-fluctuation theory and the path-integral technique are very limited in their range of validity. Furthermore, applications to pulse propagation, higher moments, probability distributions, and other observables are extremely difficult to carry out. Recent long-range, low-frequency ocean-acoustic propagation experiments have gathered precise data that cannot be quantitatively calculated with available analytic techniques. For this reason, numerical simulations have been developed. Section III describes one such experiment, and some numerical simulation results that appear to support the assumption that internal waves are the source of observed fluctuations.

Much of the interest in ocean acoustics derives from the possibility of its use in monitoring ocean processes. The measurement of large-scale, average properties of the ocean, such as its average temperature, is needed for evaluating global warming theories. The measurement of mesoscale processes, such as meandering of the Gulf Stream, or the motion of 100-km-scale eddies, is needed to evaluate models of ocean circulation. The measurement of the strength and spectra of internal waves in various parts of the world ocean is needed to understand mechanisms of heat and momentum transfer within the ocean system.

The theory of waves in random media is needed to translate measurements of acoustic fluctuations into values of internal-wave spectral parameters. This same theory is needed to calculate the uncertainties, or limitations, imposed by internal-wave-induced fluctuations on measurements of mesoscale and larger processes. Finally, in the measurement of ocean-basin average quantities like temperature or heat content, even the mesoscale processes can be regarded as a random medium, because 100 km is so much smaller than the size of an ocean basin. Section IV discusses our present qualitative understanding of the limits imposed on large-scale tomography by internal-wave fluctuations, and the possibilities of tomographic measurements of the internal-wave field.

II. Early Successes of Internal-Wave Dominance

The most important fact about sound propagation in the ocean is that it takes place in a waveguide. Sound speed decreases by a few percent as one drops from the ocean surface down to depth of 500-1500 m, and then increases again as one approaches the ocean bottom, which is at 4000-6000 m over most of the world ocean. Thus the entire depth of the ocean acts as a waveguide, allowing sound from a source anywhere in the ocean to be bent away from the surface and bottom, not touching the boundaries even in travelling for thousands of kilometers. (See Figure 2.)

Furthermore, attenuation of sound in seawater is very small at low frequencies, making very long-range experiments practical.

Several experiments carried out in the 1970's provided precise data on acoustic fluctuations with which to compare theoretical predictions. The successful early comparison of that data with the theory of internal-wave dominance has been described in detail elsewhere.[2,3] The most precise experiments were relatively short-range, high-frequency transmissions. One of these experiments was a transmission over 18 km between two fixed, underwater towers near Cobb Seamount off the coast of Washington.[4] The experiment was repeated with even higher precision several years later.[5] The key observations of these experiments were the temporal spectra of phase and log-amplitude, with observations made every 31.5 s for many days. The phase spectrum (Figure 3) showed an enhanced band in just the frequency range expected for internal waves. This result is expected if the transmission occurred in the weak-fluctuation regime. The strength and spectrum of internal waves were independently measured with *in situ* instruments, and this knowledge was used to make an absolute prediction of the phase and log-amplitude spectra.[6] The agreement with the phase spectrum was excellent (Figure 3), and this encouraged the belief in internal-wave dominance.

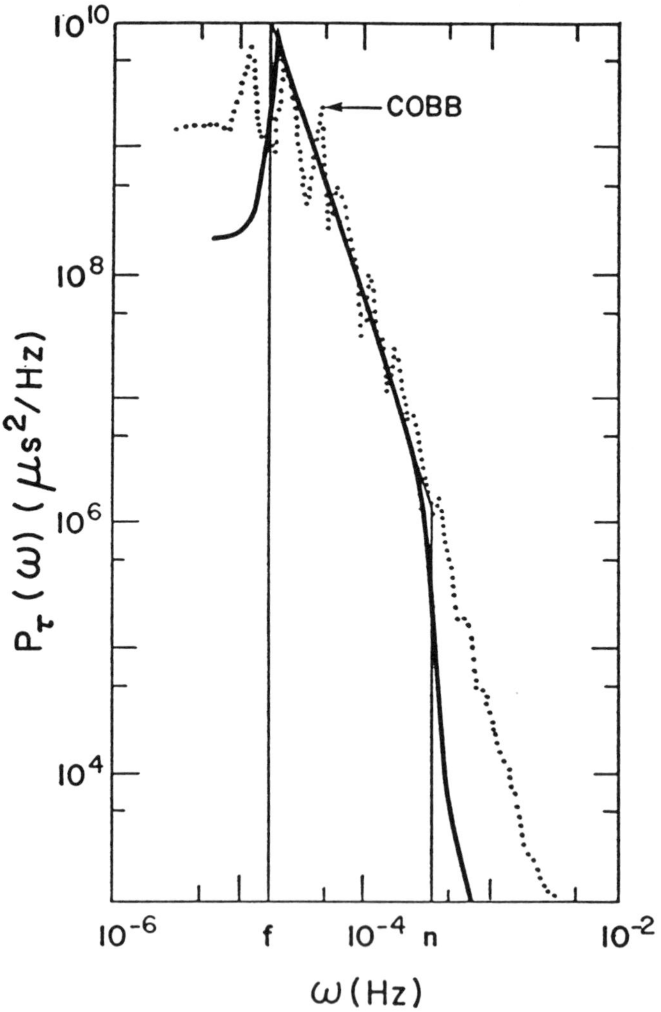

Figure 3. Phase spectrum of arriving wave in the Cobb Seamount experiment. The heavy smooth curve is the prediction of internal-wave dominance, which contributes in the frequency band between the inertial frequency f and the buoyancy frequency n.

Another experiment of that era was a 35-km transmission in water off the Azores.[3] Independently measured internal waves had strength that was higher than normal, resulting in propagation conditions that were saturated. The predictions for the phase and log-amplitude spectra were made by means of a path-integral technique[3] and gave remarkable agreement (Figure 4). The log-amplitude spectra follow a simple Gaussian prediction, but the phase spectra were a sophisticated combination of the weak-fluctuation band and a strong-fluctuation tail. Furthermore, the behavior of the fluctuations could be understood with even more precision. Figure 5 shows phasor plots from the Azores experiment;[7] the movement of the phasor in an azimuthal direction is a consequence of a combination of saturation with the dispersion relation of ocean internal waves. A quantitative fit to this motion for the Azores experiment (Figure 6) was quite successful.

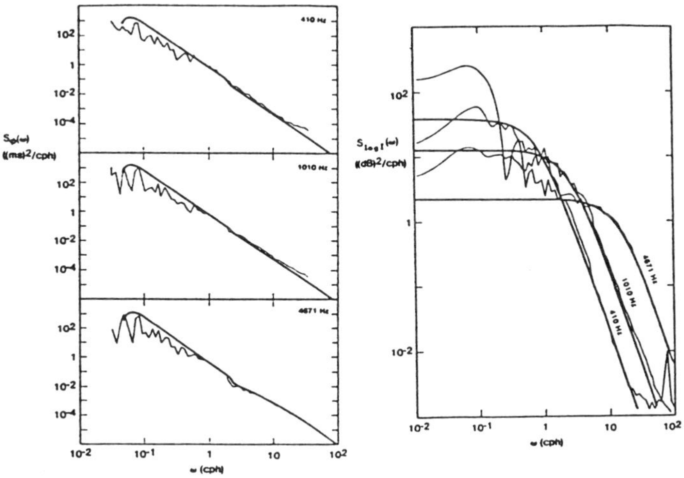

Figure 4. Spectra in the Azores experiment. The smooth curves are predictions of internal-wave dominance. The buoyancy frequency is 2 cph. (a) Phase; (b) log-intensity.

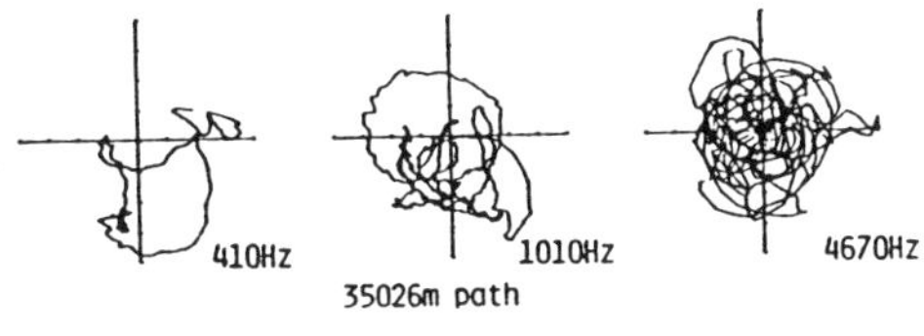

Figure 5. Phasor plots from the Azores experiment. The general trend to move in an azimuthal direction even in the saturated region is a consequence of the dispersion relation of internal waves.

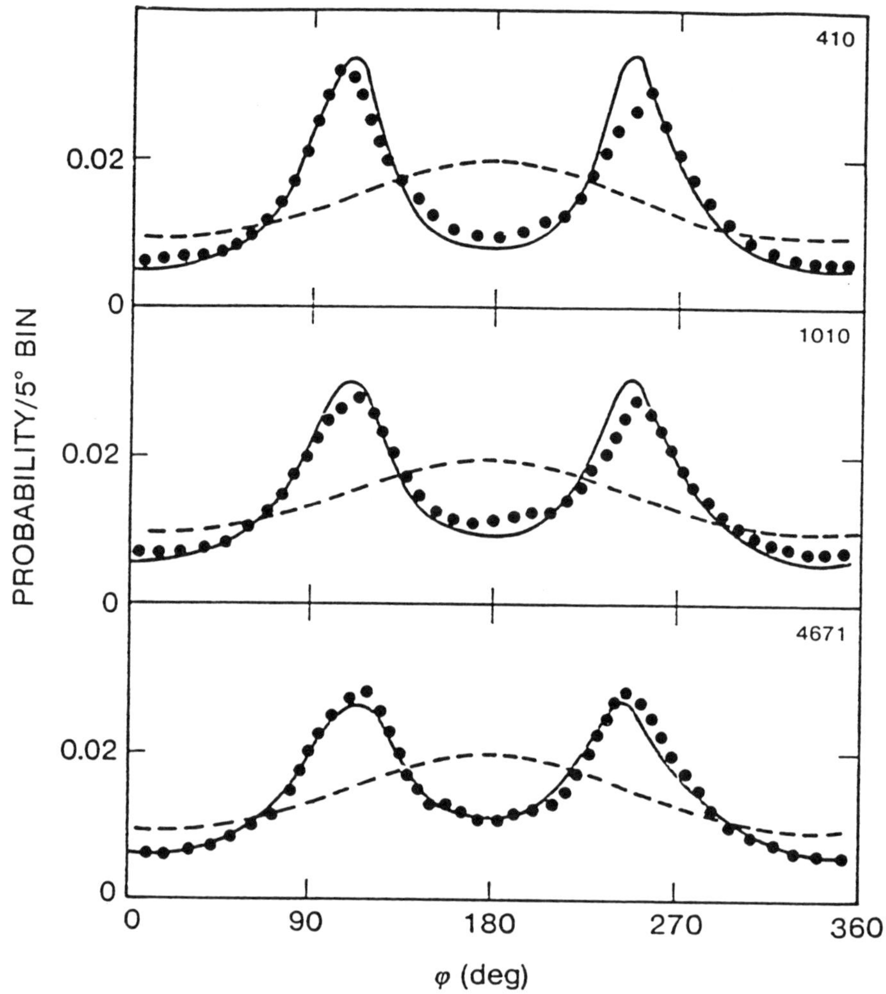

Figure 6. Histograms of the angle between the phasor position and its velocity in the complex plane. The peaks near 90 and 270 degrees are a consequence of tendency toward azimuthal motion; the solid curves are predictions of internal-wave dominance. The dashed curves are predictions for a two-dimensional Gaussian behavior.[3]

The Azores experiment included pulse transmissions in which the pulses were on the order of 1 ms in width. The prediction of the tails on the received average pulse was a triumph of the path-integral technique for calculating fluctuations (Figure 7).

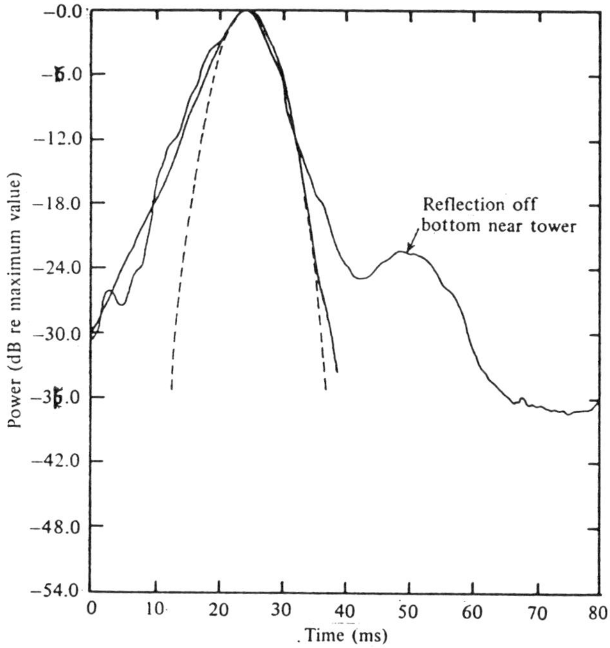

Figure 7. Average arrival pulse in the Azores experiment. The dashed curve is a Gaussian with σ predicted from internal-wave dominance. The solid smooth curve is a more precise prediction depending on a combination of internal waves and the ocean sound channel. The precursor represents early-arriving energy, and is a consequence of the sound channel.

The above was only a brief selection of results from a program of comparison of theory and experiment that included many other experiments and very detailed theoretical calculations of many observables, including temporal and spatial coherence functions, higher moments of intensity, and the probability density function of intensity. There were long-range, low-frequency experiments in these comparisons, but they typically did not have the precision or the multiple measurements necessary for a stringent test of theory.

III. Recent Long-Range Experiments

Stimulated by early success of predicting acoustic fluctuations caused by real ocean processes, a number of groups embarked on an ambitious program of using acoustics to monitor large-scale processes in the ocean. This effort goes under the name of "Ocean Acoustic Tomography".[8] At the same time it was pointed out that long-range experiments could be used to monitor the geographic distribution of small-scale ocean processes, in particular, internal waves.[9]

The first use of long-range acoustic measurements to measure internal-wave strength used a single source-receiver pair separated by 300 km in the Atlantic. The arrival time of a dozen rays were measured every hour over six weeks, and the several-millisecond rms fluctuations combined with a dominant depth associated with each ray to provide a picture of the depth distribution of internal waves in that geographical region (Figure 8). [10,11]

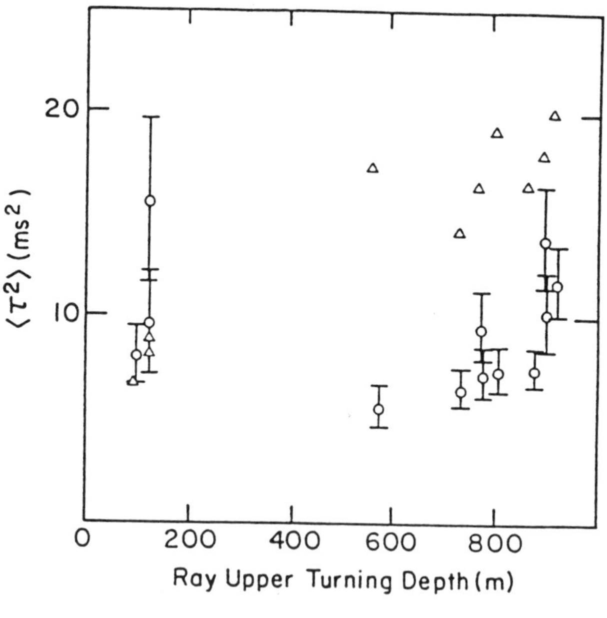

Figure 8. Internal-wave tomography from a 300-km ocean-acoustic experiment.[11] The experimental points are the variances of travel time observed along rays whose upper turning points are taken as their representative depths. The triangles are the predictions of a reference strength of internal waves. One sees that the internal waves below 500 m are only about half the reference strength.

A more recent experiment was carried out in the Pacific with a source separated by 1000 km from a receiving 3000-m vertical line array of 50 hydrophones.[12] The effective transmitted pulse was 12 ms wide, with 250-Hz center frequency, allowing measurement accuracies of order one millisecond. A sample received pulse (Figure 9) shows multiple wavefront segments spread over four seconds arrival time. The global shape of the pulse agrees well with a simple geometrical optics prediction. These pulses were transmitted every 10 minutes for 21 hours, and also every hour for nine days. Each wavefront segment is sampled by a number of hydrophones, so that the arrival time fluctuations in both time and vertical position can be measured. Figure 10 shows the variances for the segments that can be separated out; each of these segments represents a particular ray trajectory through the 1000 km from the source to the receiver array. Figure 11 shows the temporal coherence function for two of the segments; segment -36 shows a clear component with 12-hr period, while both segments show a component with coherence time less than the pulse separation time of 10 min. This type of information for all the segments can be used in an inversion scheme for determining the ocean processes that caused the fluctuations. For example, the 12-hr periods are probably due to internal tides.

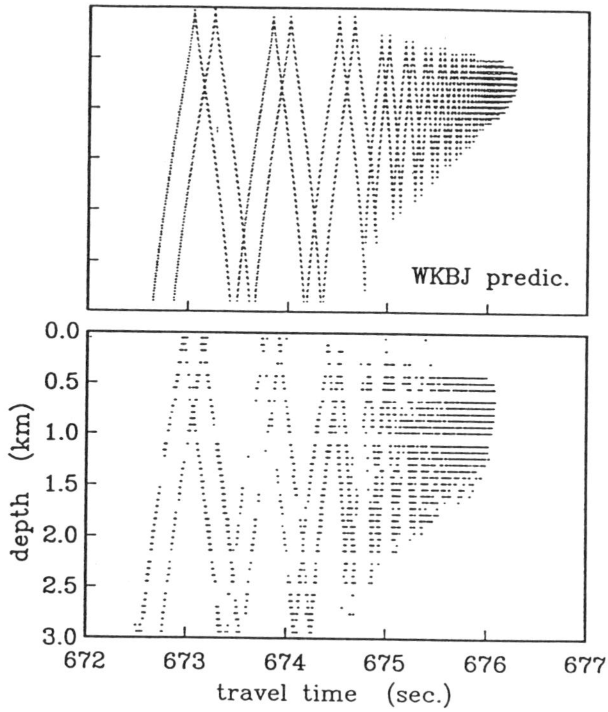

Figure 9. Arrival structure in time and depth of the intensity of a sound pulse from an 800-m-deep source traveling 1000 km to a vertical line array of 50 hydrophones spread in depth from the surface to 3000 m.[12] (a) Geometrical optics prediction using a range-independent sound-speed profile; (b) One pulse from the experiment in the Pacific Ocean.

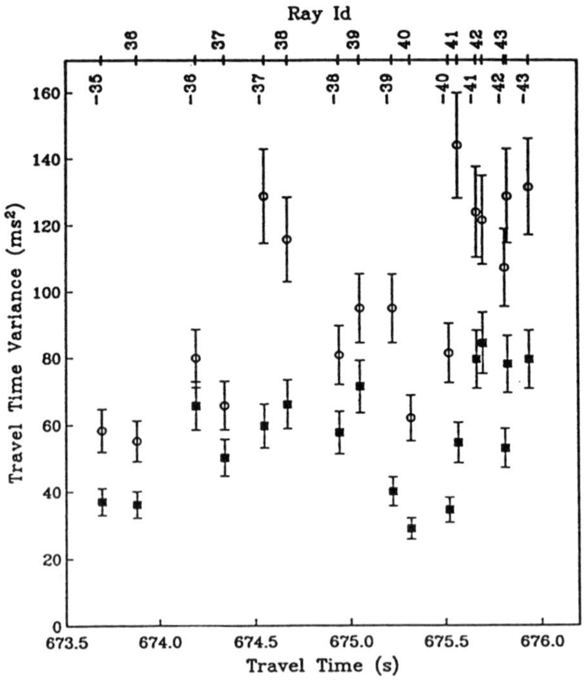

Figure 10. Variances of travel time observed along rays that arrive to particular segments in Figure 9. These fluctuations are caused by internal waves.

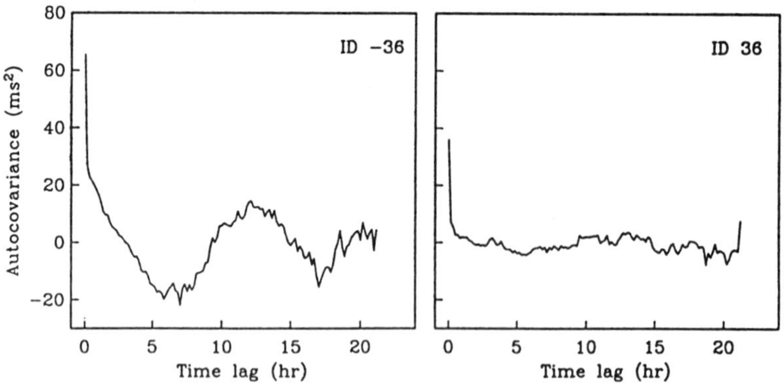

Figure 11. Temporal coherence function for two arrival segments seen in Figure 9. The spike at zero lag indicates a process with coherence time less than the time between pulses (10 min), while the 12-h period seen in one of the functions indicates the effect of an internal tide.

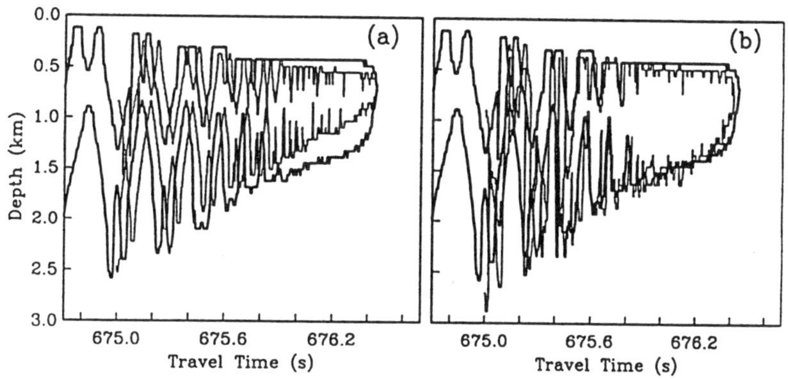

Figure 12. Numerical simulation of ocean sound propagation compared with data from the Pacific experiment. The thick solid line is a representation of the area within which sound arrives at the receiver. The thinner line represents the result from numerically simulating the sound propagation. (a) No internal waves in the simulation, and the simulated arrival is significantly narrower in depth than the real data. (b) Internal waves in the simulation, and one sees that the sound energy has diffused in depth, in qualitative agreement with the real data.

The fluctuations with coherence time less than ten minutes (and simultaneously, vertical coherence lengths less than 60 m) were unexpected. But analytic calculations of pulse fluctuations are difficult. Some work is now in progress to numerically simulate 1000-km propagation through a field of internal waves.[13] A preliminary result is shown in Figure 12. Both sections of Figure 12 show the envelope of the experimentally received pulse, averaged over all the transmissions. Figure 12a compares with the envelope from a numerical simulation without internal waves; it does not agree well. In Figure 12b, the simulation included internal waves, and it shows a clear expansion of the envelope in depth, in qualitative agreement with the observations. Therefore it appears that internal waves may very well be the cause of the diffusion in depth of the sound energy. These same simulations can be checked for the occurrence of arrival-time fluctuations with coherence times and lengths in agreement with the surprising experimental observations; this work is in progress.

IV. Mesoscale and Internal-Wave Tomography

The variance of fluctuations in arrival time due to small-scale ocean processes has been measured in one experiment to be 40-100 ms^2 at 1000 km. Future numerical simulations will be able to tell us how this number depends on range. Some of this variance has very small coherence time and coherence length, while a significant portion, probably from internal tides, is highly correlated in depth, with correlation time about 12 h.

Mesoscale tomography experiments attempt to use the arrival times of many rays to invert for ocean structure of scale about 100 km. The above fluctuations act as an irreducible noise that limits the accuracy of tomographic inversions. Many samples at a rapid rate can greatly reduce the mean of the rapid fluctuations, but the internal-tide effects have such long time and space scales that little reduction is possible. Thus tomography programs must address the effect of internal-wave limitations.

Internal waves do more than just introduce fluctuations with zero mean. It is possible that they can introduce a mean bias in the arrival time of a given set of rays.[14] This possibility must be investigated by means of numerical simulation. A mean bias of this type is of general interest in studies of waves in random media, because it means that, for any medium, small-scale statistical structure can be confused with larger scale structure.

Ocean-basin-scale tomography is becoming of interest, for example, to those studying global warming and the consequent expected rise in the mean temperature of the ocean.[15] By measuring travel times of sound traversing many thousands of kilometers of the world ocean, it is possible in principle to measure the change in ocean mean temperature. However, the noise on the measurement comes from smaller scale, statistical processes: mesoscale and internal waves. Understanding these processes statistically becomes of first order importance for global-scale measurements.

Heat and momentum flux in the ocean is of importance to our understanding of the climate system, and these quantities are affected by the small-scale processes such as internal waves. Therefore a program to measure the statistical strength of internal waves as a function of depth and geographical position is of value to our overall understanding of climate. Acoustic measurements of travel time fluctuation can be used for this purpose. It is of interest that the bias effect mentioned above is not reciprocal. That is, internal waves cause a mesoscale bias, but mesoscale does not cause a bias in the measurement of fluctuation variance. However, the recently observed unexpected contribution to the variance with very rapid space and time variability is a new limitation on the accuracy of deducing internal-wave strength from fluctuation variance. Thus our lack of understanding of the forward problem is a limiting factor to our desire to use acoustic fluctuations as a probe of internal waves.

Acknowledgments

This work has been partially supported by the Office of Naval Research Code 1125OA. I are grateful for a grant from the W.M. Keck Foundation. Discussions and help from John Colosi, Charles Bracher, Jeff Simmen, and Guang-Yu Wang are gratefully acknowledged. This is Contribution Number 193 from the Institute of Tectonics.

1. V.I. Tatarskii, in *The effects of the turbulent atmosphere on wave propagation*, National Technical Information Service, Springfield, VA, 1971.

2. S.M. Flatté, R. Dashen, W.M. Munk, K.M. Watson & F. Zachariasen, in *Sound Transmission Through a Fluctuating Ocean*, Cambridge University Press , 1979.

3. S.M. Flatté, "Wave propagation through random media: Contributions from ocean acoustics," *Proc. of the IEEE* 71, 1267–1294(1983).

4. T.E. Ewart, "Acoustic fluctuations in the open ocean — A measurement using a fixed refracted path," *J. Acoust. Soc. Am.* 60, 46–59(1976).

5. T.E. Ewart & S.A. Reynolds, "The mid-ocean acoustic transmission experiment — MATE measurement using a fixed refracted path," *J. Acoust. Soc. Am.* 75, 785–802(1984).

6. W.H. Munk & F. Zachariasen, "Sound propagation through a fluctuating ocean — Theory and observation," *J. Acoust. Soc. Am.* 59, 818–838(1976).

7. S.A. Reynolds, S.M. Flatté & R. Dashen, B. Buehler, P. Maciejewski,, "AFAR measurements of acoustic mutual coherence functions of time and frequency," *J. Acoust. Soc. Am.* 77, 1723–31(1985).

8. W.H. Munk & C. Wunsch, "Ocean acoustic tomography: A scheme for large-scale monitoring," *Deep Sea Research* 26A, 123–161(1979).

9. S.M. Flatté, "Principles of acoustic tomography of internal waves," *Proc. Oceans '83* (Aug. 29 - Sept. 1, 1983).

10. S.M. Flatté & R.B. Stoughton, "Theory of acoustic measurement of internal-wave strength as a function of depth, horizontal position, and time," *J. Geophys. Res. - Oceans* 91, 7709–7720(1986).

11. R.B. Stoughton, S.M. Flatté & B. Howe, "Acoustic measurements of internal-wave rms displacement and rms horizontal current off Bermuda in late 1983," *J. Geophys. Res. - Oceans* 91, 7721–7732(1986).

12. T.F. Duda, Stanley M. Flatté, J.A. Colosi, B.D. Cornuelle, J.A. Hildebrand, W.S. Hodgkiss, Jr., P.F. Worcester, B.M. Howe, J.A. Mercer & R.C. Spindel, "Measured wavefront fluctuations in 1000-km pulse propagation in the Pacific Ocean," *J. Acoust. Soc. Am.* 92, 939–955(1992).

13. J.A. Colosi, S.M. Flatté & C. Bracher, "Numerical simulation of pulse propagation through a random field of internal waves in an ocean waveguide," *International Meeting for Wave Propagation in Random Media*, Seattle, WA (1992).

14. J.L. Codona, D.B. Creamer, S.M. Flatté, R. Frehlich & F. Henyey, "Average arrival time of wave pulses through continuous random media," *Phys. Rev. Letters* 55, 9–12(1985).

15. A. Baggeroer & W.H. Munk, "The Heard Island feasibility test," *Physics Today* 45, 22–30(1992).

Section 2

Scintillation Measurements and Their Interpretations

Ocean acoustic propagation measurements and wave propagation in random media

Terry E. Ewart and Stephen A. Reynolds

Applied Physics Laboratory, University of Washington, Mailstop HN-10, Seattle, Washington, 98105

ABSTRACT

Over the last two decades, observations of acoustic scintillations in the ocean have been useful in testing theories that predict the statistics for WPRM. Ocean acoustic observations differ from scattering processes observed in other WPRM sub-fields in two important ways. First, the statistics of the refractive index can be determined accurately from oceanographic measurements. Hence, exacting tests of theories can be done when careful measurements of the medium and the wavefield are made synoptically. Second, the scattering of acoustic energy is predominantly influenced by the ocean internal wave field, which tends to exhibit statistical stationarity in many ocean regimes, and not by non-stationary turbulent processes, as observed in the atmosphere. We discuss space/time acoustic propagation experiments carried out in three scattering regimes: (1) weak, where the Rytov approximation is valid; (2) moderate, where multiple scattering theory is required; and (3) the case where multiple scattering in a single medium irregularity is encountered. A review is presented of the measurements and comparisons with theory.

It is clear from field and numerical experiments that 2-point statistics cannot possibly explain the observations of intensity. Because two-point statistics are the norm when interpreting both medium and wavefield measurements, the possible failure of Gaussian statistics must be addressed. This has yet to be done. We will discuss remaining issues for which we believe further study is needed.

2. INTRODUCTION

This conference addresses fluctuations in the propagation of waves in media with random index of refraction fields. This discipline is rooted in the practicalities of dealing with those fluctuations, and the beauty associated with explaining them. We present observations of acoustic scintillation statistics measured by our group over three paths in the deep ocean. The ocean environment is ideal in the sense that the scattering refractive index field may be monitored simultaneously with scintillation measurements. However, ocean paths are complicated by an anisotropic scattering field and a depth-dependent sound speed and scattering strength. The experiments we describe attempted to minimize these complications by using propagation ranges less than 20 km. In particular, the second path we discuss was nearly horizontal. Evaluations of 4th moment theory have successfully predicted the strong ($< I^2 > > 2$) scintillation levels from this path.

The complex amplitudes of the fluctuations are observed to have highly non-Gaussian behavior; hence we are "stuck" with a need to understand the phenomena at complexities beyond the usual two-point statistics. Numerical simulations are ideal for studying field statistics over a broad range of scattering conditions. The latter part of our presentation discusses parabolic equation-based numerical experiments used to study intensity probability distributions. Using our ocean acoustic measurements as a starting point, we use the generalized gamma distribution to model simulated data from weak and through strong scatter to saturation.

Since the introduction of the system of parabolic moment equations by Shishov[1] (1968), we have had the tantalizing expectation of understanding the scattering physics. It has been a long road and we highlight some of the work pertinent to the experiments presented. We would like to emphasize that understanding observations of ocean acoustic scintillation has driven the development of theory.

Using the Garrett and Munk[2] (1972) ocean internal wave model, Uscinski[3] (1980) applied 4th moment theory to the published[4] 1971 Cobb Seamount experiment temporal spectra of intensity. These were only partial solutions however, and in 1982 Uscinski[5] provided the first full range solution to the 4th moment equation for a plane wave initial condition. Uscinski et al.[6] and Ewart et al.[7] gave predictions for the Cobb experiment intensity time spectra that showed the proper acoustic frequency dependence. The scintillation index remained under-predicted. That theory included a point source initial condition. Macaskill[8] (1983) provided a correction term to the 4th moment using two-scale expansions. That theory was applied by Macaskill and Ewart[9] (1984) to "numerical experiments" based on parabolic wave field-marching through computer-generated index of refraction fields. There was good agreement between the full range theory with the correction term, and the simulated data intensity correlations for low values of the scattering parameters (for Gaussian and 4th order power law medium transverse spectra). Ewart and Reynolds[10] (1984) presented the initial results from a subset of the MATE (Mid-Ocean Acoustic Transmission Experiment) experiment data compared to zeroth order theory using an ocean model that included linear ocean internal waves and a refinement termed finestructure. Agreement between the theory and intensity correlations from the data improved, but the under-prediction remained. Ewart et al.[11] (1985) showed that the cross frequency 4th moment theory of Uscinski and Macaskill[12] (1985) severely under-predicted the cross-spectra for intensity records at differing frequencies. A full range solution to the 4th moment equation using the path integral was formulated by Uscinski et al.[13] (1986). Earlier path integral solutions were asymptotic to short and long range cases, and not applicable to the experiments at Cobb Seamount. In this paper, we include linear internal waves and finestructure as well as a correction term to the theory for the MATE predictions.

Macaskill and Ewart[14] (1984) provided a model of the intensity moments and an intensity probability distribution of the log intensity that predicted several experiments spanning a wide range of scattering parameters. Ewart and Percival[15] (1986) demonstrated that the generalized gamma distribution models both field and numerical experiment data (acceptance at the 90% confidence level). Using numerical experiments, Ewart[16] (1989) demonstrated (for 8 decades in scattering strength corresponding to 3 decades in wavenumber and 6 decades in range) that the generalized gamma distribution is a valid model for the intensity probability distribution for WPRM. In the past few years, Uscinski's Cambridge Group has produced several publications on comparisons of wavefield simulations with numerical solutions for the 4th moment equation. References are found in Leonard and Uscinski[17]. Recently, Martin and Flatte´[18, 19] extended these simulation techniques to 3-D media for a point source.

Almost all of the testing problems arise when theories are confronted with real data sets. Having full range solutions for many of the possible 2-point separations requires comparisons with observations for validation; that is the crux of this paper.

2.1. Experiments at Cobb Seamount

Ewart[4] examined ocean acoustic fluctuations, where both the phase and the amplitude (4 and 8 kHz transmissions over an 18 km path using fixed sources and receivers) were recorded synoptically (6 days) with oceanographic measurements. The observed intensity fluctuations were large ($< I^2 > > 2$). The available theories (many based on the Rytov approximation) were able to predict the phase fluctuations (Desaubies[20, 21, 22] and Munk and Zachariasen[23]). However, these same theories were hopelessly inadequate for predicting the large intensity variance, or its spectral decomposition. For that experiment, the oceanographic observations were insufficient to determine the full 3-space/time ocean model.

In a second experiment at Cobb Seamount (MATE), the frequencies were expanded to include 2 and 13 kHz. The space/time medium fluctuations along the transmission path region were sampled by many different sensors so that various projections of the 3-space/time statistics were known. Twelve days of data on a downward path and two days on an upward path were recorded (the paths were both totally refracted). As in the first experiment, the fixed source and receivers allowed determination of the travel time to within a few microseconds, and it was possible to separate the upward or downward arrivals into identifiable Fermat paths[24, 25, 26]. In this paper, we will only discuss the MATE results.

2.2. The AIWEX Acoustic Transmission Experiment (AATE)

AATE was conducted from a multi-year icefield north of Prudhoe Bay, Alaska (Ewart and Reynolds[27]). Transmissions at 2, 4, 8, and 16 kHz were made over a 6.4 km path for 12 days to a receiving array with a 153 m vertical (synthetic) aperture. As in MATE, the arrivals were isolated into single Fermat paths and the phase and amplitude results were obtained. In AATE the intensity fluctuations were very weak, and the oceanography (as determined synoptically by the Arctic Internal Wave Experiment (AIWEX) investigators) was highly non-stationary, even when viewed over the operation's short duration.

Testing the full range moment equation theories has evolved into more precise evaluations of theories and subtle increases in our understanding of ocean variability. It is our contention that the remaining tasks facing us in our quest to validate the theories will boil down to "getting the medium right," rather than any order one changes in the theories. In the atmosphere, investigators have had to contend with a severe lack of stationarity in turbulent processes. In the ocean, horizontal homogeneity and stationarity of processes are good assumptions, but the actual spectral models are complex in both wavenumber and frequency.

3. OCEAN INDEX OF REFRACTION VARIABILITY

In the late 1960's and early 1970's, it was recognized by oceanographers that submesoscale variability was dominated by the processes of internal waves - dynamic oscillations on the weak internal density gradients of the ocean. In a series of papers, Garrett and Munk[2, 28] outlined a model for the linear internal wave field of the ocean. These dynamics differ from turbulence in that they are anisotropic, and the dispersion relation requires inclusion of spatial/temporal coupling. Another complication is additional variance in the index of refraction from processes that have been termed finestructure[29]. Finestructure may represent a significant fraction of the total $< n^2 >$, and cannot in general, be ignored. Unlike the atmospheric case, internal wave velocities produce a much smaller effect on the index of refraction (as seen by a propagating wave) than do internal wave vertical displacements. To set the stage for introducing the distinctions between atmospheric turbulence and ocean variability, we present Figure 1.

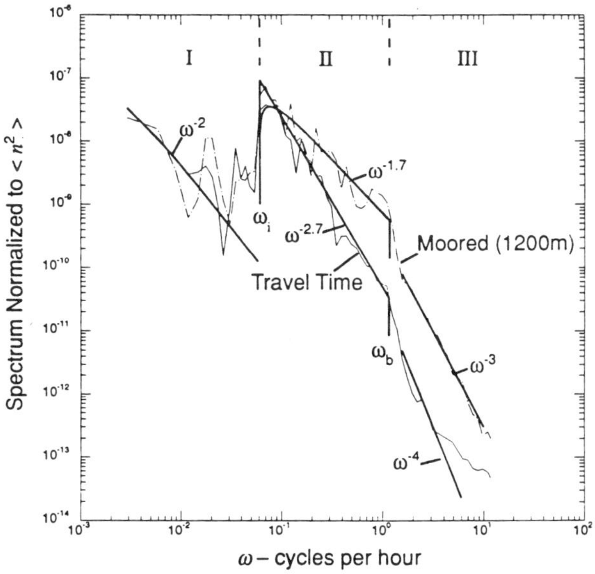

Figure 1. MATE phase and moored temperature sensor data converted to $< n^2 >$ units. (Note that cyclical units are used in the Figures.)

Here we have plotted the 2 kHz travel time spectrum of the MATE experiment, in $< n^2 >$ units. The same normalization has been done for one of the moored (fixed point) temperature spectra in the experiment, where the temperature fluctuations measured in the mooring have been converted to index of refraction fluctuations. The integrals of both spectra equal $< n^2 >$. The frequencies between the inertial frequency, w_i , and the buoyancy frequency, w_b, Regime II, is the band where linear internal wave dynamics is allowed. Regime I below w_i, that extends (unobservable in the spectrum) into Region II, is the finestructure regime, and Regime III above w_b, is the high-vertical-wavenumber microstructure with a very small total energy. In both spectra, the energy in the tide lines has been removed by combined modeling of stochastic and deterministic processes. The moored spectrum of the index of refraction is

$$\Phi_M(\omega) = \int_{-\infty}^{\infty}\int_{-\infty}^{\infty}\int_{-\infty}^{\infty} \Phi_n\left(\kappa_x, \kappa_y, \kappa_z, \omega\right) d\kappa_x\, d\kappa_y\, d\kappa_z, \tag{1}$$

where Φ_n is the 3-space/time spectrum of n. We need the medium correlation function projected (integrated) in the propagation direction, $B_{P,n}$. For straight ray propagation in z,

$$B_{P,n}\left(v_x, v_y, \tau\right) =$$

$$\int_{-\infty}^{\infty}\int_{-\infty}^{\infty}\int_{-\infty}^{\infty} \Phi_n\left(\kappa_x, \kappa_y, \kappa_z = 0, \omega\right) e^{i\left(\kappa_x v_x + \kappa_y v_y + \omega\tau\right)}\, d\kappa_x\, d\kappa_y\, d\omega, \tag{2}$$

where $(\kappa_x, \kappa_y, \kappa_z, \omega)$ and (v_x, v_y, v_z, τ) are the spectral and separation coordinates respectively. The spectrum of the phase, S, is

$$\Phi_S(\omega) = k^2\, z_R \int_{-\infty}^{\infty} B_{P,n}\left(0, 0, \tau\right) e^{-i\omega\tau}\, d\tau. \tag{3}$$

$B_{P,n}\,(0, 0, 0) = \langle n^2 \rangle L_P$, with L_P the integral scale from the integration in range, k is the acoustic wavenumber, and z_R is the range. The projected medium correlation function, $B_{P,n}\,(v_x, v_y, \tau)$, is usually called the Transverse Correlation Function, TCF. It is analogous to the integrated structure function in atmospheric scattering literature. This expression has been generalized to the curved ray case by Esswein and Flatte´[30].

For normalization, we use the standard parameters of the parabolic moment equations, where two parameters, scattering strength, Γ, and the scaled range Z, represent the scattering regimes,

$$\Gamma = k^3 \langle n^2 \rangle L_P\, L_y^2, \quad Z = \frac{z_R}{k\, L_y^2}, \text{ and } \langle S^2 \rangle = \Gamma\, Z. \tag{4}$$

L_y is the transverse correlation length. In physical terms, Γ is the "Fresnel" distance, kL_y^2, divided by the scattering (or coherence) length $(k^2 \langle n^2 \rangle L_P)^{-1}$; Z is the propagation distance scaled by the "Fresnel" distance. Thus, the phase spectrum can be written

$$\Phi_S(\omega) = \Gamma\, Z \frac{\int_{-\infty}^{\infty} B_{P,n}\left(0, 0, \tau\right) e^{-i\omega\tau}\, d\tau}{B_{P,n}\left(0,0,0\right)}.$$

We will require more detail on the TCF later in the discussion. The theoretical phase fluctuation and moored spectra plotted in Figure 1 were obtained from the MATE spectral model and (1) and (3). The steepening of the travel time spectrum in the internal wave regime arises from the dispersion relation; an extra power is subtracted in the negative power law exponent, thus effectively demonstrating the accuracy and applicability of the linear internal wave dispersion relation. In the observable finestructure regime i.e., below ω_i, the spectra are statistically the same, indicating the lack of a dispersion relation. We believe that in the microstructure regime, small scale structures are advected by the internal waves and the power law exponent changes in the same fashion between the spectrum of n and the spectrum of S.

For propagation studies in the turbulence spectrum of the atmosphere as found in Ishimaru[31], the TCF of the medium is obtained from an assumed spectrum of turbulence that is advected past the measurement apparatus. The TCF for the case of a spectrum, $\Phi_n(\kappa_x, \kappa_y, \kappa_z)$ is

$$B_{P,n(turb)}(v_x, v_y, \tau) = \int_{-\infty}^{\infty}\int_{-\infty}^{\infty} \Phi_n(\kappa_x, \kappa_y, 0) e^{i[\kappa_x (v_x - v_x\tau) + \kappa_y (v_y - v_y\tau)]}\, d\kappa_x\, d\kappa_y$$

$$= B_{P,n(turb)}[(v_x - v_x\tau), (v_y - v_y\tau)],$$

where v_x and v_y are the medium velocities transverse to the propagation. The fluctuations in time are the same as the fluctuations at a field point. For such a medium, Φ_S and Φ_M in Figure 1 would have the same spectral dependence. If the internal wave spectrum of the atmosphere plays a role in the scattering of electromagnetic waves, the effects could be detectable from the usual Taylor hypothesis models by precise measurements of the phase and the one-point index of refraction fluctuations for electromagnetic or acoustic propagation.

We turn our attention to reviewing the specific TCF's we will need in discussing ocean experimental results.

3.1. The AATE TCF

As mentioned in the introduction, the AATE propagation geometry sampled a weak scattering environment. Fluctuations in the refractive index were dominated by internal waves. Levine[32] showed that the observed environmental variability could be described by a modified GM internal wave model. The internal wave displacement variance was reduced by about a factor of 50 from open ocean values and spectral frequency dependence and modal bandwidth differed from the traditional model. For our purposes, we decompose the vertical variability denoted by mode number, j.

Using the GM formulation, the refractive index spectrum decomposed in terms of frequency and vertical mode number can be written

$$\Phi_n(\omega, j; y) = \langle n^2(y) \rangle G(\omega) H_j, \text{ where} \tag{5}$$

$$G(\omega) = G_0 \frac{\left(\omega^2 - \omega_i^2\right)^{1/2}}{\omega^p}, \; H_j = \frac{H_0}{j^2 + j_*^2}.$$

G_0 and H_0 are normalization constants such that

$$G_0^{-1} = \int_{\omega_i}^{\omega_b} \frac{\left(\omega^2 - \omega_i^2\right)^{1/2}}{\omega^p} d\omega, \text{ and } H_0^{-1} = \sum_{j=1}^{\infty} \frac{1}{j^2 + j_*^2}.$$

The level of refractive index fluctuations is related to the internal wave displacement variance, $\langle \zeta^2 \rangle$, by the relative gradient in potential sound speed. That is,

$$\langle n^2(y) \rangle = \left(\frac{1}{c_0} \frac{d c_p}{dy} \right)^2 \langle \zeta^2(y) \rangle. \tag{6}$$

The displacement variance is related in turn to the vertical profile of the internal wave buoyancy frequency. From the environmental observations, $p = 2.2$ and $j_* = 30$ for the AATE site; see Levine[32] for details. In general, because $\langle n^2 \rangle$ is a function of depth, scattering is vertically inhomogeneous along ocean rays. This was the case during AATE, and the TCF must be written as a weighted integral of (5) along the ray. We defer further discussion to the section on acoustic results.

3.2. The MATE TCF

Two ray paths were monitored along the MATE 18 km range. Unlike AATE, an effectively homogeneous scattering environment was traversed by the upward ray. However, fluctuations in the refractive index were complicated by the presence of finestructure variability roughly equal in variance to that described by the internal wave model, but apparently having a much longer time scale. This knowledge was obtained using the extensive environmental measurements made during the experiment[33]. Because only the downward path was monitored for environmental fluctuations, we predict internal wave variability along the upward path using the GM-model and test the result with the phase spectrum.

Along the MATE paths, the spectrum of the refractive index is modeled as the sum of internal wave and finestructure components

$$\Phi_n = \Phi_{n,IW} + \Phi_{n,FS}. \tag{7}$$

The covariance function of n, B_n, is defined in terms of the internal wave and finestructure correlation functions, and

$$B_n\left(v_x, v_y, v_z, \tau\right) = \langle n_{IW}^2(y) \rangle \rho_n, \text{ where } \rho_n = \left(\rho_{n,IW} + b\,\rho_{n,FS}\right). \tag{8}$$

For the downward path, $b = 1$ from observation; for the upward path we assume $b = 0$.

As mentioned earlier, we require the correlation function (8) projected along the direction of propagation. We denote this correlation function as ρ_P. For internal waves (see Equation 12 of Uscinski[3]), $\rho_{P,IW}(v_x, v_y, \tau) =$

$$2\int_{\omega_i}^{\omega_b(y)} G(\omega)\cdot\frac{\left(\omega_b{}^2-\omega^2\right)^{1/2}}{\left(\omega^2-\omega_i{}^2\right)^{1/2}}\int_{\kappa_{y_c}}^{\infty}\frac{H(\kappa_y)}{\kappa_y}\cos(\alpha\, v_x)\cos(\kappa_y\, v_y)\cos(\omega\,\tau)\,d\kappa_y\, d\omega. \quad (9)$$

$G(\omega)$ is given in (5). Because the buoyancy frequency varies slowly with depth, a WKBJ approximation is used[21]. The modal decomposition used in the AATE case, can now be written in terms of a vertical wavenumber, κ_y:

$$H(\kappa_y) = H_0^{-1}\,\kappa_{y_*}\left(\kappa_{y*}^2-\kappa_y^2\right)^{-1};\; H_0 = \int_{\kappa_{y_c}}^{\infty}\frac{\kappa_{y_*}\,d\kappa_y}{\left(\kappa_{y*}^2+\kappa_y^2\right)}; \quad (10)$$

$$\kappa_{y_*} = t\left(\omega_b{}^2-\omega^2\right)^{1/2},\; t=\frac{\pi j_*}{\omega_0 Y_0}, \text{ and } \kappa_{y_c} = \frac{\kappa_{y_*}}{j_*},$$

where $\omega_0 = \omega_b(0)$ and Y_0 is the scale depth of $\omega_b(y)$. The internal wave dispersion relationship,

$$\alpha^2 = \kappa_y^2\,\frac{\left(\omega^2-\omega_f{}^2\right)}{\left(\omega_b{}^2-\omega^2\right)} \quad (11)$$

is used in evaluating (9). The local measurements found that $p = 2.7$ and $t = 3.1\ m^{-1}s$ ($j_* = 6$). For details on the environmental modeling see Levine et al.[33].

The finestructure correlation function is modeled as

$$\rho_{n,FS}(v_x, v_y, v_z, \tau) = e^{\left[-\sqrt{\left(v_x{}^2+v_z{}^2\right)/L'^2_H + v_y{}^2/L'^2_V}\right]}\, e^{(-|\tau|/\tau_0)}. \quad (12)$$

Note that horizontal isotropy is assumed for both internal wave and finestructure fields and is well supported by the measurements. Projecting along the range direction gives for finestructure:

$$\rho_{P,FS}(v_x, v_y, \tau) = 2L'_H\cdot a\cdot K_1(a)\, e^{(-|\tau|/\tau_0)}. \quad (13)$$

K_1 is the modified Bessel function and $a^2 = \left(v_x{}^2/L'^2_H + v_y{}^2/L'^2_V\right)$.

To first order, the finestructure correlation lengths equal those for internal waves. $L_p = \rho_{P,IW}(0,0,0)$, then

$$2L'_H = L_p \; ; \; L'_V = L_V , \tag{14}$$

where $L_V = 1/\kappa_{y_*}$. Estimating τ_0 is difficult because the experiment did not last long enough; we take $\tau_0 = 500$ h.

Combining internal wave and finestructure terms, the MATE projected covariance function is

$$B_{P,n}\left(v_x, v_y, \tau\right) = < n_{IW}{}^2 > \left[\rho_{P,IW}\left(v_x, v_y, \tau\right) + \frac{b}{L_p} \cdot \rho_{P,FS}\left(v_x, v_y, \tau\right)\right]. \tag{15}$$

4. THE ACOUSTIC EXPERIMENT RESULTS

In this section we review acoustic experiment results. Predictions for the acoustic observations use the TCF's from the previous sections.

4.1. AATE

AATE provides a good example of the complications that can arise from the ocean sound channel. The sound speed profile, along with the deepest and shallowest ray paths for each receiving channel, are shown in Figure 2. There were no deterministic interfering multipaths. The predicted internal wave induced variability in the refractive index, $< n^2 (y) >$, is also shown.

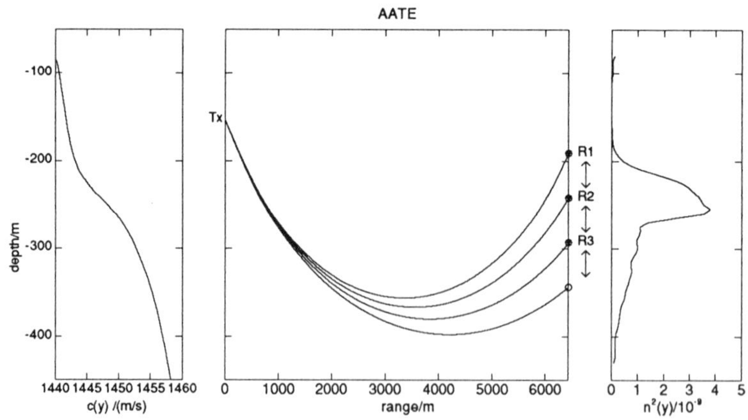

Figure 2. The AATE propagation environment. Locations of the transmitter and three depth cycling receivers are noted on the center ray trace. The rays are traced to the top and bottom positions of the depth cycling (synthetic aperture) array for each receiving channel.

Along ocean rays, the dominant scattering occurs near the turning points. Because of the very weak levels of $< n^2 >$ at the ray turning points, the fluctuations observed are very low. The measured variances in the log-amplitudes for four acoustic frequencies, as a function of depth, are shown in Figure 3. For the observed scintillation levels, weak scattering theory is appropriate. We utilize the Rytov approximation-based theory of Munk and Zachariasen[23]. The GM-formulation in (4) with parameters determined by the environmental measurements provides the spectrum of $< n^2 (y) >$ required by theory.

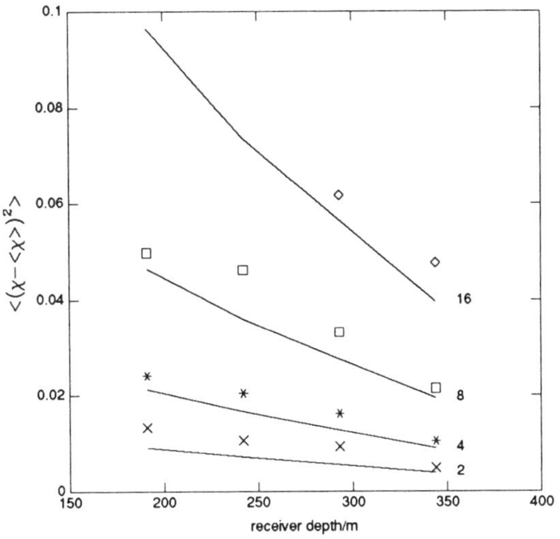

Figure 3. Predicted and observed log-amplitude variance for the receiving array depths. The predictions (solid lines) were obtained by integrating (16). The 2, 4, 8 and 16 kHz observations are shown as crosses, asterisks, boxes and diamonds respectively. (Noise estimates were subtracted from the 16 kHz variances, and are available from the bottom channel only.)

The prediction for the frequency spectrum of phase, Φ_S, and log-amplitude, Φ_χ, for a ray path with a single lower turning point can be written[23]

$$\Phi_{S/\chi}(\omega;y) = \sum_{j=1}^{\infty} \frac{k^2\,\omega_0 Y_0}{\pi} \int_0^{z_R} dz \sec^2\theta \left[1 \pm \cos\left(\beta_{SL}\, j^2\right)\right] \frac{\Phi_n(\omega, j; z)}{j\left(\omega^2 - \omega_L^2\right)^{1/2}}, \quad (16)$$

$$\text{where } \omega_L{}^2 = \omega_i{}^2 + \omega_b{}^2(y)\tan^2\theta;\ \beta_{SL} = \frac{\left(\pi\,\omega_b(y)\right)^2}{\left(\omega_0 Y_0\right)^2} \frac{z\left(z_R - z\right)}{k\, z_R},$$

and k is the acoustic wavenumber (taken constant). In the terms in brackets in (16), + corresponds to S and - corresponds to χ. ω_0 and Y_0 are scales associated with the buoyancy frequency $\omega_b(y)$, ω_i is the local inertial frequency, and θ is the ray angle. β_{SL}

for the linear profile is used here[23]. The integral is carried out along the ray path to range z_R.

To obtain the total variance of χ as a function of depth, Φ_χ in (16) is integrated over $[\omega_i, \omega_b]$. The predictions are shown in Figure 3 for the four acoustic frequencies at AATE. The symbols are the $<\chi^2>$ observations as noted in the caption. For a given depth, observations match the predicted proportional scaling with acoustic frequency. The depth dependence is similar as well. The observations uniformly exceed the predicted values, due to variability outside the internal wave bandwidth.

A better test of theory is provided by the observed frequency spectra. These are shown in Figure 4 for the 8 kHz case at the 191 m receiver. The phase spectrum peaks near ω_i because of mooring motion. (Phase and travel time are equivalent.) Although the total $<\chi^2>$ in Figure 3 exceeds the predicted value, within the internal wave band there is agreement within a factor of 2. The curves denoted GM in the figure are the predictions obtained using the traditional open ocean internal wave values. The observed spectral dependence is better matched by the AIWEX prediction of $\omega^{-2.2}$ than the ω^{-3} dependence of GM. Clearly, our ability to make reasonable predictions was possible only with the local environmental results. At low frequencies, the effect of array motion is apparent in the travel time. (This motion was on the order of 10^{-6} of the 6.4 km path.)

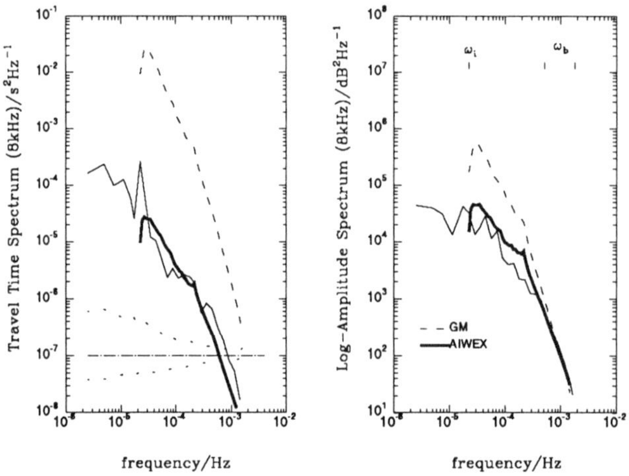

Figure 4. Observed travel time and log-amplitude spectra (solid) compared with predictions[17]. The "GM" predictions (light dashed) use open ocean parameters, whereas the AIWEX curves (bold) use the AIWEX model parameters. ω_i and ω_b are noted on the right hand graph. (The departures from regularity in the theory curves are due to the input $< n^2(y) >$ profile.) 90% confidence intervals, obtained by numerical simulation, are shown dotted about the 10^{-7} line on the left hand graph.

The two-point spatial covariances are shown in Figure 5. We have estimated the spatial covariance using the top receiver position as a common point and pairing with a series at the noted vertical distance away. Because of the severe lack of homogeneity, it is not possible to write the vertical correlations as a simple depth-time covariance. For any given time-lag, τ, the covariances fall off more rapidly with depth as the acoustic frequency increases. This behavior should be predictable by extending Munk and Zachariasen's formulation to follow the vertical inhomogeneity along two different rays. This complication has yet to be addressed.

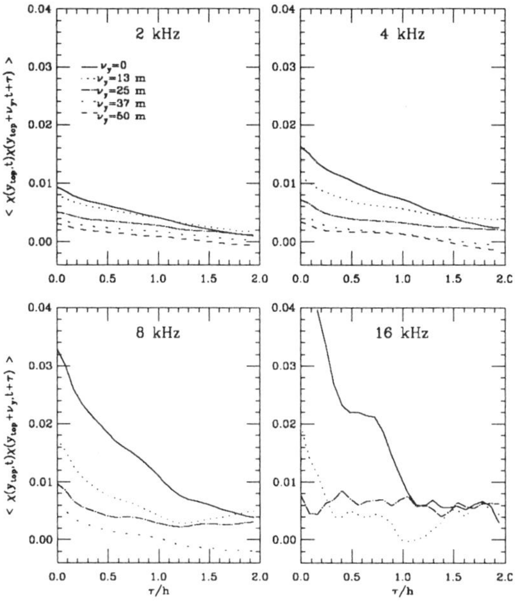

Figure 5. Two-point covariances observed at AATE for depth-separated pairs of log-amplitudes as a function of time lag τ. Depth separations are shown when they are statistically different from zero, y_{top} = 191 m.

In summary, the level of acoustic variability measured during AATE was very weak. Rytov approximation theory for acoustic scintillation due to internal waves gives a reasonably good comparison (better than a factor of 2) with the observations. Detailed knowledge of the local environment was required. Our ability to predict the observations is complicated by the vertically inhomogeneous medium, and the lack of stationarity (only 4 days of data could be included in these results). The AATE measurements provide statistics of weak ocean scattering. Comparisons with theory in this regime where the scattering environment is known are surprisingly scarce in the ocean acoustic literature.

4.2. The MATE Lower Path

Unlike the AATE case, pulses traversed this path over a nearly homogeneous scattering environment, and the paths were fixed. The average vertical sound speed profile is shown in Figure 6 along with the two ray paths. The vertical profile of refractive index fluctuations inferred from the internal wave variability is also shown.

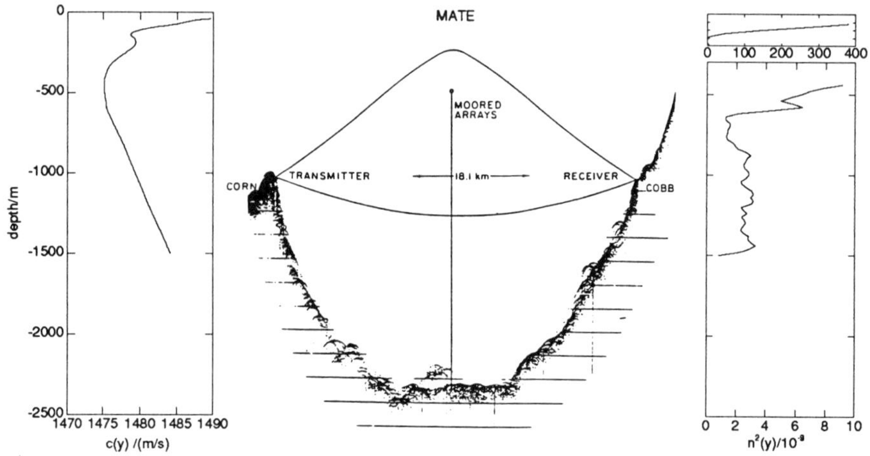

Figure 6. The MATE propagation environment. The location of the moored arrays is shown. The topography of the seamounts is an actual depth-recorder output superimposed on the ray trace. (Note the break in scale at 200 m in the plot of $n^2(y)$).

The validity of the modeled TCF (15) can be demonstrated by comparing the predicted phase spectrum with the observation at a single receiver. Using the results in (3) (neglecting ray curvature effects), the temporal spectrum of acoustic phase at scaled range z_R is given by

$$\Phi_S(\omega) = < S^2 > \left[\frac{G_1^{-1}}{\omega^{2.7}} + \frac{b}{2\pi L_p} \frac{4\,\tau_0}{1+(\omega\,\tau_0)^2} \right]; \text{ with } G_1 = \int_f^N \frac{d\omega}{\omega^{2.7}}. \tag{17}$$

The phase variance, $< S^2 >$, is obtained using the technique developed by Esswein and Flatte´[30]. (The spectral normalization used here is that the integral over positive frequencies equals the variance. (17) has an additional factor of 2 from (3).) The variance over the upward ray is roughly a factor of 5 greater than that of the downward ray. Figure 7 shows (17) compared to the observations (travel time and phase are interchangeable at MATE). Over the internal wave continuum (between ω_i and ω_b) the comparison is excellent. The lines of energy just above and below ω_i are due to deterministic tides and are not considered here. (Note that the tides are removed in Figure 1.)

We turn our attention to the intensity records. Time gating was used to separate the upward and downward path arrivals. (The instrumentation allowed only one path to be monitored at a time.) The behavior of the intensity time series is very different for the two paths. A portion of the observed intensity series during the time gating change from the downward to the upward path is shown in Figure 8. All four acoustic frequencies are plotted. This figure illustrates the dramatic changes in the intensity probability distributions and the temporal behavior in the various scattering regimes. The normalized intensity variances, $(< I^2 > - < I >^2) / < I >^2$, are listed in Table I. The levels increase with the acoustic frequency and decrease between 8 and 13 kHz. Predictions for the downward path, obtained using 4th moment theory (Ballard and Uscinski, unpublished) are listed in Table 1. The theory uses the MATE TCF (15), and includes the zeroth order solution plus a correction term. The agreement is excellent.

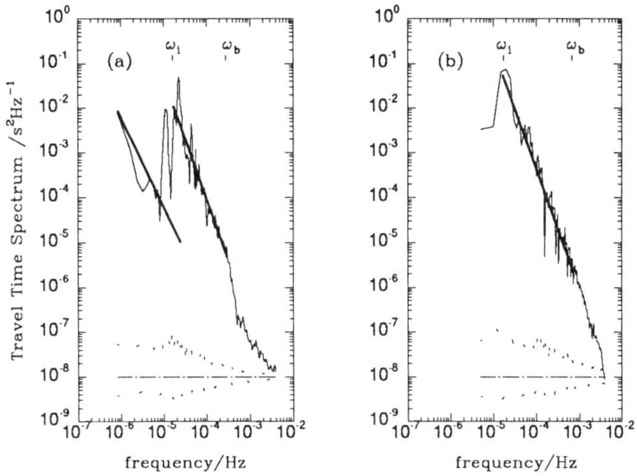

Figure 7. Observed phase spectra compared with the TCF (bold-solid) along the downward (a) and upward paths (b). The finestructure component, with correlation time-scale τ_0 = 500 *h*, is seen at frequencies below ω_i in (a). 90% confidence intervals (as in Figure 4) are shown relative to 10^{-8}.

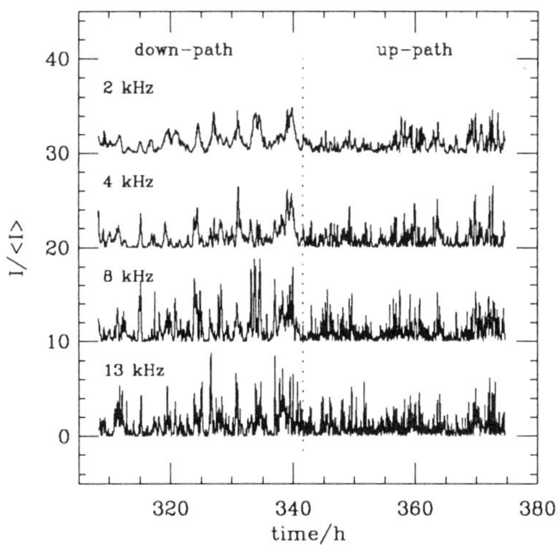

Figure 8. MATE intensities for two inertial periods before and after the instrumentation change from monitoring the downward to the upward path. The time of the change is denoted by a vertical dotted line.

Table I. **MATE** Observed Intensity Scintillation Indices

Frequencies		2	4	8	13 kHz
Down-Path					
Observed					
Tower 1	(bottom/top)	.71/.64±.08	1.28/1.21±.14	1.48/1.39±.12	1.40/1.34±.10
Tower 2	(bottom/top)	.48/.51±.06	1.07/1.05±.17	1.25/1.23±.14	1.48/1.38±.16
Predicted		.54	1.0	1.5	1.4
Up-Path					
Observed					
Tower 1	(bottom/top)	.65/.68±.07	.90/.97±.12	1.01/1.03±.11	.95/.91±.08
Tower 2	(bottom/top)	.60/.60±.07	.83/.87±.10	1.04/1.14±.11	

The intensity spectra observed at a single receiver are shown in Figure 9 for both the down and the up paths. The zeroth-order fourth moment theory predictions for four scaled ranges for 8 kHz are shown in Figure 10 along with the observed 8 kHz intensity spectrum. The theory and predictions near the inertial frequency are in excellent agreement, and we believe represent the first time that the tail of the Rytov approximation behavior has been observed or predicted in multiple scattering for the ocean TCF.

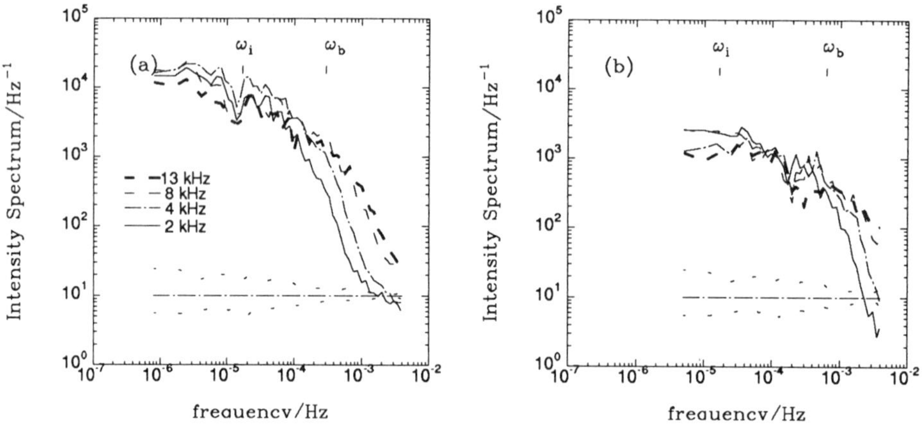

Figure 9. Intensity spectra observed over the MATE downward (a) and upward paths (b). All four acoustic frequencies are over-plotted. For temporal frequencies below 10^{-4} Hz, the spectra are statistically the same. 90% confidence levels as in Figure 4.

Two-frequency and two-point spatial intensity statistics are available from the experiment, but theory evaluation has not been attempted for the extended MATE time series. The 2-point correlations, $\rho_{1,2}(\tau)$, are shown in Figure 11 for vertical (upper graphs) and horizontal (lower graphs) separations. The solid lines are the downward path

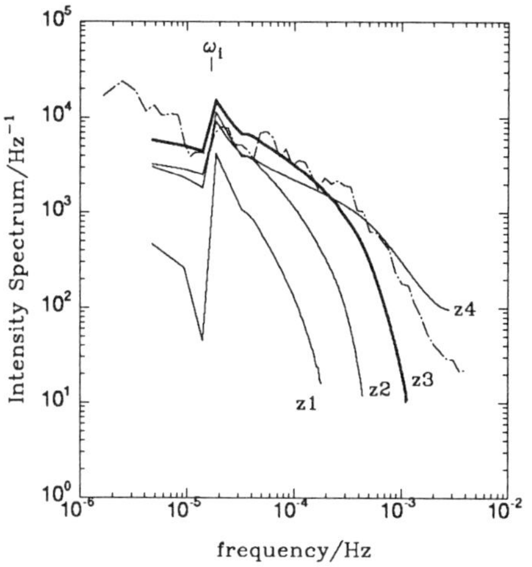

Figure 10. The zeroth-order solution to the moment equation is shown at four scaled ranges. The MATE downward path TCF was used. The observed 8 kHz intensity spectrum (corresponds to scaled range Z3) is over-plotted as dot-dashed.

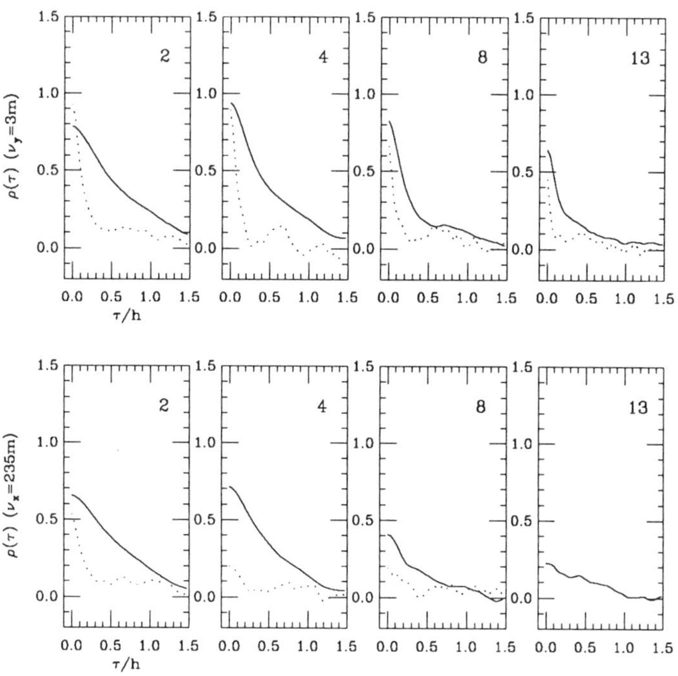

Figure 11. MATE observations of vertical (top) and horizontal (bottom) 2-point intensity correlations as functions of time lag, τ. The downward path results are shown solid and upward path shown dotted for the four acoustic frequencies.

results. The possible pairs of 2-frequency correlations of intensity for the down and up paths are plotted in Figure 12.

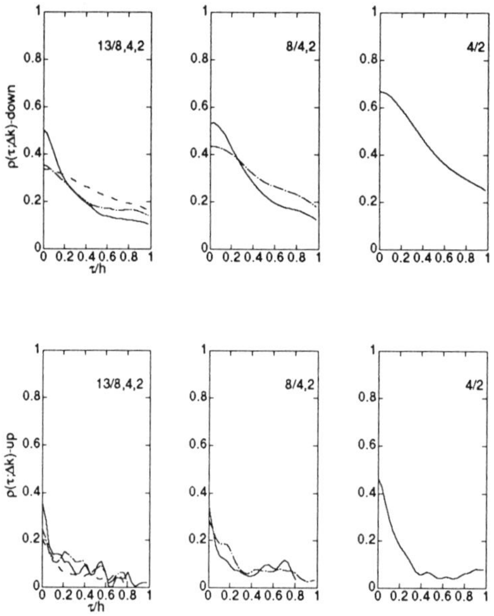

Figure 12. Pairs of 2-frequency correlations of intensity for the down (top) and up (bottom) paths at MATE. The solid, dot-dashed and dashed curves on the left correspond to 13/8, 13/4, and 13/2; solid, dot-dashed in the middle to 8/4, 8/2; and the solid curve on the right to 4/2 respectively.

4.3. The MATE Upper Path

Along the upward directed path at MATE, because $< n^2(y) >$ reaches a maximum near the turning point (see Figure 6), the scattering is largely concentrated over the narrow region of the turning point. The scintillation indices and intensity for the upward path are given in Table 1 and Figure 9(b), respectively. Spatial and cross-frequency statistics are shown in Figures 11 and 12. Note that in Figure 9(b) the absence of a roll-off below ω_i may be due to the short duration (2 days) of the time series. In any case, the evaluation of theory for the upper path has yet to be done.

Summarizing MATE, for the four pulse frequencies, on the downward path $\Gamma = (.028 - 7.7) \times 10^4$, $Z = (1.0 - 6.8) \times 10^{-2}$, and on the upward path, $\Gamma = (.20 - 56.) \times 10^4$, and $Z = (.7 - 4.8) \times 10^{-2}$. Observations of the scintillations in the 4 (Γ, Z) regimes for the MATE lower path allowed us to test full range 4th moment theory (autocorrelations) away from the asymptotic regions. The medium correlation functions are messy, but well-tested in an oceanographic sense. Predicting the observed cross-correlations of intensity for spatial and frequency two-point separations awaits further theory development.

5. THE INTENSITY PROBABILITY DISTRIBUTION

We have shown that the intensity variance can be predicted for AATE and MATE, where scattering conditions span wide ranges in Γ and Z. This is a far cry from knowing the full statistics. One might say, "If we knew the two-point correlations of the complex field, the 4th moments of the field, and the intensity probability distributions, we would know a lot about the process." The problems we face in understanding the intensity probability distribution are seen clearly in Figure 13. A one parameter distribution cannot explain the behavior for near to far range, and the true (unknown) distribution has a complicated form. Here we have plotted the 3rd and 4th intensity moments of a plane wave propagating through a medium with a 4th order power law transverse spectrum versus the 2nd moment. $\Gamma = 1000$ (a moderate ocean acoustic propagation value) and the scaled range varied from zero to the point of field saturation ($< I^2 > \approx 2$). The intensity moments of the data (circles) are those computed from many realizations. The fields were generated by Monte-Carlo methods (see Macaskill and Ewart[14] and Ewart[16]). Also plotted with various line types are the universal moments of the log normal, K, gamma, Weibull, exponential, and modified exponential[34]).

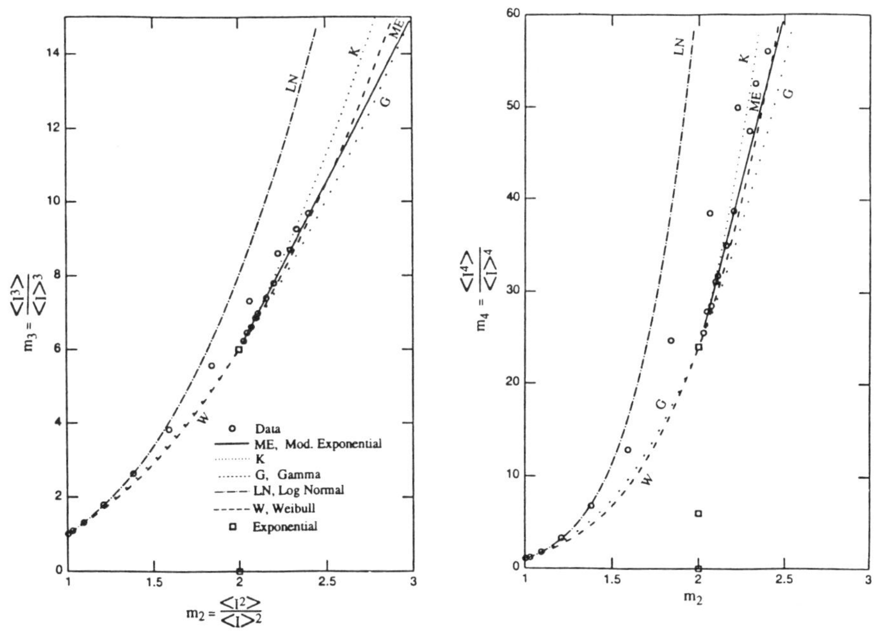

Figure 13. Moments from simulated data that has been fitted by the generalized gamma distribution plotted with several one parameter results defined by m_2.

The modified exponential distribution is a good fit for the region approaching saturation. The K distribution does not fit this region well, as was shown by Macaskill and Ewart[14]. The gamma and Weibull distributions are clearly poor models of the full-range moments. Our point is that the generalized gamma distribution fits all of the data (circles) for all of the (Γ, Z) pairs with a 90% confidence interval acceptance using theoretical-quantile-data-quantile fits for a range of Γ (10^{-1} to 10^7) and Z (10^{-4} to 10^2). Thus, it is not necessary to use asymptotic forms for the moments or the probability distributions, unless one is considering "first principle" predictions.

In Figure 14 we have plotted contours of the moments $< I^2 > / < I >^2$, (for the full suite of simulations) as a function of Γ and Z. We see lognormal moments at short ranges, an increase to a peak (the medium focus), and a drop to 2 at longer ranges. Ewart[16] showed that the peak at the medium focus follows closely the prediction of Spivack and Uscinski[35]. Using numerical solutions to the 4th moment equation, the (Γ, Z) dependence of the peaks, lognormal boundary, and exponential boundary obey the relation, $\Gamma Z^{2.3}$ = constant. It is a relatively simple exercise to generate these results for any other medium TCF. Similar results are available for the 3rd and 4th intensity moments and can be obtained by request.

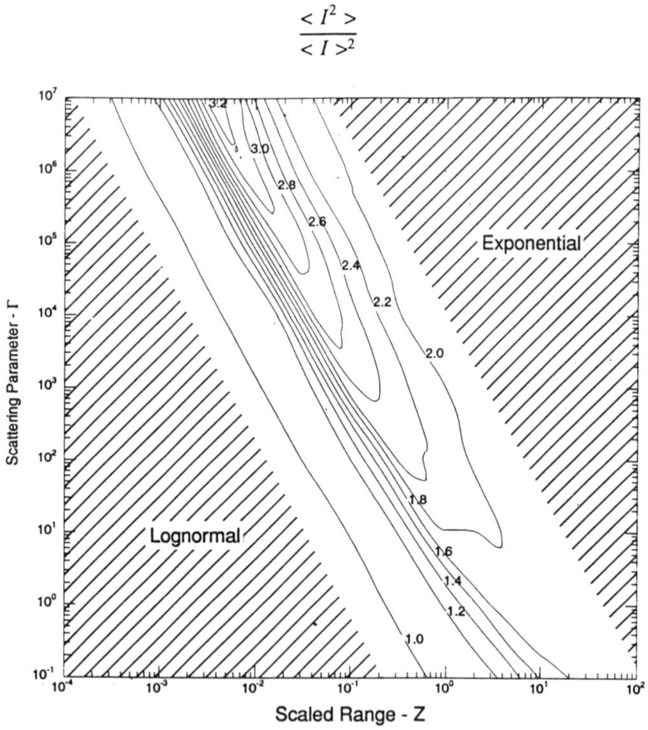

Figure 14. Contours of the 2nd moment from the simulated data.

Ewart and Ballard (unpublished) have completed a similar study for the point source initial condition for a few (Γ, Z) pairs. In this work, the Q-Q fits to the generalized gamma distribution are even better in the region of the medium focus (due to modified sampling techniques), and $< I^2 > / < I >^2$ is well predicted by moment theory for propagation from a point source. The scintillation indices at the focus increase to higher values at longer ranges for a plane wave as would be expected, but conclusions as to the results of Figure 14 will remain little changed for a point source initial condition. In order to relate these curves to a particular scattering problem, it is only necessary to recognize that the parameters, $< n^2 > L_p$ and L_y are the two parameters of the medium that define Γ and Z. Following Ewart[16], these can be interpreted in terms of k and z_R by inverting (4). Thus, given an assumed medium correlation function and its normalization, it is a simple matter to obtain predictions of observations from the simulated fields.

In the ocean case, the second moment is limited to values below ≈ 3, due to the strong frequency-dependent chemical absorption. For the same reason, the wavefields do not closely approach saturation.

5.1. MATE and AATE Intensity Probability Distributions

The hypothesis that AATE and MATE intensities are modeled with the generalized gamma distribution is well tested. The MATE data is temporal, and AATE is spatial and temporal. Uscinski and Ballard (unpublished) have modeled the MATE upper path as weak scattering near the point of inflection in the ray at 200 m and a deeply modulated phase screen at the upper turning point. The presence of a deeply modulated screen could indicate that the intensity probability distribution should be Rician; see Roberts[36]. In Figure 15, upward path intensity moments are plotted on a figure from that paper comparing the Rician normalized intensity moments. The agreement could be taken as support for the phase screen model.

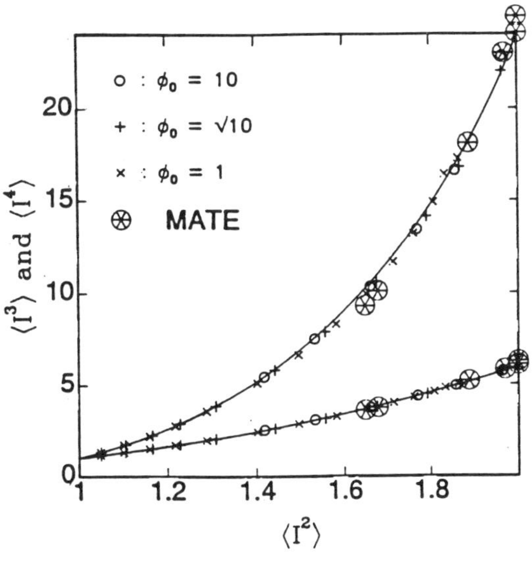

Figure 15. Observed 3rd and 4th intensity moments from the MATE upward path are over-plotted on a figure from Roberts[36]. The MATE moments are shown as circled asterisks, the other points are those from Robert's phase screen simulations, and the lines are for the Rician distribution.

6. SUMMARY AND CONCLUSIONS

We have presented results of the predicted and measured intensity correlations for three very different scattering regimes using ocean measurements. A summary of related studies of the intensity probability distribution for both ocean and simulated data has been included. Our overall conclusion is that the theoretical predictions for the 4th moment of a propagating wavefield have progressed quite far from the asymptotic region theories of a decade ago. Getting the medium TCF correct is an overriding issue, and a difficult one.

Direct 4th moment solvers now being developed will provide predictions that are accurate and relatively easy to compute. As an example, Macaskill and Ballard (unpublished) have achieved predictions superior to those presented in Ewart et al.[11] of the MATE cross frequency covariances at $\tau = 0$ using adaptive grid 4th moment solvers. Leonard and Uscinski[17] have successfully predicted cross frequency intensity correlations using adaptive grid 4th moment solvers for simulated data with Gaussian and 4th order power law TCF's up to $\Gamma = 1000$. The solvers are a reasonable approach when one "needs to get an answer," but provide little insight. The 4th moment solutions initiated by Uscinski and Macaskill in the early 1980's are difficult to evaluate for the complex ocean TCF. Clearly, more theoretical work remains. Field and numerical experiment results should be used to test these emerging theories.

6.1. AATE

The phase and intensity for this experiment have been very well predicted using the Rytov approximation theory of Munk and Zachariasen, and the TCF derived from the AIWEX ocean model. The intensity probability distributions from AATE can be modeled with a generalized gamma distribution.

6.2. The MATE Lower Path

The phase auto-spectra are well modeled using Rytov approximation theory and the MATE TCF. The intensity variances are well predicted using 4th moment theory (with zeroth order theory evaluation plus a correction). The auto-spectra of intensity are reasonably well predicted, but improvements are expected when the direct 4th moment solvers are fully tested. Direct solvers are also expected to successfully predict the two-frequency intensity cross spectra. A theoretical issue here is "what was wrong" with the 4th moment theory solutions. The spatial cross spectra have been predicted by Uscinski, but the theory has not been evaluated. The intensity probability distributions for the MATE lower path are well modeled by the generalized gamma distribution.

6.3. The MATE Upper Path

The phase auto-spectra are well modeled using the Rytov approximation theory and the MATE TCF. The sharp turning of the ray inside a horizontal correlation scale at the upper turning point has been modeled as a deeply modulated screen, with small regions of extended media on each side; initial predictions of the intensity variance with that model are not encouraging. The data intensity variance rises to 2, apparently without a peak. A new theory may be required for this case. The intensity moments are fit well using a Rician model (single phase screen assumption). The intensities are also well fit by the generalized gamma distribution; they are not close to exponential, indicating that saturation occurs at a more distant range.

The focus of WPRM studies in the ocean must include the waveguide aspects of the ocean at longer ranges and lower frequencies. A "MATE like" experiment, where the phase and the amplitude are recorded for several frequencies and ranges is needed to make further tests of theory. It is essential that such an experiment include phase and amplitude measurements over a sizable depth range. Regions of the ocean where internal waves dominate the dynamic and passive index of refraction fluctuations are well known, and could be used for such an experiment. Collaboration with many oceanographers would be essential, if one is serious about obtaining the space-time spectrum of the index of refraction and the appropriate TCF. In future intensity probability distribution studies, we will attempt to link the various irradiance distribution models, e.g. I-K distributions, with the generalized gamma model.

7. ACKNOWLEDGMENTS

The authors would like to thank John Ballard, Barry Uscinski, and Charles Macaskill for their assistance and support in preparing this paper. Nina Triffleman edited the manuscript, and Patricia Oakley, and Mike Kenney provided assistance with the computations and plotting. This research was sponsored by the Department of the Navy, Office of the Chief of Naval Research, under Grant N00014-90-J-1260 This article does not necessarily reflect the position or the policy of the Government, and no official endorsement should be inferred.

---

1. V.I. Shishov, "Theory of wave propagation in random media," *Izvestia VUZ. Radiofizika* **11**(6), 866-875 (1968).

2. C. Garrett, and W. Munk, "Space-time scales of internal waves," *Geophys. Fluid Dyn.* **2**, 225-264 (1972).

3. B.J. Uscinski, "Parabolic moment equations and acoustic propagation through internal waves," *Proc. R. Soc. Lond. A* **372**, 117-148 (1980).

4. T.E. Ewart, "Acoustic fluctuations in the open ocean--a measurement using a fixed refracted path," *J. Acoust. Soc. Am.* **60**(1), 46-59 (1976).

5. B.J. Uscinski, "Intensity fluctuations in a multiple scattering medium. Solutions of the fourth moment equation," *Proc. R. Soc. Lond. A* **380**, 137-169 (1982).

6. B.J. Uscinski, C. Macaskill and T.E. Ewart, "Intensity fluctuations. Part I: Theory," *J. Acoust. Soc. Am.* **74**(5), 1474-1483 (1983).

7. T.E. Ewart, C. Macaskill and B.J. Uscinski, "Intensity fluctuations. Part II: Comparison with the Cobb Experiment," *J. Acoust. Soc. Am.* **74**(5), 1484-1499 (1983).

8. C. Macaskill, "An improved solution to the fourth moment equation for intensity fluctuations," *Proc. R. Soc. Lond. A* **386**, 461-474 (1983).

9. C. Macaskill and T.E. Ewart, "Computer simulation of two-dimensional random wave propagation," *I.M.A. J. of Appl. Math.* **33**, 1-15 (1984).

10. T.E. Ewart and S. A. Reynolds, "The Mid-Ocean Acoustic Transmission Experiment -MATE-," *J. Acoust. Soc. Am.* **75**(3), 785-802 (1984).

11. T.E. Ewart, C. Macaskill and B.J. Uscinski, "The MATE acoustic frequency cross correlation of intensity," *J. Acoust. Soc. Am.* **77**(5), 1732-1741 (1985).

12. B.J. Uscinski and C. Macaskill, "Frequency cross-correlation of intensity fluctuations in multiple scattering," *Optica Acta* **32**(1), 71-89 (1985).

13. B.J. Uscinski, C. Macaskill and M. Spivack, "Path integrals for wave intensity fluctuations in random media," *J. Sound and Vibration* **106**(3), 509-528 (1986).

14. C. Macaskill and T.E. Ewart, "The probability distribution of intensity for acoustic propagation in a randomly varying ocean," *J. Acoust. Soc. Am.* **76**(5), 1466-1473 (1984).

15. T.E. Ewart and D.B. Percival, "Forward scattered waves in random media-the probability distribution of intensity," *J. Acoust. Soc. Am.* **80**(6), 1745-1753 (1986).

16. T.E. Ewart, "A model of the intensity probability distribution for wave propagation in random media," *J. Acoust. Soc. Am.* **86**(4), 1490-1498 (1989).

17. S.R. Leonard and B.J. Uscinski, "Accurate numerical solutions for the frequency cross-correlation of intensity fluctuations in a random medium," *App. Opt.* (in press, 1992).

18. J.M. Martin and S.M. Flatte´, "Intensity images and statistics from numerical simulation of wave propagation in 3-D random media," *App. Opt.* **27**(11), 2111-2126 (1988).

19. J.M. Martin and S.M. Flatte´, "Simulation of point-source scintillation through 3-dimensional random media," *J. Opt. Soc. Am.* **7**(5), 838-847 (1990).

20. Y.J.F. Desaubies, "Acoustic-phase fluctuations induced by internal waves in the ocean," *J. Acoust. Soc. Am.* **60**(4), 795-800 (1976).

21. Y.J.F. Desaubies, "Analytical representation of internal wave spectra," *J. Phys. Oceanogr.* **6**(6), 976-981 (1976).

22. Y.J.F. Desaubies, "On the scattering of sound by internal waves in the ocean," *J. Acoust. Soc. Am.* **64**(5), 1460-1469 (1978).

23. W.H. Munk and F. Zachariasen, "Sound propagation through a fluctuating, stratified ocean: Theory and observation," J. *Acoust. Soc. Am.* **59**(4), 818-838 (1976).

24. J.E. Ehrenberg, T.E. Ewart and R.D. Morris, "Signal-processing techniques for resolving individual pulses in a multipath signal," *J. Acoust. Soc. Am.* **63**(6), 1861-1865 (1978).

25. B. Bell, and T.E. Ewart, "Separating multipaths by global optimization of a multidimensional matched filter," *IEEE Trans. Acoust., Spch. Sig. Proc.* **ASSP-34**(5), 1029-1037 (1986).

26. B. Bell and S.A. Reynolds, "A matched filter network for estimating pulse arrival times," *IEEE Trans. Sig. Proc.* **39**(2), 477-481 (1991).

27. T.E. Ewart and S.A. Reynolds, "Instrumentation to measure the depth/time fluctuations in acoustic pulses propagated through arctic internal waves," *J. Atmos. Ocean. Tech.* **7**(1), 129-139 (1990).

28. W.H. Munk, "Internal waves and small-scale processes," *in* B. Warren and C. Wunsch (Eds.), *Evolution of Physical Oceanography*, MIT Press, Cambridge, MA, 264-291 (1981).

29. M.D. Levine and J.D. Irish, "A statistical description of temperature finestructure in the presence of internal waves," *J. Phys. Oceanogr.* **11**(5), 676-691 (1981).

30. R. Esswein and S.M. Flatte´, "Calculation of the strength and diffraction parameters in oceanic sound transmission," *J. Acoust. Soc. Am.* **67**(5), 1523-1531 (1980).

31. A. Ishimaru, *Wave Propagation and Scattering in Random Media Vol. 2*, Academic Press, New York, 389-390 (1978).

32. M.D. Levine, "Internal waves under the arctic ice pack during AIWEX: The coherence structure," *J. Geophys. Res.* **95**(C5), 7347-7357 (1990).

33. M.D. Levine, J.D. Irish, T.E. Ewart and S.A. Reynolds, "Simultaneous spatial and temporal measurements of the internal wave field during MATE," *J. Geophys. Res.* **91**(C8), 9709-9719 (1986).

34. K.S. Gochelashvili and V.I. Shishov, "Strong fluctuations of laser radiation intensity in a turbulent atmosphere-- the distribution function," *Sov. Phys. JETP* **47**(6), 1028-1030 (1978).

35. M. Spivack and B.J. Uscinski, "Accurate numerical solution of the fourth-moment equation for very large values of Γ," *J. Mod. Opt.* **35**(11), 1741-1755 (1988).

36. D.L. Roberts, "The probability distribution of intensity fluctuations due to a random phase screen with an exponential autocorrelation function," *in* B.J. Uscinski (Ed.), *Wave Propagation and Scattering*, Clarendon Press, Oxford, UK, 129-154 (1986).

On The Statistics of Equatorial Transionospheric Radio Waves

Charles L. Rino
Vista Research, Inc., Mountain View, California 94042

Abstract. Artificial earth satellites have provided unique opportunities for measuring ionospheric scintillation. This paper reviews the applications of scintillation theory to the interpretation of these data. We concentrate on the statistics of equatorial scintillation, which provide consistent examples of phenomena that are unique to the ionosphere. Equatorial scintillation at frequencies above 1 GHz was unexpected and cannot be explained with the power-law models that are used for most other media; moreover, the statistics of ionospheric scintillation are distinctly different from optical scintillation observed over extended paths that produce comparable perturbation levels.

The equatorial irregularities that are primarily responsible for the observed scintillation have been measured in situ; however, only recently has a consistent picture emerged. A two-component model, which is a direct consequence of the fundamental instability that is believed to be responsible for the irregularities, can be used to quantatively explain the measured intensity scintillation characteristics.

1. Introduction

Ionospheric scintillation was first observed in emissions from radio stars [1, 2], and for a time the phenomenon played an important role in establishing bounds on the angular extent of compact radio sources [3]. The emphasis changed dramatically following the launch of Sputnik I in 1957. Radio beacons carried by Sputnik I and the early satellites that followed it were ideally suited for measuring ionospheric scintillation. At the same time, the detrimental effects of scintillation on transionospheric communication, space-based radars, and navigation systems stimulated an extensive research effort.

Coherent dual-frequency transmissions had been used earlier to correct ionospheric errors in radio beacon measurements of rocket trajectories [4]. From the known form of its frequency dispersion, the ionospheric contribution to the signal phase at two widely separated frequencies can be determined. Thus, a dual-frequency measurement suffices to determine the ionospheric phase contribution. The technique is used on both the Navy-Navigation and the Global-Positioning-System satellites. For diagnostic measurements, one can use the highest coherent frequency to derive a phase reference against which the lower frequencies are synchronously demodulated. If the frequencies are well separated, both the amplitude and phase of the propagation disturbance at the lower frequency can then be measured.

The first systematic measurements of ionospheric phase scintillation were made by Crane [5], who used the 400/150-MHz radio beacons carried by the Navy Navigation Satellites. The review paper by Yeh [6] describes later contributions, including results from the geosynchronous Applied Technology Satellites (ATS) and the low-orbiting Wideband Satellite (see also Fremouw et al. [7]), both of which carried dedicated ionospheric propagation experiments. Davies' review [8] describes the ATS propagation experiments dedicated to measuring total electron content (TEC), which is effectively the deterministic part of the ionosphere-induced phase change upon conversion to path-integrated electron density.

A large part of the research effort in ionospheric scintillation was devoted to understanding the physical processes that structure the propagation medium. The global morphology of ionospheric scintillation and its relationships to other geophysical phenomena has been reviewed by Aarons [9]. The review papers by Ossakow [10] and Fejer and Kelley [11] provide a comprehensive survey of the physical processes that can cause ionospheric irregularities over many decades of scale sizes. Scintillation is caused by comparatively large irregularities, which generally coexist with smaller scale structures. Radar backscatter from small-scale ionospheric irregularities—Spread F—had been observed from the earliest days of ionospheric sounding [12].

This review will concentrate on the statistics of ionospheric scintillation as deduced from measurements made near the earth's geomagnetic equator. The topic was selected because of the unique interplay between the irregularity structure and signal statistics that analyses of equatorial scintillation data have revealed. The severity of scintillation activity is characterized by the intensity index often denoted $S4$ in ionospheric studies:

$$S4 = \sqrt{(< I^2 > - < I >^2} / < I >, \tag{1}$$

where $I = |u|^2$, and u is the complex signal amplitude. Scintillation is weak when $S4 < 1$ and *saturated* when $S4 \geq 1$. The spatial, temporal, and frequency coherence properties of u and/or I are important as well, but the behavior of $S4$ alone, particularly its frequency dependence, can provide considerable insight into the underlying irregularity structure.

The observation of equatorial scintillation at 3 GHz by Craft and Westerlund [13] was unexpected, although reasonable hypotheses could be advanced to explain how intense local irregularities or weaker irregularities distributed along the propagation path might account for the 3-GHz observation [14]. The fact that gigahertz scintillation is accompanied by near-Rayleigh ($S4 \approx 1$) fading at VHF, UHF, and occasionally L-Band, however, was not emphasized at the time. Most reported measurements of scintillation spectra as well as in-situ rocket measurements suggested an underlying three-dimensional power-law irregularity distribution with a spectral index greater than 4. Allowing for reasonable values of the other parameters needed to characterize the irregularities and their spatial extent, highly non-Rayleigh distributions ($S4 >> 1$) should have been expected.

Before describing how these observed characteristics of equatorial scintillation were reconciled with the in-situ measurements, it is helpful to outline the scintillation theory as it had been applied to ionospheric scintillation. The rigorous theory is computationally intensive and necessarily requires simplification for applications. Moreover, the main body of the theory addresses statistically homogeneous structures, whereas ionospheric scintillation data provide no clear demarcation between omnipresent trend-like variations and the homogeneous random fluctuations that give rise to the scintillation structures that are most easily quantified. As a consequence, critical scintillation parameters depend on the intervals over which they are measured. For example, weak intensity scintillation is generally devoid of low-frequency content whereas strong intensity scintillation can admit pronounced low-frequency structure. The low-frequency content of phase scintillation is inherently inhomogeneous at all disturbance levels.

2. Background

A theory of radiowave scintillation emerged in the United States during the early 1950s. From well-understood principles of wave propagation and scattering, a conceptual model of wave propagation in weakly irregular media was developed. The seminal paper by Booker, Ratcliffe, and Shinn [15] consolidated these ideas into an analytic theory. The developing theories of noise, stochastic processes, and signal detection provided the mathematical tools needed to characterize the random nature of the signals [16].

The synthesis of the propagation and scattering concepts that influenced ionospheric scintillation theory is eloquently summarized in the paper *Aspects of Diffraction Theory–Application to the Ionosphere* by J. A. Ratcliffe [17]. The paper contains an annotated bibliography of contributions by Booker, Briggs, Bowhill, Bramley, Fejer, Hewish, Keller, Little, Pawsey, Phillips, and Rice among others. Ratcliffe recognized that the constraint to small phase-perturbations imposed by first-order perturbation theory is rarely met in practice. He circumvented this limitation by postulating an *equivalent* phase-changing screen. The phase-changing screen is unrestricted in the magnitude and scale of the phase perturbation it imparts to the wave field. Beyond the phase screen, the disturbed wave field propagates freely as does the optical electromagnetic field beyond a lens.

The rigorous theory of wave propagation in continuous random media was developed by Russian scientists. English translations of the books by Chernov [18] and Tatarskii [19], which introduced the Rytov approximation, influenced the later development of scintillation theory in the United States[1]. The rigorous theory yields a closed hierarchy of differential equations for the moments of the complex field. The equations are difficult to solve, however, and tractable solutions have remained elusive to this day. Thus, the perturbation, Rytov, and phase-screen models provided the main conceptual and computational basis for interpreting ionospheric scintillation data.

The Rytov approximation characterizes the logarithm of the complex field, whereas most other theoretical approaches characterize the complex field itself. Even so, the Rytov and perturbation theories incorporate identical integral representations. The complex field $u(\rho, z)$ is written in one of two forms:

$$\left\{ \begin{array}{ll} \chi + i\phi & \text{MSP} \\ x + iy & \text{SPM} \end{array} \right\} = -r_e \int_0^L \iint \exp\{-i(k/2)\rho^2 L[z(L-z)]^{-1}\} \frac{L\Delta N_e(\rho, z)}{z(L-z)}\, d\rho \quad (2)$$

The upper form of (2), which is the Rytov approximation, is derived from the method of smooth perturbations (MSP). The in-phase and quadrature components of the integral represent, respectively, the log-amplitude, χ, and phase ϕ of u. In the small perturbation method (SMP), the quadrature components of the integral are interpreted directly. Because the moments of log-amplitude and phase necessarily depend on complex signal moments of all orders, and vice versa, definitive comparisons are not easily made.

The integral in (2) represents the propagation of a point emission over a path of length L in a weakly ionized medium with electron density variations represented by ΔN_e. The parameter r_e is the classical electron radius. The Fresnel scale

$$\rho_F = \sqrt{\frac{z}{k}} \quad (3)$$

marks the transition at which smaller structures produce amplitude (log-amplitude) variations when observed at distances greater than z. The larger structures contribute directly to phase, but their contribution to intensity is suppressed by Fresnel filtering. This fundamental concept can also be deduced from ray-optics arguments [21].

[1]The Russian contributions to scintillation theory are summarized in the review paper by Barabanenkov [20] and *Radio Science* Volume 10 Number 1 (1975), which is dedicated to scintillation theory.

To remove the small-amplitude-fluctuation constraint rigorously, one must ultimately use the parabolic wave equation method (PEM), but the phase-screen model captures the essential features of the strong-scatter theory. For a medium with a sharply defined characteristic scale, as represented by a gaussian irregularity spectral-density function (SDF), near- and far-zone regions correspond to propagation distances where the Fresnel scale is respectively smaller than or larger than the transverse correlation scale. Salpeter [22] developed a comprehensive summary of the scintillation characteristics of a gaussian phase screen. Singleton [23] extrapolated Salpeter's results to form closed S_4 contours, which he displayed against the Fresnel scale divided by the correlation scale (the ordinate) and the rms phase imparted by the phase screen (the abscissa).

Singleton's diagram was used extensively to qualitatively explain scintillation effects, and variants of it are still used. An important feature of the diagram is a diagonal set of contours corresponding to $S4 > 1$. Over the range of Fresnel scale ratios less than unity, increasing perturbation strength, as parameterized by the rms phase, leads to increasing $S4$, which achieves a maximum above unity and then decreases to unity. For sufficiently large Fresnel scale ratios, $S4$ increases monotonically to unity. Thus, in a gaussian medium lack of saturation sets a lower bound on the Fresnel scale parameter. More realistic physical models provide a much richer structure.

It was know at the outset of the theoretical development that naturally occurring irregularities are more accurately characterized by a power-law rather than gaussian SDF. Rufenach's [24] analysis of radio-star scintillation marked the turning point (gaussian to power-law) for ionospheric scintillation theory. Rumsey [25] was among the first to exploit the phase-screen model to determine the expected strong-scatter limiting behavior of intensity scintillation in power-law environments. Gochelashvily and Shishov [26] had developed applicable general formulae earlier. For an incident plane wave, the intensity SDF admits the deceptively simple integral representation

$$\Phi_I(\mathbf{K}) = \iint \exp\{-g(\boldsymbol{\xi}, z\mathbf{K}/k)\} \exp\{-i\mathbf{K} \cdot \boldsymbol{\xi}\}\, d\boldsymbol{\xi} \tag{4}$$

where

$$g(\boldsymbol{\xi}, \boldsymbol{\eta}) = \begin{cases} D(\boldsymbol{\xi}) + D(\boldsymbol{\eta}) - \frac{1}{2}D(\boldsymbol{\xi} + \boldsymbol{\eta}) - \frac{1}{2}D(\boldsymbol{\xi} - \boldsymbol{\eta}) \\ 8 \iint \Phi(\mathbf{q}) \sin^2(\boldsymbol{\xi} \cdot \mathbf{q}/2) \sin^2(\boldsymbol{\eta} \cdot \mathbf{q}/2) \frac{d\mathbf{q}}{(2\pi)^2} \end{cases}, \tag{5}$$

$$\Phi(\mathbf{K}) = r_e^2 \lambda^2 L \Phi_{\Delta N_e}(\mathbf{K}) \tag{6}$$

is the two-dimensional SDF of the phase path integral, and $D(\rho)$ denotes the phase structure function. For completeness,

$$< uu'^* >= \exp\{-\frac{1}{2}D(\Delta\rho)\} \tag{7}$$

is the corresponding mutual coherence function, which is invariant under free-space propagation.

Consider the generalized Kolmogorov power-law form

$$\Phi_{\Delta N_e}(\mathbf{q}) = \frac{C_s \exp\{-(q/q_i)^2\}}{[q_0^2 + q^2]^{n/2}}. \tag{8}$$

It follows from (5) that

$$g(\xi, \mathbf{K}\rho_F^2) = U8 \iint \frac{\exp\{-(\omega/Ri)^2\}}{[R_0 + \omega^2]^{n/2}} \sin^2(\omega \cdot \xi/\rho_F) \sin^2(\omega \cdot \mathbf{K}\rho_F) \frac{d\omega}{(2\pi)^2}, \tag{9}$$

where

$$l_c = [r_e^2 \lambda^2 L C_s]^{2/(n-2)} \tag{10}$$

$$R_0 = q_0 \rho_F \tag{11}$$

$$R_i = q_i \rho_F \tag{12}$$

$$U = [\rho_F / l_c]^{n/2-1} \tag{13}$$

The parameterization, although not unique, is simple and informative. For most applications, the two Fresnel parameters are ordered such that $R_0 << R_i$. The parameter l_c is a measure of the spatial coherence of the wave field, which can be demonstrated by evaluating (7) in the appropriate limit. Similarly, it can be shown that for sufficiently small U, $S4^2 \propto U$. When $R_0 << 1$ and $R_i >> 1$ intensity scintillation can be characterized in terms of the power-law index n and the perturbation strength as measured, for example, by l_c. The strict limits of this *fractal* or pure power-law regime depend on perturbation strength, but the spectral index is the critical parameter. A distinct transition occurs when $n = 4$, which is somewhat steeper than the Kolmogorov value $n = 11/3$.

From the spectral form of (5) it is readily seen that the intensity SDF is well defined in the simultaneous limits $R_0 \rightarrow 0$ and $R_i \rightarrow \infty$ as long as $2 < n < 6$. The corresponding one-dimensional phase spectral index p is one less than n. For $n < 4$, the phase structure function itself is well defined in the limit $q_0 \rightarrow 0$ and $q_i \rightarrow \infty$. In optics, this property is used to approximate the phase structure function as a fractional power multiplied by a parameter called the structure constant. Ionospheric models more often use the rms phase as a measure of irregularity strength. Either way, one finds that in the strict power-law limit, $S4$ progresses monotonically to saturation at the Rayleigh limit of unity. When n exceeds 4, however, highly non-Rician statistics are observed. The behavior of intensity scintillation in the fractal regime has been analyzed in detail by Berry [27, 28, 29, 30]. Other researchers have refined and extended the basic methodology [31, 32, 33, 34]. Numerical simulations illustrate the critical dependence of the model on the power-law index [35].

In saturation, steep power-law environments generate highly non-Rayleigh statistics whereas shallow power-law environments adhere more closely to the Rayleigh limit. Multifrequency measurements of $S4$ provide sensitive tests of the possible irregularity structures that might cause observed statistical patterns [36]. As long as the conditions $R_0 << 1$ and $R_i >> 1$ are satisfied at the reduced propagation distance $z_r = z(z_t - z)/z_t$, where z_t is the distance between the source and receiver and z is measured from the propagation disturbance, the results are applicable to the point-source conditions that prevail in real environments, although the behavior near the Fresnel limits can be very different for point and plane wave sources. The phase screen results have been verified with more fundamental computations. These include direct solutions to the fourth-order moment equations [37, 38, 39, 40], path-integrals [41, 42, 43, 44, 45], and numerical simulations [46].

2.1. First-Order Statistics

One naturally seeks a more detailed statistical description of scintillation phenomena than is provided by the even-order complex field moments. Fade distributions were routinely measured as part of early morphological studies of ionospheric scintillation [47]; moreover, a mathematical model proposed by Nakagami [48] provided a reasonable fit to the measured distributions. Nakagami's distribution was developed as a single parameter model that approximates the general distribution for jointly gaussian quadrature signal components for an extended range of $m = 1/S4^2$ values including $S4 > 1$.

In highly anisotropic media, the effects of changing propagation geometry relative to the magnetic field direction must also be accommodated. Briggs and Parkin [49] developed a simple model that predicts the angle dependence of $S4$ in weakly scattering anisotropic media. The notation $S4$ originated with their work as one of four mathematical sinctillation indices they considered. The simplicity of the model is achieved largely by assuming that the **K**-dependence of $\Phi_{\Delta N_e}$ is given by quadratic form whose principal axes lie along and transverse to the direction of the magnetic field. Directional invariance of the spectral shape is a necessary condition. With the further assumption of a gaussian spectral shape, the Briggs-Parkin formula for $S4$ is readily obtained. Similar results can be extended for power-law spectra. One finds that under weak-scatter conditions $S4$ varies with the secant of the incidence angle, with local enhancements whenever the propagation direction intercepts an axis or plane of irregularity elongation. The fade distribution itself should be sensitive to such effects, but the equatorial observing geometry does not permit field-aligned measurement. As discussed in Aarons's [9] review, the predicted geometrical enhancements are observed at high latitudes where field-aligned propagation is practical.

The form of the weak-scatter theory (2) suggests an application of the central limit theorem. From the model developed by Booker, Ratcliffe, and Shinn [15], for example, one would expect the quadrature signal components to be jointly gaussian. When the variances of the in-phase and quadrature signal components are equal, the intensity statistics are Rician, and the probability distribution admits a simple analytic form, which can be derived from the intensity moments $<I>$ and $<I^2>$ that define $S4$. Mercier [50] proved that in the limit of sufficiently large R_0, the Rice-Rayliegh limit is achieved independent of perturbation strength. The application of Mercier's theorem, however, is restricted by the combination of practical observing conditions and the irregularity structure, which typically correspond to small rather than large Fresnel-to-dominant irregularity scale sizes.

With hindsight, it is not surprising that the Rician model provides a relatively poor fit to measured ionospheric intensity distributions. Ochs and Lawrence [51] showed that log-normal statistics, which would be expected from the Rytov interpretation of the weak-scatter theory, fit optical scintillation data quite well. Even so, the general gaussian statistics do not restrict the intensity distribution to the Rice-Rayleigh forms. The difficulty is that the three parameters needed to define the unrestricted joint gaussian probability density function cannot be determined from the intensity distribution alone. A proper statistical treatment must accommodate the complex signal structure. We were able to show that, if a theoretical model is used to adjust the gaussian parameters to obtain a best fit, one can generally do as well or better than the single parameter log-normal fit [52]. The procedure is not rigorous, however, because the gaussian model is imposed arbitrarily.

Beyond Mercier's results there are few rigorous developments that predict a specific form for the intensity distribution. Heuristic models such as the K-distribution family [53] have been quite successful, but that model is not strictly consistent with the predictions of strong-scatter theory [54]. A good summary of the numerous models and model combinations that have been used for intensity scintillation can be found in Churnside and Clifford [55], who also discuss the problems associated with definitive measurement of higher-order intensity moments. The preferred method of testing distributions against models has its own pitfalls, however. As discussed in Fremouw et al. [7], quantitative measurements of scintillation require that the data be detrended to establish the local mean intensity level. In effect, the signal is arbitrarily separated into random and trend-like components. Because intensity scintillation is inherently low-pass filtered, the detrending operation is generally unobtrusive, although for sufficiently strong scintillation this might not be so. With phase scintillation one invariably removes some structure that might just as well be considered random. Figure 1 from Rino [56] illustrates this effect.

Thus, we must contend with the fact that the detrend cutoff or measurement interval influences all measured quantities to some degree. Recognizing that the detrend interval influences what is measured, it is instructive to consider how the signal statistics vary as a function of detrend cutoff. It appears that for any scintillation patch irrespective of perturbation strength, the most rapid fluctuations are gaussian distributed. We suspect that a generalized form of Mercier's theorem could be constructed that would predict this result. If the detrend interval is increased, however, a more complicated correlation develops between the quadrature signal components. A related generalization of Mercier's theorem has been developed by Jakeman and McWhirter [57].

Upon separating out the intermediate scale component by successive detrending, one finds that phase and amplitude correlation ultimately manifests itself as crescent-like trajectories on an Argand diagram. Thus, as shown by Fremouw et al. [58], the signal consists of at least two multiplicative components. Unlike the small-scale component, the intermediate-scale component is not easily characterized because of the complicated correlation between intensity and phase. In light of this discussion, we qualify our introductory remarks by stating that no evidence of strong focusing has been observed in equatorial scintillation for the largest detrend interval over which stable $S4$ estimates can be made. Small $S4$ enhancements above unity might be observed with longer detrend intervals.

3. Equatorial Scintillation

The regular occurrence of intense Spread F at the earth's geomagnetic equator has been studied for decades [59]. Radar range (height)-time-intensity maps of the disturbed regions show dramatic *plume*-like regions that extend from the base of the F-region through its peak and beyond for many hundreds of kilometers. In-situ rocket and satellite measurements have shown that the plumes coincide with large electron density depletions. The plumes are evidently initiated by a Rayleigh-Taylor instability on the steepened bottom side of the F-region after a post-sunset electric-field-driven rise.

As described in Ossakow's review [10], the physics of the plume development can be modeled in considerable detail by using numerical fluid plasma simulations. For example, the eastward drift of the neutral background causes plumes to tilt rather than to rise vertically. Also, a highly conducting (sunlight) E-layer at the terminus of the magnetic field lines that define the depleted flux tube inhibits the development of the instability and can explain the combined seasonal-longitudinal dependence of Spread

F [60].

The physics of the small-scale irregularities that cause intense backscatter and the intermediate-scale irregularities that cause scintillation are less well understood. It is nonetheless possible to simulate local conditions that can sustain various instability mechanisms, whereby the turbulence-like evolution of the structure in a small subregion can be studied. What emerges is mosaic of interrelated instability processes that can account for the occurrence of irregularity structures that span a scale-size regime covering at least seven orders of magnitude—centimeters to hundreds of kilometers.

Numerous equatorial campaigns involving coordinated ground-based, aircraft, rocket, and satellite observations have been conducted as well as synoptic sounder, scintillation, and radar measurements. The results of the major campaigns are documented in a series of topical reviews [14, 61, 62, 63, 64, 65, 66]. The synoptic data base has quantified the diurnal, seasonal, longitudinal, and solar-cycle dependencies of equatorial scintillation occurrence and associated plumes [67, 68, 69]. At any of the geomagnetic observing stations at appropriate times one can observe realizations of the same geophysical phenomena in the same place.

Prior to 1980, it was widely accepted that equatorial irregularities could be characterized by a uniform power-law spectrum from some tens of kilometers to possibly a few tens of meters where diffusion would remove the structure. Basu et al. [70] extrapolated the spectral intensity from scintillation measurements to sub-meter scale structures that cause radar backscatter. They found that the extrapolation significantly overestimates the measured backscatter intensity. They concluded that a diffusive high-frequency cutoff must exist below a few hundred meters. A uniform k^{-2} one-dimensional power-law model to the diffusive limit was reinforced by both in-situ rocket measurements and scintillation spectra; however, the early rockets [71] penetrated only the bottom side of the F-region, and the typical Fresnel scale for scintillation observations is near 1 km. Thus, structures larger than 1 km had not yet been explored in detail.

The first indication of a departure from uniform power-law behavior at large-scale sizes came from the Wideband Satellite data, which showed phase spectral indices between 1 and 2 for structures larger than 1 km. Livingston et al. [72] showed that the Wideband data were consistent with low-resolution in-situ data from the Atmospheric Explorer E (AE-E) satellite. The first direct measurements of the full scale-size range were made during the PLUMEX I experiment using a powerful well-instrumented rocket that could reach altitudes well above the F region peak. The rocket carried a multi-frequency beacon and a high-resolution in-situ probe. The probe showed a distinct two-component power-law structure throughout the region on the eastward wall of a plume (as measured by the UHF ALTAIR radar), which produced intense scintillation [73]. Upon analyzing high-resolution data from the AE-E satellite, Basu et al. [65] reported the same two-component structure. Thus, the dominant irregularity structures causing equatorial scintillation are evidently characterized by a two-component power-law spectrum; the irregularities larger than approximately one kilometer admit a shallow power-law index (< 3 for phase), while smaller irregularities are characterized by a much steeper power-law index (> 3) for phase.

Because the most intense irregularity structures populate the shallow-index regime, the lack of strong focusing could arise from dominance of the shallow power-law component. We had demonstrated this behavior by using phase-screen simulations [35], but the definitive study was performed by Franke [74] as part of his doctoral dissertation.

Franke used only multifrequency (VHF, L- and C-band) intensity scintillation data from geosynchronous satellites, which he compared to phase-screen simulations. A single *effective* phase screen was constructed whose average spectral characteristics matched the two-component power-law model discussed above. Variations along the magnetic field direction were ignored, whereby the phase screen is effectively oriented normal to the propagation direction.

Franke first showed that a single power-law model could not reproduce the frequency dependence of $S4$, as already noted, or the measured intensity correlation properties. He then showed that the two-component model not only reproduced the measured frequency dependence of $S4$ and the intensity decorrelation time but also the simultaneous intensity spectral density functions. The spatial coordinates of the simulation were converted to temporal coordinates by assuming frozen irregularity structures. Franke also inferred that the high-frequency portion of the irregularity spectrum became steeper as the irregularity region aged, as would be expected when the driving forces relaxed. Finally, Franke showed that systematic displacement of scintillation intensity features observed at different frequencies could be explained by the steep walls that are known to delineate the plume structure. Oscillatory irregularity structures were also identified with the diffraction from large but smooth irregularity structures that evidently are associated with the late phase of the structure evolution. The principal result, however, is that a dynamic two-component power-law model with its break between 500 m and 1 km can explain most of the observed equatorial intensity scintillation properties with reasonable hypotheses about the structure distribution along the propagation path.

The success of the scintillation modeling effort compels one to ask if the model is consistent with our understanding of the physics of irregularity formation. The mechanism is believed to be the generalized Rayleigh-Taylor instability, which acts to structure any electron density gradient that is being convected by an $E \times B$ drift or a neutral wind. As described in the Ossakow's [10] review, the mechanism has been demonstrated with artificial plasma clouds and numerous numerical simulations. The evolving irregularity structure has a distinct asymmetry, with the evolving structure highly elongated along the effective neutral wind direction. In effect, high density structures are drawn out of the back side of the convected plasma enhancement. One might expect the steep gradients to produce a k^{-2} in-situ one-dimensional spectrum, but Zargham and Seyler [75] used numerical simulations of randomly seeded plasma structures to show that asymmetric structures quickly evolve that have a two-component power-law spectrum when measured across the neutral wind direction, whereas a single k^{-2} power law spectrum is measured when the structure is sampled along the neutral wind direction. Because the equatorial neutral wind direction is predominantly eastward, scintillation measurements would be dominated by the two-component form as is observed.

Kelley et al. [76] showed that Zargham and Seyler's simulations were consistent with in-situ measurement that showed two-component structures. By taking into account the convection direction relative to the direction in which the in-situ structures were being sampled, he also showed that earlier single-component power-law measurements could be reconciled within the confines of a single structure mechanism. These results suggest that we have not a simple two-component in-situ irregularity structure but rather a spectrum whose one-dimensional characteristics change systematically with changes in the angle between the sampling direction and the convection direction. The systematic changes in the large-scale spectral index described by Livingston [72] might be related to systematic changes along the propagation path; however, it is clear that much more detailed simulations would be needed to explore these possibilities.

4. Discussion

We began this review by emphasizing the inhomogeneous character of the disturbed ionospheric transmission medium. The mean parameters that formally characterize the medium admit statistical variations [77]. In addition or alternatively, the complex signal admits at least three distinct multiplicative components:

$$u(\rho) = \overbrace{\exp\{i\phi(\rho)\}}^{1}\,\overbrace{\exp\{\psi(\rho)\}}^{2}\,\overbrace{u_g(\rho)}^{3} \tag{14}$$

Term 1 is a pure phase perturbation that accounts for the TEC contribution. Functionally, $\phi(\rho)$ is inhomogeneous or trend-like. Term 2 has correlated amplitude and phase components, with some characteristics of the Rytov solution. While it might seem reasonable to attribute all the inhomogeniety to the phase trends, under strong scatter conditions, the intensity itself need not be strictly homogeneous, since it will ultimately respond to the large-scale phase structure in the medium. Term 2 would formally represent this inhomogeneous intensity contribution.

From analysis of the Wideband Satellite data, it appears that a sufficiently small structure, as extracted by detrending and represented by $u_g(\rho)$, admits gaussian first-order statistics. We speculated that Mercier's theorem could be generalized to prove this conjecture and possibly to quantify the detrend cutoff. Beyond that, we do not expect the first-order statistics of $u(\rho)$ to admit a simple characterization.

By reviewing the phase-screen model, we attempted to provide some insight into the behavior of the intensity scintillation. We emphasized critical parameters such as the power-law index and the ratio of the Fresnel scale to the spatial coherence scale—Equation (13). We then reviewed Franke's results, which showed that the principal characteristics of equatorial scintillation cannot be explained by a single power-law model. Rather, a two-component power-law model, first observed by in-situ rocket measurements and later by the AE-E satellite, must be invoked. The hypothesis is strengthened further by the fact that the two-component spectral model is a characteristic of the gradient-drift instability mechanism that is believed to be the dominant structuring mechanism in the ionosphere.

There is a complicating factor in that the one-dimensional spectral characteristics are strongly dependent on the direction from which they are measured relative to the $E \times B + U$ drift direction. The full ramifications of this varying, highly anisotropic spectrum have yet to be explored. The universal nature of the instability mechanism suggests that it might play a role in other magnetically biased plasma environments. It is appropriate to conclude by reiterating the important role played by numerical simulations in our understanding of irregularity physics and in propagation through the resultant structure.

References

[1] J. S. Hay, S. J. Parsons, and J. W. Phillips. Fluctuations in cosmic radiation at radio frequencies. *Nature (London)*, 158:234–, 1946.

[2] C. G. Little and A. Maxwell. Fluctuations in the intensity of radio waves from galactic sources. *Philos. Mag.*, 42:267–278, 1951.

[3] M. H. Cohen, E. J. Gundermann, H. E. Hardebeck, and L. E. Sharp. Interplanetary scintillations II, observations. *Astrophys. J.*, 147:449–, 1967.

[4] W. W. Berning. The determination of charge density in the ionosphere by radio doppler techniques. *J. Atmos. Terr. Phys.*, 261–273, 1954.

[5] R. K. Crane. Ionospheric scintillation. *Proc. IEEE*, 65(2):180–199, 1977.

[6] K. C. Yeh and Chao-Han Liu. Radio wave scintillations in the ionosphere. *Proc. IEEE*, 70(4):324–360, 1982.

[7] E. J. Fremouw, R. L. Leadabrand, R. C. Livingston, M. D. Cousins, C. L. Rino, B. C. Fair, and R. A. Long. Early results from the DNA wideband satellite experiment – complex-signal scintillation. *Radio Sci.*, 13(1):167–187, 1978.

[8] Kenneth Davies. Recent progress in satellite radio beacon studies with particular emphasis on the ats-6 radio beacon experiment. *Space Science Reviews*, 25(4):357–430, 1980.

[9] J. Aarons. Global morphology of ionospheric scintillations. *Proc. IEEE*, 70(4):360–378, 1982.

[10] S. L. Ossakow. Ionospheric irregularities. *Rev. Geophys. Space Phys.*, 17, 1979.

[11] B. B. Fejer and M. J. Kelley. Ionospheric irregularities. *Rev. Geophys. Space Phys.*, 18:401–454, 1980.

[12] H. G. Booker and H. W. Wells. Scattering of radio waves by the F-region of the ionosphere. *Terrest. Magnetism*, 43:249–256, 1938.

[13] H. D. Craft, Jr., and L. H. Westerlund. Scintillation at 4 and 6 GHz caused by the ionosphere. AIAA 10th Aerospace Sciences Meeting, San Diego, California. January 1972.

[14] Su. Basu and M. C. Kelley. Review of equatorial scintillation phenomena in light of recent developments in the theory and measurement of equatorial irregularities. *J. Atmos. Terr. Phys.*, 37:1229–1242, 1977.

[15] H. G. Booker, J. A. Ratcliffe, and D. H. Shinn. Diffraction from an irregular screen with applications to ionospheric problems. *Philos. Trans. R. Soc. London, Ser. A*, A(242):549–, 1950.

[16] S. O. Rice. Mathematical analysis of random noise. In *Selected Papers on Noise and Stochastic Processes*, New York: Dover Publications, Inc., 1954, 133–294.

[17] J. A. Ratcliffe. Some aspects of diffraction theory and their application to the ionosphere. *Rep. Prog. Phys.*, 19:187–267, 1956.

[18] L. A. Chernov. *Wave Propagation in a Random Medium*. New York: Dover Publications, Inc., 1960.

[19] V. I. Tatarskii. *Wave Propagation in a Turbulent Medium.* New York: Dover Publications, Inc., 1961.

[20] Yu. N. Barabanenkov, Yu. A. Kravtsov, S. M. Rytov, and V. I. Tatarskii. Status of the theory of propagation of waves in a randomly inhomogeneous medium. *Sov. Phys. Usp.*, 13(5):551–680, 1971.

[21] J. W. Strobehn. Line-of-sight wave propagation through the turbulent atmosphere. *Proc. IEEE*, 56:1301–, 1968.

[22] E. E. Salpeter. Interplanetary scintillations I, Theory. *Astrophys. J.*, 147:443–448, 1967.

[23] D. G. Singleton. Saturation and focusing effects in radio star and satellite scintillation. *J. Atmos. Terr. Phys.*, 32:187–208, 1970.

[24] C. L. Rufenach. Power-law wave number spectrum deduced from ionospheric scintillation observations. *J. Geophys. Res.*, 77:4761–4772, 1972.

[25] V. H. Rumsey. Scintillations due to a concentrated layer with a power-law turbulence spectrum. *Radio Sci.*, 10(1):107–114, 1975.

[26] K. S. Gochelashvily and V. I. Shishov. Multiple scattering of light in a turbulent medium. *Opt. Acta*, 18:767–, 1971.

[27] M. V. Berry. Focusing and twinkling: critical exponents from catastrophes in non-gaussian random short waves. *J. Phys. A*, 10(12):2061–2081, 1977.

[28] M. V. Berry. Diffractals. *J. Phys. A*, 12(6):781–797, 1979.

[29] M. V. Berry and Z. V. Lewis. On the Weierstrass-Mandelbrot fractal function. *Proc. R. Soc. London, Sect. A*, 370:459–484, 1980.

[30] M. V. Berry and T. M. Blackwell. Diffractal echoes. *J. Phys. A*, 14:3101–3110, 1981.

[31] G. Parry, P. N. Pusey, E. Jakeman, and J. G. McWhirter. Focusing by a random phase screen. *Opt. Commun.*, 22(2):195–201, 1977.

[32] E. Jakeman and J. G. McWhirter. Non-gaussian scattering by a random phase screen. *Appl. Phys. B*, 26:125–131, 1981.

[33] B. J. Uscinski, H. G. Booker, and M. Marians. Intensity fluctuations due to a deep phase screen with a power-law spectrum. *Proc. R. Soc. London, Ser. A*, 374:503–530, 1981.

[34] E. Jakeman. Fresnel scattering by a corrugated random surface with fractal slope. *J. Opt. Soc. Am.*, 72(8):1034–1041, 1982.

[35] C. L. Rino and J. Owen. Numerical simulations of intensity scintillation using the power-law phase screen model. *Radio Sci.*, 19(3):891–908, 1984.

[36] C. L. Rino. On the application of phase screen models to the interpretation of ionospheric scintillation data. *Radio Sci.*, 17(4):855–867, 1982.

[37] B. J. Uscinski. Analytical solution of the fourth-moment equation and interpretation as a set of phase screens. *J. Opt. Soc. Am. A*, 2(12):2077–2091, 1985.

[38] M. Tur. Numerical solutions for the fourth moment of a finite beam propagating in a random medium. *J. Opt. Soc. Am. A*, 2(12):2161–2170, 1985.

[39] J. Gozani. Numerical solution for the fourth-order coherence function of a plane wave propagating in a two-dimensional kolmogorovian medium. *J. Opt. Soc. Am. A*, 2(12):2144–2151, 1985.

[40] A. M. Whitman and M. J. Beran. Two-scale solution for atmospheric scintillation. *J. Opt. Soc. Am. A*, 2(12):2133–2143, 1985.

[41] J. L. Codona, D. B. Creamer, S. M. Flatté, R. G. Frehlich, and F. S. Henyey. Solution for the fourth moment of waves propagating in random media. *Radio Sci.*, 21(6):929–948, 1986.

[42] J. L. Codona, D. B. Creamer, S. M. Flatté, R. G. Frehlich, and F. S. Henyey. Moment-equation and path-integral techniques for wave propagation in random media. *J. Math. Phys. (N.Y.)*, 27(1):171–177, 1986.

[43] S. M. Flatté, S. A. Reynolds, and R. Dashen. Path-integral treatment of intensity behavior for rays in a sound channel. *J. Acoust. Soc. Am.*, 82(3):967–972, 1987.

[44] S. M. Flatté, S. A. Reynolds, R. Dashen, B. Buehler, and P. Maciejewski. Afar measurements of intensity and intensity moments. *J. Acoust. Soc. Am.*, 82(3):973–980, 1987.

[45] R. Mazar, J. Gozani, and M. Tur. Two-scale solution for the intensity fluctuations of two-frequency wave propagation in a random medium. *J. Opt. Soc. Am. A*, 2(12):2152–2160, 1985.

[46] J. M. Martin and S. M. Flatté. Intensity images and statistics from numerical simulation of wave propagation in 3-d random media. *Appl. Opt.*, 27(11):2111–2126, 1988.

[47] H. E. Whitney, J. Aarons, R. S. Allen, and D. R. Seemann. Estimation of the cumulative amplitude probability distribution function of ionospheric scintillations. *Radio Sci.*, 7:1095–1104, 1972.

[48] M. Nakagami. The M-distribution: a general formula of intensity distribution of rapid fading. *Statistical Methods in Radio Propagation*, W. C. Hoffman, ed., New York: Pergammon, 3–36, 1960.

[49] B. H. Briggs and I. A. Parkin. On the variation of radio star and satellite scintillations with zenith angle. *J. Atmos. Terr. Phys.*, 25:339–365, 1963.

[50] R. P. Mercier. Diffraction by a screen causing large random phase fluctuations. *Proc. Cambridge Philos. Soc.*, 62:382–400, 1962.

[51] G. R. Ochs and R. S. Lawrence. Saturation of laser beam scintillation under conditions of strong atmospheric turbulence. *J. Opt. Soc. Am.*, 59:226–227, 1969.

[52] C. L. Rino, R. C. Livingston, and H. E. Whitney. Some new results on the statistics of radio wave scintillation: 1. empirical evidence for gaussian statistics. *J. Geophys. Res.*, 81(13):2051–, 1976.

[53] E. Jakeman and R. J. A. Tough. Non-gaussian models for the statistics of scattered waves. *Adv. Physics*, 37(5):471–529, 1988.

[54] R. Dashen. Distribution of intensity in a multiply scattering medium. *Opt. Lett.*, 10(4):110–112, 1984.

[55] J. H. Churnside and S. F. Clifford. Log-normal Rician probability-density function of optical scintillations in the turbulent atmosphere. *J. Opt. Soc. Am. A*, 4:1923–1930, 1987.

[56] C. L. Rino and J. Owen. The time structure of transionospheric radio wave scintillation. *Radio Sci.*, 15(3):479–489, 1980.

[57] E. Jakeman and J. G. McWhirter. Correlation function dependence of the scintillation behind a deep random phase screen. *J. Phys. A*, 10(9):1599–1643, 1977.

[58] E. J. Fremouw, R. C. Livingston, and D. A. Miller. On the statistics of scintillating signals. *J. Atmos. Terr. Phys.*, 42:717–731, 1980.

[59] J. R. Heman. Spread F and ionospheric F-region irregularities. *Rev. Geophys.*, 4(2):255–299, 1966.

[60] R. T. Tsunoda. Control of the seasonal and longitudinal occurrence of equatorial scintillations by the longitudinal gradient in integrated e-region pedersen conductivity. *J. Geophys. Res.*, 90:447–, 1985.

[61] J. Aarons. Equatorial scintillations: A review. *IEEE Trans. Antennas Propag.*, AP-25(5):729–736, 1977.

[62] Su. Basu and M. C. Kelley. A review of recent observations of equatorial scintillations and their relationship to current theories of F region irregularity generation. *Radio Sci.*, 14(3):471–485, 1979.

[63] Sa. Basu, J. P. McClure, Su. Basu, W. B. Hanson, and J. Aarons. Coordinated study of equatorial scintillation and in situ and radar observations of nighttime F region irregularities. *J. Geophys. Res.*, 85(A10):5119–5130, 1980.

[64] Sa. Basu and Su. Basu. Equatorial scintillations—a review. *J. Atmos. Terr. Phys.*, 43(5/6):473–489, 1981.

[65] Su. Basu, Sa. Basu, J. P. McClure, W. B. Hanson, and H. E Whitney. High resolution topside in situ data of electron densities and VHF/GHz scintillations in the equatorial region. *J. Geophys. Res.*, 88(A1):403–415, 1983.

[66] Su. Basu and Sa. Basu. Equatorial scintillations: Advances since ISEA-6. *J. Atmos. Terr. Phys.*, 47(8-10):753–768, 1985.

[67] J. Aarons, J. P. Mullen, J. P. Koster, R. F. daSilva, J. R. Medeiros, R. T. Medeiros, A. Bushby, J. Pantoja, J. Lanat, and M. R. Paulson. Seasonal and geomagnetic control of equatorial scintillations in two longitudinal sectors. *J. Atmos. Terr. Phys.*, 42:861–866, 1980.

[68] R. C. Livingston. Comparison of multifrequency equatorial scintillation: American and Pacific sectors. *Radio Sci.*, 15(4):801–814, 1980.

[69] J. Aarons, H. E. Whitney, E. MacKenzie, and Sa. Basu. Microwave equatorial scintillation intensity during solar maximum. *Radio Sci.*, 16(5):939–945, 1981.

[70] Sa. Basu, Su. Basu, J. Aarons, J. P. McClure, and M. D. Cousins. On the coexistence of kilometer- and meter-scale irregularities in the nighttime equatorial F region. *J. Geophys. Res.*, 83(A9):4219–4226, 1978.

[71] M. C. Kelley, G. Haerendel, H. Kappler, A. Valenzuela, B. B. Balsley, D. A. Carter, W. L. Eclund, C. W. Carlson, B. Hausler, and R. Torbet. Evidence for a rayleigh-taylor type instability and upwelling of depleted density regions during equatorial spread-F. *Geophys. Res. Lett.*, 3:448–, 1976.

[72] R. C. Livingston, C. L. Rino, J. P. McClure, and W. B. Hanson. Spectral characteristics of medium-scale equatorial F region irregularities. *J. Geophys. Res.*, 86(A4):2421–2428, 1981.

[73] C. L. Rino, R. T. Tsunoda, J. Petriceks, R. C. Livingston, M. C. Kelley, and K. D. Baker. Simultaneous rocket-borne beacon and in situ measurements of equatorial spread F—intermediate wavelength results. *J. Geophys. Res.*, 86(A4):2411–2420, 1981.

[74] S. J. Franke, C. H. Liu, and D. J. Fang. Multifrequency study of ionospheric scintillation at Ascension Island. *Radio Sci.*, 19(3):695–706, 1984.

[75] S. Zargham and C. E. Seyler. Collisional interchange instability: 1. numerical simulations of intermediate-scale irregularities. *J. Geophys. Res.*, 92(A9):10073–10088, 1987.

[76] M. C. Kelley, C. E. Seyler, and S. Zargham. Collisional interchange instability: 2. a comparison of numerical simulations with the in situ experimental data. *J. Geophys. Res.*, 92(A9):10089–10100, 1987.

[77] V. I. Tatarskii and V. U. Zavorotnyi. Wave propagation in random media with fluctuating turbulent parameters. *J. Opt. Soc. Am. A*, 2(12):2069–2075, 1985.

Ionospheric Structures and Scintillation Spectra

Santimay Basu
Phillips Laboratory, PL/GPIA
Hanscom AFB, MA 01731

Sunanda Basu*
Boston College
Newton, MA 02159

Abstract

Ionospheric scintillation measurements are examined in relation to the satellite in-situ measurements of electron density structures in the ionosphere and the plasma processes responsible for the generation of these structures. Plasma instabilities due to either internal or external forcing functions cause the ionospheric electron density to be structured. Radio waves from satellites during their traversal through the structured ionosphere are scattered and ionospheric scintillation is recorded on the ground in the presence of a relative motion between the satellite, the ionosphere and the observer. The structured high electron density environment obtained near the magnetic equator, the auroral oval and the polar cap causes intense scintillation activity in these regions at VHF and GHz frequencies. The two-slope electron density irregularity spectrum of extended equatorial spread F structures, generated by Rayleigh-Taylor instability process, is found to be compatible with multi-frequency satellite scintillation measurements in the framework of scattering theory. In the case of equatorial bottomside sinusoidal irregularities, that are confined in altitude, the scintillation spectra exhibit the expected signature of Fresnel oscillations. However, the power-law spectral index of scintillations are found to be considerably shallower than that expected for the gaussian type of irregularity spectra observed by satellite in-situ observations. At high latitudes, two distinct classes of plasma instabilities, namely, the gradient-drift and shear-driven instability processes are identified. Scintillation and in-situ irregularity spectra of the class of irregularities causing strong scintillations nearly conform to one another. The difference between the scintillation and in-situ spectral index is 0.5 instead of unity, a result to be expected from theoretical considerations. This apparent discrepancy may be resolved in the context of anisotropic irregularity spectrum and the observing geometry as discussed by Wernik et al. (1990).

Introduction

Solar ultraviolet radiation and X-rays weakly ionize the upper atmospheric constituents to form the ionosphere which extends from an altitude of about 60 km to several thousand kilometers. The plasma density of the ionosphere is distributed non-uniformly in several layers, the lowermost D-region being formed at an altitude of about 85 km, the intermediate E-region at about 100 km and the F-region extends

---

*Now at Atmospheric Sciences Division, National Science Foundation, Washington, DC 20035, on leave from Boston College.

from about 150 km to several thousand kilometers. The bulk of the ionospheric plasma resides in the F-region, which attains its maximum plasma density at altitudes of 300 km. In the presence of the earth's magnetic field, the ionospheric plasma becomes magnetized. From the point of view of radio wave propagation, the ionosphere is thus described as a non-uniform, weakly ionized magnetoactive plasma.

The ionosphere often develops electron density structures which cover a wide range of scalelengths from a few centimeters to several thousand kilometers as illustrated in Figure 1. The large scalelengths are induced either by internal forcing functions, such as, planetary waves and gravity waves or external forcing functions such as the magnetospheric electric field at high latitudes. The small scalelength irregularities of electron density evolve due to a variety of plasma instabilities. Figure 1 shows that the wide range of scalelengths can only be studied by employing a variety of techniques, namely, the coherent and incoherent scatter radar, the digisonde, the scintillation recording systems, as well as the rocket and satellite in-situ probes.

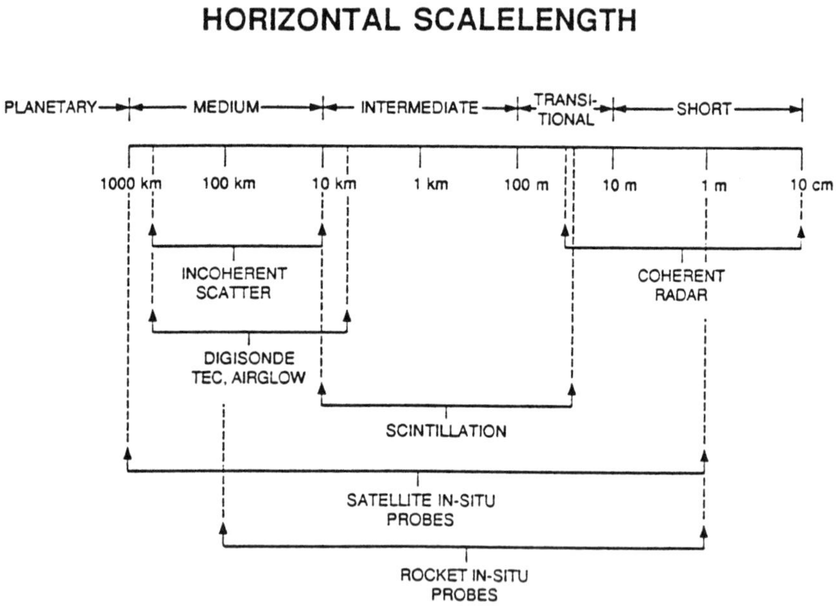

Figure 1. Variety of techniques used to study the wide range of scalelengths of electron density structures.

In the present paper we shall concentrate on the intermediate scalelength range of electron density structures which are often called electron density irregularities due to their random characteristics. Transionospheric propagation of radio waves are markedly affected by the F-region irregularities since the electron content of the F-region predominates but, on occasions, the E-region irregularities may also affect the propagation. These electron density irregularities impose random phase fluctuations across the wavefront of a satellite signal that traverses the ionosphere. The emerging wavefront with phase fluctuations develop intensity fluctuations during its propagation to the ground and, as a result, spatial variations of intensity and phase develop on the ground. In the presence of a relative motion between the satellite, the ionosphere and the observer on the ground, temporal variations of intensity and phase, known as intensity and phase scintillations are recorded by a receiver on the ground. For several decades, ionospheric scintillation observations have been made on a global scale which have contributed greatly to our understanding of the morphology of intermediate scale electron density irregularities (Aarons, 1982; Basu and Basu, 1981, 1985; Basu et al., 1988a). The theories of weak and strong ionospheric scintillation have been developed in the framework of wave propagation in a random medium (Tatarskii, 1961) by Rino (1979, 1980) and Yeh and Liu (1982 and references therein).

The objective of the present paper is to compare the electron density structures observed by satellite in-situ measurements with the structure of scintillation imposed by these irregularities on satellite signals. In addition the plasma instability mechanisms responsible for generating these ionospheric structures are discussed. In addition to the naturally occurring electron density irregularity structures, the structure of scintillation artificially induced by ground-based high-power high-frequency radio waves will also be discussed. The plan of the paper is as follows. In Section 1, the relationship between the one-dimensional irregularity spectrum obtained from satellite in-situ measurements and scintillation spectrum induced by these structures will be reviewed. In Section 2, we shall discuss the characteristics of equatorial F-region electron density structures observed in-situ by satellites in relation to the scintillation spectrum and the plasma instability mechanisms. In Section 3, the characteristics of scintillations induced by naturally occurring electron density irregularities will be compared with those artificially induced by high power, high frequency radio waves. In Section 4, the characteristics of electron density irregularity structures and scintillation in the high latitude ionosphere in the context of plasma instability mechanisms will be examined. Section 5 summarizes the results discussed in the paper.

1. Irregularity Spectrum and Scintillation Spectrum

The relationship between the three-dimensional electron density irregularity power spectrum and phase and intensity scintillation spectrum are given below. These were established by Cronyn (1970).

The three-dimensional wavenumber (k) spectrum of electron density irregularities with a power law index of q and axial ratios α, β and γ along the three components of wavenumber κ_x, κ_y, and κ_z can be expressed as

$$F_{ns}(k) \propto \left[1 + (\alpha\kappa_x)^2 + (\beta\kappa_y)^2 + (\gamma\kappa_z)^2\right]^{-q/2} \tag{1}$$

If a satellite-borne electron density probe scans through these irregularities in the x-direction with a velocity U, the temporal one-dimensional irregularity spectrum obtained from the space probe measurements may be given by

$$F_{nt}(\nu) = \frac{2\pi}{U} \iint_{-\infty}^{+\infty} d\kappa_y d\kappa_z F_{ns}(2\pi\nu / U, \kappa_y, \kappa_z) \qquad (2)$$

Under forward scatter approximation and for irregularity drift speed U in the x-direction, the temporal spectrum of phase scintillation is obtained as

$$F_{\phi t}(\nu) = 2\pi(r_e\lambda)^2(L/U) \int_{-\infty}^{+\infty} d\kappa_y F_{ns}(2\pi\nu / U, \kappa_y, 0) \qquad (3)$$

where ν - the frequency,
r_e - the classical electron radius, 2.818×10^{-15} m,
λ - the radio wavelength in meters,
L - the layer thickness of the irregularities in meters.

The corresponding intensity scintillation spectrum is

$$F_{\phi t}(\nu) = 2\pi(r_e\lambda)^2(L/U) \int_{-\infty}^{+\infty} d\kappa_y F_{ns}(2\pi\nu / U, \kappa_y, 0) \cdot \Im(2\pi\nu / U, \kappa_y) \qquad (4)$$

where $\Im(2\pi\nu/U, \kappa_y)$ - the Fresnel filter function.

Considering Eqs. (1) to (4), the high-frequency asymptotes of the spatial spectrum of the electron density irregularities and the temporal spectra of phase and intensity scintillations should be as shown in Figure 2 for q=4.

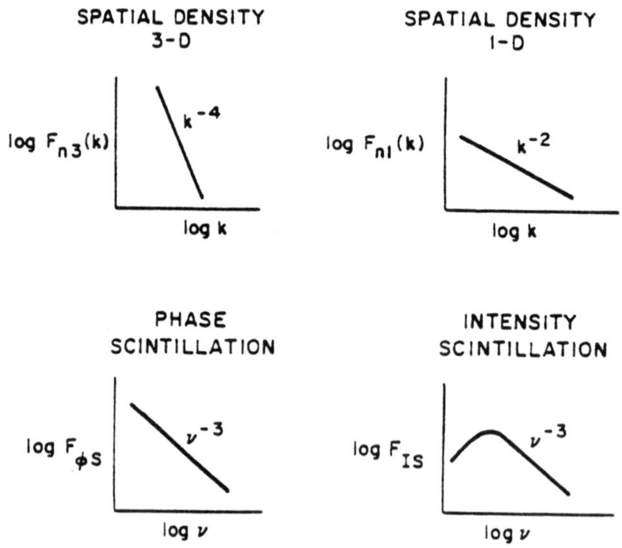

Figure 2. The relationship of the three-dimensional electron density irregularity spectrum, the one-dimensional irregularity spectrum from space probe measurements, and the phase and intensity scintillation spectra.

2. Equatorial F-Region Irregularities and Scintillation

Figure 3 (Basu et al., 1988a) illustrates, with magnetic coordinate system, the global distribution of worst-case scintillation at low GHz (L-band) frequencies during the solar maximum and minimum conditions. The irregularities of electron density in the F-region of the ionosphere cause the observed scintillation. During the solar maximum period, scintillations are much enhanced due to increased background ionization density. The irregularity amplitude ($\Delta N/N$) does not vary much with variation of solar activity, but the background ionization density varies by at least a factor of 10 from the solar maximum to the minimum period. The diagram shows that the major scintillation activity is concentrated around the magnetic equator in the pre-midnight period, in the auroral region during the nighttime period, and in the polar region at all local times.

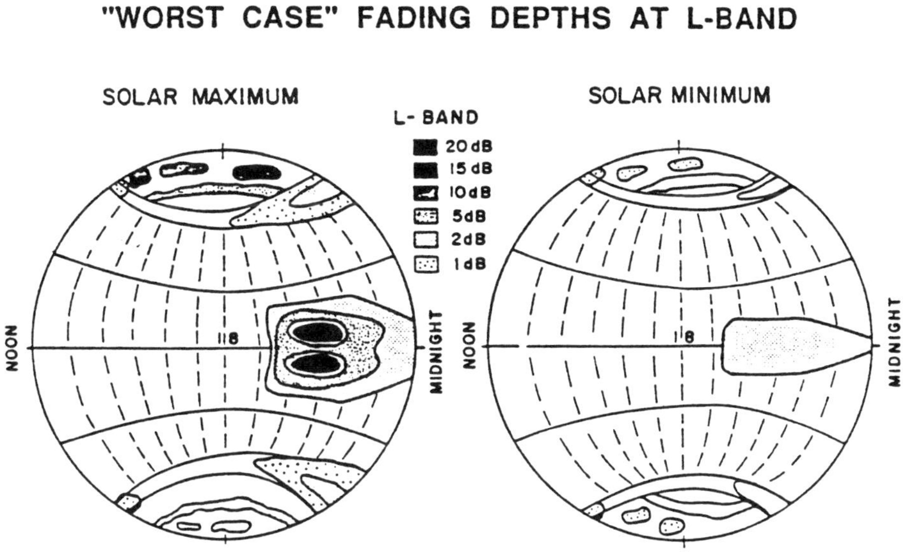

Figure 3. Global variation of scintillation fades during solar maximum and solar minimum conditions.

The equatorial F-region irregularities are generated by the Rayleigh-Taylor instability mechanism (Dungey, 1956; Basu et al., 1978; Tsunoda, 1980, 1981; Kelley, 1989) after sunset when the eastward electric field is enhanced. The irregularities evolve in plume-like structures (Woodman and LaHoz, 1976), extend in altitude to more than 1000 km and, being field-aligned, encompass a nominal latitude range of ±15° magnetic latitude. The plumes contain irregularities in the scalelength range of tens of kilometers to tens of centimeters. Figure 4 shows such plume structures being detected by radar backscatter at 50 MHz and 1.6 GHz scintillations at the magnetic equator.

The effect of these field-aligned plumes on scintillation activity is much enhanced at about 15°N and 15°S magnetic latitudes since the background ionization density is enhanced due to the transport of ionization from the magnetic equator.

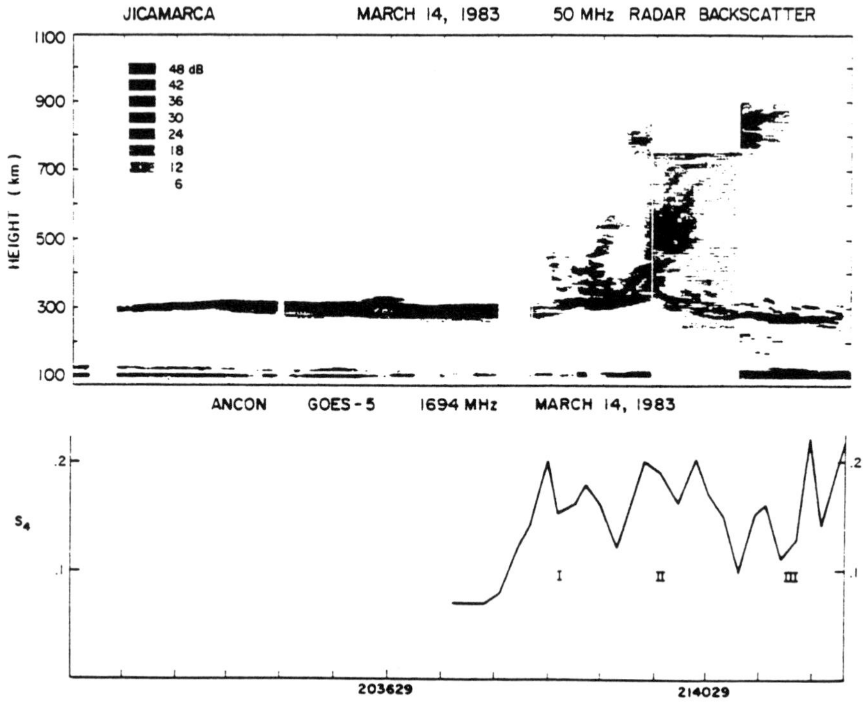

Figure 4. (Top) Temporal variation of the range and intensity of 50-MHz radar backscatter observed at Jicamarca. (Bottom) Variation of the amplitude scintillation index (S_4) of 1694-MHz signals from the geostationary satellites GOES-5 recorded at Ancon.

Scintillation measurements performed at such locations using multi-frequency transmissions from the Marisat satellite indicate (Figure 5) that even the scintillation index, S_4, at 1.54 GHz may attain a saturated level of 1.2 (Basu et al., 1987).

The satellite in-situ irregularity data acquired by the Atmospheric Explorer-E satellite at such locations of intense scintillation activity and its two slope ($p_1 = -1.5$ and $p_2 = -3.3$) spectrum with a break scalelength of 750 m are illustrated in the bottom and the top panels of Figure 6 (Basu et al., 1983). A detailed description of the two slope irregularity spectrum obtained from satellite in-situ measurements may be obtained in the forthcoming publication of Livingston and Dabbs (1992). Similar two slope spectrum has been observed in rocket in-situ irregularity measurements (Rino et al., 1981; Kelley et al., 1982). Franke and Liu (1983) utilized two-slope in-situ irregularity power spectra obtained by Basu et al. (1983), and performed simulation studies to obtain multi-frequency scintillation spectra (Figure 7). Excellent agreement between the observations (Figure 5) and the simulation results (Figure 7) establishes the consistency between the in-situ irregularity spectrum and the scintillation spectrum in the framework of the scattering theory.

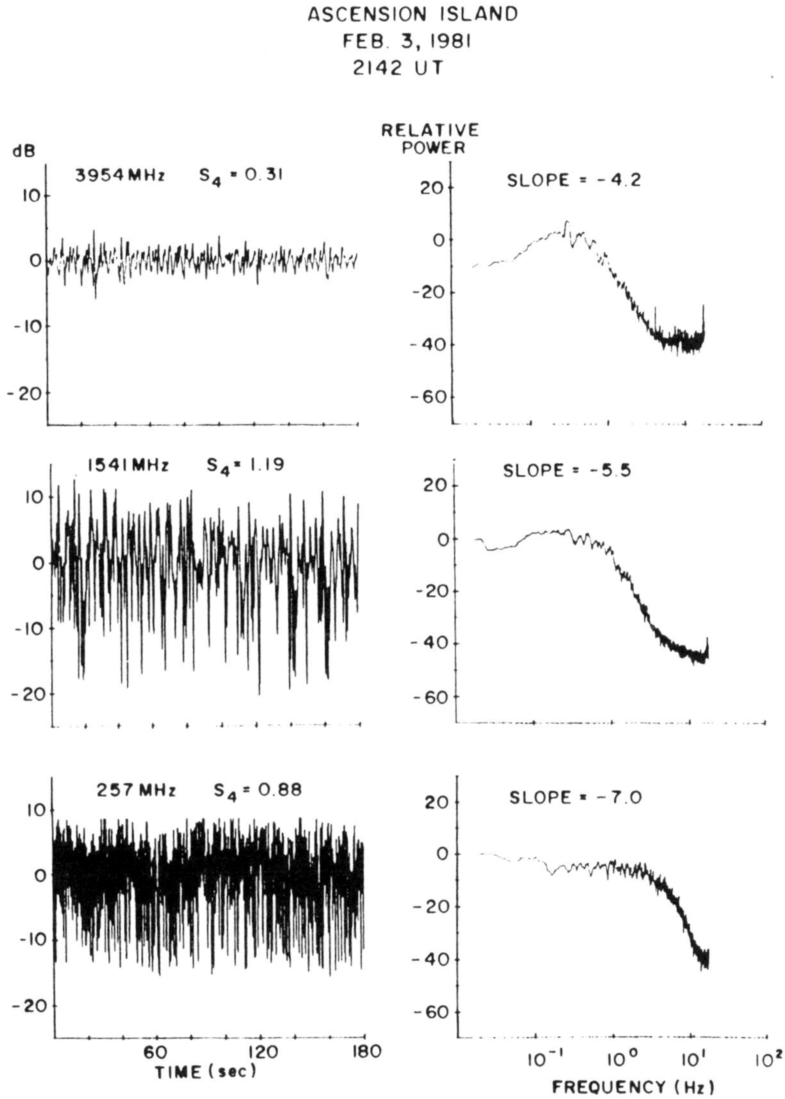

Figure 5. A 3-min data segment of scintillations observed at 3954, 1541, and 257 MHz at Ascension Island and their respective spectra.

Equatorial irregularities causing scintillations are typically distributed over a layer of 200-km thickness around a mean altitude of 350 km. However, on occasions, the irregularities which are not fully developed, become confined within a thin layer in the bottomside F-region and exhibit sinusoidal spatial variations. The spatial structure and the spectrum of such bottomside sinusoidal structures (Valladares et al., 1983) detected by AE-E satellite in-situ measurements are shown in the top panel of Figure 8. The bottom panel shows that the corresponding scintillation spectra exhibit Fresnel oscillations as is to be expected in the case of weak scattering by a thin irregularity layer. It should, however, be noted that the one-dimensional irregularity

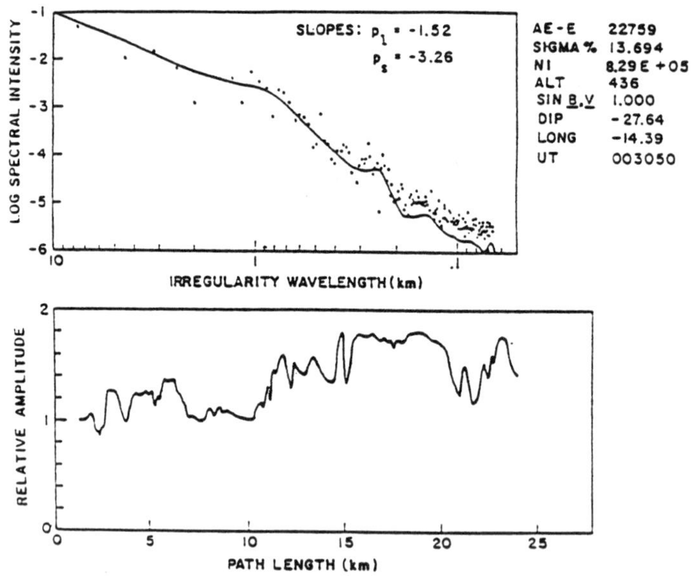

Figure 6. A 3-sec sample of high resolution RPA data of relative amplitudes obtained on AE-E orbit 22759 on Dec 17, 1979 (lower panel). The upper panel shows a power spectrum of the linearly detrended data for this sample obtained by the FFT (dots) and maximum entropy (solid line) techniques.

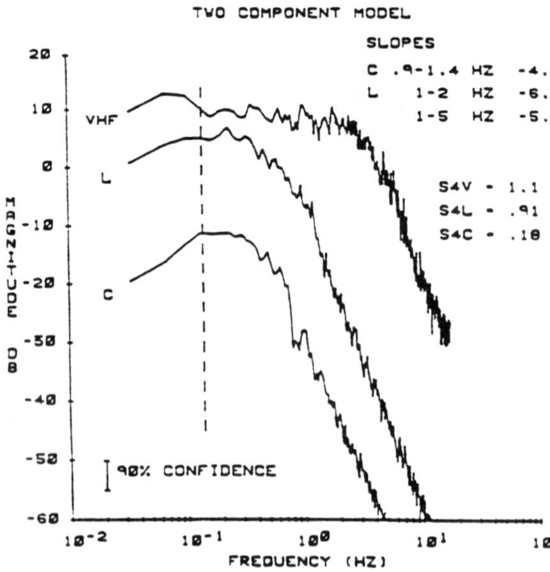

Figure 7. Multi-frequency temporal scintillation spectra computed from simulated scintillation data for the two-component power-law irregularity spectrum model (after Franke and Liu, 1983).

spectral index of 5 indicated in the top panel does not correspond to the scintillation spectral index of 2.4 as illustrated in the bottom panel.

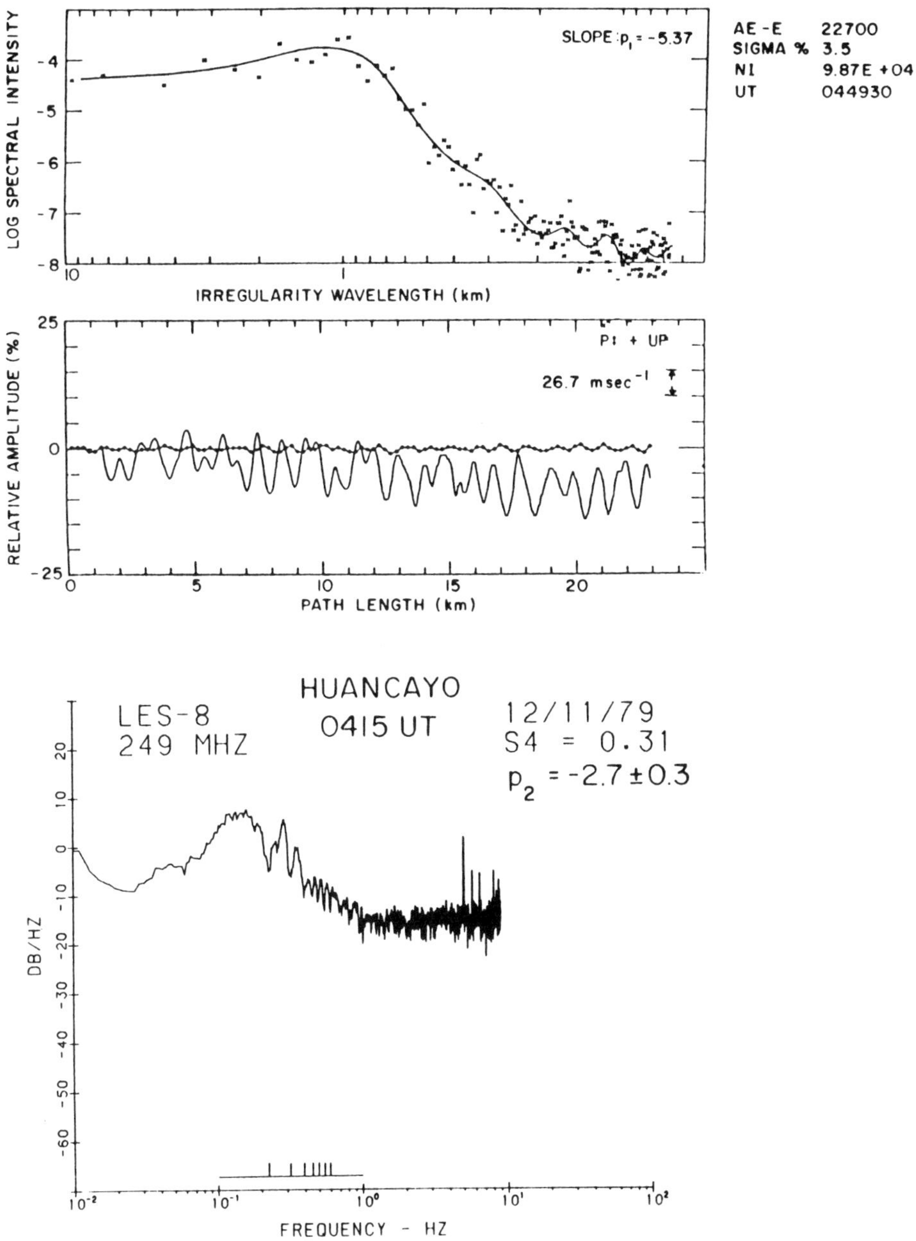

Figure 8. The top panel shows the spatial structure and spectrum of a 3-sec interval of RPA data obtained on AE-E orbit 22700 during which bottomside sinusoidal structures were observed. The bottom panel shows the spectrum of LES-8 scintillation spectrum observed at this time from Huancayo, Peru.

3. Scintillations Induced by High-Power HF Heaters

The F-region irregularities may also be artificially induced by high power HF waves commonly called heaters (Erukhimov et al., 1979; Basu et al., 1987b; Frey et al., 1984). These artificial irregularities are excited by thermal forces (Fejer, 1979) in contrast to the natural irregularities which are generated by either homogeneous or inhomogeneous electric fields. The artificial irregularities attain their maximum amplitude 30 to 40 seconds after the heater is turned on and are quenched within one or two minutes after the heater is turned off. Figure 9 shows the locus of the 250-km subionospheric point of the ray path from a quasi-stationary polar beacon satellite transmitting at 250 MHz through the heater beam of the Max-Planck Institut für Aeronomie at Ramfjordmoen, Norway. The HF heater was operated at 5.1 MHz with an effective radiated power of 200 MW and was cycled 10 min on and 10 min off. The top and bottom panels of Figure 10 show the intensity and phase scintillation spectra during the fully developed state of scintillation. The intensity scintillation index corresponds to weak scintillations with $S_4 = 0.12$. The Fresnel oscillations are very evident in both the phase and intensity scintillation spectra and the Fresnel minima in the intensity scintillation spectra coincide with the Fresnel maxima of phase scintillation spectra. It indicates that the heater generated irregularities are confined to a layer with thickness less than 50 km. Model computations indicate that the irregularity amplitude, $\delta n/n$, is of the order of only 1%.

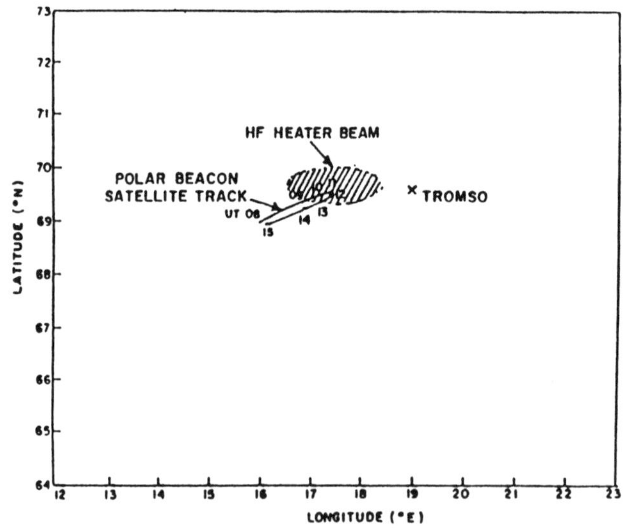

Figure 9. Geometry of observations indicating the half-power beamwidth (shaded region) of the HF heater and the locus of intersection of the ray path from Tromso to the polar beacon satellite with 250-km ionospheric height. The universal times are marked alongside the sub-ionospheric track.

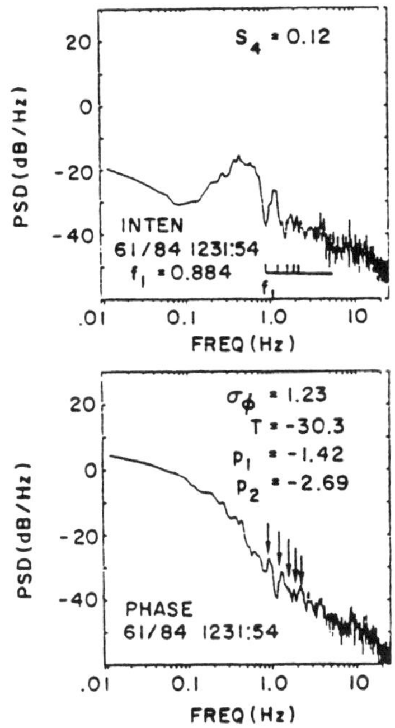

Figure 10. The intensity and phase scintillation spectra during the full developed state of scintillation with Fresnel oscillations indicated.

4. High Latitude Scintillations

Scintillations at high latitudes are commonly observed in the nightside auroral oval above 65° corrected geomagnetic latitude (CGL) and within the polar cap located between 75°CGL and the magnetic pole. In the auroral oval, scintillations may be induced by irregularities generated locally by auroral particle precipitation, but much stronger irregularity structures convected from the polar cap may cause very intense scintillations (Tsunoda, 1988; Basu et al., 1991). Within the polar cap, the interplanetary magnetic field (IMF) dictates the type of plasma structures that are detected there (Buchau et al., 1985; Weber et al., 1884, 1986). These authors established that during a southward orientation of the interplanetary magnetic field, discrete patches of ionization, 1000 km in extent, with enhanced ionization density of the order of 10^{12} m^{-3} from midlatitudes are convected into the polar cap and then move in an antisunward direction and may enter the nightside auroral oval. On the other hand, when the IMF is northward, plasma structures oriented parallel to the noon-midnight direction develop due to particle precipitation and move in the dawn-dusk direction (Carlson, 1984). The influence of IMF on plasma structures observed within the polar cap is schematically illustrated in Figure 11. Basu et al. (1988b) showed that these patches of ionization develop small scale irregularities, which cause intense scintillations, due to the gradient drift instability mechanism as illustrated in the right-hand panel of Figure 11. The bottom diagram of the right-hand panel, adapted after the numerical simulation results of Mitchell et al. (1985), shows prominent protrusions or "fingers" of enhanced density growing outward from the boundary opposite to the direction of convection. At non-linear stages of development of the instability, the entire patch of ionization gets structured. On the other hand, the left-hand panel of Figure 11 shows that for the northward configuration of IMF, plasma gets structured at the edges of the arc due to velocity shear instability. The bottom panel, adapted after Keskinen et al. (1988) illustrates how the plasma gets structured by collisional Kelvin-Helmholtz instability mechanism.

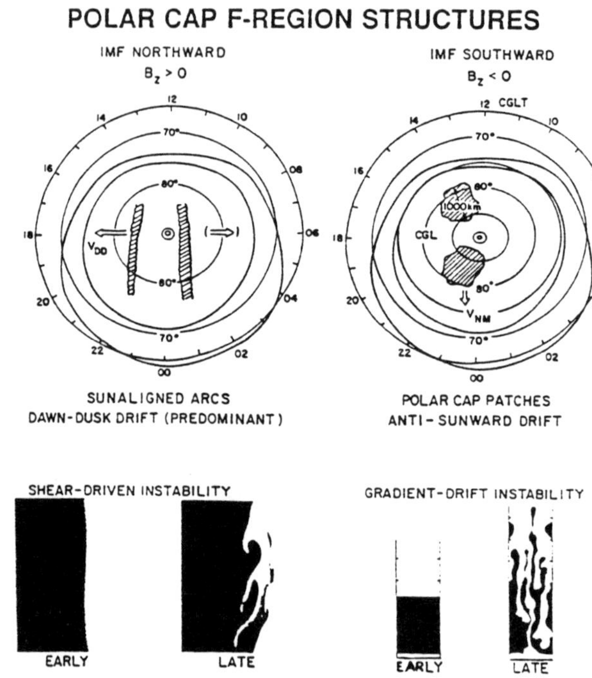

Figure 11. Influence of Interplanetary Magnetic Field (IMF) on plasma structures within the polar cap.

The electron density irregularity structure and its spectrum for polar cap patches, obtained from DE-2 satellite in-situ measurements of electron density fluctuations (Basu et al., 1990a) is illustrated by Figure 12. The spectrum shows that the one-dimensional irregularity spectrum between the scalelength range of 10 km to 250 m can be described by a spectral index of -1.9. Phase and intensity scintillation spectra for such patches of ionization obtained around the time of the in-situ measurements are shown in Figure 13. These measurements were made at Thule, Greenland, using the 250-MHz transmissions from the polar beacon satellite. The phase spectral slope, which is expected to reflect the in-situ spectra, is found to be -2.8 in good agreement with the one-dimensional in-situ spectral slope of -1.9. However, this agreement is an exception rather than the rule at high latitudes. The distribution of in-situ density irregularity and phase scintillation spectral slopes shown in Figure 14 indicates that the median value of the phase scintillation spectral index is -2.3 whereas for the one-dimensional in-situ density irregularity spectrum the index is -1.9. Such discrepancies have been examined by Wernik et al. (1990) in the context

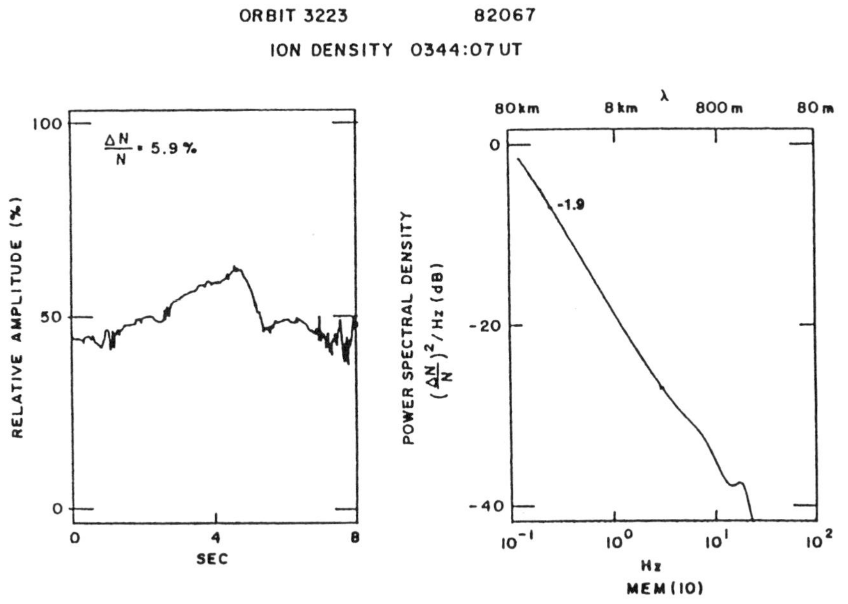

Figure 12. Spectrum of one 8-sec sample of $(\Delta N/N)_{rms}$ data from DE-2 satellite on March 8, 1982.

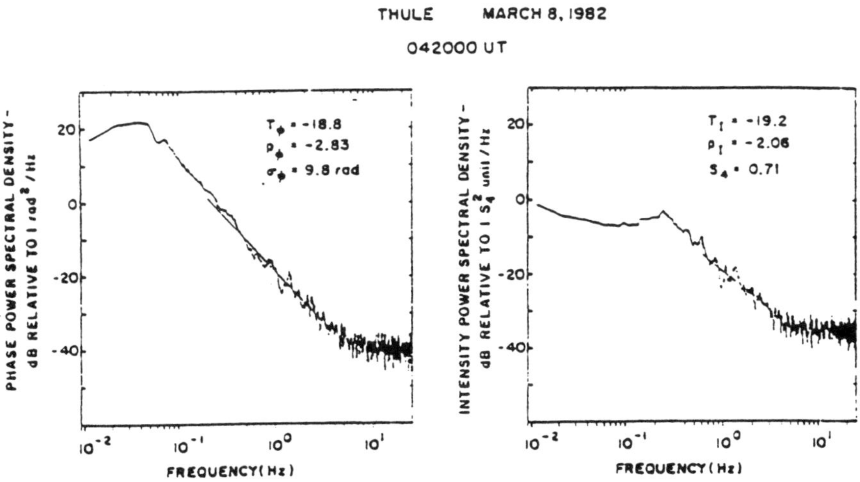

Figure 13. Spectra of rms phase deviation (left panel) and intensity scintillation index S_4 (right panel) from Thule on March 8, 1982.

of anisotropic irregularity spectra and geometry of observations. It should be noted that the patches produce very intense scintillation primarily because the ionization density is so high. The major portion of the ionization density of the patches is not locally generated but reflects the high density of the sub-auroral ionosphere which is convected into the polar cap.

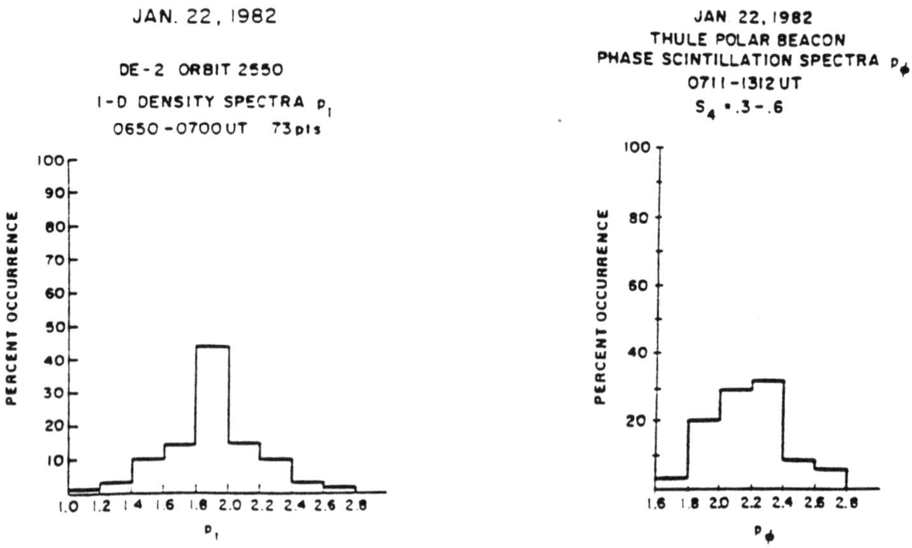

Figure 14. Histograms of one-dimensional in-situ density spectral index, p_1, (left panel) and phase spectral index, p_ϕ, (right panel).

The irregularities in the arcs generated by the velocity shear instability mechanism have overall spectral slopes similar to that of the patches as shown by Basu et al. (1990b). However, the power spectral density of shear generated irregularities is higher at short scales and are more efficient in causing scintillations in the VHF range as established by Basu et al. (1986). In spite of this increased efficiency, the ionization density of the arcs, being locally generated by particle precipitation, is usually a factor of 5 less than the patches (Carlson et al., 1984). As a result, scintillations caused by the arcs are not as intense as that caused by the patches.

5. Summary

Ionospheric scintillation studies seem to be based on a solid foundation since theoretical and experimental investigations of scintillations have been well coordinated with satellite in-situ measurements of irregularity structures and development of plasma instability theories generating such structures in the ionosphere.

Overall, there seems to be good agreement between satellite in-situ measurements of ionospheric structures, numerical simulation and field measurements. However, there remains discrepancies between in-situ irregularity and scintillation spectral slopes at high latitudes and at equatorial latitudes in the case of bottomside sinusoidal irregularities. The effects of anisotropic irregularity spectrum and observing geometry as discussed by Wernik et al. (1990) need to be investigated further.

Acknowledgment

The work at Phillips Laboratory (PL) was funded by AFOSR Task 2310G9 and the work at Boston College was partially supported by PL Contract F19628-90-K-0007.

References

Aarons, J., Global morphology of ionospheric scintillation, *Proc. IEEE*, *70*, 360, 1982.

Basu, S. and Su. Basu, Equatorial scintillations - a review, *J. Atmos. Terr. Phys.*, *43*, 473, 1981.

Basu, S., Su. Basu, J. Aarons, J.P. McClure, and M.D. Cousins, On the coexistence of kilometer- and meter-scale irregularities in the nighttime equatorial F region, *J. Geophys. Res.*, *83*, 4219, 1978.

Basu, S., E. MacKenzie, S. Basu, E. Costa, P.F. Fougere, H.C. Carlson, and H.E. Whitney, 250 MHz/GHz scintillation parameters in the equatorial, polar and auroral environments, *IEEE Journal on Selected Areas in Communications*, *SAC-5*, 102, 1987a.

Basu, S., Su. Basu, P. Stubbe, H. Kopka, and J. Waaramaa, Daytime scintillations induced by high-power HF waves at Tromso, Norway, *J. Geophys. Res.*, *92*, 11149, 1987b.

Basu, S., E. MacKenzie, and Su. Basu, Ionospheric constraints on VHF/UHF communications links during solar maximum and minimum periods, *Radio Sci.*, *23*, 363, 1988a.

Basu, Su. and S. Basu, Equatorial scintillations: Advances since ISEA-6, *J. Atmos. Terr. Phys.*, *47*, 753, 1985.

Basu, Su., S. Basu, J.P. McClure, W.B. Hanson, and H.E. Whitney, High-resolution topside in-situ data of electron densities and VHF/GHz scintillations in the equatorial region, *J. Geophys. Res.*, *88*, 403, 1983.

Basu, Su., S. Basu, C. Senior, D. Weimer, E. Nielson, and P.F. Fougere, Velocity shears and sub-km scale irregularities in the nighttime auroral F-region, *Geophys. Res. Lett.*, *13*, 101, 1986.

Basu, Su., S. Basu, E.J. Weber, and W.R. Coley, Case study of polar cap scintillation modeling using DE-2 irregularity measurements at 800 km, *Radio Sci.*, *23*, 545, 1988b.

Basu, Su., S. Basu, E. MacKenzie, W.R. Coley, J.R. Sharber, and W.R. Hoegy, Plasma structuring by the gradient drift instability at high latitudes and comparison with velocity shear driven processes, *J. Geophys. Res.*, *95*, 7799, 1990a.

Basu, Su., S. Basu, E.J. Weber, and G.J. Bishop, Plasma structuring in the polar cap, *J. Geomag. Geoelectr.*, *42*, 763, 1990b.

Basu, Su., S. Basu, E. Costa, C. Bryant, C.E. Valladares, and R.C. Livingston, Interplanetary magnetic field control of drifts and anisotropy of high latitude irregularities, *Radio Sci.*, 26, 1079, 1991.

Buchau, J., E.J. Weber, D.N. Anderson, H.C. Carlson, Jr., J.G. Moore, B.W. Reinisch, and R.C. Livingston, Ionospheric structures in the polar cap: their origin and relation to 250-MHz scintillation, *Radio Sci.*, *20*, 325, 1985.

Carlson, H.C., V.B. Wickwar, E.J. Weber, J. Buchau, J.G. Moore, and W. Whiting, Plasma characteristics of polar cap F-layer arcs, *Geophys. Res. Lett.*, *1*, 895, 1984.

Cronyn, W.M., The analysis of radio scattering and space probe observations of small-scale structure in the interplanetary medium, *Astrophys. J.*, *161*, 755, 1970.

Dungey, J.W., Convective diffusion in the equatorial F region, *J. Atmos. Terr. Phys.*, *9*, 304, 1956.

Erukhimov, L.M., V.I. Kovalev, A.M. Lerner, E.N. Myaskinov, I.N. Poddel'skii, and A.V. Rakhilin, Spectrum of large-scale artificial inhomogeneities in the F layer of the ionosphere, *Radio Phys. Quantum Electron., Engl. Trans.*, *22*, 888, 1979.

Fejer, J.A., Ionospheric modification and parametric instabilities, *Rev. Geophys.*, *17*, 135, 1979.

Franke, S.J. and C.H. Liu, Observations and modeling of multi-frequency VHF and GHz scintillations in the equatorial region, *J. Geophys. Res.*, *88*, 7075, 1983.

Frey, A., P. Stubbe, and H. Kopka, First experimental evidence of HF produced electron density irregularities in the polar ionosphere: Diagnosed by UHF radio star scintillations, *Geophys. Res. Lett.*, *11*, 523, 1984.

Kelley, M.C., J.F. Vickrey, C.W. Carlson, and R. Torbert, On the origin and spatial extent of high-latitude F-region irregularities, *J. Geophys. Res.*, *87*, 4469, 1982.

Kelley, M.C., *The Earth's Ionosphere, Plasma Physics and Electrodynamics*, Academic Press Inc., San Diego, CA, p. 121, 1989.

Keskinen, M.J., H.G., Mitchell, J.A. Fedder, P. Satyanarayana, S.T. Zalesak, and J.D. Huba, Nonlinear evolution of the Kelvin-Helmholtz instability in the high-latitude ionospher, *J. Geophys. Res.*, *93*, 137, 1988.

Livingston, R.C. and T.M. Dabbs, Observations of equatorial plasma irregularity spectrum, To be submitted to *Radio Sci.*, 1992.

Mitchell, H.G., J.A. Fedder, M.J. Keskinen, and S.T. Zalesak, A simulation of high latitude F-layer instabilities in the presence of magnetosphere-ionosphere coupling, *Geophys. Res. Lett.*, *12*, 283, 1985.

Rino, C.L., A power law phase screen model for ionospheric scintillation. 1. Weak scatter, *Radio Sci.*, *14*, 1135, 1979.

Rino, C.L., Numerical computations for a one-dimensional power law phase screen, *Radio Sci.*, *15*, 41, 1980.

Rino, C.L., R.T. Tsunoda, J. Petriceks, R.C. Livingston, M.C. Kelley, and K.D. Baker, Simultaneous rocket-borne beacon and in-situ measurements of equatorial spread-F intermediate wavelength results, *J. Geophys. Res.*, *86*, 2411, 1981.

Tatarskii, V.I., *Wave Propagation in a Turbulent Medium*, McGraw-Hill Book Co., Inc., 1961.

Tsunoda, R.T., Magnetic field-aligned characteristics of plasma bubbles in the nighttime equatorial ionosphere, *J. Atmos. Terr. Phys.*, *42*, 743, 1980.

Tsunoda, R.T., Time evolution and dynamics of equatorial backscatter plumes. 1. Growth phase, *J. Geophys. Res.*, *86*, 139, 1981.

Tsunoda, R.T., High-latitude F-region irregularities: a review and synthesis, *Rev. of Geophys.*, *26*, 719, 1988.

Valladares, C.E., W.B. Hanson, J.P. McClure, and B.L. Cragin, Bottomside sinusoidal irregularities in the equatorial F region, *J. Geophys. Res.*, *88*, 8025, 1983.

Weber, E.J., J. Buchau, J.G. Moore, J.R. Sharber, R.C. Livingston, J.D. Winningham, and B.W. Reinisch, F-layer ionization patches in the polar cap, *J. Geophys. Res.*, *89*, 1683, 1984.

Weber, E.J., J.A. Klobuchar, J. Buchau, H.C. Carlson, Jr., R.C. Livingston, O. de la Beaujardiere, M. McCready, J.G. Moore, and G.J. Bishop, Polar cap F layer patches: structure and dynamics, *J. Geophys. Res.*, *91*, 12121, 1986.

Wernik, A.W., C.H. Liu, S.J. Franke, and M. Gola, High-latitude irregularity spectra deduced from scintillation measurements, *Radio Sci.*, *25*, 883, 1990.

Woodman, R.F. and C. LaHoz, Radar observations of F region equatorial irregularities, *J. Geophys. Res.*, *81*, 5447, 1976.

Yeh, K.C. and C.H. Liu, Radio wave scintillations in the ionosphere, *Proc. IEEE*, *70*, 324, 1982.

Scintillation in the Solar Wind (IPS)

Wm. A. Coles,
Electrical and Computer Engineering,
University of California, San Diego,
La Jolla, CA 92093-0407

Introduction

Intensity scintillation in the solar wind (IPS) was discovered in 1964 by Hewish, Scott and Wills[1]. Since then it has been used extensively both to catalog the angular diameter of radio sources[2] and, with other scattering measurements, to probe the solar wind plasma. The latter application divides neatly into the study of transients, the topic of reviews by Woo[3] and Hewish[4] in this issue; and the study of quasi-static structures, which is the topic of this review. Perhaps the most important results in this area have been in velocity measurements. IPS observations[5,6], carried out regularly over 20 years, have demonstrated how the velocity structure of the solar wind varies with solar activity in a way that could not be done by any other method. Recently IPS[7] has been used to make precision measurements of the radial acceleration profile of the solar wind, that can be compared with theoretical magneto-hydrodynamic (MHD) models. Scattering measurements have also been used to study the microturbulence[8,9,10,11], but this has not advanced as far as the velocity work, primarily because of weakness in the theory of turbulence in a magnetized plasma. These IPS studies have driven theoretical work on scattering, for example in velocity estimation[12], in the effect of source coherence on intensity scintillation[13], and in two frequency cross correlations[14]. They also demonstrate some interesting effects related to the presence of Alfven waves[15] (MHD shear waves) which do not, by themselves, cause density fluctuations, but which affect ambient density fluctuations in measureable ways. In this paper I will discuss some recent work which is directly related to the presence of these Alfven waves.

The Solar Wind

The solar wind[16,17,18] is a radial outflow of completely ionized plasma, primarily hydrogen with a few % helium, originating in the corona and continuing to at least 50 AU. It begins at zero velocity at the corona, at a kinetic temperature of about 2×10^6K. It accelerates rapidly, becoming supersonic with respect to the various MHD modes in the first 10 solar radii (R_s). It carries the solar magnetic field out with it, becoming radial outside of 2.5 R_s. Above open field regions the corona is cooler, and the speed is higher, than above closed field regions. The open field regions are called coronal holes. The sun rotates with a period (near the equator) of 27 days. This causes the magnetic field of the solar wind to form an Archimedean spiral as the plasma carries it out. The polar rotation period is longer by several days and this differential rotation amplifies the magnetic field, causing bouyant flux ropes which erupt from the surface in active regions.

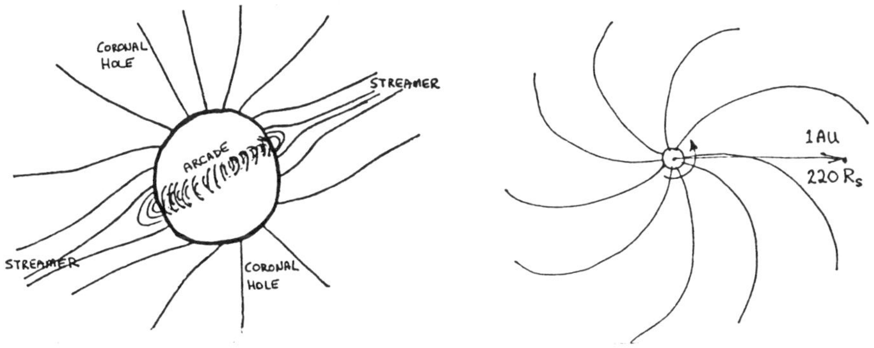

Figure 1.(a) (left panel) The Solar magnetic field as viewed from the earth. (b) (right panel) Top view showing the large scale spiral structure.

The most important question relating to the solar wind (perhaps in the whole of solar physics) is, "what causes the superheating of the corona to 2×10^6 K?" Given that the corona is this hot, a supersonic solar wind should exist. But it should not be as fast as is actually observed. The next question then is, "what causes the additional acceleration of the solar wind?" It is possible that the answers to both these questions involve wave dissipation, and that this can be detected and measured using various scintillation techniques.

Scattering Methods

The scattering phenomena which have been used to probe the solar wind include intensity scintillation, angular scattering, spectral broadening, phase scintillation and pulse broadening. The geometry of a typical observation is sketched Figure 2. The source might be a natural radio source, a spacecraft beacon, or a radar echo from a planet[19]. The scattering is concentrated where the line-of-sight is closest to the sun because the spectrum of refractive index drops quickly with distance from the sun, $\Phi_n(\kappa) \propto R^{-4}$. About half the intensity variance and half the contribution to angular scattering or spectral broadening comes from a sector which subtends about 70° at the sun.

The refractive index fluctuations in a plasma are $\delta n \propto \lambda^2$, which gives $\delta\phi \propto \lambda$ and $D_\phi(s) \propto \lambda^2$, unlike the neutral atmosphere where $D_\phi(s) \propto \lambda^{-2}$. As the sun is approached the scintillations become stronger and saturate as sketched Figure 3. To observe closer to the sun one must use a shorter wavelength. This is a fortunate situation because the sun

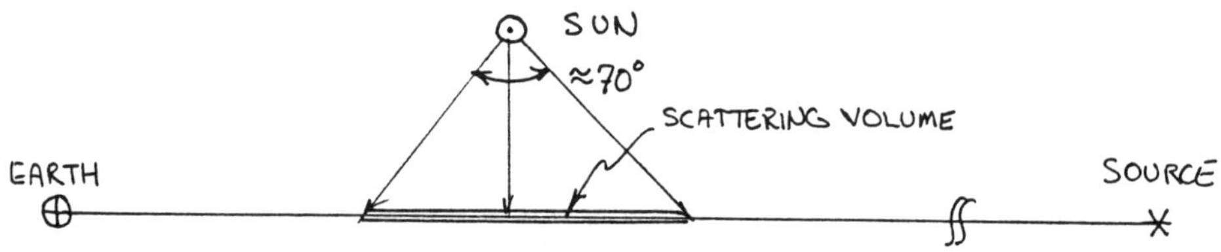

Figure 2. The geometry of an IPS observation near the sun.

is a very intense radio source. To observe near the sun one needs a narrow antenna beam, to avoid observing the sun directly or in the near sidelobes. Of course the beam width is proportional to the wavelength for a given antenna.

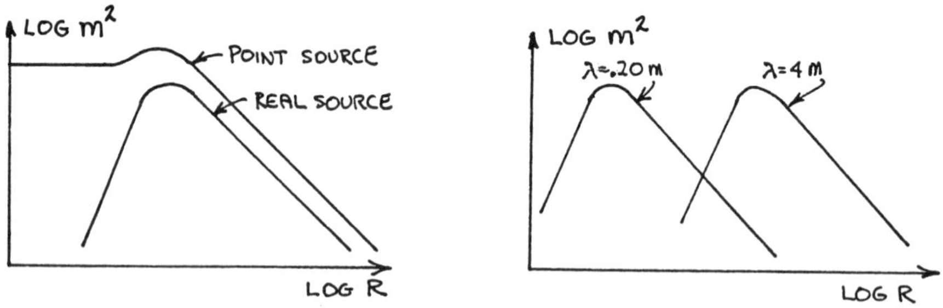

Figure 3. The behavior of the normalized variance m^2 vs distance from the sun.

The velocity is estimated by cross correlating the intensity scintillations measured at two spaced receivers[12,20,21]. If the entire scattering volume is moving with the same velocity one will see correlation functions with a well-defined time lag as shown in Figure 4. However if there is a distribution of speed withing the scattering volume (but the direction is constant) one will see a highly skewed function as shown in Figure 5.

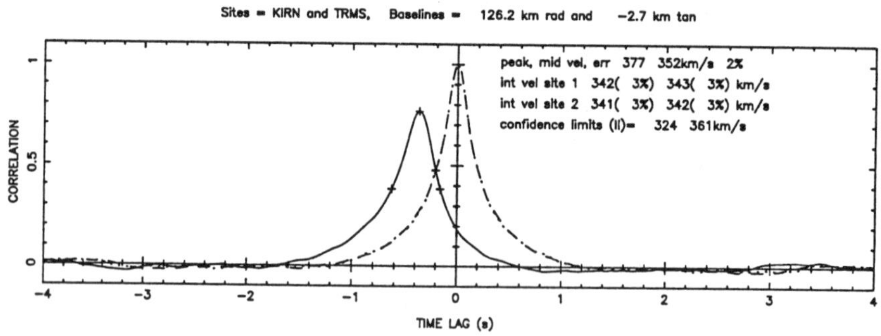

Figure 4. Typical intensity cross correlation with a well-defined velocity.

Occasionally one sees two grossly different speeds, which gives rise two a double peaked cross correlation as shown in Figure 6. If the direction of the velocity is spread over some random distribution but the speed is more or less constant, then the cross correlation is depressed but not skewed. In general the time lag at the intersection of the auto and cross correlations is a good measure of the mean velocity $V_{mean} \approx b/2\tau_{int}$. However it is necessary to fit models to the correlations functions to obtain estimates of the mean velocity and the rms velocity in the parallel and perpendicular directions. An example of such model fitting is shown in Figure 7.

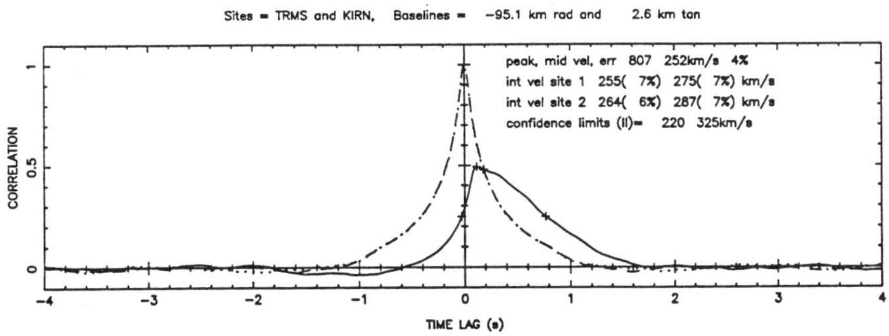

Figure 5. A cross correlation showing the effect of a large spread of radial velocity within the scattering volume.

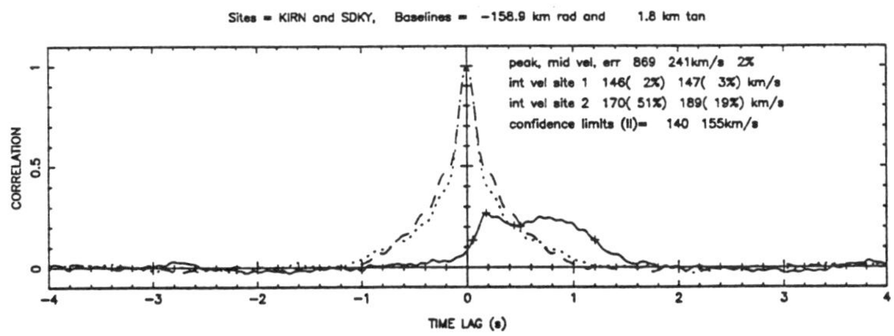

Figure 6. A cross correlation showing two distinct velocities within the scattering volume.

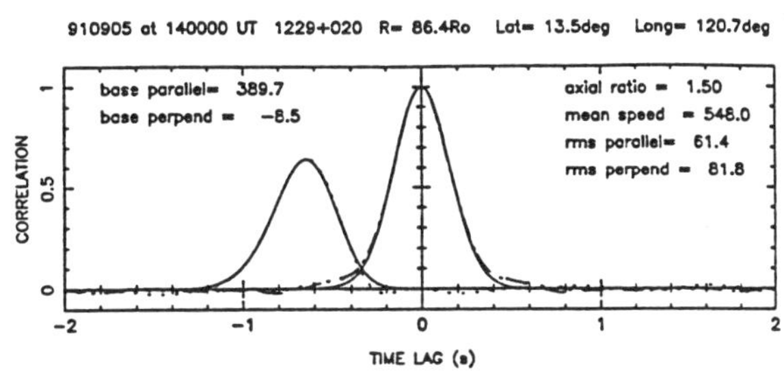

Figure 7. A cross correlation including a model fit, showing the effect of random velocity in both parallel and perpendicular directions.

When a point source passes near the sun it broadens[22,23], usually into an apparently gaussian blob with elliptical symmetry. Of course the apparent brightness distribution $B(\vec{\theta})$ is just the Fourier transform of the mutual coherence function $\Gamma(\vec{s})$ and we usually work directly in the $\Gamma(\vec{s})$ domain. A typical observation is plotted as $\Gamma(s)$ vs s in Figure 8a. However if plotted vs $Q=(\vec{s}^{\,t}\mathbf{M}\vec{s})^{1/2}$ this spread is organized into a monotonic function as shown in Figure 8b. The matrix of the transformation **M**, gives the axial ratio and orientation of this ellipse. This is usually elongated in the radial direction (i.e. field aligned). The axial ratio measured in this way, at scales of 3 to 30 km, increases rapidly inside of 10 R_s.

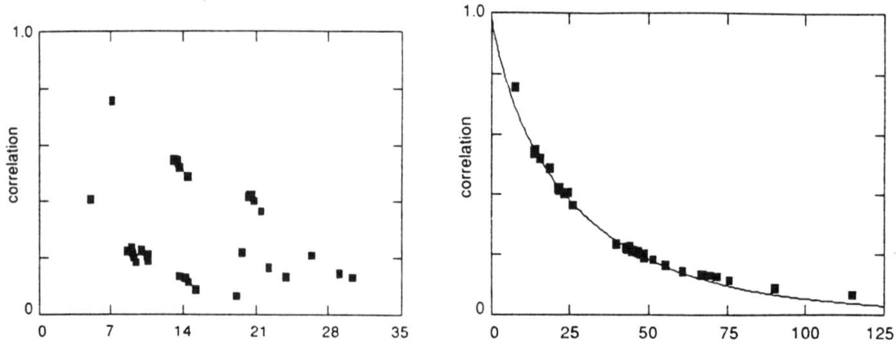

Figure 8.(a) (left) The mutual coherence function vs the magnitude of the baseline. (b) (right) The mutual coherence function vs a quadratic form in the baseline, showing elliptical symmetry.

When a monochromatic source is observed through the solar wind it is spread into a frequency spectrum by differential Doppler shifts from angular scattering[8,24]. In fact we can write the relationship between spectral broadening and the structure function more directly in the correlation domain as

$$B_e(\vec{s}=\vec{V}\tau) = exp\,[-\tfrac{1}{2}D\,(\vec{s}=\vec{V}\tau)] \quad so \quad D\,(\vec{s}=\vec{V}\tau) = -2\log B_e(\vec{s}=\vec{V}\tau).$$

Here $B_e(\tau)$ is the Fourier transform of the power spectrum $P\,(\omega)$.

A monochromatic point source also permits direct phase measurements, although this is non-trivial for a moving vehicle like a spacecraft. Here one can take advantage of the dispersion of the plasma and, by measuring the phase at two different frequencies, recover both the plasma and the free space contributions. One can show theoretically that the observed phase structure function converges to the wave structure function at large scales. By simulation we have observed that this convergence is quite good for scales larger than the Fresnel scale[11]. Thus observations of a spacecraft beacon can be used in two ways, spectral broadening provides the small scales and phase scintillation provides the large scales. An example of this is shown in Figure 9.

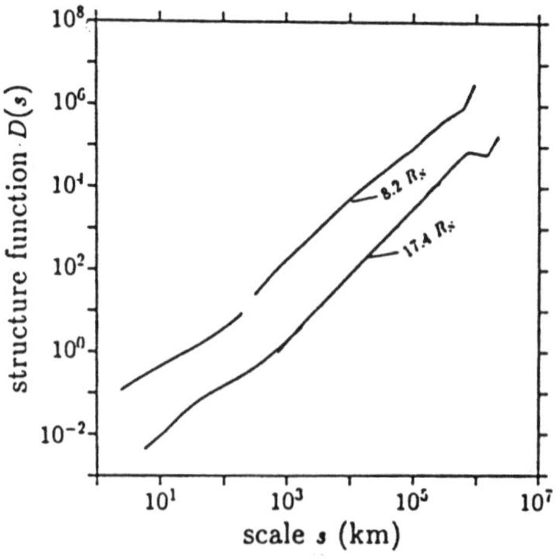

Figure 9. Estimates of the wave structure function by spectral broadening (small scales) and by phase scintillation (large scales).

Velocity Observations

It has recently become possible to make a large number of two-antenna velocity observations at distances between 12 and 120 R_s, where a significant amount of acceleration occurs. These observations[7] have been made using the European Incoherent Scatter Facility (EISCAT), a three antenna system in northern Scandinavia. This system can provide very high accuracy because the antenna baselines are much longer than the Fresnel scale. It is also very convenient for summer work because the sun is circumpolar most of the summer and sources near the sun are observable 24 hrs/day.

To study the radial velocity profile one must first sort out the substantially larger variations with heliographic latitude, longitude and time. The magnitude of this problem can be seen by the plot of velocity vs distance in Figure 10. Clearly this plot is of little value in constraining theoretical acceleration models. However one can select observations which are at almost the same latitude, longitude and time. An example of such a selection is shown in Figure 11 with an overplot of some theoretical curves. The lowest theoretical line is one in which the wind is driven by thermal conduction alone. The parameters have been adjusted to give the highest velocity consistent with other observational constraints. The higher theoretical lines show models in which the acceleration is enhanced by the ponderomotive force of the Alfven wave gradient. Although these observations do not prove that Alfven waves accelerate the solar wind, they are certainly consistent with this explanation and one can estimate the amplitudes of the Alfven waves necessary to provide this acceleration.

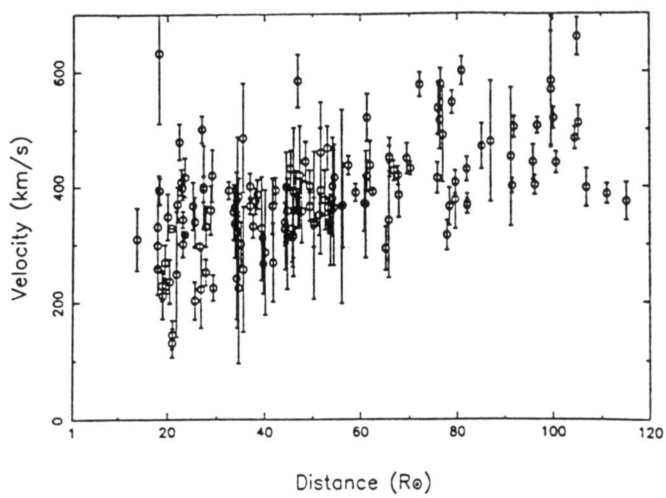

Figure 10. Estimated velocity vs distance for all data collected at EISCAT in the summer of 1990.

However we occasionally see observations like those shown in Figure 12. Clearly none of the theoretical lines comes close to the observations. Roughly 25% of the streams observed between 1990 and 1992 show this characteristic, so it is not rare. But it does seem to be correlated with the magnetic neutral line which is shown in Figure 13. Overplotted on this magnetic field map are the traces of the scattering volume projected back to the solar surface. All the late accelerating streams we have seen are similarly close to a field reversal.

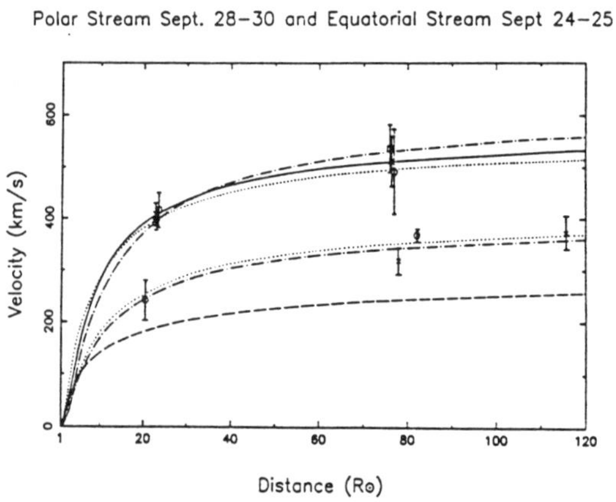

Figure 11. Estimated velocity vs distance for two different streams. The solid lines are theoretical models. The lowest is thermally driven. The higher velocities include both thermal conduction and a flux of Alfven waves.

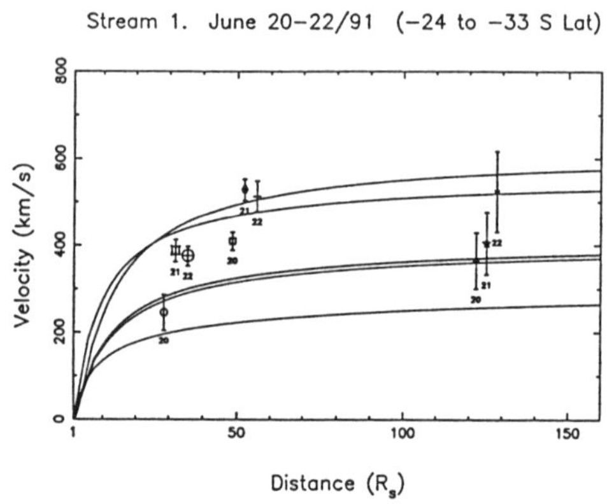

Figure 12. Estimated velocity vs distance for a stream which does not agree well with theoretical models as its acceleration is delayed.

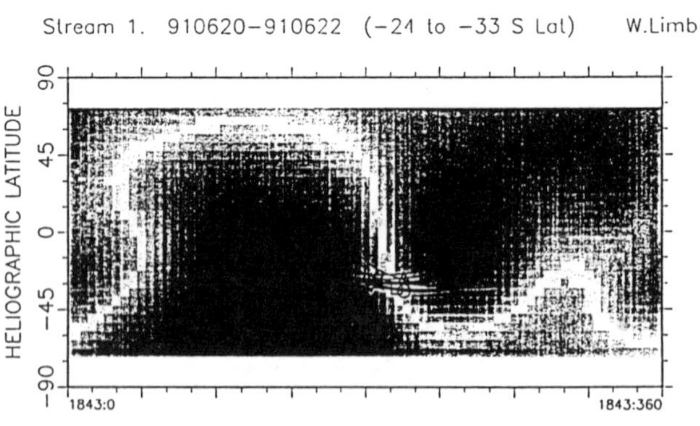

Figure 13. The magnitude of the radial component of the coronal magnetic field estimated at 2.5 R_s. The light region is zero field (the polarity reversal).

The Spatial Spectrum of Electron Density

Observations of the wave structure function[10], scaled to 13 cm wavelength, are shown in Figure 14. These plots were assembled over many years from many sources including: angular scattering, spectral broadening, intensity scintillation, phase scintillation and long baseline interferometry. They indicate strongly that the spectrum is not a simple Kolmogorov power-law, but is enhanced by approximately a factor of 10 at small scales. They also indicate strongly that there is a dissipation scale which varies roughly as $l_0\,(km) \approx R\,(R_s)$. Similar estimates of the inner scale have been obtained by other workers[9,25,26,27].

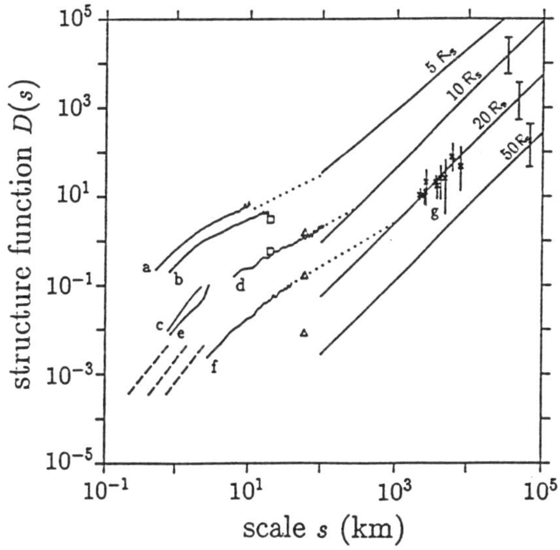

Figure 14. Estimates of the wave structure function at four different solar distances by a variety of techniques.

This assembly of different observations was not as convincing as it could be because of the time variability and the possibility of uncorrected bias in the different techniques involved. Recently[11] it was possible to analyze both spectral broadening and phase scintillation taken simultaneously. Some of these results are shown in Figure 15. This shows the same enhancement without the ambiguity in the earlier assembly.

A particularly interesting feature of this data is that the scales between 3 and 30 km are known to be highly anisotropic[23]. Angular scattering observations were used to determine the axial ratio as a function of distance. Some of these observations are shown in Figure 16. Although there are not very many observations, because it is a difficult and expensive observation to make, they clearly show the the axial ratio increases as the sun is approached.

For some time the reason for the variation with distance was assumed to be due to the underlying plasma process (which was not known either). However recently[15] another possibility has come to light. The Alfven waves that are, perhaps, necessary to accelerate the solar wind (and are also directly observed by spacecraft at greater distances), can isotropize the angular scattering. The mechanism is quite straightforward. We first assume, as appears very likely, that the anisotropy is field aligned. Now the Alfven waves which cause the magnetic field direction to vary will also cause the orientation of the microstructure to vary. The coherence length for these Alfven waves is

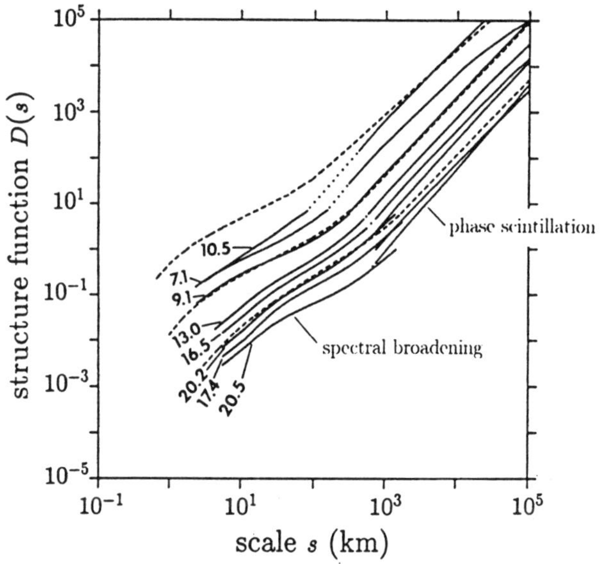

Figure 15. Estimates of the wave structure function from Voyager spacecraft data using both angular scattering and phase scintillation simultaneously.

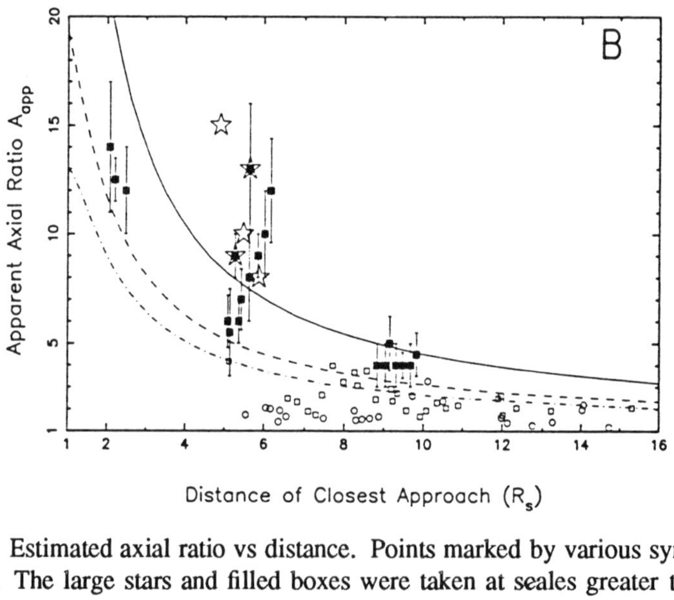

Figure 16. Estimated axial ratio vs distance. Points marked by various symbols are observations. The large stars and filled boxes were taken at seales greater than the inner scale. The small open symbols were taken at scales smaller than the inner scale. The lines drawn over the data are calculated from solar wind acceleration models, similar to those shown in Figure 11, which include a flux of Alfven waves to increase the velocity.

much shorter than the length of the scattering volume, so there are many independent waves in the scattering volume. Thus an observation of angular scattering will see a random superposition of scattering ellipses with different orientations. Integration through this distribution will produce an image with the same functional form but a smaller axial ratio. The apparent axial ratio will be inversely proportional to the wave amplitude. The amplitude of the Alfven waves is known to increase with distance near the sun (at least until they saturate). So the apparent axial ratio will decrease with distance (until the waves saturate). At saturation the waves will have reduced the axial ratio to approximately 2. The effect of wave amplitude $\delta B/B$ on the apparent axial ratio is shown in Figure 17 for intrinsic axial ratios of 5, 10, 20, 50 and 100.

The apparent axial ratio for a solar wind observation can then be predicted by integrating through a theoretical model of the solar wind. Several examples of such calculations are drawn over the raw data in Figure 16. In these calculations it was assumed that the intrinsic axial ratio was infinite. One can see from the goodness of fit that it is entirely entirely plausible that this model is correct; that the instrinsic structure is essentially two dimensional. This suggests that one should consider mechanisms, such as the parametric decay instabilities, which might produce field-aligned structures. It also indicates that estimates of turbulent dissipation rates based on insotropic turbulence could be grossly in error.

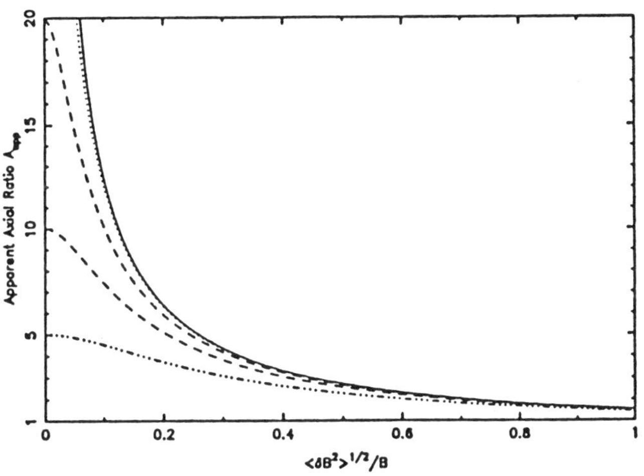

Figure 17. The effect of Alfven wave amplitude $\delta B/B$ on the apparent axial ratio.

The Effect of Alfven Waves on Intensity Scintillations

The effect of Alfven waves on angular scattering is to isotropize the scattering. Since we do not know the intrinsic axial ratio, we can only bound the wave amplitudes. However using intensity scintillation we can do a little better. Although Alfven waves do not cause density fluctuations, they will cause ambient density fluctuations to move laterally. This causes the apparent direction of flow to change. If the structure is isotropic this effect can be observed in two-antenna cross correlations as a reduction in the peak of the cross correlation as shown in Figure 7. However if the structure is highly anisotropic, as it is close to the sun, then it does not reduce the peak of the cross correlation at all, the apparent velocity is simply the sum of the flow velocity and the wave velocity. Clearly it is important to sort out this effect.

It appears that the intrinsic structure must be nearly isotropic outside of 30 R_s because we obtain good model fits to the cross correlations using expected wave amplitudes as shown in Figure 7. However observations inside of 10 R_s must be strongly affected. Apparently the intrinsic microstructure becomes more isotropic somewhere between 10 and 30 R_s. We do not have a good idea how to determine this at present, but some possibilities are suggested by the correlation functions shown in Figure 18. This displays contours of the spatial cross correlation at a two fixed time lags, for Alfven waves in an anisotropic medium travelling inwards and outwards at one third of the flow velocity. The clear difference in these correlations suggests that we may be able to sort out the wave direction by multiple baseline observations.

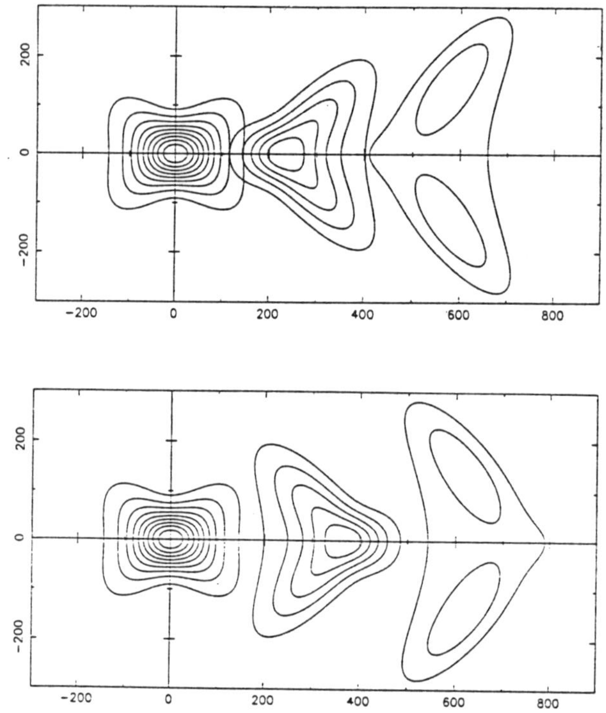

Figure 18. The spatial correlation of intensity at time lags of 1 and 2 s. The velocity was 300 km/s, the baseline 300 km, the wave velocity 100 km/s, and the wave amplitude unity. (a) (top) inwards propagation (b) (bottom) outwards propagation.

Acknowledgement

I would like to thank my colleagues at NAIC in Arecibo, the NRAO in Socorro and EISCAT, Tromso for their help in taking and interpreting the observations described here. In particular I would like to mention John Armstrong, Gabriel Bourgois, Ruth Esser, Rod Frehlich, Russ Grall, John Harmon, Masayoshi Kojima, Jussi Markkanen and Barney Rickett. Support for this work was provided by the Norwegian Research Council (NAVF grant 427-91/026) and the U.S. National Science Foundation (under grant ATM-9008608).

References

1. Hewish, A., Scott, P. F., and Wills, D., Interplanetary scintillation of small diameter radio sources, *Nature*, *203*, 1214, 1964.

2. Readhead, A.C.S., IPS of radio sources at metre wavelengths - II, *Mon. Not. Roy. Astro. Soc.*, *155*, 185, 1971.

3. Woo, R., Spacecraft radio scintillations and solar system exploration, this issue.

4. Hewish, A., Interplanetary scintillation imaging of disturbances in the solar wind, this issue.

5. Kojima, M. and T.Kakinuma, Solar cycle evolution of the solar wind speed structure between 1973 and 1985 observed with the interplanetary scintillation method, J.Geophys. Res., 92, 7269, 1987.

6. B. J. Rickett and W. A. Coles, "Evolution of the Solar Wind Structure over a Solar Cycle: IPS Velocity Measurements Compared with Coronal Observations", *J. Geophys. Res.*, **96**, 1717-1736, 1991.

7. Coles, W. A., R. Esser, U.-P. Lovhaug, and J. Markkanen, Comparison of solar wind velocity measurements with a theoretical acceleration model, *J. Geophys. Res., 96*, 13849, 1991.

8. Woo, R., and J. W. Armstrong, Spacecraft radio scattering observations of the power spectrum of electron density fluctuations in the solar wind, *J. Geophys. Res., 84*, 7288, 1979.

9. Scott, S. L., W. A. Coles, and G. Bourgois, Solar wind observations near the sun using interplanetary scintillation, *Astron. Astrophys., 123*, 207-215, 1983.

10. Coles, W. A., and J. K. Harmon, Propagation observations of the solar wind near the Sun, *Astrophys. J., 337*, 1023, 1989.

11. Coles, W. A., W. Liu, J. K. Harmon and C. L. Martin, The solar wind density spectrum near the sun: Results from Voyager radio measurements, *J. Geophys. Res., 96*, 1745, 1990.

12. Bourgois, G., W. A. Coles, G. Daigne, J. Silen, T. Turunen, and P. J. Williams, Measurements of the solar wind velocity with EISCAT, *Astron.Astrophys., 144*, 452--462, 1985.

13. Codona, J. L. and R. G. Frehlich, Scintillation from extended incoherent sources, *Radio Science, 22*, 469-480, 1987.

14. Codona, J. L., D. B. Creamer, S. M. Flatte, R. G. Frehlich and F. Henyey, Two-frequency intensity cross-spectrum, *Radio Science* **21,** 805-814, 1986.

15. Coles, W. A. and R. Esser An Observational Limit to the Amplitude of Alfven Waves in the Solar Wind and Comparison With an Acceleration Model, *J. Geophys. Res., in press,* 1992.

16. Brandt, J. C.,*Introduction to the solar wind*, Freeman, 1970.

17. Hundhausen, A. J., *Coronal Expansion and Solar Wind*, Springer-Verlag, New York, 1972.

18. Marsch, E. and R. Schwenn, *Solar Wind Seven*, Pergammon, Oxford, 1992.

19. Harmon, J. K., and W. A. Coles, Spectral broadening of planetary radar signals by the solar wind, *Astrophys. J., 270*, 748, 1983.

20. Briggs, B. H., G. J. Phillips, and D. H. Shinn, The analysis of observations on spaced receivers of the fading of radio signals, *Proc. R. Soc. London, Ser. B., 63*, 106, 1950.

21. Coles, W.A. and J.J.Kaufman, Solar wind velocity estimation from multistation IPS, *Radio Science, 13*, 591, 1978.

22. Blesing, R. G., and P. A. Dennison, Coronal broadening of the Crab nebula 1969-1971, *Proc. Astron. Soc. Aust., 2*, 84, 1972.

23. Armstrong, J. W., W. A. Coles, M. Kojima, and B. J.Rickett, Observations of field-aligned density fluctuations in the solar wind, *Astrophys. J., 358*, 685, 1990.

24. Woo, R., F. C. Yang and A. Ishimaru, Structure of density fluctuations n the solar wind near the sun deduced from Pioneer 6 spectral broadening measurements, *Astrophys. J. 210*, 593, 1976.

25. Yakovlev, O. I., A. I. Efimov, V. M. Razmanov and V. K. Shtrykov, Inhomogeneous structure and velocity of the circumsolar plasma based on data of the Venera-10 station, *Sov. Astronomy, 24*, 454, 1980.

26. Martin, J. M., Voyager microwave scintillation measurements of solar wind plasma parameters, *Sci. Rep., D208-1985*, Cent. for Radar Astron., Stanford Univ., Stanford, Calif., 1985.

27. Manoharan, P. K. and S. Ananthakrishnan, Determination of solar-wind velocities using single-station measurements of interplanetary scintillation, *Mon. Not. R. Astr. Soc., 244*, 691, 1990.

RADIO ASTRONOMICAL SCINTILLATION IN THE INTERSTELLAR PLASMA

Barney Rickett
Electrical and Computer Engineering Department
University of California, San Diego

Introduction

In astronomy there are various irregular media that act to distort and perturb the signals received. However, the perturbations can be recorded and used to investigate the various irregular media. Such inverse studies become increasingly valuable with increasing distance to the medium, since direct sampling then becomes increasingly impractical. This paper will concentrate on radiowave propagation through the interstellar medium, for which there is no hope of direct measurements. However, I will start by reviewing the various media through which astronomical signals travel.

The approximate scales relevant to scattering in the Earth's atmosphere, ionosphere, the solar wind and the interstellar medium are listed in Table 1. The Fresnel scale ($R_F = \sqrt{z/k}$) is given in terms of the typical distance z to the scattering medium and wavenumber k. The field coherence scale ρ_0 is defined as the transverse scale over which there is an rms phase difference of one radian. Here we characterize the strength of scattering by the ratio W:

$$W = [R_F/\rho_0]^2 = [R_S/R_F]^2 = R_S/\rho_0 \tag{1}$$

Note that W can also be expressed in terms of the radius R_S of the scattering disc, which is the region at the distance of the scattering medium that makes the main contribution to the field at a single observing point. Thus $R_S = z\,\theta_0$, where $\theta_0 = 1/(k\rho_0)$ is the characteristic scattering angle. Evidently $W > 1$ corresponds to strong scattering and vice versa. The Born variance is often used as a measure of strength of scattering; $m_B^2 \propto W^{5/6}$ for a Kolmogorov medium.

Table 1. The Scattering Media in Radio Astronomy

	Atmosphere		Ionosphere		Solar wind	Interstellar	Intergalactic
Z km	3	3	350	350	1.510^8(1AU)	10^{16}(300pc)	10^{22}(300Mpc)
λ m	0.1	10^{-6}	10	0.1	1	1	1
R_F m	7	0.02	750	75	1.510^5	1.310^9	1.310^{12}
ρ_0 m	10^4	0.01	710^3	210^6	10^6–10^4	510^7	?
$\sqrt{W}$	710^{-4}	2	0.1	410^{-5}	0.15 - 3	25	?
V m/s	1	1	100	100	400	10^5	10^5 - 10^8?
τ_{pat}sec	7	0.01	7	0.7	0.3 - 0.1	10 min (dif)	?
						5 day (ref)	1-100 day ?

Irregularities in the neutral atmosphere cause the familiar optical twinkling of stars, which is the bain of optical astronomers and is minimized by siting large telescopes at high elevations and is now combatted with increasing success by "active optics". In optical propagation through the atmosphere W is typically within a factor of 5 of unity, corresponding to substantial scintillation effects. However, since the atmospheric scattering is non-dispersive, it becomes extremely weak at radio wavelengths and can be ignored. In contrast, the various ionized media cause phase perturbations that increase linearly with wavelength, and scattering becomes strong at meter wavelengths. For the ionsophere W depends on the state of the ionosphere, which is strongly dependent on local time and latitude; typical values are given in Table 1. For the solar wind, W increases as the elongation angle (between the sun and the source of the radiowaves) decreases. The elongation at which W=1 decreases as the radio wavelength decreases. For the interstellar medium, W increases with distance through the Galactic disc, and for paths toward the center of the Galaxy. The value (W=625) is given for a relatively short galactic path; for longer paths at longer wavelengths W is greater by many orders of magnitude. For sources outside of our Galaxy there is also the possibility of scintillation in the intergalactic medium. The properties of this medium are only tentative at present and, with one possible exception[1], no intergalactic scintillation effects have been found. In recent years examples of gravitational lensing have been found, for lines of sight that pass very close to a massive galaxy. In principle the random distribution of galaxies causes random gravitational effects for waves from extremely distant sources, and also the stars in our galaxy can cause a similar effect. These effects are too small to observe at present; see the paper by Erukhimov and Shpiro (these proceedings).

I will now concentrate on scintillation due to the random distribution of plasma density in the interstellar medium (ISM), with emphasis on what has been learned from scintillation studies about the ISM, rather than on the nuisance that such scintillations can be to radio astronomers. Even for typical observing conditions at cm wavelengths, the level of scattering is strong (W ~ 625) and we rely heavily on theory for the asymptotic strong scattering limit ; at meter wavelengths values of W can be as large as 10^7. In strong scattering "asymptopia" (as Stan Flatté calls it), the phenomena are separated into the diffractive scale near ρ_0 and the refractive scale near R_S, which differ by a factor of W.

The angles of scattering are very small (10^{-8}–10^{-5} radians) and the parabolic wave equation is appropriate. In the nearly 25 years since the phenomena were first observed, we have determined that the plasma density can be described by a wave structure function that is a power law from scales as small as 10^7 m up to scales as large as 10^{12} m. The exponent α of the structure function is very close to 5/3, the outer scale lies above the largest values of R_S that have been used. The inner scale is in the same region as ρ_0 for many observing conditions, though there is some disagreement on which is bigger in a typical obervation[2,3]. Figure 1 is a schematic structure function, illustrating the various scales, as we believe them to be ordered in the interstellar medium. See the proceedings of the 1988 San Diego Workshop 1988[4] and Rickett[5] for reviews. The conclusion that "turbulence" in the plasma density behaves very much like that of the earth's neutral atmosphere has caused the theoreticians of astrophysical plasmas considerable difficulty. The ISM densities are so low that the scales involved are much smaller than the mean free path for collisions. Thus any inertial range of turbulence must be supported by a variety of collisionless processes, most of which cannot operate on scales as small as those observed.

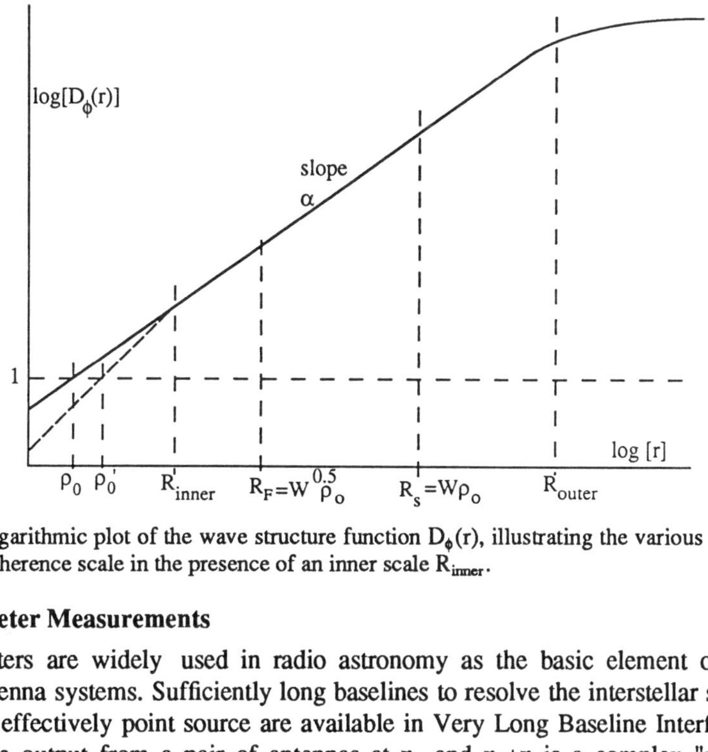

Figure 1. Logarithmic plot of the wave structure function $D_\phi(r)$, illustrating the various scales. ρ_0' is the field coherence scale in the presence of an inner scale R_{inner}.

Interferometer Measurements

Interferometers are widely used in radio astronomy as the basic element of image-forming antenna systems. Sufficiently long baselines to resolve the interstellar scattering (ISS) of an effectively point source are available in Very Long Baseline Interferometry (VLBI). The output from a pair of antennas at $\mathbf{r}_1$ and $\mathbf{r}_1+\mathbf{r}$ is a complex "visiblity": $V(\mathbf{r}_1,\mathbf{r},\omega) = u(\mathbf{r}_1,\omega)\, u^*(\mathbf{r}_1+\mathbf{r},\omega)$, which is averaged over times of a few minutes; here u is the reduced scalar field. The ensemble average of this quantity is the second moment, well-loved by scintillators for its simple and robust theoretical description.

$$< V(\mathbf{r},\omega) > \; = \; \gamma_2(\mathbf{r},z,\omega,\omega) \; = \; \exp[\,-0.5D_\phi(\mathbf{r},z)\,] \qquad (2)$$

Here $D_\phi(r)$ is the structure function for geometric phase, calculated on a straight line through the medium. The field coherence scale is defined by $D_\phi(\rho_0) = 1$ and the radius of the scattered angular spectrum is defined as $\theta_0 = 1/(k\rho_0)$. For a density spectrum of power law form with a wavenumber exponent of α+2 and an inner scale of l_0, for $\rho_0 > l_0$

we have:

$$\gamma_2 \approx \exp[-(r/\rho_0)^\alpha] \text{ for } r > l_0 \,; \quad \gamma_2 \approx \exp[-(r/l_0)^2 (lo/\rho_0)^\alpha] \text{ for } r < l_0 \quad (3a)$$

For $\rho_0 < l_0$

$$\gamma_2 \approx \exp[-(r/l_0)^\alpha (lo/\rho_0)^2] \text{ for } r > l_0 \,; \quad \gamma_2 \approx \exp[-(r/\rho_0)^2] \text{ for } r < l_0 \quad (3b)$$

Hence with an array, making a set of interferometers, we could estimate ρ_0 and α and l_0 - if sufficient averaging could be achieved. Practical integration times can include several characteristic diffractive times, which for a pattern moving at velocity v is ρ_0/v. A common assumption has been that this approximates an ensemble average, from which the scintillation parameters have been estimated[6–11]. However, the integration times have been very much shorter than the refractive scintillation times and have not included effects of image wander, since an image centering algorithm is typically used in VLBI. The result is a flattening of the inferred structure function and hence a bias to the estimates of α and hence of l_0. Goodman and Narayan[12] have studied the statistics of V in this snapshot regime, in which an estimate is made over a time long compared to diffractive times and short compared to refractive times. They did not, however, calculate the effect of image centering.

Measurements have also be made of the image wander due to ISS using VLBI. In the snapshot regime a typical angle of wander θ_r can be estimated as:

$$\sigma_{\theta_r} \approx \frac{[D_\phi(R_S)]^{0.5}}{kR_S} \approx \theta_0 W^{\alpha/2-1} \approx \theta_0 W^{-1/6} \quad (4)$$

The surprise is that even for values of W as large as 10^4, a Kolmogorov spectrum will still provide image wander by an rms angle that is 20% of the size of the scattered image. Thus it appears that tests for image wander might be made and used as a sensitive diagnostic for $[D_\phi(R_S)]$. In practical VLBI, absolute position measurements are less precise than relative position measurements, but since the latter depend on the curvature of phase they effectively probe smaller scales - as do intensity scintillations. Gwinn et al.[13] successfully used this method to constrain α on a line of sight toward the galactic center. See also Fey and Mutel[14].

Pulse Broadening and Arrival Time Fluctuations

Pulsars are radio sources in our Galaxy that emit pulsed radio signals over 0.1 to 10 GHz. Each pulsar has a characteristic period, ranging from 0.0016 to 4 sec. These spinning neutron stars have become a very valuable tool for study of the ISM. First we consider the effects of the dispersion in the medium and then go on to the scattering.

In radio astronomy the observing frequency ω is always much greater than the plasma frequency ω_p (and the cyclotron frequency) in interstellar space. Thus the refractive index can be written to a high degree of accuracy as:

$$n = 1 - \omega_p^2/2\omega^2 = 1 - N_e r_e \lambda^2/2\pi, \quad (5)$$

where N_e is the electron density and $r_e = 2.82\ 10^{-15}$m is the classical electron radius. Defining the scalar field U in terms of a reduced field u involves the mean refractive index, which is dispersive:

$$U(\mathbf{r},\omega) = u(\mathbf{r},\omega) \exp[-ik\langle n\rangle z] \quad (6)$$

For a narrow band receiver $\Delta\omega$ centered at ω_o we can expand the mean phase $k\langle n\rangle z$

about ω_o

$$k\langle n\rangle z = \phi_o + (\omega-\omega_o)\langle T_g\rangle + \cdots, \tag{7}$$

where the group delay $\langle T_g\rangle$ is:

$$\langle T_g\rangle = \frac{\partial}{\partial\omega}\Big[k\langle n\rangle z\Big]_{\omega_o} = \frac{z}{c} + \int_0^z \langle N_e\rangle dz \, \frac{r_e\, 2\pi c}{\omega_o^2} \tag{8}$$

Here the "light travel time" (first term) is, say, 1000 years and the dispersive delay (second term) is, say, 2 sec for a nearby pulsar at 1 GHz. The latter can be measured with great precision for each pulsar. It depends on the integrated electron column density or "dispersion measure" and can be used to estimate the distance z, since the average electron density is approximately known.

Consider a pulsar measurement in which the pulse intensities are accumulated, according to the (known) period, to form an average pulse profile. In practice the result is the convolution of the emitted profile and several broadening effects. After removing the instrumental resolution, the dispersion of arrival times across the receiver bandwidth and the intrinsic profile, one can estimate the effect of the interstellar medium as $J(t,\mathbf{r})$, where $\mathbf{r}$ is the antenna position. $J(t,\mathbf{r})$ represents the two most important propagation effects: the wander in arrival time and the broadening of the pulse profile due to interstellar scattering. A true ensemble average $\langle J(t)\rangle$ will be wider than a typical sample $J(t,\mathbf{r}_1)$, and it can be written as a convolution (*):

$$\langle J(t)\rangle = \exp[-(t/\sigma_g)^2] * J_d(t) \tag{9}$$

Here σ_g is the rms wander in the group delay T_g. J_d is the diffractively broadened profile, of width τ_d (at 1/e). This broadening is due to the extra travel time for waves that are scattered into an angular spectrum of radius θ_0. This idea gives the approximate relation:

$$\omega_o \tau_d \approx \omega_o \frac{z\theta_0^2}{2c} = \frac{z}{2k}(k\theta_0)^2 = 0.5W \tag{10}$$

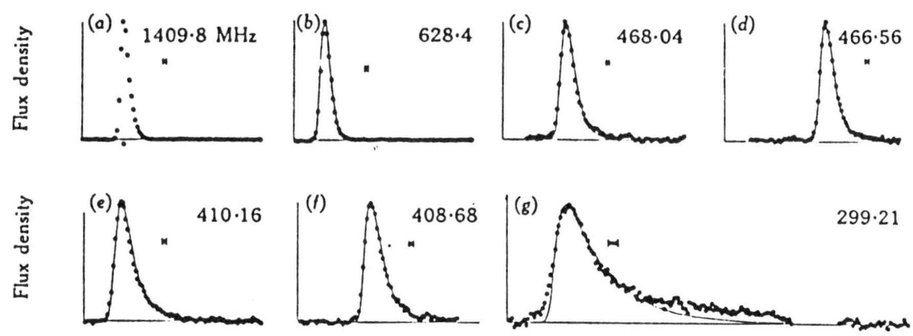

Figure 2. The average pulse shape J(t) of pulsar PSR0833-45 at the 7 frequencies indicated in MHz. The intrinsic shape is thought to be nearly independent of frequency, since the asymmetrical broadening, which increases rapidly with decreasing frequency accurately follows the scaling law expected for τ_d. The time scale, which is the same for each, is 89 msec across the horizontal axis of panel (g).

Figure 2 shows pulse profiles for the Vela pulsar (PSR0833-45) at 7 frequencies[15]. At 1410 MHz the profile reflects the intrinsic pulse shape, which changes only slowly with frequency. As the frequency is decreased, there is very pronounced broadening of the

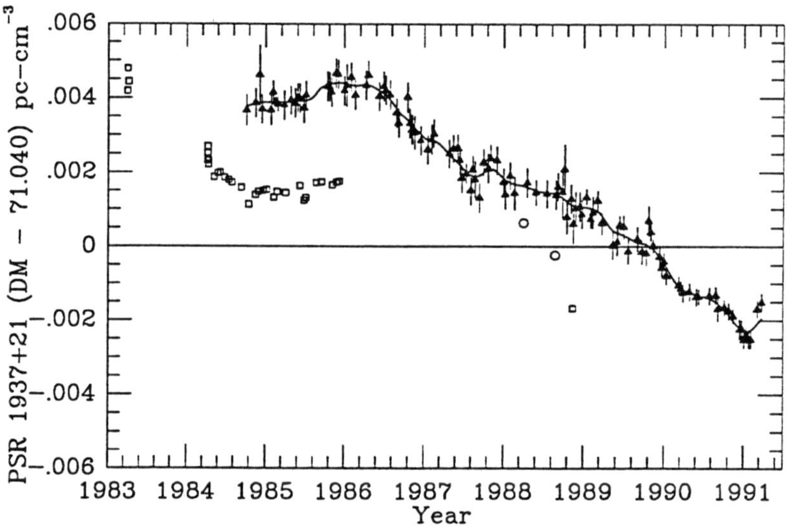

Figure 3 . Changes in the integrated electron column density (DM) from 1983-1991 for pulsar PSR1937+21. Triangles come from differential arrival time measurements between 1.41 and 2.38 GHz, made by Taylor and Ryba using the Arecibo telescope [18]. Open squares are from Cordes et al.[17] between 0.43 and 1.41 GHz. The extraordinary precision is indicated by noting that 0.0006 DM units correspond to an arrival time difference of 1 microsecond.

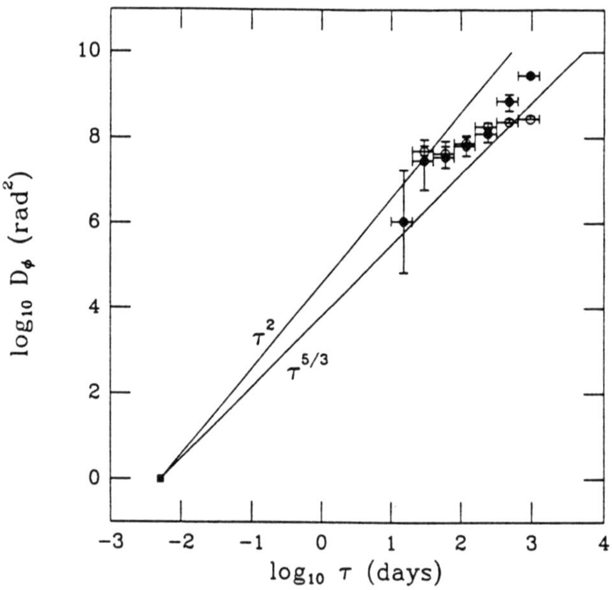

Figure 4 . The structure function D_ϕ estimated for the line of sight toward PSR1937+21 at 1.41 GHz. The filled circles are from 0.43 to 1.41 GHz, and the open circles from a 3-year subset of the 1.41 to 2.38 GHz data in Figure 3. The differences are not significant. The point at $D_\phi = 1$ is from diffractive scintillation of the same pulsar. There is a power law behavior over 5 decades; the time scale of 1000 days maps to a spatial scale of $4.3 10^{12}$ m (using a velocity of 50 Km/s).

profile due to ISS. Referring to eq (10) and noting that $\theta_0 \propto \omega^{-1-2/\alpha}$, we obtain $\tau_d \propto \omega^{-2-2/\alpha} \approx \omega^{-4.4}$ for a Kolmogorov medium. The observed frequency dependence is in good agreement with the exponent -4.4. There are now about 500 pulsars known and for many of them τ_d has been measured[16]. This can be used to estimate $\rho_0 = R_F/\sqrt{W} \approx R_F/\sqrt{2\omega_o\tau_d}$ for each object. The results of such measurements are addressed below.

Turning now to arrival time wander, an individual realization $J(t,\mathbf{r}_1)$ is approximately $J_d(t-T_g)$, ie its center is offset by the applicable group delay T_g due to the electron column density at the epoch of the observations. Here we ignore any statistical deviations of the pulse shape due to the finite averaging time. Variations of T_g due to changes in the electron column density are small and slow. Nevertheless in recent years they have been measured for a few pulsars. The most intriguing is PSR1937+214, the original millisecond pulsar (period = 1.6ms). This pulsar is an extraordinarily accurate clock, and has been studied extensively[17,18]. In order to measure the clock properties, the changes in group delay have to be removed. This has been done by concurrent measurements at 1408 and 2380 MHz by Joe Taylor at Princeton over 8 yrs. Figure 3 shows the results for electron column density, displayed as changes in the dispersion measure (in suitable astronomical units parsec cm^{-3}). There is a close link between the group delay and the "geometric phase" ϕ in the ISM, since they both depend on straight line integrals through the medium:

$$\delta T_g = \frac{\partial\phi}{\partial\omega}\Big|_{\omega_o} = -\frac{\phi}{\omega_o} \tag{11}$$

Thus by taking the structure function of the group delay, we can estimate $D_\phi(vt) = \omega_o^2 D_{T_g}$, where t is the time lag ranging up to, say, 4 years and v is the appropriate velocity for passage of the ISM through the line of sight. The resulting log/log plot is shown in Figure 4 for a 3-year subset of these data, scaled to a frequency of 1408 MHz[17]. This method probes scales from 10^{10}m to 10^{13}m. On the same plot we can place ρ_0 where the structure function equals one. The value used is for the same pulsar and frequency and is derived from measurement of the pulsar's dynamic spectrum, as discussed below. The result is a remarkable plot of the structure function over 5 orders of magnitude, showing a mean exponent between 5/3 and 2. The detailed analysis of the complete data has yet to be published, but should lead to useful constraints on the inner scale.

Diffractive Intensity Scintillations versus Time and Frequency

The left panel of Figure 5 is a dynamic spectrum of the pulsar PSR1937+214 at 430 MHz[17], showing fluctuations due to diffractive intensity scintillation, which decorrelate over a frequency range of about 10 KHz and over a time scale of about 100 sec. We write the intensity deviation as

$$\delta I(t,\omega) = |\,u(r{=}vt,\omega)\,|^2 - \langle I\rangle\,, \tag{12}$$

where u is the reduced scalar field and in practice <I> is a local average intensity. Such fluctuations have a modulation index of 100% and can be characterized by a correlation function versus frequency and time:

$$\gamma_{2I}(r{=}vt,\omega,\omega{+}\delta\omega) = \langle\,\delta I(t',\omega)\;\delta I(t'{+}t,\omega{+}\delta\omega)\,\rangle \tag{13}$$

For a true ensemble average in asymptotic strong scattering the statistics of u become Gaussian, and the fourth moment in eq (13) becomes the square of the second moment:

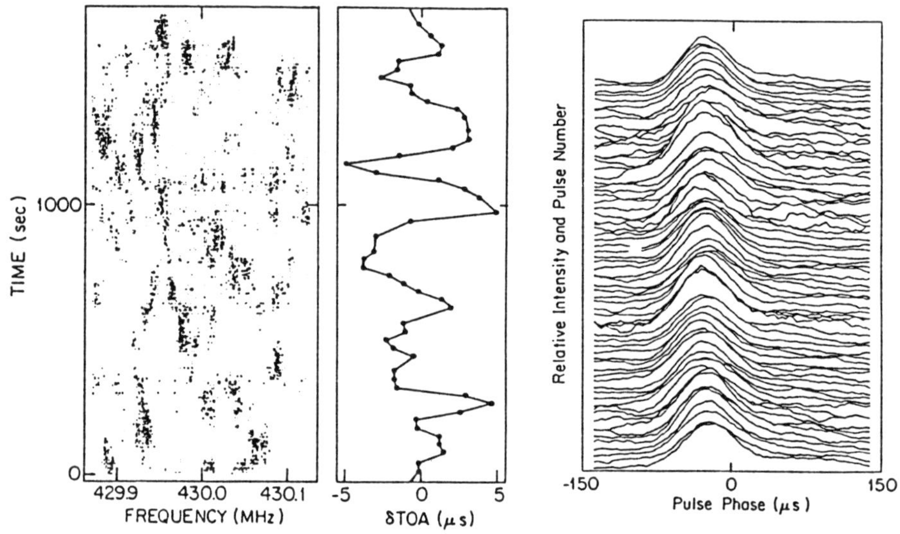

Figure 5 . Pulse signals from PSR1937+21 (period 1.6 msec) near 430 MHz recorded at Arecibo observatory. Left panel shows the pulse amplitude versus frequency evolving over time. Darker regions representing stronger pulses. The pulse power, averaged over 32 sec (ie over 20000 pulses), is displayed in grey scale as a single horizontal line covering 250 KHz centered at 430 MHz. Successive spectra are plotted for 1600 sec. The right hand panel shows the successive average pulse shapes for the same data. Each one is the convolution of the emitted pulse shape and the scattering response $J(t, \mathbf{r})$. The vertical axis gives the antenna position $\mathbf{r}$, as the pattern crosses the earth. The middle panel gives the fluctuations of the estimated arrival time T_g from each average.

$$\gamma_{2I}(r,\omega,\omega+\delta\omega) = |\gamma_2(r,z,\omega,\omega+\delta\omega)|^2 \tag{14}$$

γ_2 can then be factored into a refractive term and a diffractive term[19].

$$\gamma_2(r,z,\omega,\omega+\delta\omega) = \exp[-(\delta\omega\sigma_g/2)^2]\ \gamma_d(\delta\omega) \tag{15}$$

A useful analysis by Codona et al.[20] examines these approximations for a phase screen presents improved expressions, valid for intermediate levels of scattering.

In practice the average in eq (13) will be over many diffractive scintillation times, but less than one refractive time. Thus what is estimated approximates $\gamma_d(\delta\omega)$. This correlation function has a characteristic width $\Delta\omega_d$, at which $\gamma_d(\Delta\omega_d) = 0.5$. It is of interest here to note the connections with pulse broadening. Still in strong sattering, $\gamma_2(\delta\omega)$ is the Fourier transform of the ensemble average pulse shape $<J(t)>$, and the product in eq (15) corresponds to the convolution in eq (9). In addition, γ_d is the Fourier transform of the instantaneous pulse shape $J_d(t)$ [19]. An important consequence of the Fourier transform relation is the connection between the pulse broadening time τ_d and the diffractive decorrelation bandwidth $\Delta\omega_d$:

$$\Delta\omega_d\, \tau_d \approx 1 \tag{16}$$

The relation with the pulse shape is illustrated in the middle panel of Figure 5, in which each of the pulse spectra in the upper panel are presented by their corresponding pulse shapes. In these plots the intrinsic pulse profile is included and would have to be "deconvolved" to reveal the Fourier relation precisely.

There are now measurements of $\Delta\omega_d$ for very many pulsars and it has become a valuable diagnostic for ISS effects[16,21]. If the scattering medium were spatially homogeneous, the scaling laws would be:

$$\Delta\omega_d \;\propto\; z^{-1-2/\alpha}\,\omega_o^{-2-4/\alpha} \;=\; z^{-2.2}\,\omega^{4.4} \tag{17}$$

with the latter form applicable to a Kolmogorov medium.

Galactic Distribution of Density Turbulence

I use the phrase density "turbulence" to describe the local strength of the inhomogeneities in electron density, even though they do not necessarily imply the presence of an active turbulent process. Radio astronomers have adopted the following model for the inertial range of the electron density wavenumber spectrum:

$$P_{3N}(\kappa) \;=\; C_N^2\,\kappa^{-\beta} \quad \text{where } L_o^{-1} \ll \kappa \ll l_o^{-1} \tag{18}$$

This convention[22] is to be contrasted with that in the optical literature where C_n^2 is defined as the multiplier in the structure function of refractive index. From the measured value of $\Delta\omega_d$ for a pulsar, the line of sight integral (scattering measure) is obtained:

$$SM \;=\; \int_0^z C_N^2\,dz' \quad ; \quad \text{we also know:} \quad DM=\int_0^z \langle N_e\rangle dz' \tag{19}$$

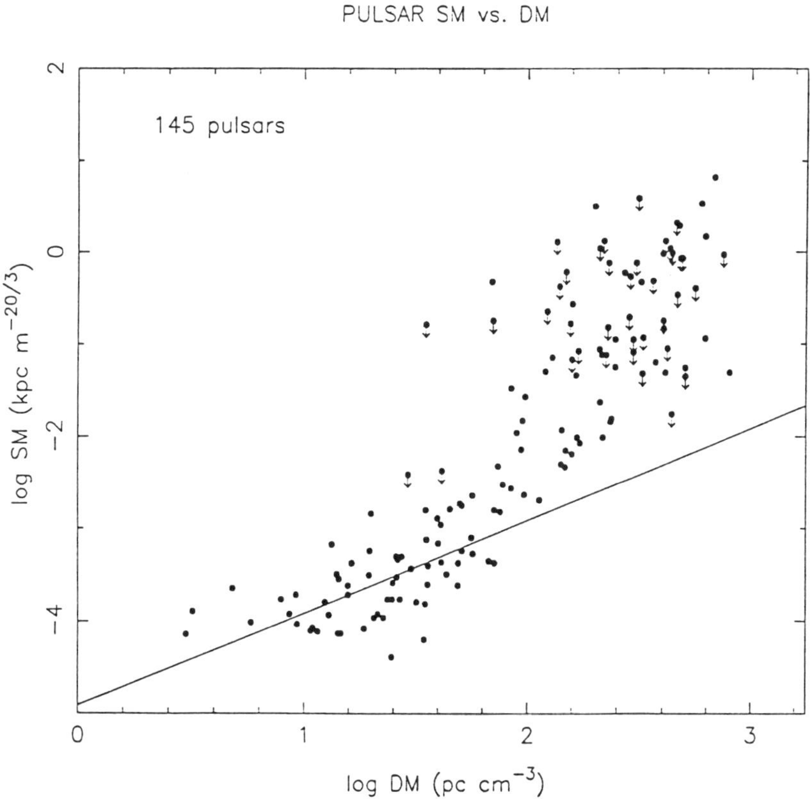

Figure 6. Each point represents a particular pulsar in a scatter plot of log SM (kpc $m^{20/3}$) versus log DM (pc cm^{-3}). For a uniform distribution of turbulence we expect a straight line of unit slope.

If C_n^2 and $<N_e>$ were the same throughout the galaxy SM should be linearly related to DM. A recent analysis of 145 pulsars[23] has been completed and is shown in Figure 6. The straight line, fitted from 0 to 80 pc cm^{-3}, supports the approximate constancy of the density parameters in the local ISM. If the density were uniform at 0.025 cm^{-3} this range would correspond to distances out to 3 Kpc from the earth over which the strength of turbulence is approximately constant. The plot shows at greater distances C_n^2 increases more quickly than $<N_e>$. In some cases SM lies a thousand times above DM. Each pulsar probes a different line of sight across the Galaxy, and Cordes et al.[23] derive a three component model for the distribution of these parameters with distance from the center of the Galaxy. Their model for the mean density agrees well with their model for density turbulence. This result provides important evidence that the electrons that cause the dispersion also cause the scattering. One component of their model is a sparse distribution of very strongly enhanced density turbulence, which has proved to be hard to identify astrophysically. This work illustrates some of the ways in which radio scintillation studies have helped probe the Galaxy, but it strays well beyond the realm of this meeting.

Refractive Intensity Scintillations

For intensity scintillations, the two dimensional wavenumber spectrum of the intensity fluctuations - $\Phi_{\delta I}(\kappa)$ is used to describe the spatial statistics of the scintillations. Under asymptotic strong scattering, $\Phi_{\delta I}(\kappa)$ can be usefully expanded in separate approximations for low and high wavenumbers. These lead to fluctuations on a large refractive scale (R_S) and small diffractive scale (ρ_0), respectively. For strong scattering of a plane wave by a screen the low wavenumber expansion is:

$$\Phi_{\delta I}(\kappa) \;=\; \frac{\kappa^4 z^2}{k^2}\; \Phi_\phi(\kappa)\; \exp[-D_\phi(\frac{\kappa z}{k})] \qquad (20)$$

Here $\Phi_\phi(\kappa)$ is the power spectrum for phase emerging from the screen and, in general, the first term is the $\sin^2$ "fresnel filter". However in very strong scattering, the exponential term sets an upper wavenumber cut-off at $1/R_S$, which restricts κ to where the sine can be approximated by its argument. Extensions to spherical waves in an extended scattering medium, which are readily obtained, do not change the basic scale. An integral over the wavenumber spectrum then gives the refractive variance m_r^2, which is expressed as a function of the strength of scattering for a Kolmogorov model in eq (21).

The refractive contribution to interstellar scintillation was not recognised, until slow fluctuations of pulsars were shown to be distance-dependent[24] and the fluctuation time (many days to years) was shown to be correctly given by R_S/v[25]. Since refractive variations happen so slowly, only a few time series have been accumulated. Figure 7 shows 430 days of intensities from a pulsar PSR1911-04 recorded at 610 MHz by Kaspi and Stinebring[26]. Each point is an average over very many diffractive scales and shows the refractive scintillations clearly. The upper plot is the structure function of the scintillations, from which a time scale of 10-30 days and an rms modulation m_r greater than 0.2 can be inferred. For the same pulsar the diffractive scintillation has also been observed and $\Delta\omega_d$ measured, from which W can be inferred. Thus we can test the observations of m_r against predictions, from various models for the spectrum of the medium. It is convenient to display the results versus W, rather than distance to the object, in order to suppress effects due to non-uniformity in the lines of sight for different objects[27,28]. Figure 8 is thus a logarithmic plot of m_r against W. For a Kolmogorov spectrum we expect

$$m_r^2 \propto W^{\alpha-2} = W^{-1/3} \quad \text{for } \rho_0 > l_o \qquad (21)$$

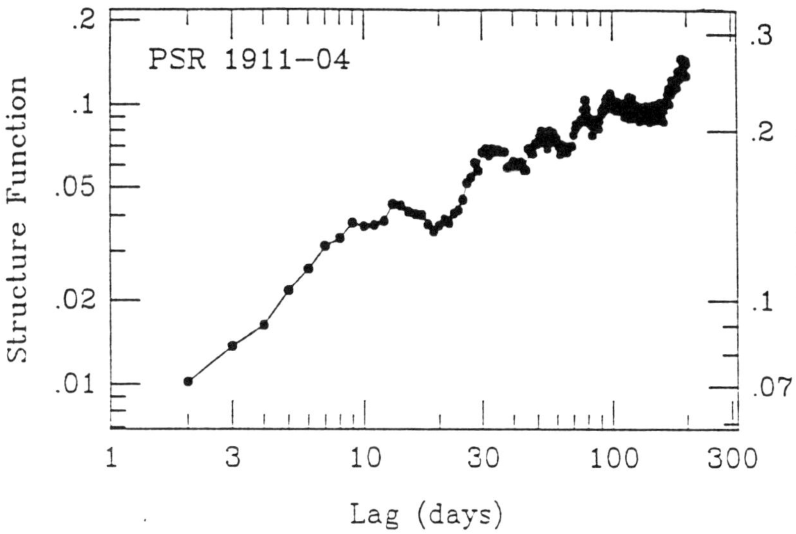

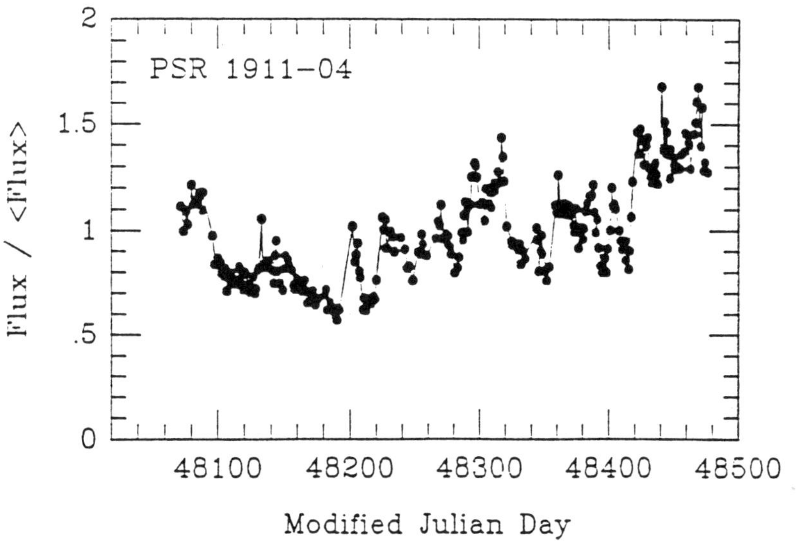

Figure 7. The Bottom panel shows the time series of refractive intensity variations for the pulsar PSR1911-04 at 610 MHz[26]. The top panel shows the structure function of these variations, revealing a time scale of about 20 days, to be compared with a diffractive time scale of minutes.

$$m_r^2 \propto (l_o/R_S)^{-1/3} = (l_o/R_F)^{-1/3}\, W^{-1/6} \quad \text{for } \rho_0 < l_o$$

These models are illustrated in the plot, in which the solid line has no inner scale and the dashed lines show the effects of increasing inner scale. The majority of the data lie above the Kolmogorov model and favor a substantial inner scale; however, the data from many of the more distant pulsars (plotted as stars), suggest no inner scale. On the same graph I have plotted the approximate form of an improved asymptotic theory for m_r due to Wang and Dashen [29] A comparison has also been made of the measured and predicted time scales, which reveals measured values too small by factors of 2 or more. In summary, there is a first order agreement between the observations and a Kolmogorov spectrum, but in second order there are discrepancies that are still under review, and may indicate to a modified spectrum.

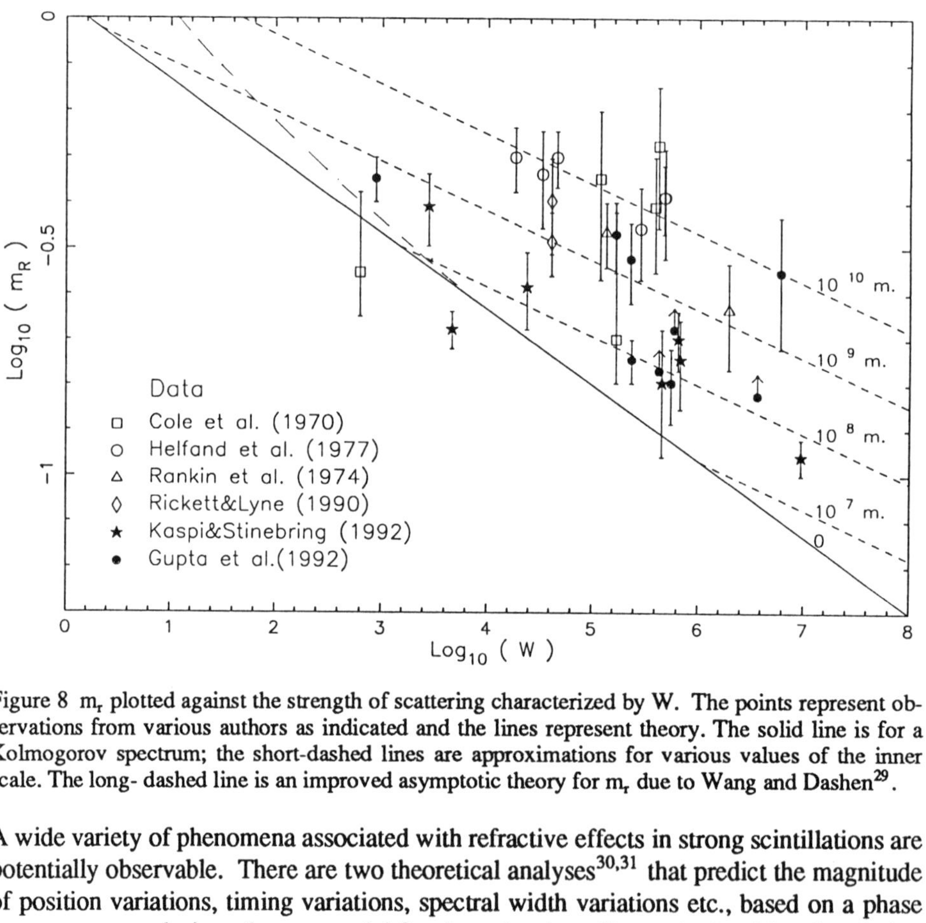

Figure 8 m_r plotted against the strength of scattering characterized by W. The points represent observations from various authors as indicated and the lines represent theory. The solid line is for a Kolmogorov spectrum; the short-dashed lines are approximations for various values of the inner scale. The long- dashed line is an improved asymptotic theory for m_r due to Wang and Dashen[29].

A wide variety of phenomena associated with refractive effects in strong scintillations are potentially observable. There are two theoretical analyses[30,31] that predict the magnitude of position variations, timing variations, spectral width variations etc., based on a phase screen wave analysis and a ray model for the refractive effects. An intuitive description of the refractive variations starts with a phase screen at distance z from an observer. The small scale variations in the screen broaden the waves into an angular spectrum. At a given location the mean angle of arrival of this spectrum $(\theta_{rx}, \theta_{ry})$ is determined by the average phase gradient over the scattering disk R_S, that lies above the observer, with an rms value given by eq (4) for a power law spectrum. The variations in the screen over scale R_S cause the refraction angle to vary about zero, on a spatial scale R_S. This refractive image wander also causes a lateral displacement of the diffractive pattern by a vector

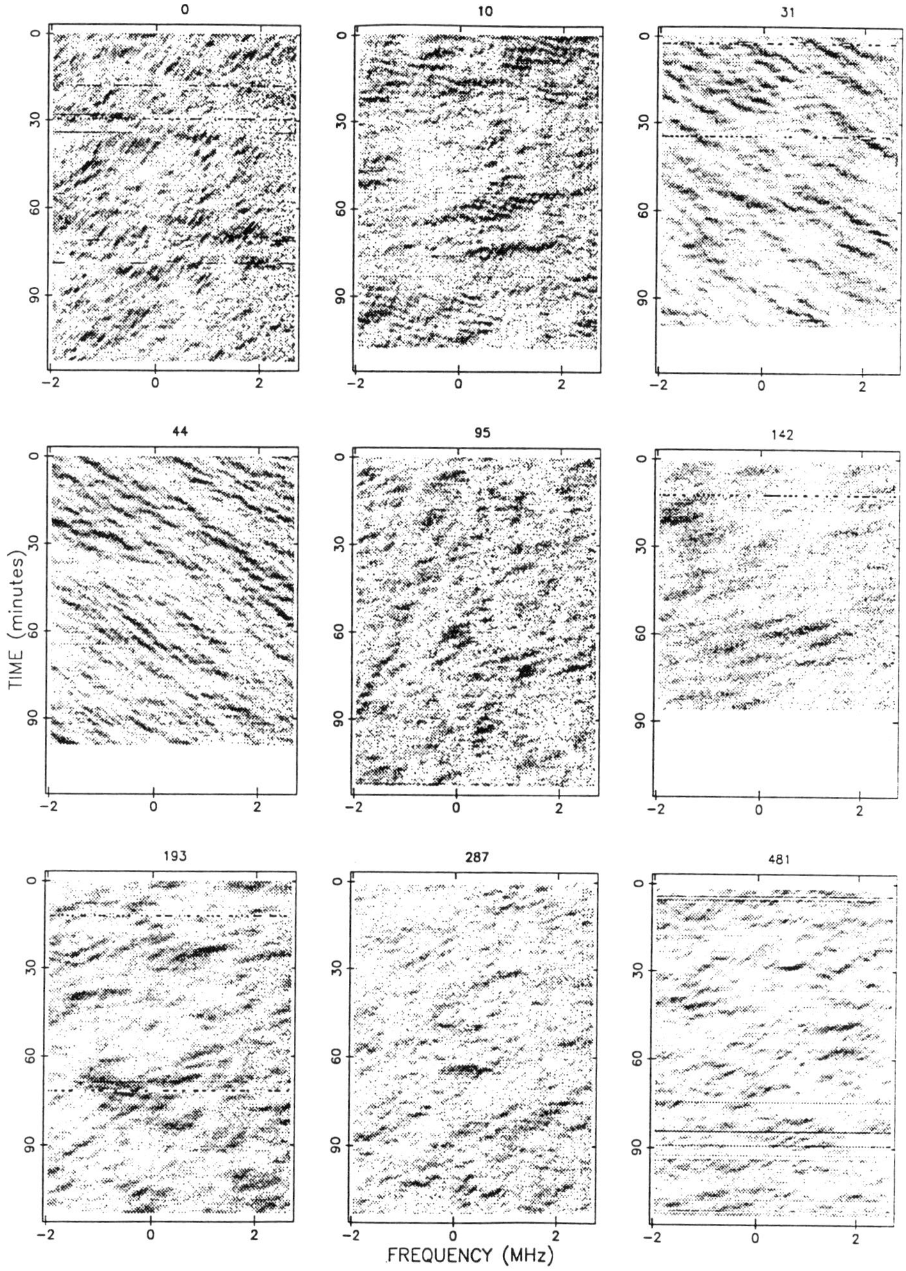

Figure 9. The dynamic spectra of diffractive scintillations from pulsar PSR0823+26 at 9 epochs spanning 481 days. The density of the shading is proportional to the average pulse intensity as in Figure 5. The spectrum covers 5 MHz centered at 408 MHz. The horizontal scale is in MHz and the vertical scale is in minutes.

$(X,Y) = (z\theta_{rx}, z\theta_{ry})$. If we think of a local peak in the diffractive interference pattern, it is displaced laterally by (X,Y) from its position in the absence of refraction. Since the refractive shifts vary smoothly on a scale of R_S, the diffractive peaks are not only shifted but also crowded together or spread apart by large scale curvature of the phase. This is essentially a ray description of the steering of the diffraction pattern by the large scales in the screen. In an extended medium the term "ray bundle" is sometimes used. Where the ray bundles are converging, there is a local maximum of the refractive scintillations and vice versa where they are diverging. A condition for this description to be even approximately correct is that the associated intensity variations should be small amplitude, or $X,Y << R_S$, which will be satisfied in asymptotic strong scattering.

Refractive Drifts in Dynamic Spectra

In Figure 9 the dynamic spectra of a pulsar due to diffractive interstellar scintillation show persistently sloping features[32]. These are another manifestation of irregular refraction. Since the medium is dispersive, the angles of refraction are strongly dependent on frequency f: $\theta_{rx} \propto f^{-2}$ This causes frequency-dependent spatial offsets, which translate to frequency-dependent time offsets $X \propto f^{-2}$ for an individual diffractive peak and correspond to sloping features in Figure 9. Such sloping features are seen in a substantial fraction of the dynamic spectra that are recorded (eg Smith and Wright[33]). They change magnitude and sign over a time scale that has not been well measured observationally. The nine panels for Figure 9 are sampled for 1-2 hours each from pulsar PSR0823+26 near 408 MHz over a total span of 481 days, with separations from 10-190 days. From the foregoing approximate description, one would expect the time scale for the slope variations to be R_S/v. (For more detailed theoretical discussion see references 30 and 31). Clear reversals of the slopes are apparent between panels 1 and 3 of Figure 9 with oppositely sloping features. The second panel shows patches of periodic patterns and superimposed slopes of opposite signs. Hewish et al.[34] explained the periodicities as interference between two dominant angles of arrival. This has been elaborated by Cordes and Wolszczan[35,36].

The notion of interference between refracted ray bundles begins to break the distinction between refractive and diffractive effects. In trying to understand these phenomena, I have been puzzled by the inferred prevalence of only two dominant directions of arrival, for such long paths through the scattering medium. The most promising analysis that I have seen is by Goodman et al.[37], who analysed this question for a medium with an inner scale. They noted that in strong scattering the effect of an inner scale is to produce occasional caustics, which will tend to dominate the signal arriving at a given location. On occasion there could be two such caustics, corresponding to the "dominant directions of arrival" above. From a qualitative point of view I expect such effects to be important near the peak of the scintillation index curve, ie strong focussing from an inner scale sized irregularity. I estimate that this corresponds to inner scales that are on the order of 0.1-$1R_S$. However, inner scale estimates in the ISM have been typically smaller than this by more than three orders of magnitude, so that one might expect an extremely large number of caustics simultaneously. A complete theory has yet to be proposed that answers the question of under what circumstances would one expect only a few caustics and what fraction of the time should such effects be visible. The awkward fact is that we are talking about the variability of a fourth moment quantity, needing an eighth moment description in principle.

Summary and Conclusions

A wide variety of radio observations have been used to study scattering and scintillation caused by plasma inhomogeneities in the interstellar medium. The strength of scattering is extremely strong, involving diffractive and refractive scales that differ by 2 to 6 decades. A notable feature of this regime is that refractive effects are still important and can be readily separated from diffractive scintillation. There is substantial first order agreement between the observations and a Kolmogorov spectrum for the electron density over about 6 decades in wavenumber. The strength of the density turbulence, however, includes a highly non-uniform component, having local enhancements by up to 5 orders of magnitude. These are particularly concentrated toward the inner part of the Galaxy.

When examined in finer detail the observations show several puzzling features, which are topics of active research.

1) The structure functions estimated from VLBI studies indicate an inner scale on the order of 10^5m. The data-processing involved an image centering algorithm, whose effect has not been properly studied.

2) Detailed measurements for the amplitude and time scale of refractive scintillations show larger amplitudes and shorter time scales than are predicted by the asymptotic Kolmogorov model. The amplitude can be explained by an inner scale near the Fresnel scale ($\sim 10^8$m), in disagreement with the estimate above. The time scale discrepancy points to an improper model for the distance and velocity of the inhomogeneities.

3) The dynamic spectra of diffractive scintillations, when sampled over times short compared to the refractive scintillation times, is not well understood theoretically. This regime is the same as that for VLBI observations in that estimates of a fourth moment quantity are made on a time scale too short for an ensemble average. Numerical simulations of moderately strong scattering exhibit many of the observed effects, but the simulations are not yet feasible for the extremely strong scattering of ISS, nor have scaling laws been formulated to map the results to this regime.

4) Occasional discrete "propagation events" have been identified for particular radio sources (Fiedler et al[38]). These could be isolated examples of determinstic structures crossing a line of sight or they could be an extreme example of a continuum of scattering phenomena, which might be related to the observed departures from a Kolmogorov spectrum.

References

1 Gopal-Krishna, "An intergalctic origin for the low-frequency flux variations of extrta-galactic radio sources," *Current Science*, **60**, 117-120 (1991)

2 S.R. Spangler & C.R. Gwinn,"Evidence for an inner scale to the density turbulence in the interstellar medium," *Astrophys. J.*, **353**, L29-32 (1990)

3 W. A. Coles, , R.G. Frehlich, B. J. Rickett, and J. L. Codona, "Refractive Scintillation in the Interstellar Medium," *Astrophys. J.*, **315** 666-674 (1987)

4 J. M. Cordes, B. J. Rickett and D. C. Backer, (Editors), "*Radio wave Scattering in the Interstellar Medium*", UC San Diego, Jan. 1988, AIP Conference Proceedings No. 174 (1988)

5 B. J. Rickett, "Radio propagation through the turbulent interstellar plasma," *Ann. Rev. Astron. & Astrophys.*, **28**, 561-605 (1990)

6 B. Dennison, M. Thomas, R. S. Booth, R. L. Brown, J. J. Broderick, and J. J. Condon, "A VLBI survey of interstellar broadening of extragalactic radio sources seen at low galactic latitudes," *Astron. Astrophys.*, **135**, 199-212 (1984)

7 S. R. Spangler, R. L. Mutel, J. M. Benson and J. M. Cordes, "Interstellar scattering of compact radio sources near supernova remnants," *Astrophys. J.*, **301**, 312-319 (1986)

8 P. N. Wilkinson, R. E. Spencer and R. F. Nelson, "Interstellar scattering towards Cyg X-3," *The Impact of VLBI on Astrophysics and Geophysics*, Proc. IAU Symp. 129 305-306 (1988)

9 R. L. Mutel and J-F. Lestrade, "VLBI measurements of Interstellar turbulence spectra," *AIP Conference Proceedings No. 174*, 122-128 (1988)

10 M. J. Reid and J. M. Moran, "Masers" *Ann. Rev. Astron. Astrophys.* **19**, 231-276 (1981)

11 P. J. Diamond, A. Martinson, R. S. Booth and A. Winnberg, "Preliminary Results from a 7-station VLBI survey of OH Masers in the Galactic plane," *AIP Conference Proceedings No. 174*, 195-199 (1988)

12 J. Goodman and R. Narayan, "The shape of a scatter-broadened image II. Interferometric Visibilities," *Mon. Not. Roy. Astro. Soc.*, **238**, 995-1028 (1989)

13 C. R. Gwinn, J. M. Moran, M. J. Reid and M. H. Schneps, "Limits on refractive interstellar scintillation towards Sagittarius B2" *Astrophys. J.*, **330**, 817-827 (1988)

14 A. L. Fey and R. L. Mutel, 608 MHz VLBI observations of the compact double radio source 2050+364: constraints on interstellar scattering" *Astrophys. J.*, in press (1992)

15 M. M. Komesaroff, P. A. Hamilton and J. G. Ables, "Linear polarization and spectrum of PSR0833-45 and the effects of scattering," *Austr. J. Phys.*, **25**, 759-777 (1972)

16 J. M. Cordes, J. M. Weisberg and V. Boriakoff, "Small scale density turbulence in the interstellar medium" *Astrophys. J.*, **288**, 221-247 (1985)

17 J. M. Cordes, A. Wolszczan, R. J. Dewey, M. Blaskiewisicz and D. R. Stinebring, "Timing and scintillations of the millisecond pulsar 1937+214" *Astrophys. J.*, **349**, 245-261 (1990)

18 M. Ryba, "*High precision timing of millisecond pulsars*," Ph.D. thesis, Princeton University, (1991)

19 L. C. Lee and J. R. Jokipii, "Strong scintillations in astrophysics. II A theory of temporal broadening of pulses," *Astrophys. J.*, **201**, 532-543 (1975)

20 J. L. Codona, D. B. Creamer, S. M. Flatte, R. G. Frehlich and F. S. Henyey, "Two-frequency intensity cross-spectrum" *Radio Science*, **21**, 805-814 (1986)

21 J. M. Cordes, "Space velocities of radio pulsars from interstellar scintillations," *Astrophys. J.*, **311**, 183-196 (1986)

22 B. J. Rickett, "Interstellar scattering and scintillation of radio waves," *Ann. Rev. Astron. & Astrophys.*, **15**, 479-504 (1977)

23 J. M. Cordes, J. M. Weisberg, D. A. Frail, S. R. Spangler and M. Ryan, "The galactic distribution of free electrons," *Nature*, **354**, 121-124 (1991)

24 W. Sieber, "Causal relationship between pulsar long-term intensity variations and the interstellar medium," *Astron. and Astrophys.*, **113**, 311-313 (1982)

25 B. J. Rickett, W. A. Coles and G. Bourgois, "Slow scintillation in the interstellar medium," *Astron. and Astrophys.*, **134**, 390-395 (1984)

26 V. M. Kaspi and D. R. Stinebring, "Long term pulsar flux monitoring and refractive interstellar scintillation," *Astrophys. J. Lett.* in press, (1992)

27 B. J. Rickett and A. G. Lyne, "Intensity variations of the Crab pulsar and refractive interstellar scintillation," *Mon. Not. Roy. Astro. Soc.*, **244**, 68-75 (1990)

28 Y. Gupta, B. J. Rickett and W. A. Coles, "Refractive interstellar scintillation of pulsar intensities," *Astrophys. J.*, in press January 1993

29 G-Y. Wang "*Statistics for waves through random media*", Ph.D. Thesis, University of California, San Diego (1991)

30 J. M. Cordes, A. Pidwerbetsky and R. V. E. Lovelace, "Refractive and diffractive scattering in the interstellar medium," *Astrophys. J.*, **310**, 737-767 (1986)

31 R. W. Romani, R. Narayan and R. D. Blandford, "Refractive effects in pulsar scintillation," *Mon. Not. Roy. Astro. Soc.*, **220**, 19-49 (1989)

32 Y. Gupta, B. J. Rickett and A. G. Lyne, "Refractive scintillation in pulsar dynamic spectra" in preparation (1992)

33 F. G. Smith and W. C. Wright, "Frequency drift in pulsar scintillation," *Mon. Not. Roy. Astro. Soc.*, **214**, 97-107 (1985)

34 A. Hewish, A. Wolszczan and D. A. Graham, "Quasi-periodic scintillation patterns of the pulsars PSR1133+16 and PSR1642-03," *Mon. Not. Roy. Astro. Soc.*, **21**, 167-179 (1985)

35 J. M. Cordes and A. Wolszczan, "Multiple imaging of pulsars by the interstellar medium," *Astrophys. J. Lett.* **307**, L27-L31 (1986)

36 A. Wolszczan amd J. M. Cordes, "Interstellar interferometry of the pulsar 1237+25," *Astrophys. J. Lett.* **320**, L35-L39 (1987)

37 J. J. Goodman, R. W. Romani, R. D. Blandford and R. Narayan, "The effects of caustics on scintillating radio sources," *Mon. Not. Roy. Astro. Soc.*, **229**, 73-102 (1987)

38 Fiedler, R. L. Dennison, B. Johnston, K. J., Hewish, A. "Extreme scattering events caused by compact structures in the inerstellar medium," *Nature*, **326**, 675-678 (1987)

Section 3

Reconstruction of Source and Medium Parameters Using Scintillation

Astronomical Speckle Masking, Speckle Spectroscopy and Long-Baseline Interferometry

K.-H. Hofmann and G. Weigelt

Max-Planck-Institut für Radioastronomie
Auf dem Hügel 69, D-5300 Bonn 1, Germany

1. Introduction

The atmosphere of the earth restricts the angular resolution of large, ground-based telescopes to about 0.5 arcsec. Fortunately, it is possible to overcome atmospheric image degradation by various interferometric methods. The speckle interferometry method[1] can reconstruct the diffraction-limited autocorrelation. True images with diffraction-limited resolution (e.g., 0.02arcsec) can be reconstructed by the Knox-Thompson method[2], by the speckle masking method[3–5] and by the nonredundant mask technique[6–8]. More references for image reconstruction methods can be found in the review articles on interferometric imaging by Dainty[9], Roddier[10], and Weigelt[11]. The nonredundant mask technique and speckle masking can measure the so-called "closure phases" which are required for optical long-baseline interferometry. The great advantage of optical long-baseline interferometry with three or more telescopes is the fact that it can yield images and spectra with fantastic angular resolution. For example, with a baseline of 150m a resolution of 0.001arcsec can be obtained. Speckle spectroscopy methods have the advantage that they yield the spectrum of each resolution element of the object in addition to the high-resolution image. The raw data for the nonredundant mask technique are Michelson interferograms. Image reconstruction from the Michelson interferograms can be performed by various radio phase closure methods. The raw data for the Knox-Thompson and for the speckle masking method are speckle interferograms. The intensity distribution $i_n(x)$ of the n-th speckle interferogram can be described by

$$i_n(x) = o(x) \otimes p_n(x), (n = 1, 2, 3, \ldots, N \sim 10^3 - 10^5), \tag{1}$$

where $o(x)$ is the object, $\otimes$ denotes convolution, $p_n(x)$ is the psf of the atmosphere/ telescope, x is a 2-D space vector, and N is the total number of speckle interferograms. In the text below, the index n of the random functions i_n and p_n and of their Fourier transforms will be omitted.

2. Speckle Masking Bispectrum Processing

Speckle masking consists of the following image processing steps 1 to 4:
Image processing step 1: Calculation of the *ensemble average triple correlation*

$$\langle i^{(3)}(x,y)\rangle = \langle \int i(x')\, i(x'+x)\, i(x'+y)\, dx' \rangle, \tag{2}$$

or calculation of the *ensemble average bispectrum*

$$\langle I^{(3)}(u,v)\rangle = \langle I(u)\, I(v)\, I^*(u+v)\rangle \tag{3}$$

of many interferograms $i(x)$. u and v are 2-D coordinate vectors. Since x, y, u, and v are 2-D vectors, the triple correlation $i^{(3)}(x, y)$ and the bispectrum $I^{(3)}(u, v)$ are *4-dimensional.*

Image processing step 2: Compensation for additive noise bias terms in the average bispectrum [12,13]. Bias terms in the average bispectrum are caused, for example, by photon noise in the speckle interferograms of faint astronomical objects.

Image processing step 3: Compensation of the speckle masking transfer function (SMTF). With $I = OP$ we get for the average bispectrum $\langle I^{(3)}(u,v)\rangle$:

$$\langle I^{(3)}(u, v)\rangle = O(u)O(v)O^*(u + v)\,\langle P(u)P(v)P^*(u + v)\rangle. \tag{4}$$

$\langle P(u)P(v)P^*(u + v)\rangle$ is called the SMTF. We can divide by the SMTF and obtain for the bispectrum $O^{(3)}(u, v)$ of the object $o(x)$,

$$O^{(3)}(u, v) \equiv O(u)O(v)O^*(u + v) = \langle I(u)I(v)I^*(u + v)\rangle / \langle P(u)P(v)P^*(u + v)\rangle. \tag{5}$$

Image processing step 4: Derivation of modulus and phase of $O(u)$ from $O^{(3)}(u, v)$. We denote the phase of the Fourier transform of the object by $\phi(u)$ and the phase of the bispectrum of the object by $\beta(u, v)$, i.e., $O(u) = |O(u)|exp[i\phi(u)], O^{(3)}(u, v) = |O^{(3)}(u, v)|exp[i\beta(u, v)]$. Inserting in Eq. 5 yields

$$|O^{(3)}(u, v)|exp[i\beta(u, v)] = |O(u)|exp[i\phi(u)]\ |O(v)|exp[i\phi(v)]\ |O(u + v)|exp[-i\phi(u + v)],$$

$$\Rightarrow exp[i\beta(u, v)] = exp[i\phi(u)]\ exp[i\phi(v)]\ exp[-i\phi(u + v)]$$

$$\Rightarrow \phi(u + v) = \phi(u) + \phi(v) - \beta(u, v). \tag{6}$$

This equation is a recursive equation for calculating the phase ϕ at coordinate $w \equiv u + v$ if ϕ is known at coordinates u and v. The above equation is called a "closure phase relation". Since speckle masking can measure closure phases it can be applied to optical long-baseline interferometers. In this case the Knox-Thompson method cannot be applied.

Fig. 1 shows a speckle masking observation of R 136, the central object in the 30 Doradus nebula[13]. Fig. 1a is one of the 6000 speckle interferograms of R 136 taken with the 1.5-m ESO/Danish telescope. Fig. 1b is a long-exposure image reconstructed from the re-centered speckle interferograms (re-centered to overcome image motion). Fig. 1c is the diffraction-limited speckle masking image of R 136 reconstructed from the 6000 speckle interferograms. This image shows 46 stars in the whole 4.9 arcsec field. The magnitudes of the stars are in the range of 12.6 to 16.4. The separation of the closest double stars is 0.07 arcsec.

Fig. 2 shows a speckle masking observation of the nearby Seyfert II galaxy NGC 1386[14]. Fig. 2a is one of the 7000 speckle interferograms recorded at the 2.2-m ESO/MPG telescope (H-α filter). Each speckle interferogram of NGC 1386 consists of about 145 photoevents per frame (the black dots are the single photoevents). Fig. 2b is a long-exposure image reconstructed from the re-centered speckle interferograms. Fig. 2c is the high-resolution speckle masking image of NGC 1386 reconstructed from the 7000 speckle interferograms.

3. Iterative Image Reconstruction from the Bispectrum

From the object bispectrum $O^{(3)}(u,v)$ obtained by speckle masking (Eq. 5) a diffraction-limited image can be derived by (a) the recursive method (Eq. 6) or (b) an iterative least-squares method. The iterative building block method[15–17] discussed below is an extension of the gridfit method[18] used in radio interferometry. It searches for that high-resolution image which has the best agreement with the measured object bispectrum. In many cases the iterative method yields higher SNR than the recursive method.

The image of iteration step k+1 is produced by adding a (k+1)th diffraction-limited point spread function (psf) to the image of step k. The psf (i.e. one "intensity building block") is added at such a position x' that the distance between the bispectrum of the (k+1)th image and the bispectrum measured is as small as possible. To this end we have to determine the position x'_{min} of the minimum in the distance function

$$d_{k+1}(x') = \int |O^{(3)}(u,v) - O^{(3)}_{k+1}(u,v)|^2 \, du \, dv, \qquad (7)$$

where $O^{(3)}_{k+1}(u,v)$ denotes the bispectrum of the (k+1)th image $o_{k+1}(x) = o_k(x) + h(x-x')$, and x' is the position of the (k+1)th psf $h(x-x')$. For increasing the speed of calculating the distance function an approximated expression

$$\begin{aligned} d_{k+1}(x') \;\; &\approx \;\; \text{const} + \frac{6}{k} \int [O^{(3)}_k(-u,-v) - O^{(3)}(-u,-v)] \\ &\qquad \times O_k(u) O_k(v) \, exp\{2\pi i (u+v) x'\} \, du \, dv, \end{aligned} \qquad (8)$$

of $d_{k+1}(x')$ is used. The approximation is based on the fact that for a large number k of iteration steps (k≫1) the changes introduced by a further psf are small. After a large number of such iteration steps (~1000 to 5000) the high-resolution image of the object is obtained.

Fig. 3 shows an application of the combination speckle masking and building block method to computer-generated speckle data[16]. Fig. 3a is the artificial object, a star cluster consisting of 5 stars. Fig. 3b is the long-exposure image of the 10 000 simulated speckle interferograms. Fig. 3c shows the diffraction-limited image reconstructed from the average bispectrum (10 000 interferograms with ~100 photoevents/frame) by the building block method. Fig. 3d is the diffraction-limited image obtained by the recursive method from the same average bispectrum. Figs. 3c and 3d show typical results of three independent experiments using 10 000 simulated speckle interferograms.

Fig. 4 shows a speckle masking / building block observation of Eta Carinae. The images in Fig. 4 show: (a) a single speckle interferogram of Eta Carinae, (b) a long-exposure image obtained by averaging 300 re-centered speckle interferograms, and (c) the diffraction-limited image reconstructed from the same 300 speckle interferograms by speckle masking and the building block method.

4. Speckle Spectroscopy

The three methods *objective prism speckle spectroscopy*[19,20], *slit speckle spectroscopy*[21], and *projection speckle spectroscopy*[22] can yield both high-resolution images and the spectrum of each resolution element.

Fig. 5 shows the first astronomical application of projection speckle spectroscopy[23,24].

Fig. 5a is one of 2000 projection speckle spectrograms of the close double star GC 6771 taken with the 2.2-m ESO/MPG telescope. Fig. 5b shows the long-exposure image reconstructed from the projection speckle spectrograms (Fig. 5a). Fig. 5c is the high-resolution object/spectrum image $o(x,\lambda)$ reconstructed from the same spectrograms by speckle masking. $o(x,\lambda)$ containes diffraction-limited resolution in x-direction and wavelength information in y-direction. The separation of the double star (Fig. 5c) is 0.14 arcsec in horizontal direction.

5. Optical Long-Baseline Interferometry and Aperture Synthesis

The great advantage of long-baseline interferometry is that it can yield images and spectra with fantastic angular resolution. Possible image reconstruction methods are the phase closure method (for the 'single-speckle' case)[25,26,6–8] and the combination of speckle masking and building block method (for the 'multi-speckle' case; large telescopes at visible wavelengths). Fig. 6 is a complete computer simulation of optical long-baseline interferometry with the ESO VLT Interferometer in the multi-speckle case[27]. Fig. 6a is the pupil function (similar to the ESO VLT Interferometer; four 8-m telescopes). Fig. 6b shows the *uv*-coverage of the whole computer experiment: data recording at 13 different rotation angles of the earth during 8 hours was simulated (aperture synthesis by earth rotation). Fig. 6c is the object, a triple star. Fig. 6d shows one of 10 000 multi-speckle long-baseline interferograms of a point source (without photon noise). The interferogram consists of about 16 speckles with interference fringes in each speckle. Speckle with fringes were obtained since many turbulence cells in front of each telescope were simulated. Fig. 6e shows one of 10 000 interferograms of the object with photon noise corresponding to 10 000 photoevents per interferograms. Fig. 6f is the diffraction-limited image reconstructed from the 10 000 interferograms by speckle masking and the building block method (10 000 photoevents per interferogram; 1000 iterations of the building block method; star separation ~5 milli-arcsec for λ=700nm and 100m diameter of the array; an experiment with only 300 photoevents/frame is reported in Ref. 28).

Acknowledgements:

We thank ESO for observing time.The results shown in Figs. 1, 2, 4 and 5 are based on data collected at the European Southern Observatory, La Silla, Chile.

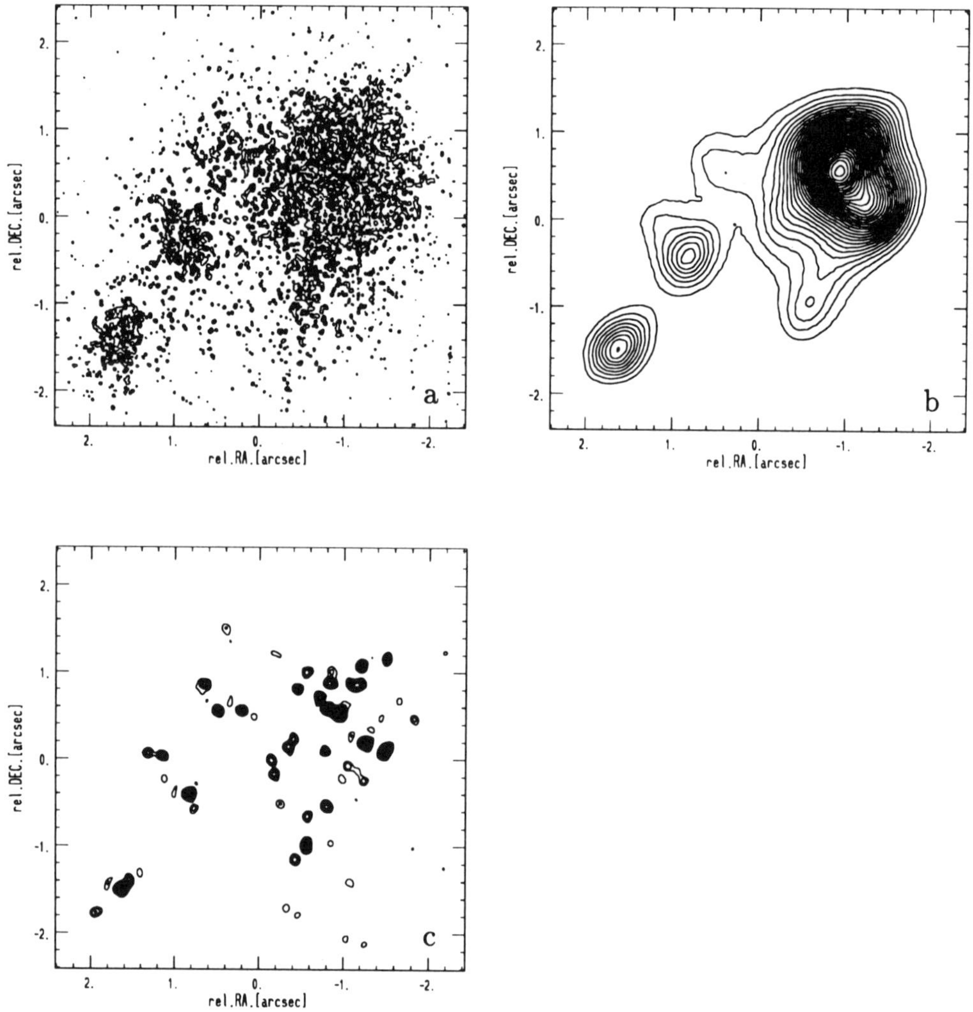

Fig. 1: Speckle masking observation of R136

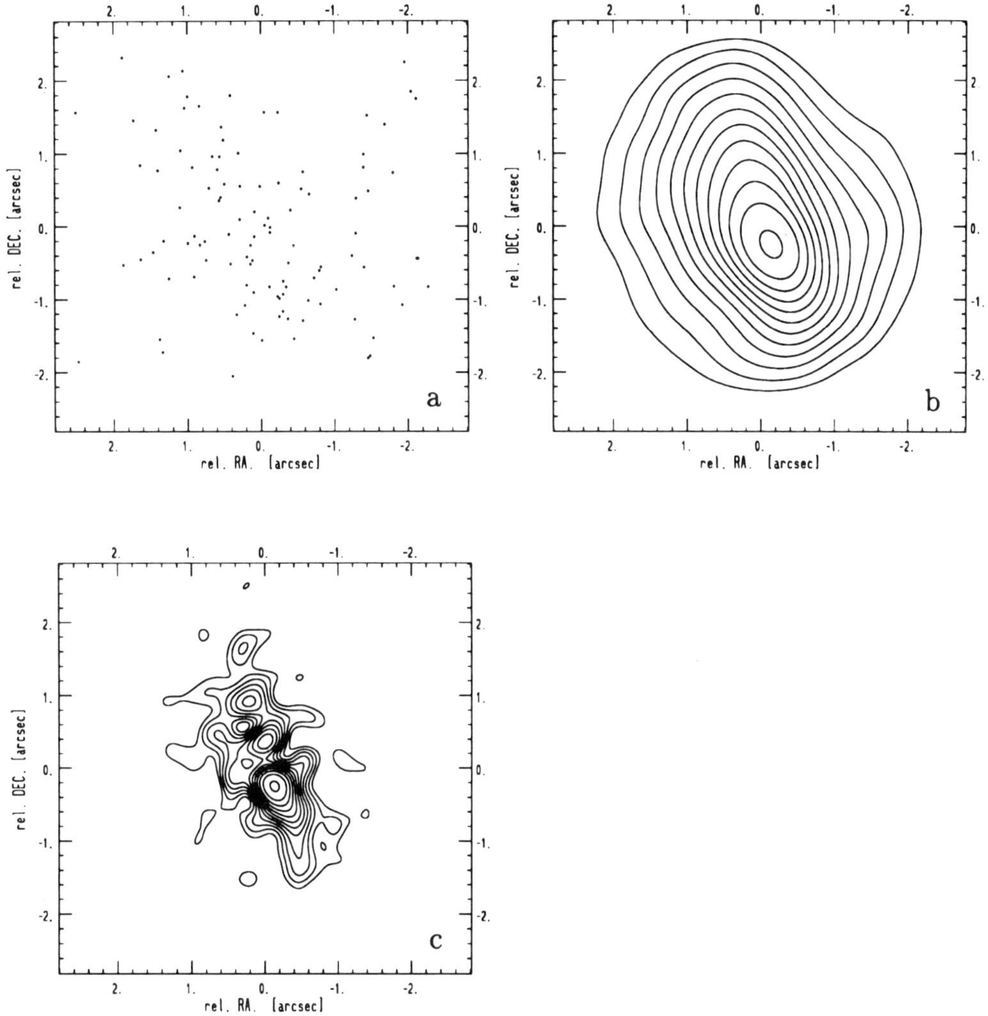

Fig. 2: Speckle masking observation of NGC 1386

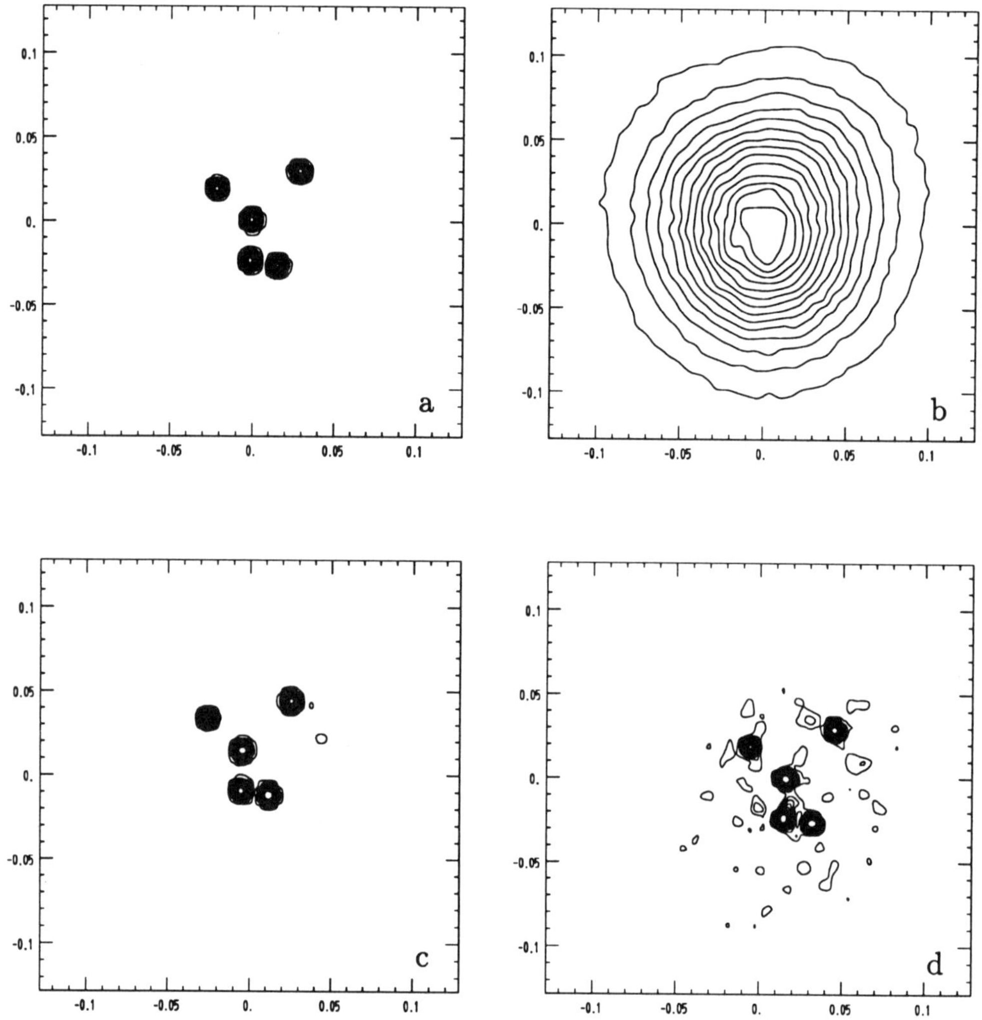

Fig. 3: Computer simulation of diffraction-limited imaging using the combination of speckle masking and building block method

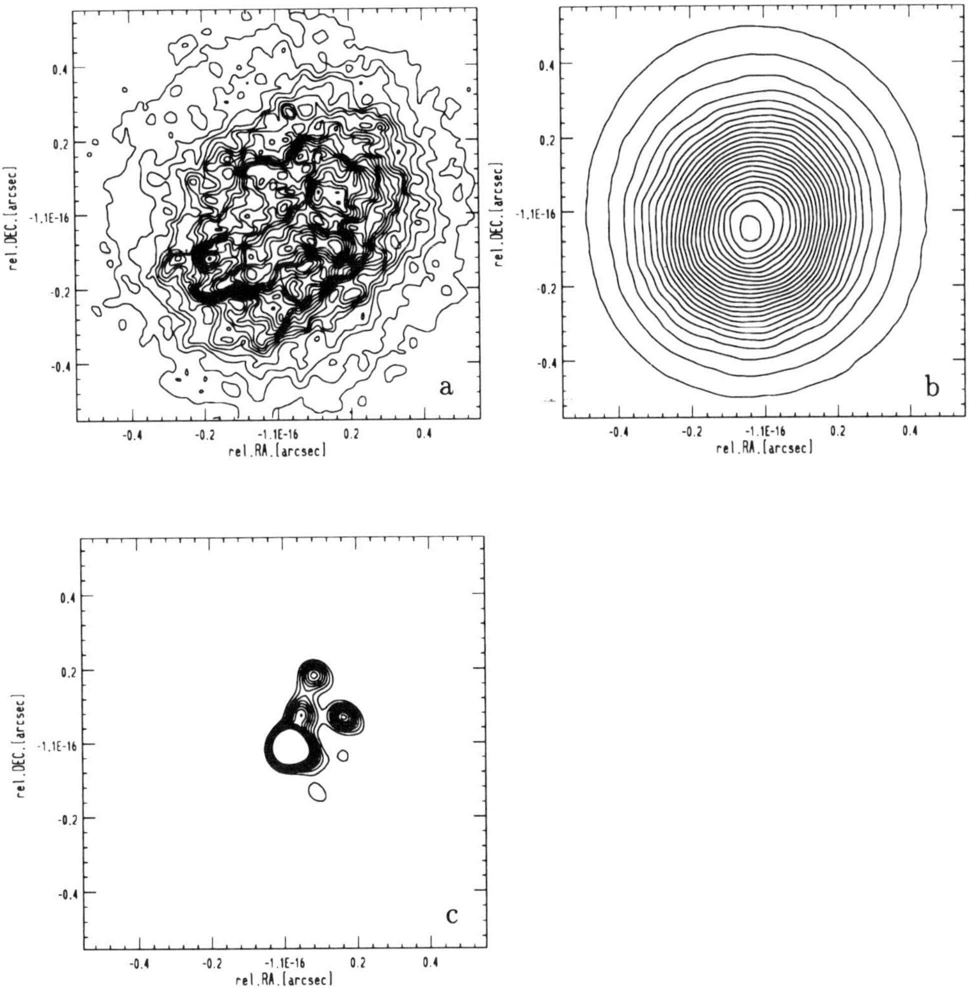

Fig. 4: Speckle masking/building block observation of Eta Carinae

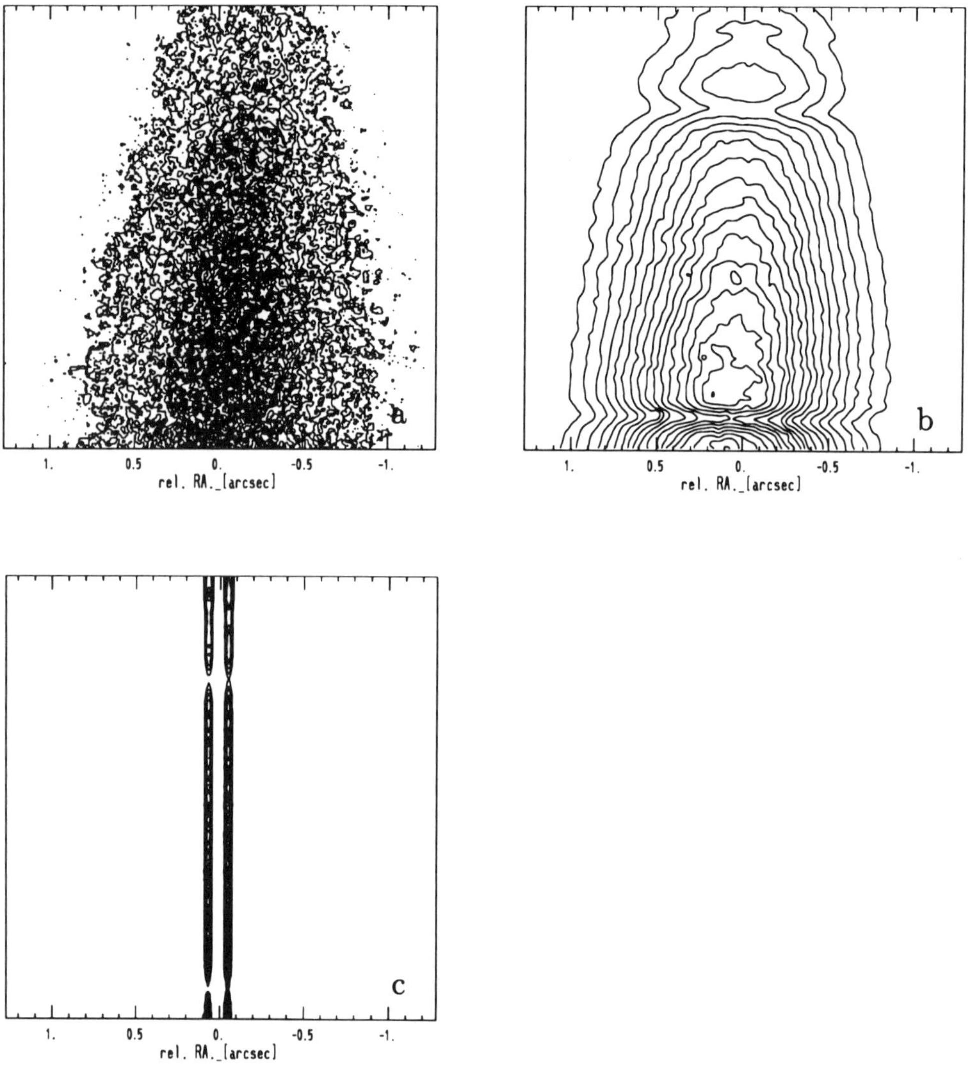

Fig. 5: Projection speckle spectroscopy of the double star GC 6771

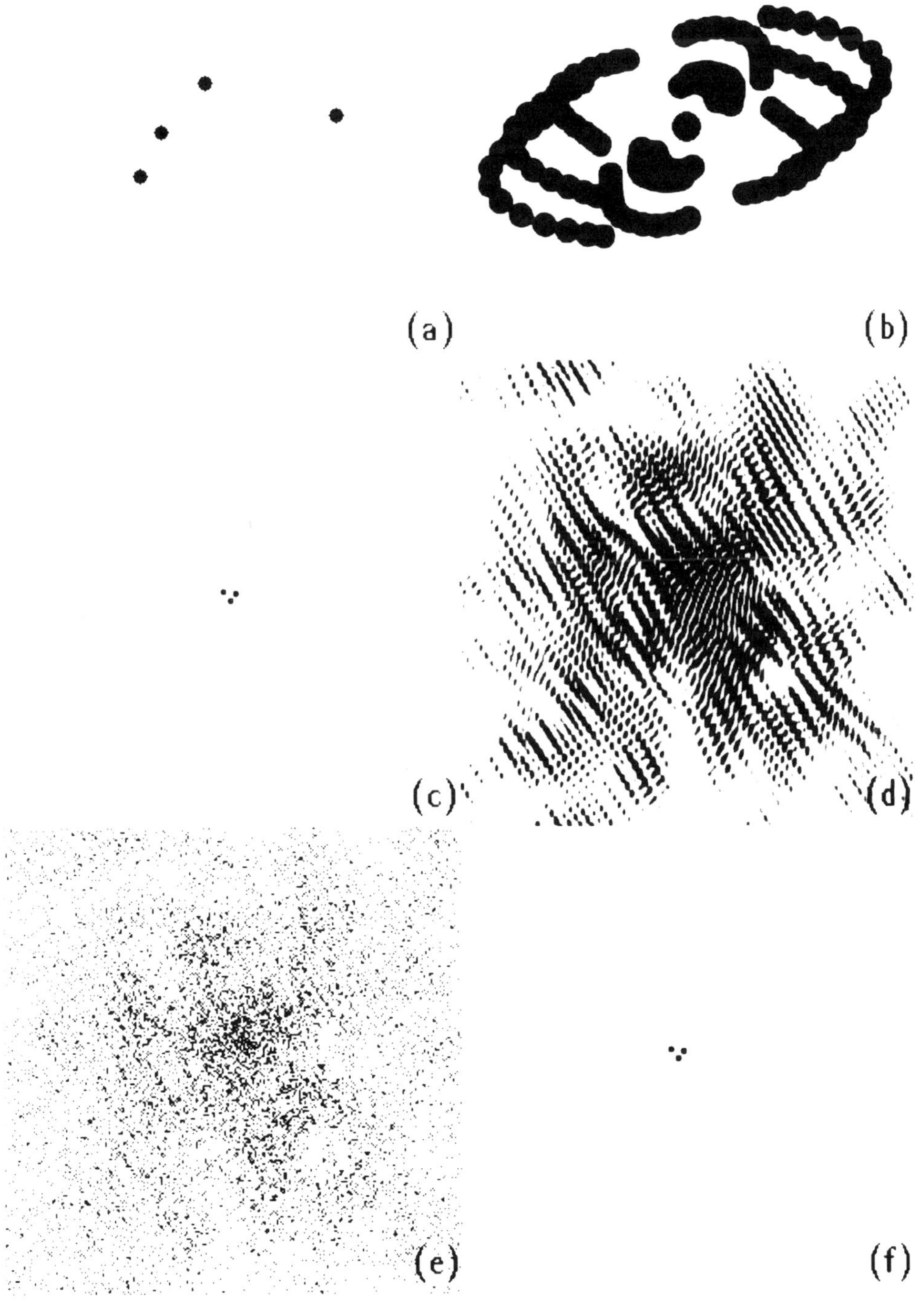

Fig. 6: Computer simulation of optical long-baseline interferometry with the ESO VLT Interferometer in the multi-speckle mode

6. References

1) A. Labeyrie, "Attainment of diffraction-limited resolution in large optical telescopes by Fourier analysing speckle patterns in star images", *Astron. Astrophys.* **6**, 85-87 (1970).
2) K.T. Knox, B.J. Thompson, "Recovery of images from atmospherically degraded short-exposure images", *Astrophys. J.* **193**, L45-L48 (1974).
3) G. Weigelt, "Modified astronomical speckle interferometry, speckle masking", *Optics Commun.* **21**, 55-59 (1977).
4) G. Weigelt, B. Wirnitzer, "Image reconstruction by the speckle masking method", *Optics Lett.* **8**, 389-391 (1983).
5) A.W. Lohmann, G. Weigelt, B. Wirnitzer, "Speckle masking in astronomy: triple correlation theory and applications", *Appl. Opt.* **22**, 4028-4037 (1983).
6) J.E. Baldwin, C.A. Haniff, C.D. Mackay, P.J. Warner, "Closure phase in high-resolution optical imaging", *Nature* **370**, 595-597 (1986).
7) A.C.S. Readhead, T.S. Nakajima, T.J. Pearson, G. Neugebauer, J.B. Oke, W. Sargent, "Diffraction-limited imaging with ground-based optical telescopes", *Astrophys. J.* **95**, 1278-1296 (1988).
8) T. Nakajima, S.R. Kulkarni, P.W. Gorham, A.M. Ghez, G. Neugebauer, J.B. Oke, T.A. Prince, A.C.S. Readhead, "Diffraction-limited imaging. II. optical aperture-synthesis imaging of two binary stars", *Astrophys. J.* **97**, 1510-1520 (1989).
9) J.C. Dainty, Stellar Speckle Interferometry, in: Laser Speckle and Related Phenomena, Topics in Applied Physics, Vol. **9**, ed. J.C. Dainty (Springer Verlag), ch. 7 (1984).
10) F. Roddier, "Interferometric imaging in optical astronomy", *Physics Reports* **170**, 97 (1988).
11) G. Weigelt, "Triple-Correlation Imaging in Astronomy", *Progress in Optics*, Vol. **29**, ed. E. Wolf, 24 (1991).
12) B. Wirnitzer, "Bispectral analysis at low light levels and astronomical speckle masking", *J. Opt. Soc. Am. A* **2**, 14-21 (1985).
13) E. Pehlemann, K.-H. Hofmann, G. Weigelt, "Photon bias compensation in triple correlation imaging and observation of R 136", *Astron. Astrophys.* **256**, 701-451 (1992).
14) W. Mauder, I. Appenzeller, K.-H. Hofmann, S. Wagner, G. Weigelt, P. Zeidler, "Speckle masking observations of the Seyfert II galaxy NGC 1386", *Astron. Astrophys.* **264**, L9 (1992).
15) K.-H. Hofmann, G. Weigelt, "Image reconstruction from the bispectrum using an iterative algorithm and applications of the method to astronomical objects", *Digital Image Synthesis and Inverse Optics*, A.F. Gmitro, P.S. Idell, I.J. LaHaie (eds.), Proc. SPIE 1351 pp. 522 (1990).
16) K.-H. Hofmann, G. Weigelt, "Building block method: image reconstruction from the bispectrum using an iterative algorithm, in: *High Resolution Imaging by Interferometry II, 14-18 Oct. 1991*, J.M. Beckers, F. Merkle (eds.), European Southern Observatory, Garching, Germany (in press).
17) I.-K. Morita, "The bispectrum analysis technique im Millimeter Interferometry", *Radio Interferometry: Theory, Techniques and Applications*, IAU Coll. 131, ASP Conference Series, T.J. Cornwell, R.A. Perley (eds.), Vol. **19**, pp. 197 (1992).
18) I.I.K. Pauliny-Toth, E. Preuss, A. Witzel, K.I. Kellermann, D.B. Shaffer, G.H. Purcell, G.W. Grove, D.L. Jones, M.H. Cohen, A.T. Moffet, J. Romney, R.T. Schilizzi,

R. Rinehart, "High resolution observations of NGC 1275 with a four-element international radio interferometer", *Nature* **259**, 17-20 (1976).
19) G. Weigelt, "Speckle Interferometry, Speckle Holography, Speckle Spectroscopy, and Reconstruction of High-Resolution Images from ST Data", in: *Conf. Proc. Scientific Importance of High-Angular Resolution at IR and Optical Wavelengths (ESO)*, pp.95 (1981).
20) G. Weigelt, G. Baier, J. Ebersberger, F. Fleischmann, K.-H. Hofmann, R. Ladebeck, "High resolution speckle methods for overcoming image degradation caused by the atmosphere and telescope aberrations", *Opt. Eng.* **25**, 706 (1986).
21) G. Weigelt, F. Grieger, K.-H. Hofmann, R. Pausenberger, "Slit speckle spectroscopy", in: *High Resolution Imaging by Interferometry II, 14-18 Oct. 1991*, J.M. Beckers, F. Merkle (eds.), European Southern Observatory, Garching, Germany (in press).
22) F. Grieger, F. Fleischmann, G. Weigelt, "Objective Prism Speckle Spectroscopy and Projection Speckle Spectroscopy", in: *Proc. High-resolution Imaging by Interferometry (ESO)*, p. 225 (1988).
23) F. Grieger, G. Weigelt, "Projection Speckle Spectroscopy and Objective Prism Speckle Spectroscopy", in *Proc. 15th Congress of the ICO* (SPIE Vol. **1319**), p. 440 (1990).
24) F. Grieger, G. Weigelt, "Projection speckle spectroscopy", in: *High Resolution Imaging by Interferometry II, 14-18 Oct. 1991*, J.M. Beckers, F. Merkle (eds.), European Southern Observatory, Garching, Germany (in press).
25) R.C. Jennison, "A phase sensitive Interferometer technique for the measurement of the Fourier transforms of spatial brightness distribution of small angular extent", *MNRAS* **118**, 276 (1958).
26) W.T. Rhodes, J.W. Goodman, "Interferometric techniques for recording and restoring images degraded by unknown aberrations", *J. Opt. Soc. Am.* **63**, 647-657 (1973).
27) T. Reinheimer, K.-H. Hofmann, G. Weigelt, "Computer simulations of interferometric imaging with the VLT Interferometer", in: *High Resolution Imaging by Interferometry II, 14-18 Oct. 1991*, J.M. Beckers, F. Merkle (eds.), European Southern Observatory, Garching, Germany (in press).
28) T. Reinheimer, K.-H. Hofmann, G. Weigelt, "Interferometric imaging with arrays of large optical telescopes in the multi-speckle mode", submitted to *Astron. Astrophys.* (1993)

DOUBLE PASSAGE IMAGING THROUGH TURBULENCE

C.J. Solomon

Blackett Laboratory, Imperial College,
London SW7 2BZ, U.K.

1.1 INTRODUCTION

There is a large body of literature concerned with the study of and compensation for the effects of atmospheric turbulence on incoherent imaging. Most of this has been directed at the astronomical problem. Roddier's review[1] remains an excellent overview of the main principles and compensation techniques. In contrast, relatively little attention has been given to the possibility of coherent imaging through turbulence. In this paper, we attempt to monitor the progress which has been made in this area and to outline both the limitations and possibilities of this approach.

The analysis of imaging through turbulence when the object is coherently (i.e. laser) illuminated is more complicated than the incoherent case. This is because one must specify how the object is illuminated and whether the object is deterministic or statistical. As regards the illumination, this also may be deterministic (for example, a plane wave) or statistical (for example, a wave that has already passed through turbulence). Although we shall briefly refer to other cases, for most of this paper we shall be concerned with one case in particular. Specifically, this is where a deterministic (mirror-like) object is both illuminated and viewed through the same inhomogeneities and we refer to this as the double passage case.

1.2 DOUBLE PASSAGE - A BRIEF HISTORY

There has been a good deal of investigation into the double passage of waves through random media over the last fifteen years. This work has been largely concerned with the propagation of waves through random media rather than imaging per se. Much of it has originated from the former Soviet Union[2–12]. The basic effect that is predicted and experimentally observed is that the intensity of radiation in the backscatter direction is significantly higher than in the regions surrounding it. This is closely related to the enhanced backscattering (EBS) effect observed for light scattering from rough surfaces and dense volume media. The key to understanding the effect (in whatever context it appears) *lies in the occurrence of coherence channels whenever the receiver coincides with the position of the transmitter*. Theoretical analyses were given by Kravtsov and Saichev[4,5] and by Jakeman[10,11] whilst experimental confirmation was provided by the experiments of Gurvich and Kashkarov[3]. For an excellent and exhaustive review of the EBS effect in optics, the reader is referred to the article by Barabanenkov, Kravtsov, Ozrin and Saichev [13].

Analysis of double passage *imaging* as distinct from propagation has been examined in a general way by Fante[14] who discussed the problem of imaging through a random phase screen using light of arbitrary coherence. Jakeman [15] has investigated the problem of double passage imaging through turbulence when the object is incoherent. A significant step was made by Mavroidis and Dainty [16] who showed that when the object is deterministic, the time-averaged double passage image contains diffraction-limited information on the object reflectivity. This has provided the stimulus for a growth of interest in the question of double passage imaging over the last two years. Although their analysis assumed a somewhat idealised model, subsequent analyses and experiments suggest that their conclusions may be extended to more general situations[17–21]. Solomon et al[18] proposed the use of non-redundant and Michelson apertures to improve the S/N ratio and perform a reasonably successful reconstruction of a complex (although still deterministic) object. A further refinement proposed was the use of polarisation coding of the illumination. This theoretically eliminates the need for time averaging over the statistics of the turbulence[20]. Experimental confirmation of both the basic coherence enhancement and the efficacy of polarisation coding has been provided[19,21]. Other papers exploring aspects of the problem from a theoretical viewpoint have also been forthcoming recently[22–24].

In the next section, we explore the connection between the EBS effect and double passage imaging. To do this, we will assume two very simple and closely related situations. The first will serve to explain the EBS effect while the second, a trivial extension of the first, will demonstrate the existence of diffraction-limited information in a time-averaged double passage interferogram.

1.3 EBS AND DOUBLE PASSAGE - PHYSICAL BASIS

Figure 1 depicts a point transmitter situated at $\underline{r}_T$ with a point scatterer located at point $\underline{r}_S$ embedded in random inhomogeneities (turbulent or otherwise). Let the point transmitter produce a primary scalar field described by the Green's function $G(\underline{r}_T, \underline{r})$. When this wave is incident on the point scatterer at $\underline{r}_S$, the backscattered field at the general point $\underline{r}$, $u(\underline{r})$, is given by[13],

$$u(\underline{r}) = \alpha G(\underline{r}_T, \underline{r}_S) G(\underline{r}_S, \underline{r}) \tag{1}$$

where α is the amplitude of scattering. The scattered intensity at the general point $\underline{r}$ is then,

$$I(\underline{r}) = |\alpha|^2 I(\underline{r}_T, \underline{r}_S) I(\underline{r}_S, \underline{r}) \tag{2}$$

where $I(\underline{r}_S, \underline{r}) = |G(\underline{r}_S, \underline{r})|^2$. We now normalise the scattered intensity $I(\underline{r})$ to the intensity at the point $\underline{r}$ obtained in the absence of the random inhomogeneities, $I^o(\underline{r})$. We have,

$$I^o(\underline{r}) = |\alpha|^2 I(\underline{r}_T - \underline{r}_S) I(\underline{r}_S - \underline{r}) \tag{3}$$

so that the normalised intensity $J(\underline{r}) = \frac{I(\underline{r})}{I^o(\underline{r})}$ is simply,

$$J(\underline{r}) = J(\underline{r}_T, \underline{r}_S) J(\underline{r}_S, \underline{r}) \tag{4}$$

where $J(\underline{r}_S, \underline{r}) = \frac{I(\underline{r}_S, \underline{r})}{I^o(\underline{r}_S - \underline{r})}$.

The key point now follows. From the conservation of energy flux for the intensity of a wave emitted by a point isotropic source we have $\langle I(\underline{r}_S, \underline{r}) \rangle = I^o(|\underline{r}_S - \underline{r}|)$, so that

$$\langle J(\underline{r}_S, \underline{r}) \rangle = 1 \tag{5}$$

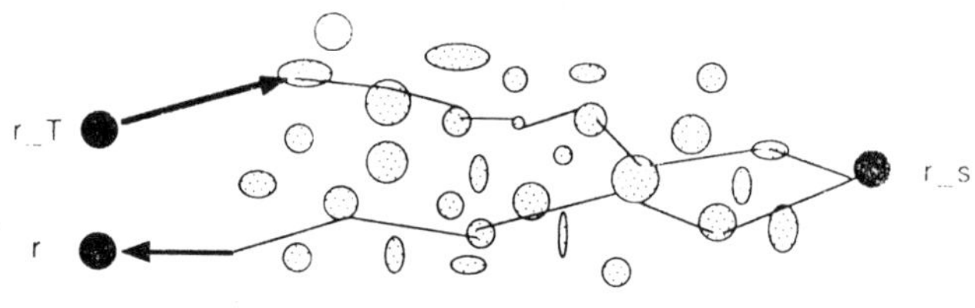

Figure 1: The enhanced backscattering effect (EBS) occurs when the points of transmission and reception coincide.

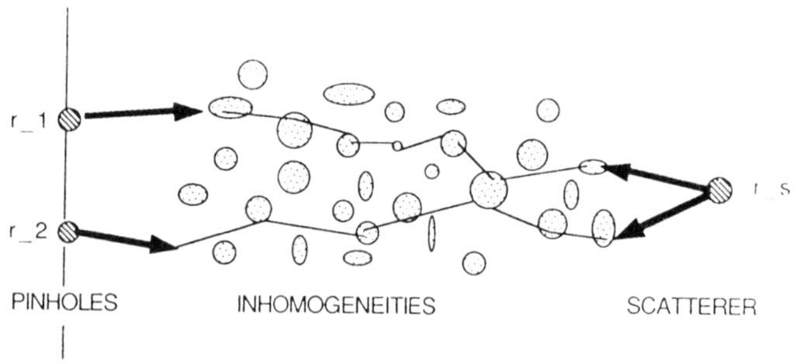

Figure 2: The double passage effect - a simple extension of EBS.

However, if we observe the average intensity at the point $\underline{r}_T$ the location of the transmitter, we have from eq. 4,

$$\langle J(\underline{r}_T)\rangle = \langle J^2(\underline{r}_S, \underline{r}_T)\rangle \qquad (6)$$

Since the *mean square will always exceed the square of the mean (= 1 from (5))*, the backscatter enhancement factor will always be larger than unity. Thus

$$\langle J(\underline{r}_T)\rangle = \langle J^2(\underline{r}_S, \underline{r}_T)\rangle > \langle J(\underline{r}_S, \underline{r}_T)\rangle^2 = 1 \qquad (7)$$

Let us now consider a simple extension of figure 1. In figure 2, we consider an aperture containing two pinholes located at points $\underline{r_1}$ and $\underline{r_2}$. The illumination consists of a coherent wave which passes through both pinholes and propagates through the random medium to the point object at $\underline{r}_S$. The backscattered light is received at the *same two pinholes.* In other words, source and receiver again coincide. The situation this time however, is slightly different - not only do we have the '*straight-back*' paths (i.e. $\underline{r_1} \rightarrow \underline{r}_S \rightarrow \underline{r_1}$ and $\underline{r_2} \rightarrow \underline{r}_S \rightarrow \underline{r_2}$) but we have the 'cross-over' paths ($\underline{r_1} \rightarrow \underline{r}_S \rightarrow \underline{r_2}$ and $\underline{r_2} \rightarrow \underline{r}_S \rightarrow \underline{r_1}$). Due to the reciprocity theorem, these two cross-over paths are equivalent - they involve propagation through the same inhomogeneities - and this contribution to the optical field at the pinholes therefore *remains coherent despite the presence of the inhomogeneities.* This, of course, is true for a medium which is randomly evolving in time provided its characteristic correlation time is large compared with the round-trip time of the light - a situation which would certainly prevail in many situations of interest. The straight-back paths on the other hand, are randomly phased with respect to each other having travelled different paths through the medium.

If we interfere the backscattered light at the two pinholes, we essentially have two contributions to the interferogram. The coherent reversible paths generate a stable fringe pattern which is unaffected by the time-evolving random medium whilst the straight-back paths will also generate a fringe pattern but one which randomly translates in the image plane as the medium evolves. As we time average over the statistics of the inhomogeneities, the randomly translating fringes wash out and average to a constant background leaving the stable fringe pattern superimposed. In other words, the effect of the straight-back light is to reduce the contrast of the time-averaged fringe pattern. Viewed in this way, the straight-back light is really a noise-like contribution and needs to be minimised. An approach to doing this is discussed later.

1.4 DOUBLE PASSAGE INTERFEROMETRY

In order to arrive at quantitative results, we shall assume an idealised imaging model in which the inhomogeneities can be represented by a random screen lying in the pupil plane and the deterministic object lies in the Fraunhofer region (fig. 3). Further, we shall take the statistics of the screen to be a circular complex gaussian process. The limitations of this model are clear but, as we hope to show, it leads to quantitative results and some interesting physical insights. We shall also be interested in how far the conclusions which we reach can be extended to more complex situations and this will be discussed later.

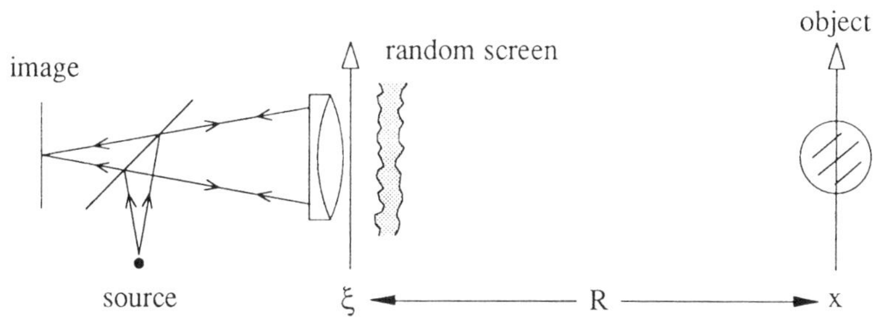

Figure 3: A schematic of the double passage imaging model.

Imagine that the object lying in the Fraunhofer region in fig. 3 is a simple point scatterer and that we illuminate and view it through a two pinhole aperture. If the scattered light returns through a *different* screen (i.e. an independent realisation), then on the basis of the simple argument given in the previous section, we would not expect the time-averaged image to show any significant fringes. (This implicitly assumes, of course, that the separation of the pinholes exceeds the correlation length of the screen - a condition enforced in both the analysis and experiment). On the other hand, if the reflected light passes through the same screen, we expect a non-zero visibility. A straightforward analysis[17] shows that under the assumption of complex gaussian statistics, we expect the visibility of the cosine fringes to be 1/3. This corresponds to a Fourier spectrum having side peaks of magnitude 1/6 relative to the central peak (i.e. zeroth frequency). In other words, we expect a definite non-zero transmission over the time-average of the single frequency defined by the two-pinhole aperture. In the case of a pure phase screen, the side peak to central ratio is 1/4. Fig. 4 shows the results of a computer simulation confirming these results. In figure 4a we see the effect of illuminating through one random screen but viewing through an uncorrelated screen - the fringes have washed out due to their random translation in the image plane. However, when illumination and viewing take place through the same screen, we observe fringes of reduced but finite visibility (figs. 4b and 4c) in agreement with the analytical predictions.

Figure 4: Computer simulations of double passage Michelson interferograms.

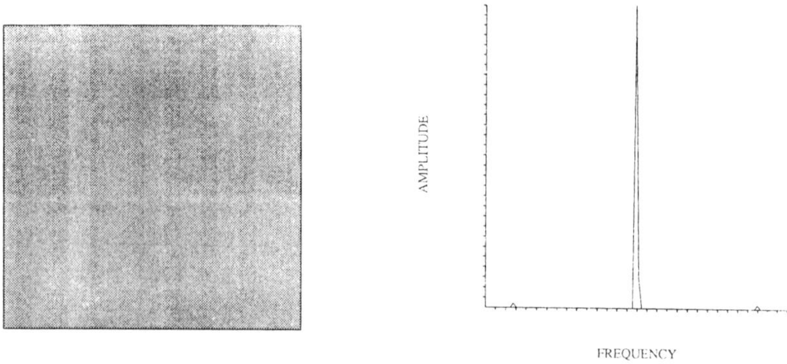

(a) Interferogram obtained when illumination and viewing take place through statistically independent screens (2000 realisations)..

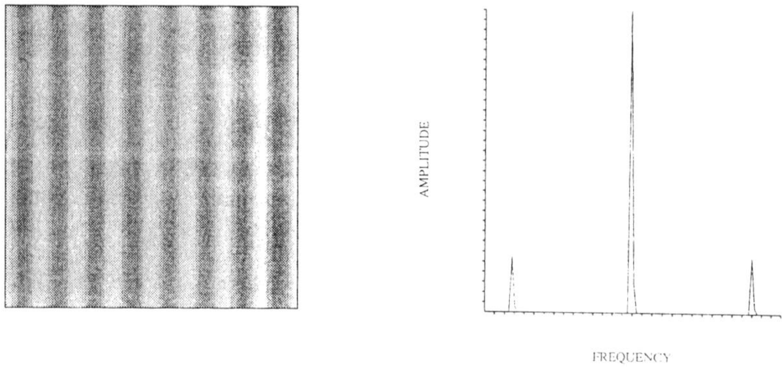

(b) Double passage interferogram and Fourier spectrum of a point source for a complex-gaussian screen (2000 realisations).

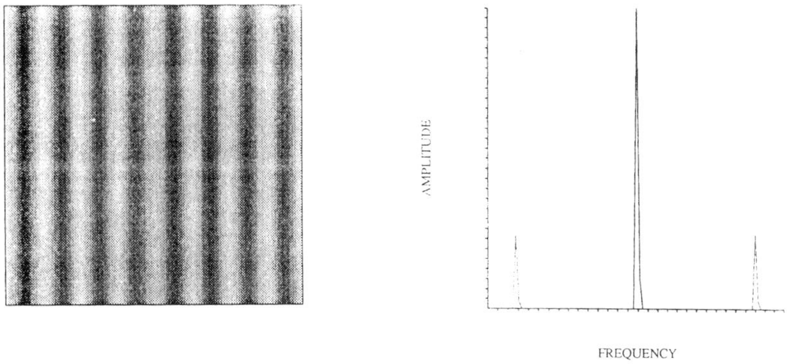

(c) Double passage interferogram and Fourier spectrum of a point source for a Kolmogoroff phase screen (2000 realisations).

Figure 5 shows the experimental apparatus used in an attempt to confirm this basic effect[19]. A collimated laser was split into two beams and illuminated the system through the two pinholes A and B. After passage through the turbulent cell to the plane mirror and back again, the reflected light passed through the two-pinhole aperture to a CCD line-scan array placed in the focal plane of the lens. The CCD array, controlled by a computer, captured sequential scans of the one-dimensional intensity in the image plane. When illumination and viewing take place through both pinholes, we expect the double passage effect to occur. If, however, illumination takes place through only one of the pinholes, though instantaneous fringes are observed, we do not expect fringes to be present in the time-averaged image. Figs 6a–6c show the recorded time-averaged intensity for the cases of a) no turbulence b) turbulence but single pinhole illumination c) turbulence with two pinhole illumination. For b) and c) the turbulent cell was positioned close to the two pinholes (i.e. in the pupil plane). Figs 6d–6f show similar plots of the recorded intensity when the turbulent cell was positioned midway between the pupil and the plane mirror. Clearly, we see that a degree of coherence is preserved in the double passage case. It is interesting to note that this also takes place when the turbulent cell is located at some distance from the aperture.

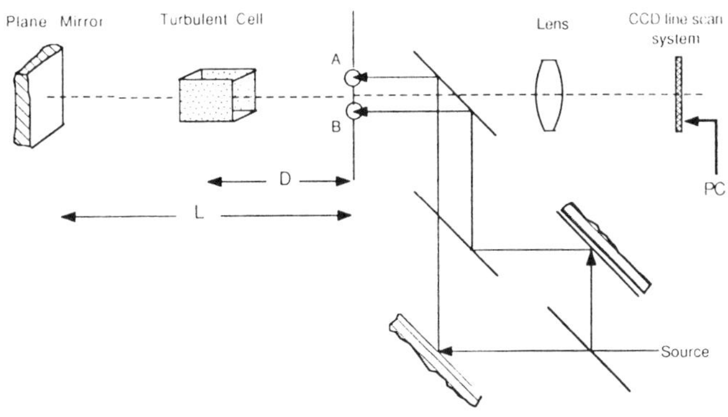

Figure 5: Schematic of an experiment performed to demonstrate the enhancement of coherence after double passage through turbulence.

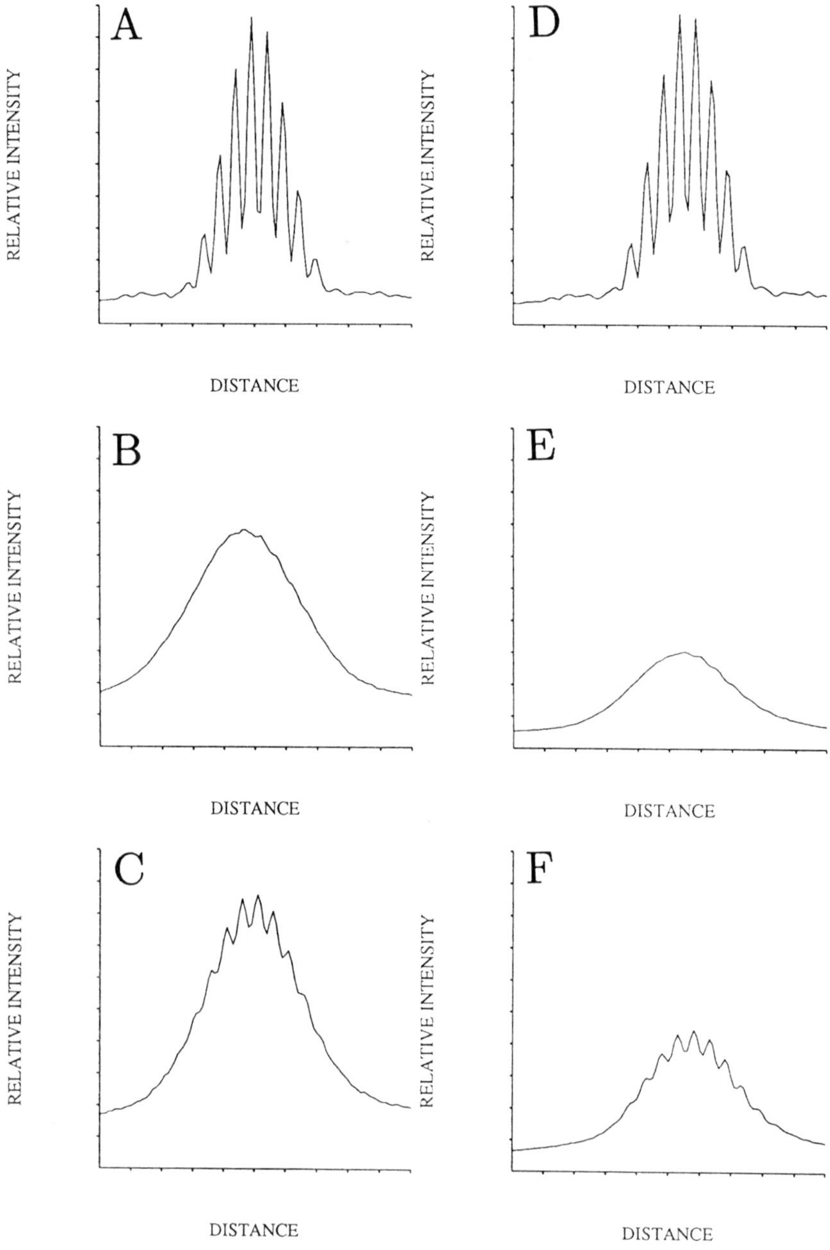

Figure 6: Experimentally observed one-dimensional intensities.
(a) No Turbulence. (b) Turbulence: Single pinhole illumination. (c) Turbulence: Two pinhole ilumination. (d) No Turbulence. (e) Turbulence: Single pinhole illumination. (f) Turbulence: Two pinhole ilumination.

2.1 IMAGING WITH NONREDUNDANT APERTURES

It is easy to show that the image does in fact contain diffraction-limited information on the object. In the particularly simple case discussed above, the image (i.e. the single frequency defined by the Michelson aperture) contains information on a related but different object frequency. Imaging may be considered as an interferometric process and it is possible to extend this to a general (although still deterministic) object and a non-redundant aperture[18]. Representing the nonredundant aperture by a sum of N delta functions,

$$h_0(\xi) = \sum_{j=1}^{N} \delta(\xi - \xi_j) \tag{1}$$

analysis yields an expression for the instantaneous image spectrum,

$$\begin{aligned} I(\eta) = \sum_4 a(\xi_k)a^*(\xi_k+\eta)a(\xi_j)a^*(\xi_l) \\ \times\ \tilde{O}(\xi_k+\xi_j)\tilde{O}^*(\xi_k+\xi_l+\eta)\quad \delta(\xi_k-\xi_m+\eta) \end{aligned} \tag{2}$$

where $\sum_4$ denotes summation over the four indices j, k, l and m ; $a(\xi)$ is the amplitude transmission of the screen ; $\tilde{O}(\xi)$ is the Fourier transform of the object amplitude reflectivity and ξ_i denotes the general position vector of the ith pinhole.

Time averaging eq. (2) under the assumption of complex gaussian statistics yields -

$$\begin{aligned} \langle I(\eta)\rangle = \sum_4 [C(\eta)C(\xi_l-\xi_j) + C(\xi_l-\xi_k)C(\xi_k-\xi_j+\eta)] \\ \times\ \tilde{O}(\xi_k+\xi_j)\tilde{O}^*(\xi_k+\xi_l+\eta)\quad \delta(\xi_k-\xi_m+\eta) \end{aligned} \tag{3}$$

where $C(\eta)$ is the coherence function (autocorrelation) of the random screen. We now consider the non-redundant aperture to comprise N pinholes *whose vector separations all exceed the correlation length* of the screen. Such an aperture will have the property that no significant correlation of the field exists between any two pinholes. In this case, we can legitimately replace the coherence function, C, by the corresponding delta function since $C(\xi_l - \xi_j)$ is only significantly different from zero when $\xi_l = \xi_j$. In other words, we purposely devise the aperture to exclude low image frequencies (ie those $< \frac{r_0}{\lambda R}$). This is quite distinct from the unrealistic assumption of a delta correlated random screen. Making the delta function substitution in eq. (3), we end up with two contributions - one in the zeroth frequency and one in the non-zero frequencies. The zeroth frequency term is given by setting j = l and k = m in the first term of (3) and by setting j = k = l = m in the second term to give,

$$\langle I(0)\rangle = \sum_{l=1}^{N}\sum_{m=1}^{N} \mid O(\xi_l+\xi_m)\mid^2 \ + \ \sum_{k=1}^{N} \mid O(2\xi_k)\mid^2 \tag{4}$$

The non-zero frequencies are obtained by setting k = l and j = m in the second term. This gives,for $\eta \neq 0$,

$$\langle I(\eta)\rangle = \sum_{l=1}^{N} \sum_{m=1,m\neq l}^{N} \mid O(\xi_l + \xi_m) \mid^2 \;\; \delta(\eta - [\xi_m - \xi_l]) \qquad (5)$$

If we examine (5), we see that *for a non-redundant aperture only one* of the $N^2 - N$ possible combinations of ξ_l and ξ_m can reproduce the frequency η. We can therefore omit the summation to get the simple result that a non-zero frequency in the image defined by the specific pupil pinhole vectors ξ_F and ξ_B is related to the object frequency $\xi_F + \xi_B$ by,

$$\langle I(\xi_F - \xi_B)\rangle \;\; = \;\; \mid O(\xi_F + \xi_B) \mid^2 \;\; \delta(\eta - [\xi_F - \xi_B]) \qquad (6)$$

Eq. (6) represents the main result of this section. We see that frequencies in the time-averaged imaged can be directly mapped to the square modulus of frequencies in the object spectrum. Note also that *difference* coordinates in the image spectrum map to *sum* coordinates in the object spectrum.

2.2 SIGNAL-TO-NOISE

Eq. (6) obtained above is strictly valid only in the limit of infinite time-averaging. Analysis[18] shows that under the practical limitations of finite time-averaging the object term that we seek to evaluate has to dominate $N^2 - 1$ other terms (which contribute a spurious background of different object frequencies) if a reasonable estimate of the object term is to be obtained. A non-redundant aperture comprising N sub-apertures will simultaneously image ${}^N C_2$ frequencies but each of these frequencies must compete with $N^2 - 1$ other terms which gradually diminish in magnitude in proportion to the exposure time. An heuristic analysis shows that optimal signal-noise is actually obtained for the simple two pinhole or Michelson aperture. Put in simple language, we find that it is better to recover the frequencies one by one (but quickly) than to recover many frequencies simultaneously (but correspondingly more slowly).

2.3 COMPUTER IMAGE RECONSTRUCTION

In an attempt to verify the analysis, computer-simulated double passage images of test objects were obtained[18]. In fig. 7a, we have a diffraction-limited image of Tigger * taken with a full aperture and no random screen - it represents the best possible reconstruction. Figure 7b is a long-exposure image taken by illuminating through one random screen and viewing through an uncorrelated one. Figure 7c is a double passage image, reconstructed by sequentially probing the central (lower) frequencies with two pinhole apertures and then using a phase retrieval algorithm to reconstruct the object from its Fourier magnitude; although the reconstruction is far from perfect it represents a considerable improvement on fig. 7b.

---

* Tigger is the tiger in A.A Milne's children's book "Winnie the Pooh".

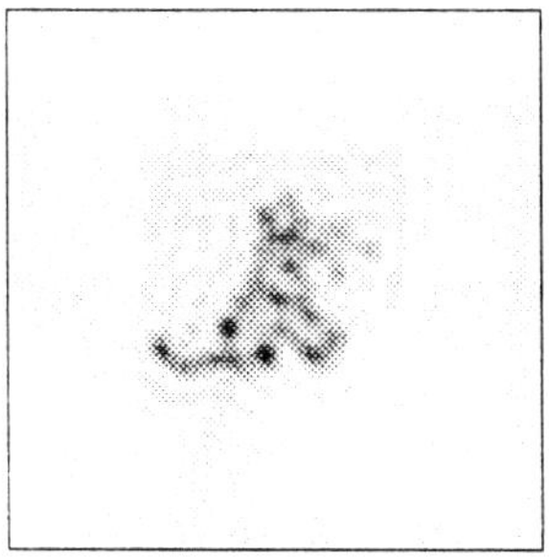

(a) Diffraction-limited image.

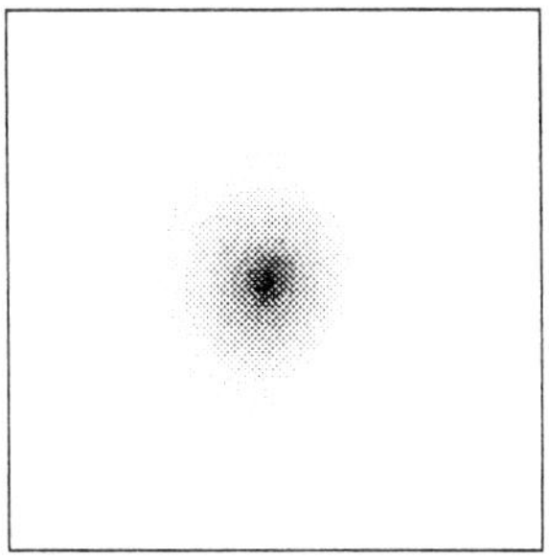

(b) Image through uncorrelated screens.

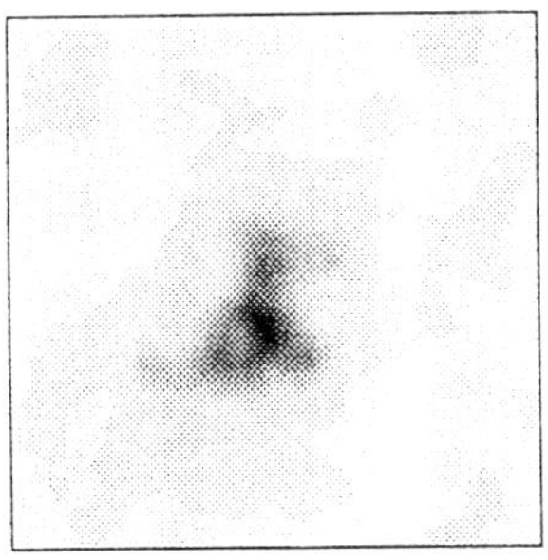

(c) Double passage Image.

Figure 7: Double passage imaging.
(a) Diffraction limited image of Tigger.
(b) Image of Tigger after forward and reverse passage through statistically independent screens.
(c) Reconstruction of Tigger achieved by double passage recovery of the Fourier modulus followed by phase recovery.

2.4 POLARISING NONREDUNDANT APERTURES

Referring to figure 3 , it has been demonstrated that *crossover* paths AOB and BOA contribute the desired information to the image whilst the *straight-back* paths do not contribute useful information. The time averaging process is necessary so that the stable fringes will be superimposed on a background which is sufficiently uniform that it does not significantly affect the estimate of the object spectrum. This can be viewed as a desirable, coherent contribution (paths AOB and BOA) and an unwanted, incoherent contribution (paths AOA and BOB).

Fig. 8 is a schematic of a system which would only allow light travelling on the desired, reversible paths AOB and BOA to contribute to the image. The laser illumination passing through A and B in fig. 8 is assumed to be linearly polarised (electric vector lies parallel to the x axis of the coordinate system). In front of pinhole A, is placed a quarter wave plate with its fast axis at 45°, in front of pinhole B is placed a similar quarter wave plate with its fast axis at -45°. Light passing through the pinholes on the return passage passes through a linear, horizontal polariser (transmission axis parallel to the x axis).

A formal analysis using the Mueller calculus yields the desired result[20]. However, it is instructive and simple to analyse this system in a qualitative manner. For simplicity, we make two idealised assumptions. Firstly, propagation through the random screen is not considered to depolarise the incident light. Secondly, reflection from the object does not significantly affect the polarisation state of the incident beam. Consider first the polarisation state of the light following the path AOB. The quarter wave plate acting on the incident, horizontally polarised light produces right, circularly polarised light *. Reflection from the object, in accordance with our second assumption, does not change the polarisation state. Passage through the quarter wave plate at B, changes the polarisation state back to horizontal which is passed by the horizontal polariser. A similar argument shows that light on the path BOA is transmitted through the system. The paths AOA and BOB have the following polarisation sequences,

AOA : horizontally polarised (incident) $\rightarrow$ right circular ($\frac{\lambda}{4}$ at 45°) $\rightarrow$ vertically polarised ($\frac{\lambda}{4}$ at 45°).

BOB : horizontally polarised (incident) $\rightarrow$ left circular ($\frac{\lambda}{4}$ at -45°) $\rightarrow$ vertically polarised ($\frac{\lambda}{4}$ at -45°).

Light travelling along paths AOA and BOB is therefore not transmitted by the horizontal polariser. This has two important consequences. Firstly, having eliminated (at least in theory) the incoherent contribution we obtain an estimate of the object frequency from the *short-exposure* image. Furthermore, the conclusion in[18] that the Michelson-Fizeau aperture was superior to multiple pinhole non-redundant apertures in the double passage method relied on the assumption that the unwanted paths (such as AOA and BOB) could not be eliminated. Free from this restriction, we can extend the approach to a multiple pinhole aperture and thereby obtain estimates of many spatial frequencies simultaneously from the short-exposure image. In fig. 9, we depict a multiple pinhole aperture with a quarter wave plate placed in front of each pinhole. Quarter wave plates placed at 45° are denoted by $+$ and those positioned at -45° are represented by $-$. It

* We use here the convention adopted by Shurcliff [25]. If looking towards the source, the electric vector at a fixed point in space is rotating clockwise, the polarisation state is right circular.

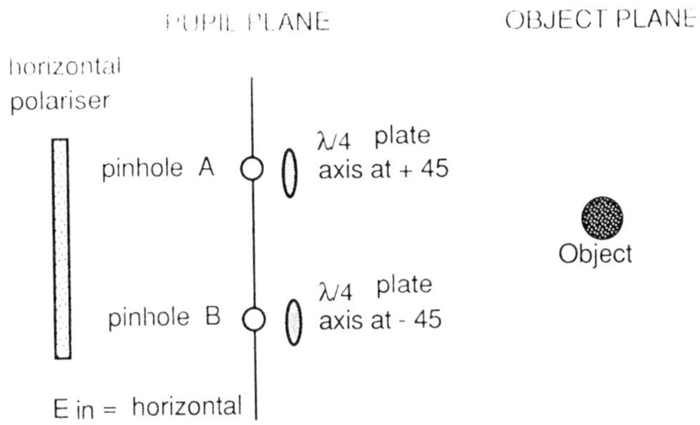

Figure 8: Schematic of a polarising aperture which eliminates the incoherent contribution to the interferogram.

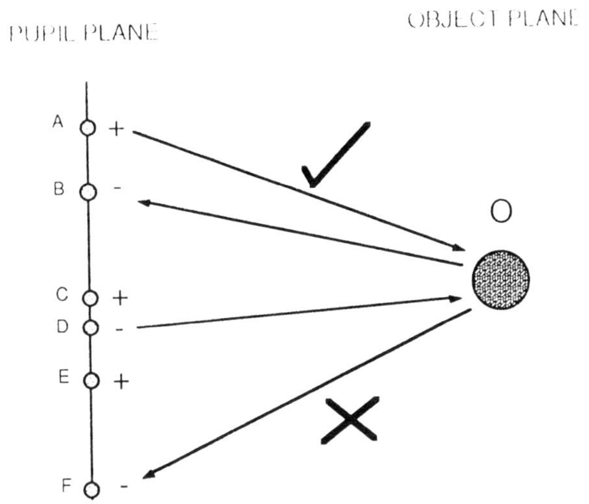

Figure 9: The concept of a polarising nonredundant aperture.

is clear, from our preceding analysis, that those spatial frequencies (defined by the separation between the pinholes) having *different* signs at the two defining pinholes will be imaged by the system, whereas those defined by *like* signs will be eliminated. It is straightforward to show that the number of spatial frequencies defined by this polarising aperture is $(\frac{n}{2})^2$ for n even or $\frac{n+1}{2}\frac{n-1}{2}$ for n odd. For an ordinary n pinhole non-redundant aperture we obtain $\frac{n(n-1)}{2}$ frequencies. Clearly in the limit that n is large, the polarising aperture images 1/2 as many frequencies.

The use of a polarising Michelson aperture to eliminate the incoherent portion of the light has been experimentally demonstrated[21]. The experimental arrangement was essentially the same as that depicted in fig. 5 with the addition of the quarter wave plates and polariser. In fig. 10, the results are summarised. Fig 10a shows the time-averaged one-dimensional intensity observed in the focal plane with no turbulence present and no quarter wave plates. In fig 10b, we see the time-averaged, double passage intensity with turbulence but no wave plates - we see that some degree of coherence is preserved as expected. In fig 10c, we see the effect of adding the quarter wave plates and polariser - the visibility of the fringes is significantly increased. Due to experimental logistics, the visibility is somewhat less than that obtained with no turbulence present but under better conditions it should be possible to approach the theoretical value of 1 which, of course, is equivalent to that obtained with no turbulence present. Fig 10d shows the effect of trying to interfere the two incoherent (i.e. straight-back) beams (AOA and BOB). As expected there are no discernible fringes. This experiment has thus shown that we can separate the coherent and incoherent parts of the light by manipulating the polarisation. It has also shown that we can have enhanced backscattering of intensity without an enhancement of coherence and an enhancement of coherence without enhanced backscattering !

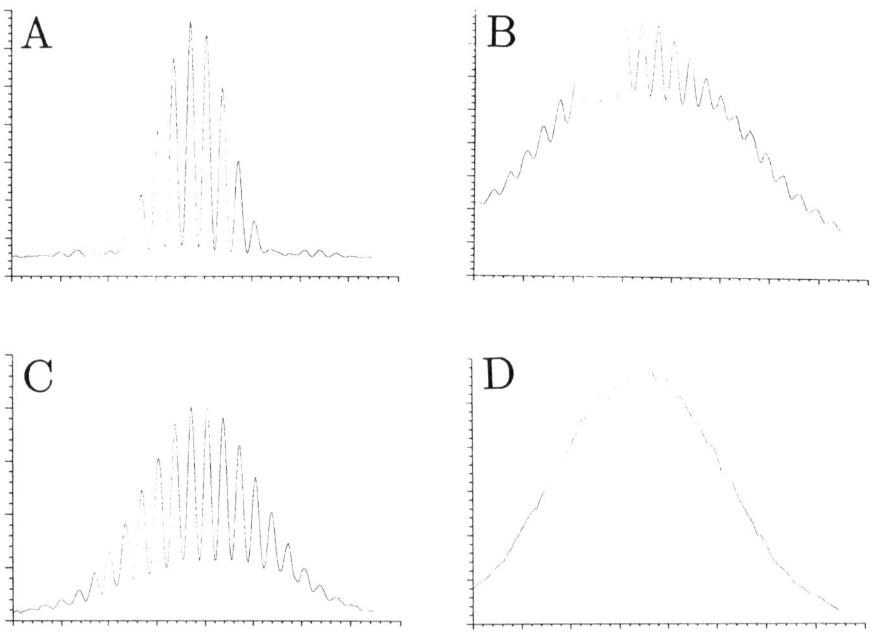

Figure 10: Time averaged, one-dimensional intensities observed in the focal plane of the system depicted in Figure 8.

(a) No Turbulence present: No quarter wave plates: Visibility of fringes is 0.91.
(b) Turbulence present: No quarter wave plates: Visibility of fringes is 0.2.
(c) With turbulence and quarter wave plates: Visibility of fringes is 0.56.
(d) Turbulence and quarter wave plates present but with polariser rotated to the horizontal. No interference pattern is visible.

3.1 DISCUSSION

We have reviewed recent work on the double passage imaging of deterministic objects obscured by a random screen. It would seem appropriate now to examine the limitations and assumptions inherent in this work and also to ask whether there is any ground for extending the idea to more demanding and general situations. Some of this will be speculative and we trust the reader to take it as such.

Firstly, let us reiterate that the analytical results quoted in this review have been obtained for a deterministic object under the idealised model of a random screen situated in the pupil plane. Certain cases may indeed correspond to this restricted situation. However, many situations transparently do not. Indeed, if we are considering a statistical object and average over the object ensemble, we cannot obtain diffraction-limited information about the object from the time-averaged image. In this case, we must examine the higher-order statistics of the image to have any possibility of retrieving this information. This becomes quite complex due to the non-linearity in the measured quantity, intensity.

The question of whether or not we can extend the argument to a more general case (i.e. an extended random medium or a screen located at some distance from the pupil) is an important one. Primarily, this is because the majority of interesting situations come under this category. In addition, in the special case of a phase screen located in the pupil, it is possible to obtain the same information by a simpler technique of simple point illumination - point reception. The key question then is: - " Does the coherence enhancement (due to the reciprocity of optical paths) which characterises the double passage effect always tells us something about the object ? Analysis of the most general situation is difficult and the results obtained do not easily lend themselves to clear physical interpretation. However, the recent paper of Yoon and Roy[22] has shown that in the case of an extended random medium, the time-averaged image still contains diffraction-limited information on the Fourier modulus of the object. The question of retrieval in this case has yet to be examined. We may also note the experimental result[19] that the visibility of fringes and thus degree of coherence does not depend (at least not significantly) on the location of the turbulence. In other words, the basic physical effect remains the same.

Most real objects are statistical (i.e incoherent) and the possibility of obtaining information about statistical objects obscured by a random medium under coherent illumination is a question of real interest. Using a simple model, it is possible to show that when an object illuminated by a plane wave scatters through turbulence (i.e single passage) the time-averaged energy spectrum can theoretically yield diffraction-limited information on the object spectrum[26] . Whether anything can be done with double passage remains an open question - certainly higher-order statistical methods would be necessary.

We may finally note that the approach discussed here does not yield phase information and this, of course, is a serious limitation. It is physically intuitive on the basis of the phase cancellation effect between two points in the pupil which are connected by crossover paths. In other words, the relative phases of two points in the pupil connected by reciprocal paths is determined only by their relative phase at the illumination points and is independent of the object. This remains true even when the object vibrates or randomly translates. Paradoxically, it may even be that practical application of the double passage effect lies not so much in imaging as in it's very insensitivity to object phase !

REFERENCES

1. F. Roddier, " The effects of atmospheric turbulence in optical astronomy " in Progress In Optics XIX North-Holland (1981).

2. A. G. Vinogradov, Yu A. Kravtsov and V.I. Tatarski, " The effect of intensification of back-scattering by bodies that are situated in a medium having random inhomogeneities ", Radiophys Quant Electron **16**, 818–823 (1973)

3. A. S. Gurvich and S.S. Kasharov, " Problem of enhancement of scattering in a turbulent medium ", Radiophys Quant Electron **20**, 547–549 (1977)

4. Yu A. Kravtsov and A.I. Saichev, " Effects of partial wave-front reversal during the reflection of waves in randomly inhomogeneous media ", Sov Phys JETP **56**, 291–294 (1982)

5. Yu A. Kravtsov and A.I. Saichev, " Effects of double passage of waves in randomly inhomogeneous media ", Sov Phys Usp **25**, 494–508 (1982)

6. V. P. Aksenov, V. A. Banakh, V. M. Buldakov, V. L. Mironov and O. V. Tikhomiriva, " Intensity fluctuations of light in the focal plane of a receiving telescope caused by round-trip propagation through a turbulent atmosphere ", Opt lett **10**, 107–109 (1984)

7. Yu A. Kravtsov and A.I. Saichev, " Properties of coherent waves reflected in a turbulent medium ", J. Opt. Soc. Am. A **2**, 2100–2105 (1985)

8. I. B. Episov and V. V. Zosimov, " Observation of partial phaseconjugation in the case of reflection in a randomly inhomogeneousmedium ", Opt Spectrosc **60**, 234-236 (1986)

9. V. A. Banakh, V. M. Buldakov and V. L. Mironov, " Effect of light-source spatial coherence in the reflected radiation intensity distribution in a telescope focal plane ", Opt Spectrosc **58**, 68–70 (1985)

10. E. Jakeman, " Enhanced backscattering through a deep random phase screen ", J. Opt. Soc. Am. A **5**, 1638–1648 (1988)

11. E. Jakeman, P. R. Tapster and A. R. Weeks, " Enhanced backscattering through a deep random phase screen", J. Phy. D: Appl Phys **21**, S32–S36 (1988)

12. P. R. Tapster, A. R. Weeks and E. Jakeman, " The observation of backscattering enhancement through atmospheric phase screens", J. Opt. Soc. Am. A **6**, 517–522 (1989)

13. Yu. N. Barabanenkov, Yu. A. Kravtsov, V.D. Ozrin and A.I. Saichev, " Enahanced backscattering in optics ", in Progress In Optics XXIX (Elsevier Science Publishers B.V. (1991)

14. R. L. Fante, " Imaging of an object behind a random phase screen using light of arbitrary coherence ", J. Opt. Soc. Am. A **2**,2318–2329 (1985)

15. E. Jakeman, " Active imaging through a random phase screen " J. Phys. D: Appl Phys. Vol 24, 227 - 232 (1991)

16. T. Mavroidis and J.C. Dainty, " Imaging after double passage through a random screen ", Opt Lett, August 1st (1990)

17. T. Mavroidis, C.J. Solomon and J.C. Dainty, " Imaging a coherently illuminated object after double passage through a random screen ", J. Opt. Soc. Am. A (**8**, 1003 - 1013,) (1991).

18. C.J. Solomon, R.G. Lane, T. Mavroidis and J.C. Dainty, " Double passage imaging through a random screen using a non-redundant aperture ",J. Mod. Opt, Vol. 38, No. 10, 1993-2008 (1991)

19. A.N. Bogaturov,A.A.D. Cañas, J.C. Dainty,A.S. Gurvich, V.A. Myakinin, C.J. Solomon, N.J. Wooder, " Observation of the enhancement of coherence by backscattering through turbulence ", Optics Communications, Vol 87 No. 1, 1-4 (1992).

20. C.J. Solomon and J.C. Dainty, " Use of polarisation in double passage imaging through a random screen", Optics Communications, Vol 87 (1992), 207-211.

21 A.N. Bogaturov, A.S. Gurvich, V.A. Myakinin, J.C. Dainty, C.J. Solomon, N.J. Wooder, " Use of polarisation in interferometry after double passage through turbulence ", Optics letters May 15th (1992).

22. D. G. Roy and G. Yoon, " Double passage recovery of an object embedded in a random medium ", Opt Lett, Vol. 17, no. 8 (April 1992)

23. I. Freund, " Time reversal symmetry and image reconstruction through multiple scattering media ", J. Opt. Soc. Am. A, vol. 9, no. 3 (1992)

24. W.T. Rhodes and G. Welch " Determination of a coherent wave field after double passage through a diffuser " J. Opt. Soc. Am. A, vol. 9, no. 2 (1992)

25. W. A. Shurcliff, " Polarised Light ", Oxford University press (1962)

26. C.J. Solomon and J.C. Dainty, " Imaging a coherently illuminated object through a random screen using a dilute aperture ", J. Opt. Soc. Am. A, Vol. 9, No. 8 (1992)

IMPLEMENTATION OF ADAPTIVE OPTICS

G. Rousset

Office National d'Etudes et de Recherches Aérospatiales (ONERA),
BP 72, 92322 Châtillon Cedex, France

1 Introduction

Atmospheric turbulence severely limits the angular resolution of the ground-based large telescopes. Whatever their aperture diameter, the angular resolution is seldom better than an arcsecond in good astronomical sites.[1] In the past twenty years, new techniques have come to light to overcome this natural limitation and reach the diffraction limit.[2] They are based on either a posteriori digital processing, as in Speckle Interferometry[3,4] and Deconvolution from Wavefront Sensing[5], or real-time wavefront correction as in Adaptive Optics[6] (AO). It is nevertheless agreed that AO brings the most desired improvement for astronomical observations.[7] Although the AO technique was initially proposed by Babcock in 1953,[8] only a few applications have been carried out in astronomy.[9,10,11] Most of the developments in this field have been supported by defense programs.[6,12] Recently, important results have been declassified concerning the laser guide star technique.[12,13,14] Stimulated by the coming new generation of very large telescopes, several programs are now undertaken to implement AO at a number of astronomical sites.[13,14]

Figure 1 outlines the basic components of an AO system, set at a telescope focus.[15] Light from a star is collected by the telescope and is reflected on a deformable mirror. The surface of this mirror is adjusted in real-time to compensate for the turbulence-induced wavefront (WF) distortions. Part of the light is sent to an imaging plane and part to a WF sensor. This sensor analyses the beam in order to derive the WF errors. A control system processes the signal delivered by the WF sensor and deduces the phase values required to drive the deformable mirror. By this way, the loop is closed and the incoming WF is flattened. Therefore, the resulting image is near diffraction-limited. An AO system is in fact a multichannel servo-loop system. The deformable mirrors are very powerful WF correctors. They are generally made of a continuous thin plate mirror and discrete piezoelectric actuators, pushing and pulling on the rear face of the thin mirror.[7] Different types of WF sensor have been proposed.[6,7] The Shack-Hartmann WF sensor is commonly used. It measures the WF slopes within a number of subapertures sampling the telescope aperture.[15] Recently, F. Roddier has proposed a new concept of AO system using WF curvature sensing and bimorph deformable mirror.[16]

Section 2 is devoted to the study of the parameters to take into account for designing an AO system. A modal expansion of the WF is used for a detailed analysis of this problem. The two fundamental limitations of AO are presented in Section 3. The modal analysis is also used for the discussion. The COME ON program is presented in Section 4. The reported results allow the evaluation of the system performance. The main points of this paper are summarized in Section 5.

2 Design of an AO system

2.1 First considerations

The requirements for an AO system are directly deduced from the characteristics of the turbulence to correct. The two main atmospheric turbulence parameters, scaling the system, are:[1]

- r_0 the Fried's diameter which is a measure of the effect of turbulence on a WF.[17] It is related to the field coherence length ρ_0 by $r_0 = 3.18\ \rho_0$. It is obtained by:

$$r_0 = \left[0.423\ k^2 \int_0^\infty C_n^2\ (z)\ dz\right]^{-3/5} \tag{1}$$

where $C_n^2(z)$ is the structure constant of the refractive index fluctuations which is a measure of the turbulence strength along the line of sight. For typical turbulence conditions, r_0 is of the order of 10 cm for visible light. Since r_0 is proportional to $\lambda^{6/5}$ (Eq.(1)), it is much larger for infrared (IR) light: for instance 60 cm at 2.2 μm.

- $\bar{v}$ the average wind velocity which determines the correlation time of the WF distortions. $\bar{v}$ is defined by:[18]

$$\bar{v} = \left[\left(\int_0^\infty C_n^2(z)\ v^{5/3}(z)\ dz\right) / \left(\int_0^\infty C_n^2\ (z)\ dz\right)\right]^{3/5} \tag{2}$$

where v(z) is the wind velocity perpendicular to the line of sight. For typical turbulence conditions, $\bar{v}$ is of the order of 10 m/s.

Related to these two atmospheric parameters, the basic design characteristics for an AO system are:

- The number of spatial degrees of correction N, i.e. the number of actuators of the deformable mirror or the number of subapertures of the WF sensor.
- The correction bandwidth f_{AO} which is a measure of the speed of the servo-loop.

In order to specify N, the fitting WF error can be used. The fitting error is the difference between the turbulence-induced WF aberration and the correction made by a deformable mirror with N actuators, assuming an AO system of infinite bandwidth. This spatial error is due to the finite number of degrees of correction while the turbulence is theoretically described by an infinite number of degrees of freedom. The expression of the fitting WF error variance σ_N^2 for an optimized system is:[10,19]

$$\sigma_N^2 = 0.3\ N^{-5/6}\ (D/r_0)^{5/3} \qquad (rd^2) \tag{3}$$

where D is the telescope diameter. A currently used rule of thumb is to specify N to be of the order of (D/r_0),[2] e.g. on a 4 m telescope in visible N = 1600 for typical conditions.[7] From Eq. (3), this leads to a fitting error σ_N = 0.55 rd (or $\sigma_N = \lambda/11$). Because of the dependence of r_0 on $\lambda^{6/5}$ (Eq. (1)), a number of projects are pursued to implement AO on astronomical telescopes in IR, hence limiting N.[13,14]

For bandwidth specification, the temporal rejection error can be used. The rejection error is the residual WF error after filtering the turbulence distortions by the closed-loop servo response, assuming an AO system with an infinite number of degrees of correction. This temporal error is due to the finite speed of the servo-loop: finite correction bandwidth. The expression of the rejection error σ_f^2 has been calculated by Greenwood:[20]

$$\sigma_f^2 = \left(0.4\ \frac{\bar{v}}{r_0}\ \frac{1}{f_{AO}}\right)^{5/3} \tag{4}$$

A currently used rule of thumb is to specify f_{AO} to be of the order of $\bar{v}/r_0$, e.g. in visible f_{AO} = 100 Hz in typical conditions.[7] From Eq. (4), this leads to a rejection error σ_f = 0.47 rd (or $\sigma_f = \lambda/13$).

The quality of the compensated image is frequently expressed in terms of Strehl ratio *Sr*. The Strehl ratio is defined as the ratio of the image peak intensity to the peak intensity of a perfect diffraction-limited image. In first approximation when the residual WF error variance σ_S^2 is relatively small, *Sr* is given by:

$$Sr = \exp\ (-\ \sigma_S^2) \tag{5}$$

For the two above-mentioned cases, i.e. $\sigma_S^2 = \sigma_N^2$ and $N = (D/r_0)^2$, and, $\sigma_S^2 = \sigma_f^2$ and $f_{AO} = \bar{v}/r_0$, *Sr* is equal to 0.74 and 0.8 respectively.

2.2 Modal expansion

For a more detailed analysis of AO and turbulence correction, WF modal expansion is a useful tool. Such an expansion allows to characterize the spatial and temporal behavior of the turbulence relevant when designing or optimizing an AO system. For convenience, only the Zernike polynomials are used in this paper. Other sets of modes have already been considered in the literature such as the Karhunen Loeve polynomials,[21] the mirror deformation eigenmodes[22] or optimal correction

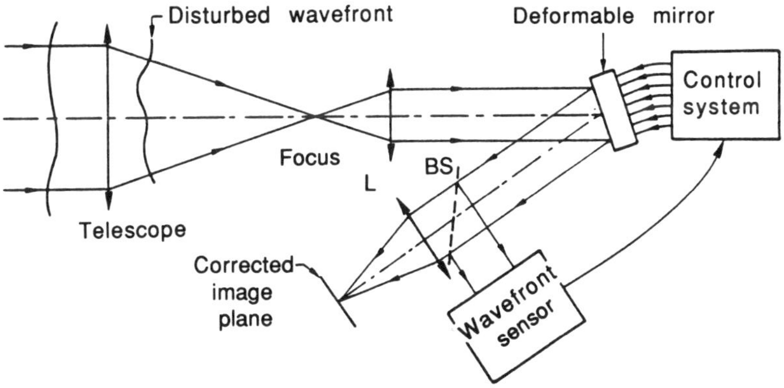

Fig. 1. Principle of an adaptive optics system. L: relay optics, BS: beam splitter.

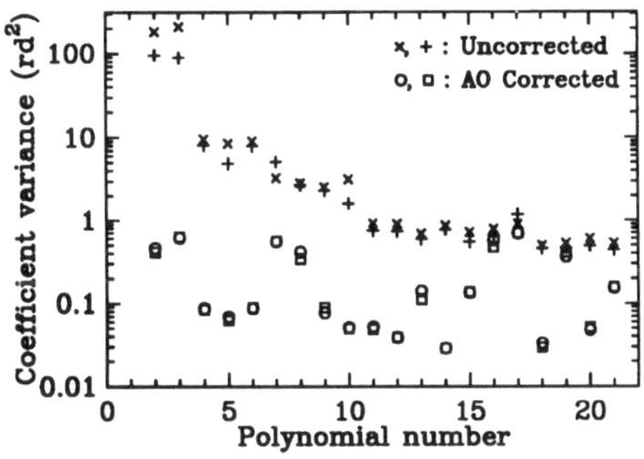

Fig. 2. Variance of the Zernike coefficients $<a_i^2>$ versus the polynomial number i, calculated at 0.5 μm with 2000 samples. (×,+): two sets of uncorrected WF, (O,□): two sets of AO corrected WF. $r_0(0.5\ \mu m) \approx 10$ cm, $\bar{v} \approx 5$ m/s.

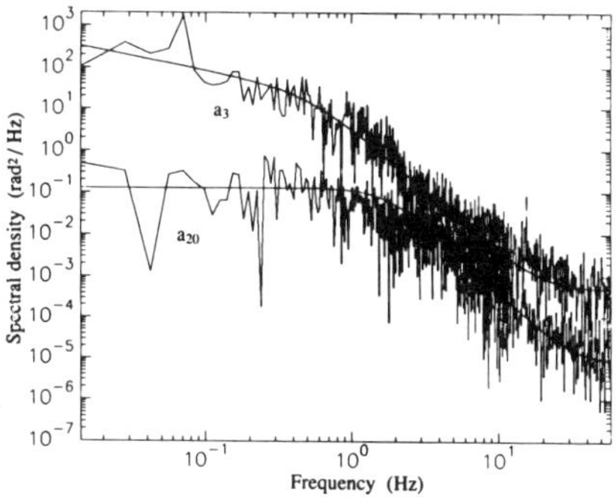

Fig. 3. Measured and fitted PSDs of two Zernike coefficients: a_3 a tilt, a_{20} a 5th order polynomial. Computed at 0.5 μm. $r_0(0.5\ \mu m) = 9$ cm, $\bar{v} = 4$ m/s.

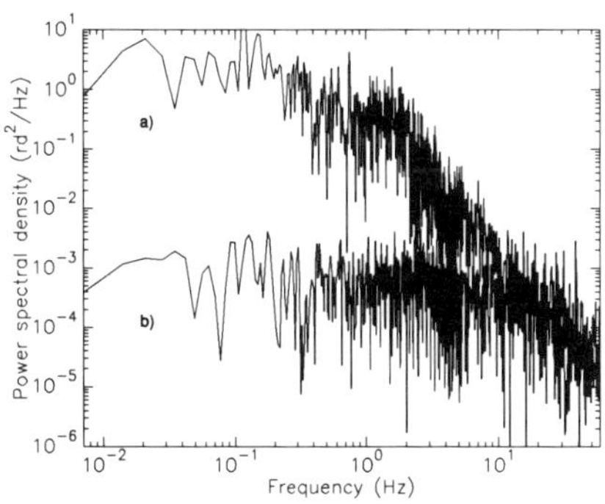

Fig. 4. Measured PSDs of Zernike coefficient a_4 (defocus). a) from an uncorrected WF set, b) from an AO corrected set, computed at 0.5 μm. $r_0(0.5\ \mu m) = 10$ cm, $\vec{v} = 5$ m/s, $f_{AO} = 25$ Hz.

functions.[23] The main advantage of the Zernike polynomials is their analytic expression which allows the derivation of their properties in the turbulence.[19,21,24,25]

Let us consider an AO system controlling directly the first N Zernike polynomials. In order to understand how to choose N, the main properties of the turbulent phase S(x,y), expanded on the Zernike polynomials $Z_i(x,y)$, are presented first. The expansion is expressed as:

$$S(x,y) = \sum_{i=1}^{\infty} a_i \, Z_i \, (x,y) \tag{6}$$

where the a_i's are the expansion coefficients obeying Kolmogorov statistics.[19] The Zernike polynomials are usually classified by their radial degree n and for each radial degree by their azimuthal frequency m ($m \leq n$). Figure 2 shows measures of coefficient variances $<a_i^2>$ versus the polynomial index i, for the first 21 Zernike polynomials (made with the COME ON AO system described in Section 4).[26] Crosses (+,×) are two sets of turbulence-induced WF distortions, without any correction by the servo-loop. The other points in Fig. 2 will be discussed later. As theoretically predicted,[19] the low order polynomials are dominant in the perturbation of the WF: the two tilts together (Z_2 and Z_3, n = 1) represent about 87% of the WF variance while the defocus and the two astigmatisms (Z_4, Z_5 and Z_6, n = 2) about 7% only. It is clear that the compensation of the low order modes must be first well done before thinking to extend efficiently the compensation to higher order modes. Very low order instruments have already been developed as: fast tip-tilt correction (limited to Z_2 and Z_3, n = 1),[27,28] or, curvature sensing and correction (limited to the first 13 Zernike polynomials).[16] They have been successfully used to improve image quality.

In Fig. 2, Crosses are a spatial representation of the turbulent WF on the Zernike polynomials: a Zernike spectrum of the turbulence. The radial degree n corresponds to the spatial frequency. The coefficient variances can be expressed as:

$$\left\langle a_i^2 \right\rangle = c(n) \left(\frac{D}{r_0} \right)^{5/3} \tag{7}$$

where c(n) depends on the radial degree n only. For n sufficiently high (typically n > 5), c(n) is fitted by the asymptotic form $(n + 1)^{-11/3}$ which is the classical expression of the Kolmogorov spectrum. This behavior is clearly observed in Fig. 5 discussed in Section 3. When controlling the first N Zernike polynomials by an AO system, the fitting error is given by $\sum_{N+1}^{\infty} \left\langle a_i^2 \right\rangle$.[19] Taking into account the $(n+1)^{-11/3}$ behavior of $<a_i^2>$ and the number of polynomials per radial degree (equal to n+1), Eq. (3) is derived. It is very close to the form given by Noll.[19]

Associated to this spatial representation, the temporal power spectral density (PSD) of Zernike coefficients provides complementary information on the temporal characteristics of the turbulence. Figure 3 displays for comparison the PSDs of two Zernike coefficients: a_3 a tilt (n = 1), and a_{20} a polynomial of radial degree n = 5.[29] Note that the tilt fluctuations (140 rd^2 at 0.5 μm) have much higher strength than the 20th polynomial fluctuations (0.63 rd^2), as already seen in Fig. 2. The general form of these two spectra has been fitted. The behavior is summarized by a slow decrease at low frequency (a_3: $f^{-2/3}$, a_{20}: f^0) and after a knee a fall-off at high frequency ($f^{-10/3}$).[24,29] An increase of the knee frequency f_i with the increase of the polynomial number is also observed: f_i follows the approximated law $f_i = 0.3\ (n+1)\ \bar{v}/D$. This behavior suggests the requirement of the increase of f_{AO} when increasing N in order to obtain an efficient correction of the higher order modes. Notice that f_i does not depend on r_0 but on D. The high frequency behavior of the PSD is due to the filtering induced by the telescope aperture.

Let us now consider how the correction is made by the AO system. The compensation can be understood as a filtering process reducing the PSD of the WF. The servo-loop bandwidth f_{AO} represents in fact the maximum frequency at which the AO system is still able to compensate for the distortions. Figure 4 is an example of two measured PSDs of the defocus mode (a_4, n = 2) without and with correction.[26] The turbulence knee frequency f_4 is around 2 Hz. The system attenuates the PSD at low frequency up to 20 Hz, value close to the bandwidth of the COME ON system in open loop at 0 dB. The corrected PSD is in fact the uncorrected PSD multiplied by the error transfer function of the servo-loop. The attenuation in variance is around 100 (ratio of $<a_4^2>_{uncor.}$ and $<a_4^2>_{cor.}$). At low frequency, the corrected PSD is mainly limited by the noise level of the WF sensing. Decreasing r_0 and/or f_{AO} would modify the corrected PSD adding more turbulence residuals (Fig. 3 of Ref. 30).

For each controlled polynomial, the same type of compensation can be observed. As a result, in addition to the fitting error σ_N^2, a rejection error has to be considered for the N controlled polynomials. This rejection error is no more expressed by Eq. (4): σ_f^2 is now composed by the errors of N degrees of correction only. Fig. 2 displays for comparison spatial expansion of uncorrected and corrected sets of WF. Points (□,o) represent the rejection error of the system. Large attenuation of the low order polynomial fluctuations are observed. f_{AO} being the same for all the modes, the low order modes are much more attenuated relatively to the high orders, because of their lower knee frequency. In Fig. 2, the measured WF variance $\left(\sum_{2}^{21} <a_i^2>\right)$ is reduced from 450 (×) and 230 (+) rd^2 to the rejection error $\sigma_f^2 = 4.5\ rd^2$ at 0.5 μm. The corresponding fitting error σ_{21}^2 is evaluated to 8 rd^2 from Eq. (3) for r_0 = 10 cm.

To summarize, the residual WF error σ_S^2 can be expressed by:

$$\sigma_S^2 = \sigma_N^2 + \sigma_f^2 \tag{8}$$

- σ_N^2 is the fitting error generally given by Eq.(3). It represents the contribution of all the degrees of freedom of the turbulent WF distortions orthogonal to the N degrees of correction.
- σ_f^2 is the rejection error due to the servo-loop bandwidth limitations and is composed by the N degrees of correction. This terms will be completed in the next

section by the noise contribution and the anisoplanatism effect.

To conclude this section let us discuss the required trade-off between N and f_{AO} in the design of a system. For infinite bandwidth, i.e. ideal system, increasing N increases the quality of the image. In practice, f_{AO} is never infinitely large. With small bandwidth, low order compensation will only be possible. For large bandwidth, different systems can be designed depending on the goals, e.g. the image quality to achieve. Furthermore depending on the turbulence conditions, different number of degrees of correction can be implemented in order to optimize the correction capability of the system. Finally, note that increasing N increases the computing load by a factor proportional to N^2 while increasing f_{AO} increases it by a factor proportional to f_{AO}.

3 Fundamental limitations

Whatever the number of spatial degrees of correction and the servo-loop bandwidth, the correction brought by AO is limited by two fundamental phenomena independent of the system: the photon noise and the anisoplanatism effect.

3.1 Photon noise

This issue is addressed when the guide star, used for sensing the WF distortions, provides a limited number of photons. Hereafter is only considered the case of WF sensing made at visible wavelength by a Shack-Hartmann WF sensor equipped with a photon-noise-limited detector. For a more complete treatment of the noise problem, the reader can refer to the literature.[5,31,32] The basic relation expresses the dependence of the noise-induced variance of the measures $\sigma_m{}^2$, at the wavelength of the WF sensing λ_{WFS}, on the photon flux and the characteristics of the experiment:[15]

$$\sigma_m^2 = \frac{\pi^2}{2}\frac{1}{p}\left(\frac{w}{w_d}\right)^2 \qquad (rd^2) \qquad (9)$$

where p is the number of photons detected per subaperture, w the width of the guide star image at λ_{WFS} and w_d the width of the diffraction Airy pattern of a subaperture. As p, $\sigma_m{}^2$ is related to the quantum efficiency of the detector and the optical throughput of the system. Moreover, the extension of the image w must be as limited as possible: e.g. diffraction-limited image and unresolved guide star. Such requirements are seldomly achieved. In particular for astronomy, AO is implemented for IR wavelengths keeping the WF sensing at visible. In consequence, the WF sensor subaperture size d is generally of the order of $r_0(\lambda_{Im})$ where λ_{Im} is the IR wavelength at which the imaging is done. Therefore, d is much larger than $r_0(\lambda_{WFS})$ and the width of unresolved guide star is limited by the seeing. $\sigma_m{}^2$ expressed at λ_{Im} is given by:

$$\sigma_m^2 = \frac{\pi^2}{2}\frac{1}{p}\left(\frac{d}{r_0(\lambda_{WFS})}\right)^2\left(\frac{\lambda_{WFS}}{\lambda_{Im}}\right)^2 \qquad (rd^2) \qquad (10)$$

p/d^2 is in fact the photon density per unit of exposure time in the WF sensing.

Therefore, the required guide star magnitude is directly related to σ_m^2, the WF quality to achieve. In this regime where the WF sensing is limited by the seeing, the guide star magnitude does not depend on the aperture diameter neither on the subaperture size but only on $r_0^2(\lambda_{WFS})$.[31,32] Using large λ_{Im} allows to reach higher magnitude for same WF quality. The dependence of the guide star magnitude on λ_{WFS} is very low as deduced from Eqs. (10) and (1).

How is this measurement error propagated on the degrees of correction ? Figure 5 illustrates the noise expansion on the Zernike polynomials. The two sets of theoretical points present the turbulence-induced WF coefficient variance $<a_i^2>$ (×) and the noise-induced WF error coefficient variance $<b_i^2>$ (□). This figure displays two spatial spectra expressed in terms of Zernike polynomials. The turbulence spectrum (×) has the asymptotic form $(n+1)^{-11/3}$. The dependence is $(n+1)^{-2}$ for the noise spectrum.[33] Indeed, the noise of the measures is white and Shack-Hartmann sensor delivers WF slopes.[34] For an AO system having N degrees of correction, the noise is only propagated on the N degrees. The total noise induced error variance is given by $\sum_{2}^{N} <b_i^2>$ and is proportional to Log(N).[34] Figure 5 allows to understand a critical point in AO control at low level. Calling signal $<a_i^2>$ and noise $<b_i^2>$ a decrease of the signal to noise ratio (SNR) is observed in Fig. 5 when increasing the polynomial number i, i.e. the radial degree n. This is directly due to the behavior of the two spectra: $(n+1)^{-11/3}$ and $(n+1)^{-2}$. Depending on the seeing, i.e. the value of r_0, and on the guide star magnitude, the SNR per polynomial changes significantly. Polynomials concerned by very low SNR, will not be controllable, leading to a necessary spatial filtering of the WF. Hence, the guide star magnitude imposes the number of useful degrees of correction N_u for any AO system: $N_u \leq N$.

Fig. 6 displays two measured temporal PSDs of Zernike polynomials Z_3 and Z_{20}, without any correction by AO, in the case of low light level.[35] In this experiment, only an average number of 9 photons were detected per subaperture and per exposure time. The deduced SNR is: 9 for Z_3 and 0.5 for Z_{20}. For Z_{20}, the turbulence and noise contributions to the PSD are not simply distinguishable. In such a case, a specific servo-loop control algorithm must be used in order to filter the noise before driving the deformable mirror.[35] This can be done using a criterium which minimizes the residual error variance for each degree of correction. For such a task, the modal analysis is very helpful because each mode can be optimized independently, resulting in better noise rejection than in zonal control.[36] In zonal control, all the degrees of correction (the deformable mirror actuators) are mainly equivalent in terms of SNR. Let us notice that, in any case, the noise in the measurement is not fully propagated to the deformable mirror because of the finite value of the servo-loop bandwidth.

3.2 Anisoplanatism

The second fundamental limitation in AO is due to anisoplanatism effect.[37] This problem occurs for instance when the object of interest for astronomical observations is off-axis from the guide star used for the WF sensing.[1] In this case, the two light beams do not exactly go through the same disturbances, in particular when crossing the high altitude turbulence layers. The difference between the two WF is much larger when increasing the angular separation between the object and the guide star. Typical

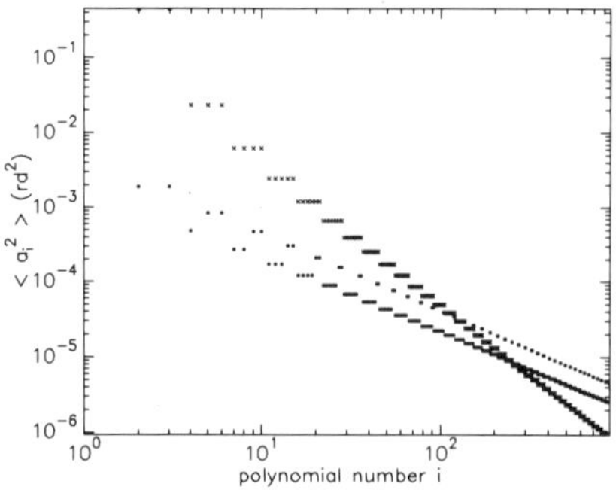

Fig. 5. Theoretical expansion of the turbulence-induced WF (×) and of the noise-induced WF error (□) on the Zernike polynomials.

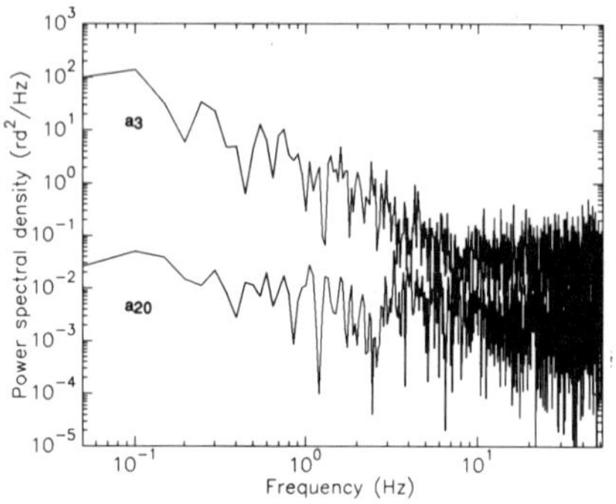

Fig. 6. Measured PSDs of two Zernike coefficients: a_3 and a_{20}, in the case of low light level: 9 detected photons per subaperture ($r_0 \simeq 12$ cm).

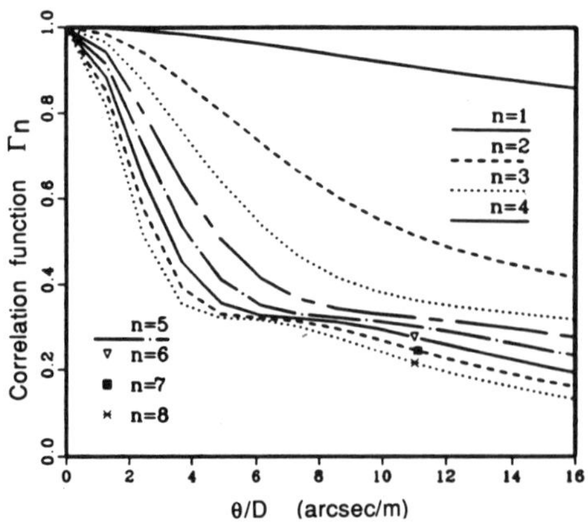

Fig. 7. Angular correlation function Γ_n, for n ranging from 1 to 8, derived from the Hufnagel C_n^2 profile.

values of isoplanatic patch size are a few arcseconds at visible, and a few tens of arcseconds at IR.[1,7] Once gain implementing AO for IR is much more favorable.

A modal expansion of the WF is once again a good way to analyse this problem.[25,38] Anisoplanatism means: after propagation the two WF are partially correlated. This partial correlation is better understood by modal expansion, e.g. Zernike expansion. Figure 7 shows the angular correlation functions Γ_n for a number of radial degrees n.[39] Γ_n is the average of the Γ_i's, the angular correlation functions of the polynomials having the same radial degree:

$$\Gamma_i = \langle a_{iob}\, a_{igs} \rangle \,/\, \langle a_i^2 \rangle \tag{11}$$

where a_{iob} and a_{igs} are the expansion coefficients for object and guide star, respectively. These Γ_n were calculated using the Hufnagel C_n^2 profile,[40] and integrating the propagation between 20 and 1 km. As a first remark, this figure reveals the strong dependence of Γ_n on the angular separation θ and n. If the decorrelation of the tilts (n = 1) is slow when increasing θ, for other n the decorrelation is steep. In addition, the 50% correlation angle decreases very fast for the first radial degrees. The peculiar shape of the curves is related to the Hufnagel profile. The first part of the curves (steep decorrelation) is due to the high altitude layer while the second part (the plateau) is related to the low altitude layers.

For a given angular separation θ, the only Zernike polynomials which are useful in the AO correction, are the polynomials having their 50% correlation angle larger than θ.[38,39] Indeed considering N polynomials in the correction and the sole anisoplanatism effect, the residual WF error is expressed by:

$$\sigma_S^2 = \sigma_N^2 + 2 \sum_{2}^{N} \langle a_i^2 \rangle \, (1 - \Gamma_i) \tag{12}$$

The second term of the left member of the equality is the anisoplanatism error for an AO system. This error, as for the rejection error and the noise induced error, is only present on the N degrees of correction. For each polynomial, the anisoplanatism error is $2 \langle a_i^2 \rangle (1 - \Gamma_i)$. A polynomial has to be considered for the compensation only if $\Gamma_i > 0.5$. Otherwise, the AO system amplifies the turbulence-induced WF distortions associated to the concerned polynomial. Hence, for each angular separation θ there is an optimal number of degrees of correction to consider. In such an analysis, no isoplanatic patch is defined, the system will only do its best. If required, the isoplanatic patch can be defined by the choice of an image quality to achieve, i.e. a given residual WF error.[38] But for astronomical applications, such a definition is not a key issue. Because of the use of the criterium $\Gamma_i > 0.5$ for the selection of the useful degrees of correction, the compensated image will be of better quality than without any compensation. Even a poor improvement may bring substantial gain in the observation, e.g. better SNR in the IR image.

As a summary of this section, photon noise and anisoplanatism limit the quality of the correction brought by AO. The both limitations are well analysed and managed by WF modal expansion. Indeed, for a given AO system with N degrees of correction, the useful number of degrees of correction N_u will be reduced by noise and anisoplanatism effects: $N_u \leq N$. This number has to be adapted in function of the observing conditions.[41] Optimized modal control allows to obtain the best achievable results.

4 The COME ON program

The COME ON program is a joint France-European Southern Observatory (ESO) project to implement AO on large astronomical telescope. The first phase is now accomplished: it was the construction and test of the COME ON AO prototype system. This system was aimed to compensate images at IR wavelengths. The second phase of the program is already well advanced: it concerns the upgrade of the COME ON system in a new system called COME ON PLUS.

4.1 The COME ON AO prototype system

The COME ON system was installed at the Cassegrain focus of the ESO 3.6 m telescope at La Silla (Chile) for 5 observing runs during the years 90 and 91.[13,26,30,42] The goal was the demonstration of the potential of AO for astronomy. In October 1989, first diffraction-limited images had been obtained at 2.2 μm on the 1.52 m telescope of the Observatoire de Haute Provence (France).[11,43]

The system components and characteristics are described in details in Ref. 44. Table 1 summarizes its main features. The deformable mirror is a continuous facesheet mirror with 19 actuators, developed at Laserdot (France). A separated two-axis plan mirror compensates for the overall WF tilt. Both mirrors are driven by a digital control loop. The Shack-Hartmann WF sensor uses 20 subapertures. The system has therefore 20 degrees of correction. The WF computer processes the signal delivered by the WF sensor and provides in real-time 40 WF slopes to the control computer. This one is in charge of the calculation of 21 commands sent to the two adaptive mirrors. The command algorithm is based on a modal analysis of the zonal control, allowing the filtering of the system eigenmodes.[45] The bandwidth of the servo-loop was first 9 Hz at 0 dB in open-loop. It has been extended in April 1991 to 25 Hz at 0 dB in open-loop.[26] The IR imaging camera is a 32 × 32 InSb array with a 0.1 or 0.05 arcsec pixel size on the sky. The camera is read-out noise limited. Images were recorded at the standard astronomical photometric bands.[30] The exposure time depends on the stellar magnitude and ranges from 16 ms to a few seconds. Long exposure images sum up about a hundred of elementary images.

In the five observing runs at La Silla, three different cameras have been tested for the WF sensing channel.[46] For visible WF sensing, the intensified Reticon was limited to magnitude $m_v = 10$. Therefore, it was replaced for astronomical observations by an Electron-Bombarded CCD (EBCCD). It was developed at Laboratoire d'Electronique Philips (LEP) in France, and allows to reach the photon-noise limit. Guide stars of magnitude, $m_R = 11.5$ in full correction and $m_R = 13$ in tilt correction only, were used to close the loop. IR WF sensing at 2.2 μm was also tested on the sky using a HgCdTe IR CCD.[46] Guide stars of magnitude $m_K = 1.5$ were used to close the loop. All these magnitude values were far from optimum as the system optical

Table I - Main characteristics of the COME ON and COME ON PLUS systems

	COME ON	COME ON PLUS
Deformable mirror	19 piezoelectric actuators ± 7.5 μm stroke ± 1500 V high voltage 3.5 kHz bandwidth	52 ± 5 μm ± 420 V 12 kHz
Tip-tilt mirror	Gimbal mount, 140 Hz bandwidth, 6 marsec sensitivity, 7 arcsec dynamic	
WF sensor (Shack-Hartmann)	20 subapertures 100 × 100 pixels 120 Hz sampling frequency Two stage intensified Reticon array	32 56 × 56 200 Hz (max.) EBCCD (low flux)
Control system	Hardwired WF computer 68020 based control computer Zonal control f_{AO} = 9 to 25 Hz	id. COME ON 56000 DSP board Optimized modal 35 Hz (max.)

Fig. 8. a) compensated and b) uncompensated long exposure images at λ_{Im} = 1.68 μm. r_0(0.5 μm) ≈ 12 cm, $\overline{v}$ = 4 m/s, f_{AO} = 25 Hz.

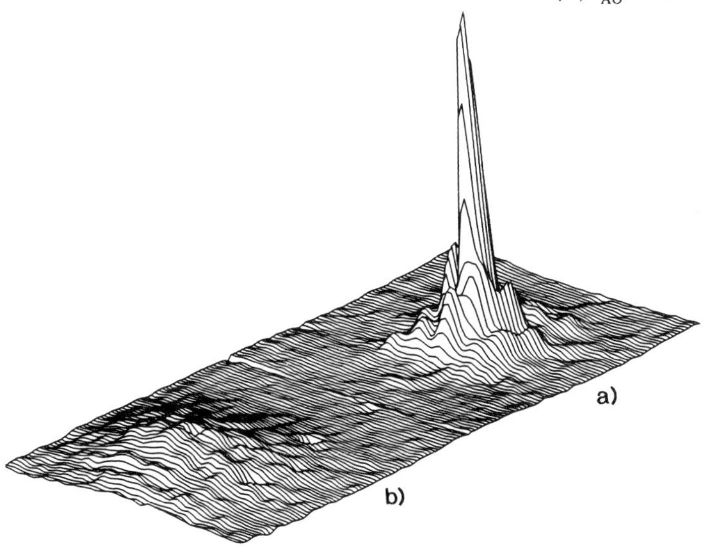

throughput was measured to be only 0.03.

For a complete investigation of the system capabilities, WF slopes, as measured by the WF sensor, were recorded in the cases of: turbulence-induced WF distortions (without any correction) and AO correction residual errors. The data reduction aims to reconstruct the WF phase from its slopes. The modal reconstruction performs the expansion of the phase on the Zernike polynomials.[5] Due to the finite number of subapertures only the first 21 coefficients are determined from the measures. Examples of result have been presented in Fig. 2 of Section 2. Fitting the measured Zernike coefficient variances with the theoretical behavior (Eq. (7)) allows to determine r_0. This was regularly done in the data reduction. For temporal analysis, PSDs are calculated for each Zernike coefficient a_i. Examples have already been given in Figs. 3 and 4 of Section 2 and Fig. 6 of Section 3. v was also determined during the data reduction by fitting the Zernike PSDs with their theoretical law (see Fig. 3).[29]

4.2 Performance

The results obtained on the ESO 3.6 m telescope have already been reported together with a detailed analysis of the correction performance.[13,26,30,42,46] Only a few selected results are presented here, they correspond to relatively bright stars, m_v ranging from 6 to 9. Figure 8 displays compensated (a) and uncompensated (b) long exposure images of a single star which demonstrate the system capability. The images are recorded at 1.68 μm (H band). For the uncompensated image, the telescope aberrations are removed by a static deformation of the deformable mirror. The measured experimental conditions are given in the caption. The quality of the images is measured by their Strehl ratio *Sr* and their full width at half maximum (FWHM). In Fig. 8 for the compensated image $Sr_c = 0.24$ and $FWHM_c = 0.12''$ and for the uncompensated image $Sr_u = 0.02$ and $FWHM_u = 0.65''$.

The most obvious gain brought by AO is the energy concentration measured by *Sr*. As a direct consequence, energy concentration leads to a gain in SNR in all circumstances where a source is imaged in a background or detector-noise-limited mode, e.g. at IR. The gain in SNR is given by Sr_c/Sr_u. In Fig. 8, Sr_c/Sr_u is about 10. In April 91, Sr_c ranging from 0.1 to 0.9 have been measured, depending on the turbulence conditions (r_0(0.5 μm) and v), and on the spectral bands.[26,42] These results were a significant improvement, from the first obtained in 90,[30] by the increase of the servo-loop bandwidth. At the longest wavelengths where $D/r_0(\lambda_{Im}) \le 6$, Sr_c is always larger than 0.35, corresponding to a rms residual WF error smaller than $\lambda_{Im}/6$. While at shorter wavelengths where $D/r_0(\lambda_{Im}) > 6$, Sr_c may be smaller than 0.3.

Another important gain brought by AO is angular resolution, measured by FWHM. Interesting resolution can be achieved even if only partial correction is provided by the system. As in Fig. 8, the compensated image FWHM is close to the 0.096" diffraction limit. In this case, only a fraction of the image total energy is gathered in the central core. Figure 9 shows the FWHM of the compensated images versus λ_{Im}. FWHM is nearly diffraction-limited for $\lambda_{Im} \ge 2.2$ μm whatever the r_0(0.5 μm) value is. The best resolution is obtained at a critical wavelength: 1.7 μm. For $\lambda_{Im} \le 1.7$ μm, AO brings a significant resolution gain. This gain is very dependent on the turbulence conditions. In this λ_{Im} range, Sr_c is mostly smaller than 0.3. In 90 with

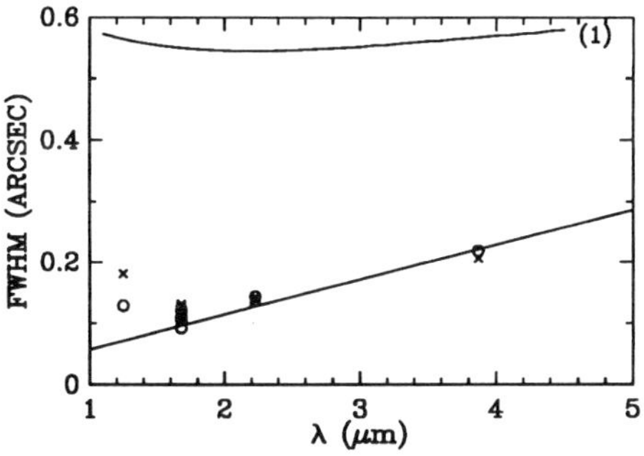

Fig. 9. FWHM versus λ_{Im}. Curve 1: theoretical long exposure uncorrected image for r_0(0.5 μm) = 16 cm. Straight line: telescope diffraction limit. Measured long exposure corrected images for r_0(0.5 μm) ranging: from 9 to 11 cm (□), 11 to 15 cm (×) and 15 to 19 cm (○). $\bar{v}$ ranges from 3 to 6 m/s. f_{AO} = 25 Hz.

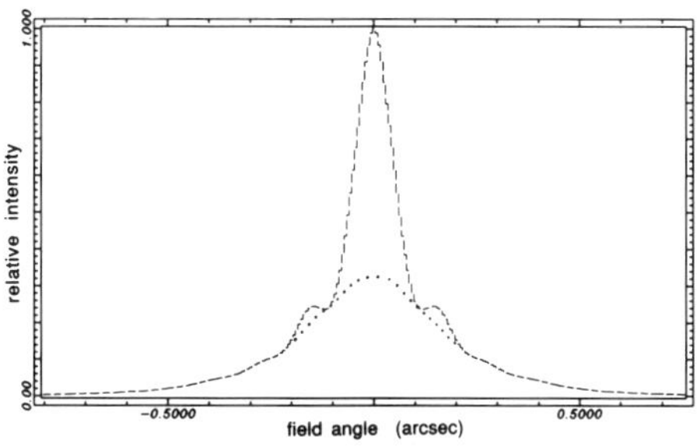

Fig. 10. Long exposure profile of a compensated image at λ_{Im} = 1.68 μm. r_0(0.5 μm) = 18 cm, $\bar{v}$ = 6 m/s, f_{AO} = 25 Hz.

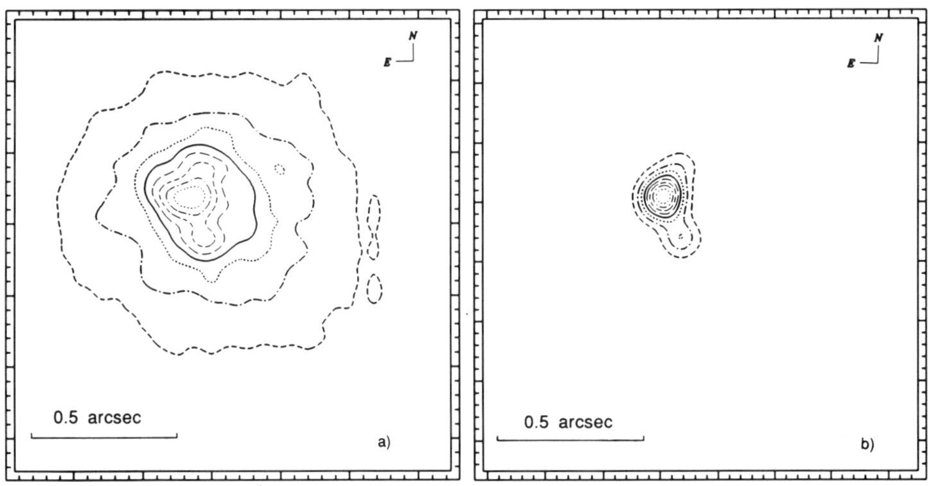

Fig. 11. Images of the low mass binary star Gl570B at λ_{Im} = 1.68 μm: a) the raw compensated image, b) the compensated image after deconvolution.

a 9 Hz bandwidth and $\bar{v}$ = 20 m/s, the critical wavelength was only 2.2 μm.[30]

Figure 10 shows a typical long exposure image profile taken at λ_{Im} = 1.68 μm.[42] A common feature, observed on a large λ_{Im} range (1.25 μm to 2.2 μm) in the compensated images, is the presence of two components in the profile: namely a sharp diffraction-limited core on top of a broad halo as foreseen by the theory.[47,48] In Fig. 10, a fit of the halo is superimposed on the image profile. It was deduced from the substraction of the theoretical diffraction pattern to the image.[26,42] The central core is in fact fully coherent while the halo is an incoherent superimposition of residual speckles.[47,49] In Fig. 10, the fraction of the image total energy in the core is about 16%. Therefore 84% of the energy is incoherently spread around in the halo. FWHM of the core is 0.096". FWHM of the halo is 0.36" while the uncompensated image has a FWHM of about 0.5". Even if the amount of energy concentrated in the central core is relatively small, the available coherence is of paramount importance for interferometric combination of large telescopes.

The characterization of partial correction in AO is of great importance in astronomy.[48] Partial correction is not only related to the fitting error[47,48] but also to a number of additional errors (described in Sections 2 and 3).[49] In front of such a problem, the key function to study is the structure function D_S of the residual phase error.[42,49] The long exposure optical transfer function (OTF) after AO correction is the product of the OTF of the aberration free telescope and of exp(- $D_S/2$). Therefore, the correction strategy must address the minimization of D_S. This approach was proposed by Tatarskii a few years ago.[23] At high spatial frequency, the phase S is decorrelated and as a result $D_S = 2\,\sigma_S^2$. Hence, the fraction of energy in the central core is given by exp(- σ_S^2).[49] The minimization of σ_S^2 maximizes the fraction of energy in the central core.

To demonstrate the astronomical value of the COME ON AO system, double stars and complex objects have been observed.[30,50] Figure 11 shows compensated images of a binary star Gl570B at 1.68 μm: a) as recorded, b) after deconvolution by a CLEAN algorithm.[50] A reference star was used to record the point spread function to be used in CLEAN. Note in the raw image the halo around the binary. The two component separation was derived: 0.15".

4.3 The COME ON PLUS system

The upgrade version of COME ON has been constructed during 1992 and is going to be set up on the ESO 3.6 m telescope in December 1992. This second prototype is dedicated to routine astronomical observations. It will provide insights for the design of the AO system of the ESO Very Large Telescope (VLT). This upgraded system has already been presented in several conferences.[13,14,35] Table 1 summarizes its characteristics. The main improvements are:

- a better correction capability with 52 degrees of correction as well as a larger servo-loop bandwidth of 35 Hz;
- a better sensitivity of the WF sensor by the use of an EBCCD camera, allowing to reach the photon noise limit at low light level. For bright guide stars, the intensified Reticon will still be used;
- a new modal control allowing the optimization of the correction in function of the

observing conditions: turbulence and guide star magnitude;

- an optical throughput of 30%.

The expected performance has been estimated using the experience obtained with the COME ON system. For bright guide star, the servo-loop bandwidth is sufficient to achieve a very good correction with all the 52 modes for $\bar{v} \leq 20$ m/s. This will allow to achieve $Sr_c = 70\%$ at 1.7 µm and partial correction at visible wavelengths. For guide star of $m_R \geq 11$, the optimized modal control will filter the high order modes. A limiting magnitude of $m_R = 16$ is expected for tilt correction only. The goal of the December run is the evaluation of the system performance on the sky.

5 Conclusion

In this paper, both the design issue and the performance evaluation has been addressed. It has been shown the importance of modal analysis for system design and operation. The good compensation of the low order modes is of great interest because they represent the most part of the WF distortions. In addition, they have the largest correlation time and correlation angle and the highest SNR. Designing a system with a large number of degrees of correction N requires also a large servo-loop bandwidth f_{AO}. A trade-off has to be found between N and f_{AO} which depends on the goals to attempt.

High image quality requires high SNR, i.e. a lot of photons. Such cases are generally seldom in astronomy. Reaching guide star magnitude $m_R \geq 12$ leads unavoidably to the partial correction regime. Moreover, the anisoplanatism limits the field where a guide star can be found around the astronomical object. For all these problems, the modal analysis allows to optimize the system in function of v, photon noise and angular correlation.

The european COME ON program has been presented. The diffraction limit of the ESO 3.6 m telescope at 1.68 µm, i.e. 0.096" resolution, has been achieved. The importance of the servo-loop bandwidth was clearly identified. In 1991, Sr_c larger than 0.35 has been measured for $D/r_0(\lambda_{Im}) \leq 6$. Recorded image profiles in partial correction confirmed the theoretical behavior: a diffraction-limited core on top of a diffuse halo. For partial correction characterization, the structure function of the residual WF errors is the key function to study and optimize. The upgraded system called COME ON PLUS has also shortly been presented.

The next step in AO for astronomy is the reduction of the fundamental limitations: photon noise and anisoplanatism. A promising way is the implementation of artificial guide star on astronomical sites. The demonstration of the capability of laser guide star system has already been done for military purposes.[12,13,14] The future trend would be the development of such a system for astronomy.[13,14] By this way, the limiting magnitude concept would perhaps disappear and the anisoplanatism problem would be hopefully reduced.

Acknowledgements

Firstly, I wish to thank all the people of the "COME ON team" who worked a lot for the success of the experiment. Particular thanks are addressed to J.C. Fontanella in addition to my colleagues at ONERA. For 6 years, the COME ON program has joined together ONERA, as prime contractor, Observatoire de Paris-Meudon, the companies Laserdot and LEP, and ESO. It has been carried out under contracts from ESO and Ministère de la Recherche et de la Technologie (France). Early developments of AO components were independently supported by Direction des Recherches Etudes et Techniques of Ministère de la Défense (France).

References

1. F. Roddier, "The effects of atmospheric turbulence in optical astronomy," Progress in Optics (E. Wolf Ed.) **XIX**, North Holland, 281-377 (1981).
2. D.M. Alloin, J.M. Mariotti (Eds.), "Diffraction-limited imaging with very large telescopes," NATO ASI Series C **274**, Kluwer Academic Publishers (1989).
3. A. Labeyrie, "Attainment of diffraction-limited resolution in large telescopes by Fourier analysing speckle patterns in star images," Astron. Astrophys. **6**, 85-87 (1970).
4. G. Weigelt, "Modified astronomical speckle interferometry, speckle masking," Opt. Commun. **21**, 55-59 (1977).
5. J. Primot, G. Rousset, J.C. Fontanella, "Deconvolution from wavefront sensing: a new technique for compensating turbulence-degraded images," J. Opt. Soc. Am. A **7**, 1598-1608 (1990).
6. J.W. Hardy, J.E. Lefebvre, C.L. Koliopoulos, "Real-time atmospheric compensation," J. Opt. Soc. Am. **67**, 360-369 (1977).
7. F. Merkle, "Adaptive optics developments at ESO," ESO Conference Proceedings **30**, 639-656 (1988).
8. H.W. Babcock, "The possibility of compensating astronomical seeing," PASP **65**, 229-236 (1953).
9. A. Buffington, F.S. Crawford, R.A. Muller, C.D. Orth, "First observatory results with an image-sharpening telescope," J. Opt. Soc. Am. **67**, 304-305 (1977).
10. J.W. Hardy, "Active optics in astronomy," ESO Conference Proceedings "Scientific importance of high angular resolution at infrared and visible wavelengths," 25-39 (1981).
11. G. Rousset, J.C. Fontanella, P. Kern et al., "First diffraction-limited astronomical images with adaptive optics," Astron. Astrophys. **230**, L29-L32 (1990).
12. R.Q. Fugate, D.L. Fried, G.A. Ameer et al., "Measurement of atmospheric wavefront distortion using scattered light from a laser guide star," Nature **353**, 144-146 (1991).
And C.A. Primmerman, D.V. Murphy, D.A. Page et al., "Compensation of atmospheric optical distortion using a synthetic beacon," Nature **353**, 141-143 (1991).
13. OSA Conference on "Adaptive optics for large telescopes," Technical Digest Series **19**, Maui (August 1992).
14. ESO Conference on "Progress in telescope and instrumentation technologies," Garching (April 1992), in press.
15. J.C. Fontanella, "Analyse de surface d'onde, déconvolution et optique active," J. Optics (Paris) **16**, 257-268 (1985).
16. F. Roddier, M. Northcott, J.E. Graves, "A simple low-order adaptive optics system for near-infrared applications," PASP **103**, 131-149 (1991).
17. D.L. Fried, "Statistics of a geometric representation of wavefront distortion," J. Opt. Soc. Am. **55**, 1427-1435 (1965).

18. F. Roddier, J.M. Gilli, G. Lund, "On the origin of speckle boiling and its effects in stellar speckle interferometry," J. Optics (Paris) **13**, 263-271 (1982).
19. R.J. Noll, "Zernike polynomials and atmospheric turbulence," J. Opt. Soc. Am. **66**, 207-211 (1976).
20. D. Greenwood, "Bandwidth specification for adaptive optics systems," J. Opt. Soc. Am. **67**, 390-393 (1977).
21. J.Y. Wang, J.K. Markey, "Modal compensation of atmospheric turbulence phase distortion," J. Opt. Soc. Am. **68**, 78-87 (1978).
22. J.P. Gaffard, G. Ledanois, "Adaptive optical transfer function modeling," SPIE Proc. **1542**, 34-45 (1991).
23. V.I. Tatarskii, "Adaptive systems and coherence," Radiophys. and Quantum Electronics **24**, Part I 590-597 and Part II 598-606 (1981).
24. C.B. Hogge, R.R. Butts, "Frequency spectra for the geometric representation of wavefront distortions due to atmospheric turbulence", IEEE Trans. Antennas Propagat. **AP-24**, 144-154 (1976).
25. G.C. Valley, S.M. Wandzura, "Spatial correlation of phase-expansion coefficients for propagation through atmospheric turbulence," J. Opt. Soc. Am. **69**, 712-717 (1979).
26. F. Rigaut, E. Gendron, P. Léna et al., "Partial correction with the adaptive optics system COME ON," ESO Conference on "High resolution imaging by interferometry," Garching (October 1991).
27. R. Racine, R.D. McClure, "An image stabilization experiment at the Canada-France-Hawaii telescope", PASP **101**, 731-736 (1989).
28. F. Maaswinkel, F. Bortoletto, R. Buonanno et al., "Image stabilization with DISCO: results of first observing run", ESO Conference Proceedings **30**, 751-760 (1988).
29. P.Y. Madec, J.M. Conan, G. Rousset, "Temporal characterization of atmospheric wavefront for adaptive optics," in Ref. 14.
30. F. Rigaut, G. Rousset, P. Kern et al., "Adaptive optics on a 3.6 m telescope: results and performance," Astron. Astrophys. **250**, 280-290 (1991).
31. M. Séchaud, G. Rousset, V. Michau et al., "Wavefront sensing in imaging through the atmosphere: a detector strategy," SPIE Proc. **1543**, 479-490 (1990).
32. B.M. Welsh, C.S. Gardner, "Performance analysis of adaptive-optics systems using laser guide stars and slope sensors", J. Opt. Soc. Am. A **6**, 1913-1923 (1989).
33. F. Rigaut, E. Gendron, "Laser guide star in adaptive optics," Astron. Astrophys., in press.
34. R.J. Noll, "Phase estimates from slope-type wavefront sensors," J. Opt. Soc. Am. **68**, 139-140 (1978).
35. E. Gendron, J.G. Cuby, F. Rigaut et al., "The COME-ON PLUS project: an upgrade of the COME-ON adaptive optics prototype system," SPIE Proc. **1542**, 297-306 (1991).
36. W.H. Southwell, "Wavefront estimation from wavefront slope measurements," J. Opt. Soc. Am. **70**, 998-1006 (1980).
37. D.L. Fried, "Anisoplanatism in adaptive optics," J. Opt. Soc. Am. **72**, 52-61 (1982).
38. F. Chassat, G. Rousset, J. Primot, "Theoretical evaluation of isoplanatic patch size," SPIE Proc. **1114**, 14-22 (1989).
39. F. Chassat, "Calcul du domaine d'isoplanétisme d'un système d'optique adaptative fonctionnant à travers la turbulence atmosphérique," J. Optics (Paris) **20**, 13-23 (1989).
40. R.E. Hufnagel, "Variations of atmospheric turbulence," in OSA Digest of the "Topical Meeting on optical propagation through turbulence", Boulder (1974).
41. P. Léna, "Perspectives de l'optique adaptative en astronomie," in "Compte-rendus

du forum optique adaptative et VLT," Observatoire de Paris (April 1990).
42. F. Rigaut, P. Léna, P.Y. Madec et al., "Latest results obtained with the COME ON adaptive optics prototype system," in Ref. 14.
43. P. Kern, F. Rigaut, P. Léna et al., "Adaptive optics prototype system for IR astronomy II: first observing results," SPIE Proc. **1237**, 345-355 (1990).
44. G. Rousset, J.C. Fontanella, P. Kern et al., "Adaptive optics prototype system for IR astronomy I: system description," SPIE Proc. **1237**, 336-340 (1990).
45. C. Boyer, V. Michau, G. Rousset, "Adaptive optics: interaction matrix measurements and real-time control algorithms for the COME ON project," SPIE Proc. **1237**, 406-415 (1990).
46. F. Rigaut, J.G. Cuby, M. Caes et al., "Visible and infrared wavefront sensing for astronomical adaptive optics," Astron. Astrophys. **259**, L57-L60 (1992).
47. R.C. Smithson, M.L. Peri, R.S. Benson, "Quantitative simulation of image correction for astronomy with a segmented active mirror," Applied Optics **27**, 1615-1620 (1988).
48. J.M. Beckers, "The VLT interferometer IV. The utility of partial adaptive optics," SPIE Proc. **1236**, 154-163 (1990).
49. J.M. Conan, P.Y. Madec, G. Rousset, "Evaluation of image quality obtained with adaptive optics partial correction," in Ref. 14.
50. F. Rigaut, M. Combes, C. Dougados et al., "Astrophysical results with COME-ON," in Ref. 14.

OPTICAL REMOTE SENSING

James H. Churnside
NOAA Wave Propagation Laboratory
325 Broadway
Boulder, CO 80303
(303) 497-6744

Abstract

The statistics of optical scintillation in the atmosphere can be used to measure several atmospheric parameters. The most obvious parameter is the refractive turbulence strength parameter, C_n^2, which is proportional to the normalized variance of irradiance fluctuations in weak path-integrated turbulence conditions. The inner scale of turbulence can be inferred under the same conditions using a number of techniques. These include measuring the correlation function of the irradiance fluctuations, comparing the variance of aperture-averaged scintillations with nonaveraged values, and comparing scintillations at two wavelengths. The outer scale can be inferred most directly from structure-function measurements. Cross-wind speed can be obtained from the spatial-temporal correlation function of scintillation. Spatial filtering techniques can be applied to obtain values at a number of different points along the propagation path. Inner-scale values obtained with visible or near-infrared light also provide information about heat flux from the surface into the atmosphere, averaged along the propagation path. Simultaneous measurements can be made at other wavelengths where the index of refraction depends on water vapor concentration, and humidity flux can also be estimated.

1. Introduction

The idea that scintillations can be used to infer information about the atmosphere should not come as a surprise to anyone who has observed heat shimmer of the horizon on a sunny day. The shimmering is sometimes more and sometimes less violent; this is clearly related to the strength of the refractive turbulence along the optical path from the source scene to the observer. This is related, as one would suspect, to the rate at which the atmosphere is being heated by the surface. The shimmering of the image also seems to drift to the left or right, sometimes slowly and sometimes more rapidly. This drift is clearly related to the speed of the cross wind along the path. Thus, relative estimates of three atmospheric parameters, turbulence strength, heat flux, and cross wind, are available to an unaided observer.

With the aid of instrumentation, quantitative estimates of these parameters and several others can be obtained. Among the others are the inner and outer scales of turbulence and fluxes of water vapor and other chemical species.

These parameters are of two types. The first type are directly related to the propagation of light. They include the refractive turbulence strength and the inner and outer scales of turbulence. The other type are the meteorological parameters, including cross wind, heat flux, and species fluxes. In this paper, we discuss some of the scintillation techniques that have been used to measure both types of parameters.

Optical techniques that do not use scintillation have also been considered, but are not considered here. Examples include measurements of turbulence strength by image spreading and inner scale measurements by angle-of-arrival fluctuations.

2. Propagation parameters

Optical propagation through the atmosphere is determined by the spectrum of the random refractive index fluctuations in the air. For many propagation geometries, the important scales are much less than the turbulent outer scale, L_0, and the spectrum can be written as[1,2]

$$\Phi_n(K) = 0.033\, C_n^2\, K^{-11/3}\, F(K l_0)\,, \tag{1}$$

where K is the turbulent wavenumber, C_n^2 represents turbulence strength, and F is a function of K times the inner scale, l_0. This function is unity at small wavenumbers and drops to zero for very large wavenumbers.

In principle, the simplest quantity to measure using scintillation is C_n^2. The normalized irradiance variance for a point source with optical wavenumber k propagating a distance L through homogeneous turbulence is given by

$$\sigma_I^2 = 0.5 k^{7/6}\, L^{11/6}\, C_n^2 \tag{2}$$

as long as this quantity is much less than one and the inner scale is much less than the Fresnel zone size $(L/k)^{1/2}$. Thus, C_n^2 can be measured by looking at a point source with a small detector and comparing the variance of the detector output to this expression as long as the conditions for its validity are met.

Near the ground, the conditions required to use Eq. (2) are difficult to meet. We would like to use a long path so that inner-scale effects can be neglected. However, on a long path, fluctuations become large, and the perturbation approach that was used to derive Eq. (2) is no longer valid. At our test range, we typically see l_0 values of about 1 cm and C_n^2 values near 10^{-12} m$^{-2/3}$ during the day. For a wavelength of 800 nm, we require a path much longer than 800 m if inner-scale effects are to be neglected. However, Eq. (2) predicts a variance of almost 12 for an 800-m path under these conditions, and thus cannot be used.

The solution is to use large, incoherent transmitter and receiver apertures. If the apertures are much larger than the inner scale, the system is not sensitive to inner-scale size eddies. If the apertures are much larger than the Fresnel zone, the signal is averaged and the variance is reduced so that the perturbation theory is valid over a much longer path. The normalized signal variance for this case can be written as[3]

$$\sigma_s^2 = 16\pi^2 k^2 L \int_0^1 du \int_0^\infty dKK\,\Phi_n(K)\,\sin^2\left[\frac{K^2 Lu(1-u)}{2k}\right] \times \left\{\frac{2J_1[0.5\,KDu]}{0.5\,KDu}\,\frac{2J_1[0.5\,KD(1-u)]}{0.5\,KD(1-u)}\right\}^2, \tag{3}$$

where u is the distance from the transmitter normalized by the path length, J_1 is the Bessel function, and D is the aperture diameter. If C_n^2 is constant along the path, it can be pulled outside of the integrals, which then can be evaluated. If the aperture diameter is large compared to the inner scale, the Bessel function cuts off the integral at relatively small wavenumbers, and the result is not sensitive to inner scale.

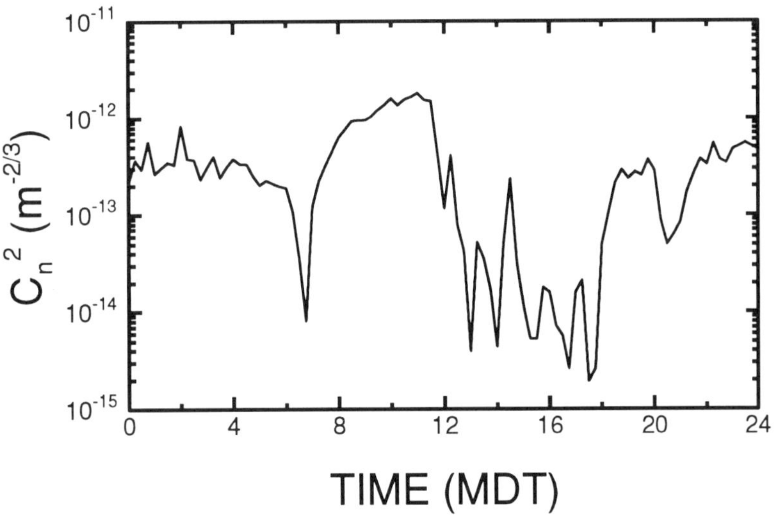

Fig. 1. Time series of 15-min averages of C_n^2 for a typical Colorado summer day.

We have used large-aperture scintillometers to measure C_n^2 during optical propagation experiments. A typical configuration employs a 100-mW light-emitting diode collimated with a 15-cm mirror as a transmitter. A second mirror 500 m away collects the light onto a photodiode, and the variance of the signal is calculated. This instrument provides good estimates of turbulence strength for values from about 10^{-16} to more than 10^{-12} $m^{-2/3}$. A typical summer diurnal cycle of 15-min averages of C_n^2 obtained with this instrument is presented in Fig. 1, which clearly shows a drop in turbulence when convective clouds form at noon.

If turbulence strength is not constant along the path, the result is a weighted average of C_n^2, where the weighting function is determined by the sine function and the two aperture filter functions $[2J_1(x)/x]$. By changing the transmitter and receiver apertures, various averages of turbulence strength can be obtained.

One application of this principle uses sinusoidal spatial filters at the transmitter and receiver.[4,5] Each combination of filters has a fairly sharp weighting function at some particular path position. If the transmitter has a filter wavelength of Λ_T and the receiver has a wavelength of Λ_R, the system will respond to turbulent fluctuations whose wavelength Λ is given by

$$\frac{1}{\Lambda} = \frac{1}{\Lambda_T} + \frac{1}{\Lambda_R} . \tag{4}$$

It will only respond to fluctuations near a distance z from the transmitter, where

$$z = \frac{\Lambda}{\Lambda_R} L . \tag{5}$$

Enough different combinations provide a profile of turbulence strength along the path.

Vertical profiles can be obtained using starlight and a spatially filtered receiver,[6,7] although the resolution is not as good as a system with spatial filters at both ends of the path.

Profiles of turbulence can also be made using crossed optical beams.[8] The covariance of scintillations of two crossed beams is measured. The correlated scintillations arise from turbulence in the common propagation volume where the two beams cross. By changing the propagation geometry, beams can be made to cross at several points along the path to provide a profile. To obtain vertical profiles, binary stars can be used as a source of the two beams.[9]

Making a scintillation measurement of inner scale involves combining two measurements that are sensitive to different scales in the turbulent spectrum and then fitting the results to the theoretical spectrum.

From Eq. (2), we can see that this can be accomplished with two scintillometers with different aperture sizes.[3,10] The ratio of the signal variances from a small-aperture system and a large-aperture system depends on inner scale but not on C_n^2. Figure 2 presents the theoretical relationship for a 4.4-cm-diameter incoherent system and a 1-mm-diameter laser system on a 260-m path. The points in the figure are 5-min averages of the optical measurements plotted against 5-min averages of inner scale obtained from hot-wire probes.[10] The agreement is as good as can be expected for a comparison of a path-averaging instrument with a point probe.

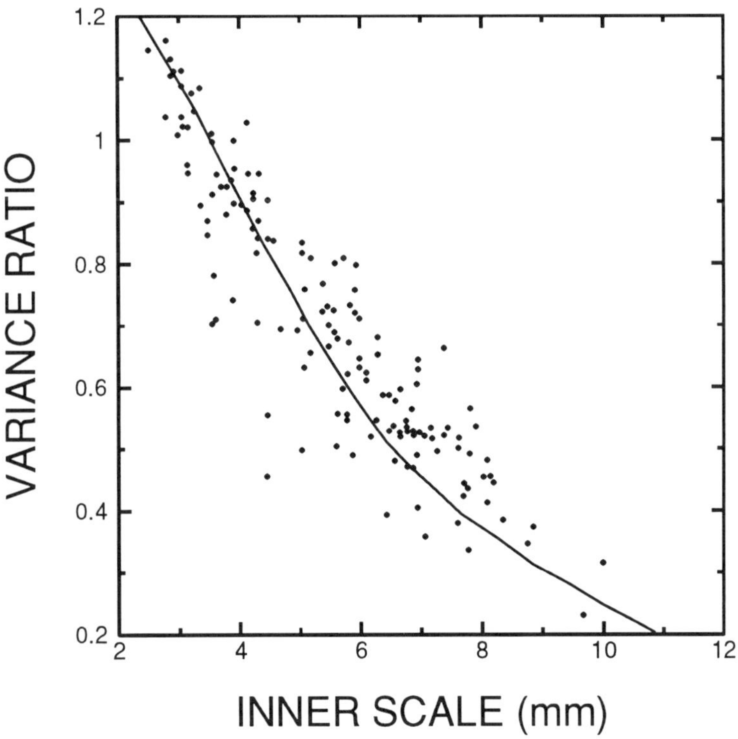

Fig. 2. Plot of variance ratio as a function of inner scale for large (4.4 cm) and small (1 mm) apertures on a 260-m path. Points are 5-min averages of data and the curve is the theoretical relationship.

We can also obtain sensitivity to two scale sizes by using two different optical wavelengths.[11] For weak scintillations, the irradiance variance is most sensitive to scales near the Fresnel zone size $(L/k)^{1/2}$. Inner scale can be obtained from the ratio of the variances at two wavelengths, as in the two-aperture-size method. Alternatively, inner scale can be obtained from the correlation of the fluctuations at the two wavelengths. The two-color-variance-ratio method is sensitive to errors because of refractive dispersion, and the correlation method is probably the better of the two.[12]

Another technique for obtaining two scale sizes is to use two different path lengths.[13] However, this method is seriously limited by any small differences in the turbulence along the two paths, and has not been used recently.

C_n^2 and l_0 can be measured simultaneously by measuring the spatial covariance of scintillations and inverting the resulting set of simultaneous equations.[14] For weak path-integrated turbulence, the covariance of the observed irradiance from a diverging source is

$$C(\rho) = 16\pi^2 k^2 L \int_0^1 du \int_0^\infty dKK \;\; \Phi_n(K)\, J_0(Ku\rho)\, \sin^2\left[\frac{K^2 Lu(1-u)}{2k}\right], \tag{6}$$

where ρ is the separation of the two points. An equation of this type is obtained for each separation used in the covariance measurement. This technique has also been used to infer profiles of turbulence strength.[15,16]

Less has been done in the area of outer-scale measurements using scintillation. Coulman et al.[17] have used stellar scintillation to infer vertical profiles of outer scale. They report outer-scale values of less than 5 m to altitudes of 18 km. Other techniques produce larger values, and the scintillation results are still controversial.

3. Meteorological parameters

Any technique for measuring C_n^2 can be easily extended to measure the component of the wind velocity across the path. Two spatially separated receivers are used, and the drift of scintillation patterns across them is measured using spatio-temporal correlation methods.

One of the simplest techniques is to use two receiver apertures in the large-aperture technique. For receivers whose centers are separated by a distance ρ, the spatio-temporal covariance of the output signals with a time delay τ is given by[18]

$$C(\rho,\tau) = 16\pi^2 k^2 L \int_0^1 du \int_0^\infty dKK\ \Phi_n(K)\ \sin^2\left[\frac{K^2 Lu(1-u)}{2k}\right] J_0[K(u\rho - v\tau)]$$
$$\times \left\{\frac{2J_1[0.5\ KDu]}{0.5\ KDu}\ \frac{2J_1[0.5\ KD(1-u)]}{0.5\ KD(1-u)}\right\}^2, \tag{7}$$

where v is the velocity of wind velocity that is aligned with the two detectors. Note that this velocity will generally be a function of the path position, and the result is some average along the path. The autocovariance can also be found from Eq. (7) by setting the separation equal to zero.

Five techniques for obtaining cross wind from measured covariance functions have been suggested. In the first of these, the velocity is obtained from

$$v = \rho/\tau_p, \tag{8}$$

where τ_p is the time delay at the peak of the covariance curve. This technique would work well if the scintillation pattern were not evolving as it translated across the receivers. Evolution of the pattern forces the peak of the covariance to smaller time delays, and this technique tends to overestimate wind speed.

The second technique recognizes that the slope of the covariance function $\delta C/\delta\tau$ at zero time delay is proportional to the cross-wind speed. The constant of proportionality is easily obtained from Eq. (7). The difficulty with this technique is that the derivative must be approximated by a finite difference measurement. For large finite differences, the function is not linear and the slope is not accurately estimated. For small finite differences, the differences are small and the system is very sensitive to system noise.

The third technique recognizes that the width of the autocovariance function is proportional to the inverse of cross-wind speed. Again, the constant of proportionality is found from Eq. (7). The major drawback of this technique is that the direction of the cross wind is not obtained.

In the next technique, the covariance and the autocovariance are both measured, and the time delay at which the two are equal (τ_c) is found. The cross wind is given by

$$v = 0.5\ \rho/\tau_c. \tag{9}$$

This technique works well, but it is necessary to measure the covariance and autocovariance at a fairly large number of time delays to obtain the crossing point accurately. Under these conditions, it may be better to measure a similar number of points across the entire covariance curve and infer the wind from the value that provides the best fit between those measurements and the theoretical curve. A detailed comparison of these five methods is presented in Ref. 18.

Velocities can also be measured using the spatial filtering techniques. If two receiver filters are used over the same aperture but shifted horizontally by $\Lambda_R/4$, the spatial responses of the two filters are sine and cosine, respectively. Thus, the covariance function will be zero at zero time delay and maximum when the time delay is equal to a quarter of the transit time of the turbulent wavelength at the path position being interrogated. In this case, because of the filtering, looking for the delay to the peak works very well. One can also infer speed from the peak of the power spectrum of the received signal in this case, although this does not provide direction information.

As in the case of turbulence measurements, profiles of cross-wind speed along the propagation path can be obtained by using several transmitter and receiver spatial wavelengths. The spatial resolution obtained by this technique is roughly given by the path length divided by the total number of cycles in the transmitter and receiver spatial filters. Figure 3 is an example of the expected and measured path-weighting functions from Ref. 5. With 10 cycles on both transmitter and receiver, the measured full width at half maximum is about 4.5 m on a 110-m path.

Crossed-beam techniques also can be used to obtain profiles of wind speed. As above, a second receiver is added and the covariance of the two signals is measured. Inference of the wind speed from the covariance is similar to the other methods.

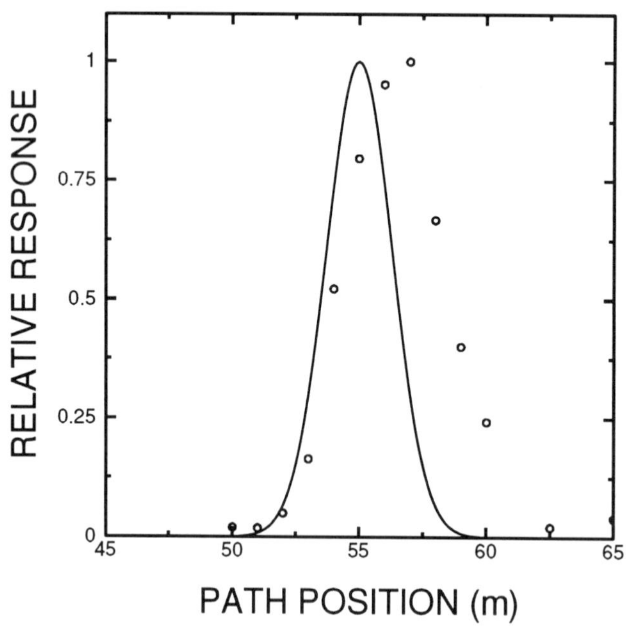

Fig. 3. Path-weighting function of spatial-filtered scintillometer over 100-m path. Curve is theoretical response and circles are measured values.

Scintillation measurement of fluxes is less direct. The theory necessary to obtain fluxes from structure parameters[19] is based on Monin-Obukhov similarity theory and can only be done in the surface boundary layer, which includes the lowest few tens of meters of the atmosphere, and only over terrain that is fairly homogeneous horizontally, such as farmland or oceans.

Surface fluxes of heat and momentum can be obtained from a measurement of C_n^2 and l_0 and the atmospheric pressure and temperature[20] as long as C_n^2 is measured in the visible or near-infrared, where refractive fluctuations are determined only by temperature fluctuations. In this case, the structure parameter of temperature fluctuations can be found from the structure parameter of refractive index fluctuations using the relationship

$$C_T^2 = \left[7.8 \times 10^{-5}\left(1 + \frac{7.5 \times 10^{-15}}{\lambda^2}\right)\frac{P}{T^2}\right]^{-2} C_n^2, \tag{10}$$

where P is atmospheric pressure and T is temperature. The energy dissipation rate is found from the inner scale using the relationship

$$\epsilon = 3000\,\frac{\nu^3}{l_0^4}, \tag{11}$$

where ν, the kinematic viscosity of air, also depends on pressure and temperature.

In Monin-Obukhov similarity theory, a length scale is defined as

$$Z = \frac{2.5\, C_P T(-M_0^3)^{1/2}}{g\rho^{1/2} H_0}, \tag{12}$$

where C_P is the constant-pressure heat capacity, g is the gravitational acceleration, and ρ is the mass density. The fluxes can be found from the relationships

$$C_T^2 = -\frac{H_0^2}{C_P^2\,\rho M_0 z^{2/3}}\, f\left(\frac{z}{Z}\right) \tag{13}$$

and

$$\epsilon = -\frac{2.5 M_0^{3/2}}{\rho^{3/2} z}\, g\left(\frac{z}{Z}\right), \tag{14}$$

where z is the height above the surface and the functions f and g are empirical functions given by

$$\begin{aligned} f(x) &= 4.9(1-x)^{-2/3} && \text{for } x \le 0 \\ &= 4.9(1+2.4x^{2/3}) && \text{for } x \ge 0 \end{aligned} \tag{15}$$

and

$$\begin{aligned} g(x) &= (1 + 0.5\,|x|^{2/3})^{3/2} && \text{for } -2 < x \le 0 \\ &= (1 + 2.5x^{3/5})^{3/2} && \text{for } 0 \le x < 1\,. \end{aligned} \tag{16}$$

Hill et al.[20] concluded that this technique works very well. Figure 4 is a plot of their data comparing temperature flux obtained by the scintillation technique with that obtained by the in situ technique of eddy correlation. Good agreement was obtained; the small bias in the data is probably due to a small calibration error in the scintillometer.

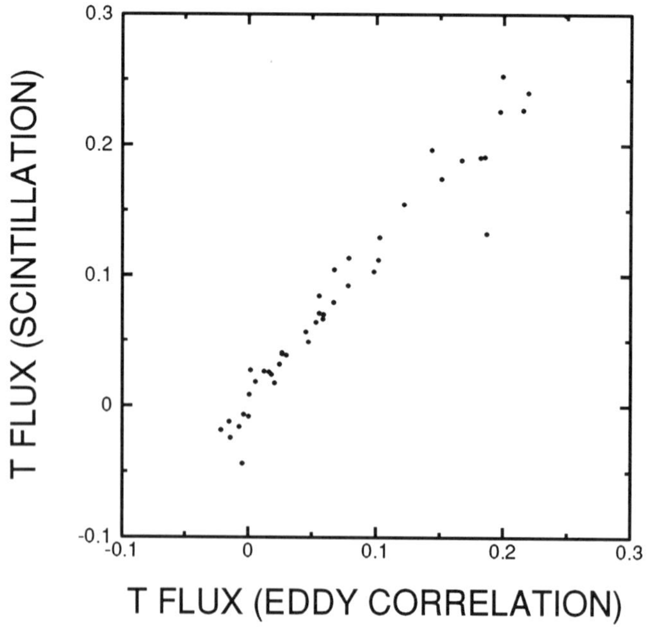

Fig. 4. Comparison of temperature flux (in units of K m s^{-1}) measured by scintillation with that measured by eddy correlation.

At wavelengths in the far-infrared, millimeter, and radio regions of the spectrum, the index of refraction of air depends on the humidity as well as the temperature. Thus, measurements of C_n^2 at these wavelengths contain information about the structure parameters of humidity C_Q^2 and the cross parameter C_{TQ}. Simultaneous measurements of inner scale and of C_n^2 at three wavelengths that have different humidity sensitivities (e.g., visible, far-infrared, and millimeter) can be used to infer heat, humidity, and momentum fluxes.[21,22] By assuming that temperature and humidity fluctuations are highly correlated (either positively or negatively), one of the three wavelengths can be eliminated.[21,23] Experimental evidence supports the use of this assumption.[24]

In principle, the scintillation technique could also be extended to obtain fluxes of other nonreactive chemical species such as methane. Hill[23] showed that the concentrations of passive additives are highly correlated under Monin-Obukhov similarity theory. This implies that differences in the fluxes of different chemical species in the boundary are because of differences in average concentration gradient. A path-averaged measurement of the concentration gradient of water vapor and of species X, combined with a scintillation measurement of average water vapor flux over the same path, can provide an estimate of the average flux of X over the path.

4. Conclusions

Scintillation techniques can be used effectively to measure turbulence strength, inner scale, wind velocity across the path, and surface fluxes. By controlling the spatial distribution of the source and the receiver sensitivity, one can control the dependence of the response on various parameters. One application of this is a system that is sensitive to a limited portion of the path; a combination of several such systems provides a profile of the parameter of interest along the path.

Scintillation techniques have several advantages over in situ techniques. Scintillation is generally not very sensitive to turbulence in the immediate vicinity of the transmitter or receiver. Thus measurements can be made with minimal disruption by the instrument. Scintillation measurements inherently represent averages over some region of the atmosphere. This reduces questions of representativeness of in situ measurements. Scintillation instruments are ideal for long-term unattended operation. They can operate through windows and therefore be mounted in sealed housings out of the weather.

5. References

1. R. J. Hill, "Models of the scalar spectrum for turbulent advection," *J. Fluid Mech.* **88**, 541-562 (1978).

2. J. H. Churnside, "A spectrum of refractive turbulence in the turbulent atmosphere," *J. Mod. Opt.* **37**, 13-16 (1990).

3. R. J. Hill and G. R. Ochs, "Fine calibration of large-aperture optical scintillometers and an optical estimate of inner scale of turbulence," *Appl. Opt.* **17**, 3608-3612 (1978).

4. R. W. Lee, "Remote probing using spatially filtered apertures," *J. Opt. Soc. Am.* **64**, 1295-1303 (1974).

5. J. H. Churnside, R. J. Lataitis, and R. S. Lawrence, "Localized measurements of refractive turbulence using spatial filtering of scintillations," *Appl. Opt.* **27**, 2199-2213 (1988).

6. J. Vernin and F. Roddier, "Experimental determination of two-dimensional spatiotemporal power spectra of stellar light scintillation. Evidence for a multilayer structure of the air turbulence in the upper troposphere," *J. Opt. Soc. Am.* **63**, 270-273 (1973).

7. G. R. Ochs, T. Wang, R. S. Lawrence, and S. F. Clifford, "Refractive turbulence profiles measured by one-dimensional spatial filtering of scintillations," *Appl. Opt.* **15**, 2504-2510 (1976).

8. T. Wang, S. F. Clifford, and G. R. Ochs, "Wind and refractive-turbulence sensing using crossed laser beams," *Appl. Opt.* **13**, 2602-2608 (1974).

9. A. Rocca, F. Roddier, and J. Vernin, "Detection of atmospheric turbulent layers by spatiotemporal and spatioangular correlation measurements of stellar-light scintillation," *J. Opt. Soc. Am.* **64**, 1000-1004 (1974).

10. G. R. Ochs and R. J. Hill, "Optical-scintillation method of measuring turbulence inner scale," *Appl. Opt.* **24**, 2430-2432 (1985).

11. E. Azoulay, V. Thiermann, A. Jetter, A. Kohnle, and Z. Azar, "Optical measurement of the inner scale of turbulence," *J. Phys.* **D21**, 541-544 (1988).

12. R. J. Hill and R. J. Lataitis, "Effect of refractive dispersion on the bichromatic correlation of irradiances for atmosphere scintillation," *Appl. Opt.* **28**, 4121-4125 (1989).

13. P. M. Livingston, "Proposed method of inner scale measurement in a turbulent atmosphere," *Appl. Opt.* **11**, 684-687 (1972).

14. R. G. Frehlich, "Estimation of the parameters of the atmospheric turbulence spectrum using measurements of the spatial intensity covariance," *Appl. Opt.* **27**, 2194-2198 (1988).

15. D. L. Fried, "Remote probing of the optical strength of atmospheric turbulence and of wind velocity," *Proc. IEEE* **57**, 415-420 (1969).

16. J. W. Strohbehn, "Remote sensing of clear-air turbulence," *J. Opt. Soc. Am.* **60**, 948-950 (1970).

17. C. E. Coulman, J. Vernin, Y. Coqueugniot, and J. L. Caccia, "Outer scale of turbulence appropriate to modeling refractive-index structure profiles," *Appl. Opt.* **27**, 155-160 (1988).

18. T. Wang, G. R. Ochs, and R. S. Lawrence, "Wind measurements by the temporal cross-correlation of the optical scintillations," *Appl. Opt.* **20**, 4073-4081 (1981).

19. J. C. Wyngaard and S. F. Clifford, "Estimating momentum, heat, and moisture fluxes from structure parameters," *J. Atmos. Sci.* **35**, 1204-1211 (1978).

20. R. J. Hill, G. R. Ochs, and J. J. Wilson, "Measuring surface-layer fluxes of heat and momentum using optical scintillation," *Boundary-Layer Meteor.* **58**, 391-408 (1992).

21. W. Kosiek and M. H. A. J. Herben, "Evaporation derived from optical and radio-wave scintillation," *Appl. Opt.* **22**, 2566-2570 (1983).

22. E. L. Andreas, "Three-wavelength method of measuring path-averaged turbulent heat fluxes," *J. Atmos. Oceanic Technol.* **7**, 801-814 (1990).

23. R. J. Hill, "Implications of Monin-Obukhov similarity theory for scalar quantities," *J. Atmos. Sci.* **46**, 2235-2244 (1989).

24. R. J. Hill, R. A. Bohlander, S. F. Clifford, R. W. McMillan, J. T. Priestley, and W. P. Schoenfeld, "Turbulence-induced millimeter-wave scintillation compared with micrometeorological measurements," *IEEE Trans. Geosci. Remote Sens.* **26**, 330-342 (1988).

Atmospheric Turbulence Profiles

Jean Vernin
Département d'Astrophysique
Université de Nice-Sophia Antipolis
F-06108 Nice Cedex France

Abstract

The statistical analysis of stellar scintillation is a powerful tool for remote optical turbulence profiling, however, layers near the observer are not detectable. The light propagation theory depends mainly on the smooth perturbation assumption. But many hypotheses are made about the statistical properties of the turbulent medium : Does it follow a Kolmogorov behavior ? Can the thermal field be considered as passive and conservative ? Within which scale range are these properties satisfied including isotropy? Scintillation is a multidimensional process which depends on the position of the observer, the angular direction of the light source, the time and the wavelength. One dimensional analysis mixes together the contribution of all the turbulent layers in such a way that it is almost impossible to locate them. But playing with different pairs of these parameters allows us to retrieve not only optical turbulence profiles, but also wind speed profiles and wind dispersion, in many different ways.

Recent interferometric, scintillation and *in situ* measurements reopen questions related to the fundamental assumptions.

Warning

The problem treated here deals with the retrieval of the physical properties of a medium traversed by a light wave coming from a punctual source at infinity (a star). We will limit ourselves to the vertical propagation case since the horizontal path is developed elsewhere. Emphasis will be put on scintillation and other important aspects such as angle of arrival and wavefront distortion will not be considered. The analysis will be foccused on the visible range of wavelength, discarding radio and acoustic waves. Finally, we do not pretend to write a review paper but rather give an overview of the various techniques we have developed in order to retrieve atmospheric profiles.

Setting the problem

A star emits spherical waves approximated here by plane waves, the phase and amplitude of which are assumed to be constant when they reach the top of our atmosphere. This medium is perturbed by micro fluctuations of the temperature field which induce refractive index inhomogeneities. The distorted phase at level h depends on the optical path given by :

$$\varphi(\mathbf{r}) = k \int_h^\infty n(\mathbf{r}, z)\, dz \qquad (1)$$

Where k is the light wavenumber and $n(\mathbf{r},z)$ is the refractive index at the position $(\mathbf{r},z)$. Z axis refers to the vertical direction. The iso-phase contour defined by equal optical paths is no longer plane but corrugated. The light rays, which propagate perpandicularly to the wave surface, are deviated and converge or diverge when they recede from the turbulence zone. At ground level, an inhomogeneous luminous flux is observed which is called scintillation and is described by the Fresnel diffraction formalism. Wave propagation through turbulent media is detailed by Tatarski [1] whose formalism has been rewritten and simplified by Roddier [2] and applied to relevant astronomical imaging problems. Fig.1 sketches the light ray tracing throughout a turbulent atmosphere, coming from two distinct sources, at two different wavelengths in the case of a single turbulent layer.

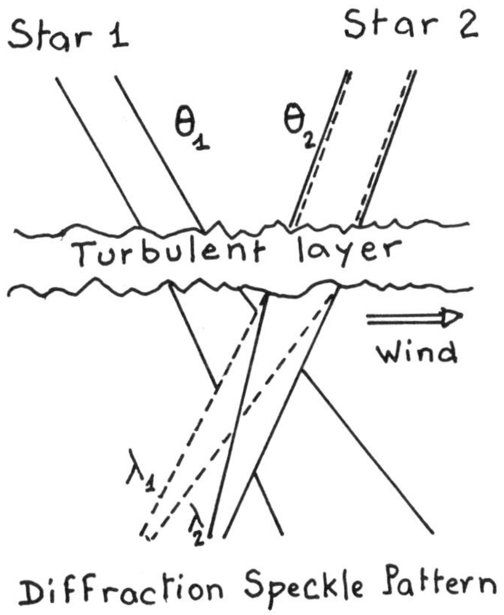

Fig.1 - Scintillation originates from converging or diverging "lenses" drawn away with the wind. Flux fluctuates depending upon the angular position of the source, wavelength, time and observer position.

The scintillation could be imagined as the converging or diverging effects of a set of lenses of random focal length, lying within horizontal sheet moving at the wind speed **V** corresponding to this altitude.

The observed diffraction pattern at ground level is the scintillation phenomenon or the so-called shadow band pattern or atmospheric speckles. The random light flux I is a function of the observer position $\mathbf{r}$, wavelength λ , time t and the angular position θ of the source. Thus, the scintillation is a multidimensional stochastic process $I(\mathbf{r}, \lambda, t, \theta)$.

The question here is : Is it possible to extract information corresponding to the medium traversed - for example what is the turbulence "strength", what is the wind speed - at every altitude along the luminous path ? Since the ground scintillation is the compounded effect of many different atmospheric layers (an integral along the optical path) the naive reader could reasonably think the answer is "no". But the courageous reader who opens this book will more likely guess that the answer is "yes, but ...".

Physical hypotheses

The theory of scintillation formation can be summerized with a small set of equations which establishes the expression for the spectral density of irradiance fluctuations. Few hypotheses concerning the physical properties of the velocity field within the turbulent layers have been assumed by Kolmogorov [3] and then extended to the temperature field by Obuklov [4] and Yaglom [5].

In order to have a simpler insight into the propagation process it is supposed that turbulence develops in one horizontal slab of thickness δh , at altitude h, where refractive index fluctuations originate, n (**x** , z), and induce phase distortions given by :

$$S_h(\mathbf{r}) = k \int_h^{h+\delta h} n(\mathbf{r}, z)\, dz \ . \tag{2}$$

First of all, the thickness dh is assumed to be small enough to insure that no diffraction effect could take place in such a thin screen and thus, that the amplitude is not perturbed. The complex amplitude after having crossed this screen is :

$$\Psi_h(\mathbf{r}) = \exp\left[i\, S_h(\mathbf{r})\right] \tag{3}$$

Then the second hypothesis requires that phase variations are weak enough (Small perturbation approximation) to allow us to develop expression (3) as :

$$\Psi_h(\mathbf{r}) = 1 + i\, S_h(\mathbf{r}) \tag{4}$$

At ground level, the real part of the complex amplitude is written as a convolution with the Fresnel filter :

$$\chi(\mathbf{r}) = S_h(\mathbf{r}) * \frac{1}{\lambda h} \cos\left(\frac{\pi \mathbf{r}^2}{\lambda h}\right) . \tag{5}$$

The spectral density of the relative irradiance fluctuations then becomes :

$$\Phi_I(K) = 0.83\, k^2\, |K|^{-11/3}\, C_N^2(h)\, \delta h \left(\sin \frac{h K^2}{2k}\right)^2 \tag{6}$$

where K is the spatial wave number and C_N^2 (h) represents the structure constant of the refractive index at altitude h. But before being able to establish expression (6) one needs to assume the statistical description given by Kolmogorov of a turbulent flow : the existence of an energy cascade from large eddies to smaller ones, without any viscous loss; the existence of a spatial domain (the inertial range) where turbulence is isotropic, homogeneous and fully developed; and finally, temperature is a passive and conservative additive.

Multidimensional statistical analysis

In this section it is demonstrated that turbulent vertical profiles cannot be assessed from a zero or one dimensional statistical analysis of scintillation. For example, scintillation variance σ_I^2 (zero-dimension) is expressed through an integral of the optical turbulence over the whole atmosphere :

$$\sigma_I^2 = 19.12 \; |\lambda|^{-7/6} \int_0^{\infty} h^{5/6} C_N^2(h) \, dh \; . \qquad (7)$$

It is obvious that an infinite set of C_N^2 (h) profiles can account for a measured scintillation index.

In the case of a one dimensional spatial analysis of the scintillation of a single star, for example a spectral density, the following expression can be written :

$$\Phi_I(K) = 0.83 \; k^2 \; K^{-11/3} \int_0^{\infty} C_N^2(h) \left(\sin \frac{h K^2}{2k} \right)^2 dh \; . \qquad (8)$$

as a summation of expression (6) over a continuous distribution of turbulence. A two dimensional Fourier transform of equation (8) leads to the scintillation spatial covariance of a single star (∗) which takes a form similar to (8) :

$$B_*(\mathbf{r}) = \int_0^{\infty} C_*(\mathbf{r}, h) \, C_N^2(h) \, dh \; . \qquad (9)$$

In this formula B_* (**r**) is experimentally measured, C_* (**r**,h) is the kernel of this integral equation and is numerically known through the derivation of equ.(8), for example, and finally C_N^2 (h) is the vertical profile one wants to retrieve. Inverting (9) belongs to the class of "ill posed" problems for which a considerable literature exists. Furthermore, the covariance B_* (**r**) has a statistical nature and is computed from a low-level light flux, both contributing to add some noise to formula an (9). The reader familiar with this kind of inversion will better understand that it is almost impossible to solve it, noticing the absence of diagonal elements in the kernel, the contours of which are plotted in Fig. 2.

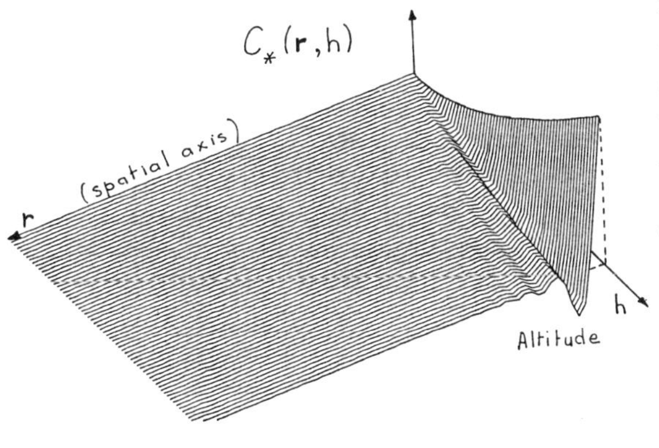

Fig.2. Representation of the kernel. C_*(r,h) corresponding to a single star. Each line shows the spatial covariance that would be given by a single turbulent layer at altitude h and with a unity C_N^2.

Once more, a one dimensional analysis will not allow us to assess a turbulence profile. This is the reason why we soon introduced the principle of multidimensional analysis[6] which consists of investigating the scintillation I(**r**, λ, t, θ) by combining several variables. In order to explain the multidimensional analysis principle we will first examine the Scidar technique (spatio-angular variable) although the spatio-temporal approach[7] was the first to show the power of this concept. The Scidar technique[8] combines both variables (**r**, θ) observing a double star as depicted in Fig. 3.

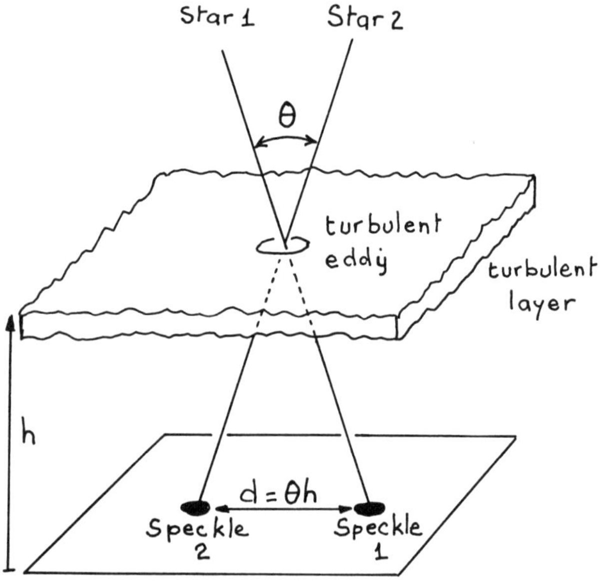

Fig.3 - Assuming a particular turbulent eddy to act as a converging lens, each star casts onto the ground a different bright speckle. Knowing θ , the measurement of d leads to the assessment of the altitude of the turbulent layer.

Each star casts onto the ground a similar Fresnel diffraction pattern of the same turbulent layer, but displaced one from another by a distance proportional to the height of the layer :

$$d = \theta\, h \tag{10}$$

because the angular separation θ between the two stars is very small, i.e. few arc seconds. The covariance of such a double pattern is the sum of the covariance of each individual pattern and the two cross-correlations between one structure and the other one. Those two latter cross-correlations carry all the information relevant to the altitude of the turbulent layer since they lie at ± θh on each side of the origin. This property completely supplies the information which was lacking in the single variable kernel $C_*(\mathbf{r}, h)$ as represented in Fig. 4.

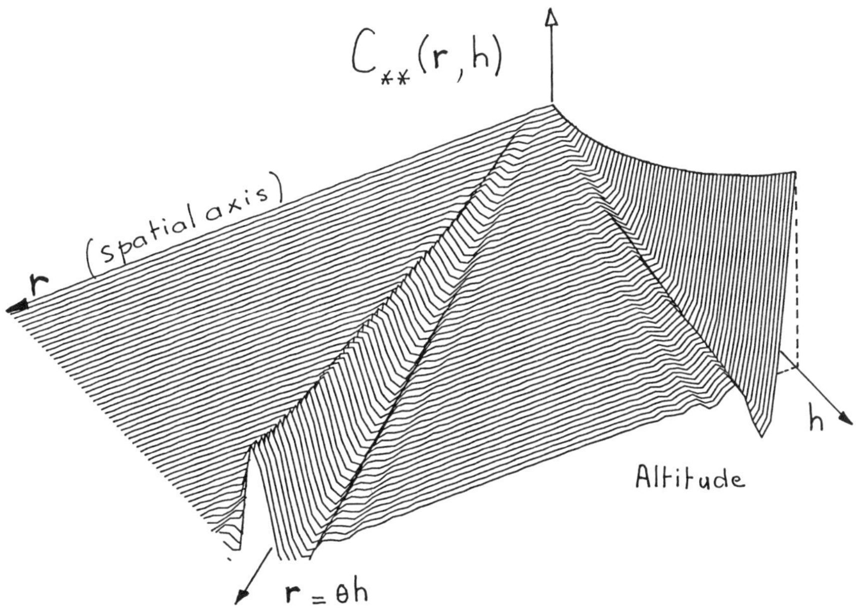

Fig.4 - Representation of the kernel C_{**} (r,h) corresponding to a double star. As in Fig. 2 each line corresponds to a single layer. Secondary bumps lie on the r = θh straight line.

One can see now some "energy" lying on the diagonal, giving hope to inverte more easily equation (9). Actually, the scintillation covariance of a double star C_{**} (r) is plotted on Fig.5 (a) which shows a secondary bump. Its inversion leads to the recovered C_N^2 (h) profile, Fig.5 (b), in which a strong turbulent layer is seen. The mathematical tool used to solve equ. (9) is an adaptation of the maximum entropy method for image restoration given by Frieden [9], with suits well for positive object reconstruction.

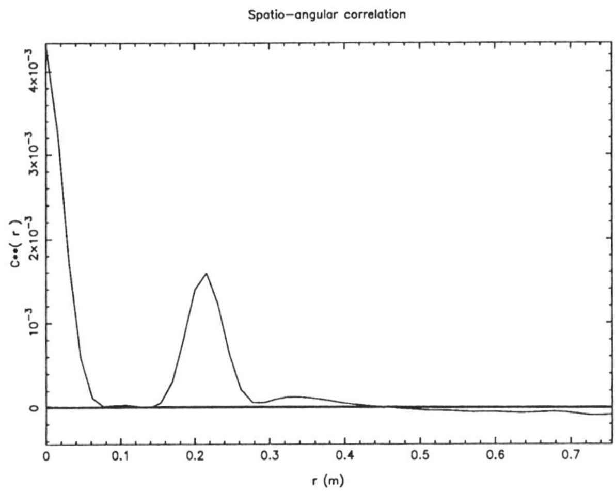

Fig.5(a) : Spatio-angular correlation of the scintillation of double star γ Ari, the angular separation of which is 7.7 arc sec.. The secondary bump is located at 21 cm from origin, which corresponds to an altitude of d / θ = 5.6 km above ground level.

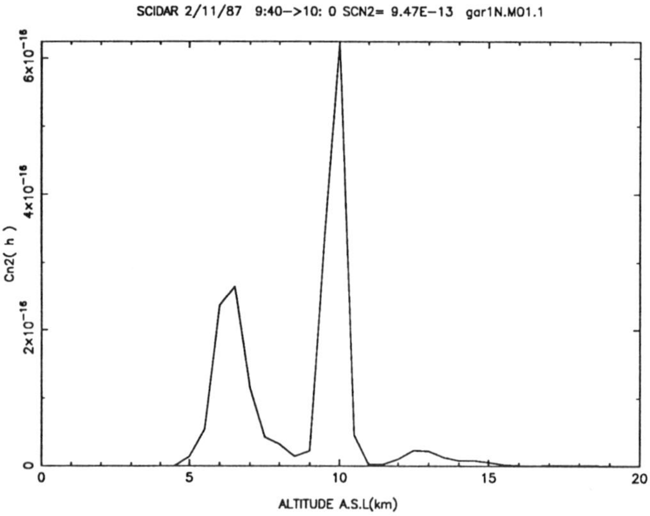

Fig.5 (b) : Reconstructed $C_N{}^2$ (h) profile corresponding to Fig. 5 (a) correlation. The strong peak located at 10 km corresponds to the secondary bump, if one knows that the observatory (Mauna Kea, Hawaii) is 4.2 km above sea level.

In other words, the multidimensional analysis consists of playing with several variables, taken in such a judicious way that the information corresponding to each altitude helps to regularize the general inversion problem stated in (9).

If one plays with the $(\mathbf{r}, \lambda)$ couple instead of $(\mathbf{r}, \theta)$ it is possible, as well as the Scidar technique, to retrieve an optical turbulence profile. This technique is named "two-color stellar scintillation" by Caccia [10], and analyses two scintillation patterns given by a single star, but at two different wavelengths. Actually, the differential refraction phenomenon allows us to separate the two speckle patterns by a distance related to the two different wavelengths, the altitude of the layer and the zenith angle of the star. The wider this angle the larger the image separation, as one can see in Fig. 6, leading to a better vertical resolution.

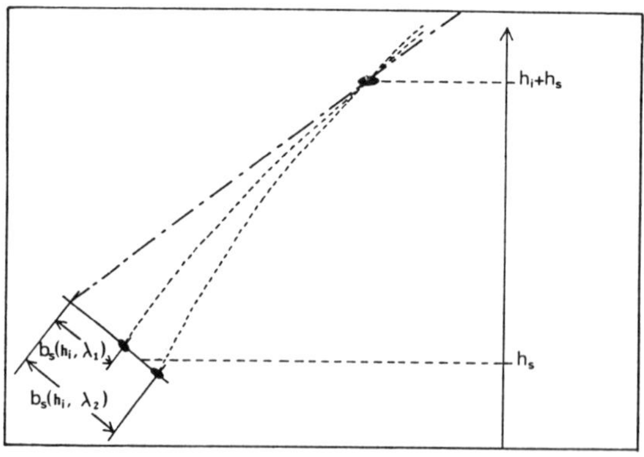

Fig.6 - Schematic and exaggerated description of the atmospheric dispersion phenomenon at two wavelengths λ_1 and λ_2 and its effects on the light refracted down to the ground by a turbulent layer at the altitude $h_i + h_s$. The distance $b_s(h_i,\lambda_1)$ is the intrinsic separation of the two-color speckle pattern induced by the turbulent layer. To give an idea of the experimental situation, note that this intrinsic separation is several tens of centimeters, whereas h_i is several kilometers.

Now, if one is interested in the speed of each turbulent layer, the couple (**r**, t) can be chosen, to compute the spatio-temporal cross-correlation function of the scintillation of a single star as proposed by Rocca [11]. Two atmospheric speckle patterns are recorded at two instants τ apart, and, assuming the frozen turbulence hypothesis, the cross-correlation of those two frames will show a bump at a distance **r** from the spatial origin such as :

$$\mathbf{r} = \mathbf{V}\,\tau \qquad (11)$$

if **V** is the speed of the layer. Fig. 7 shows such a cross-correlation measured for $\tau = 6$ ms.

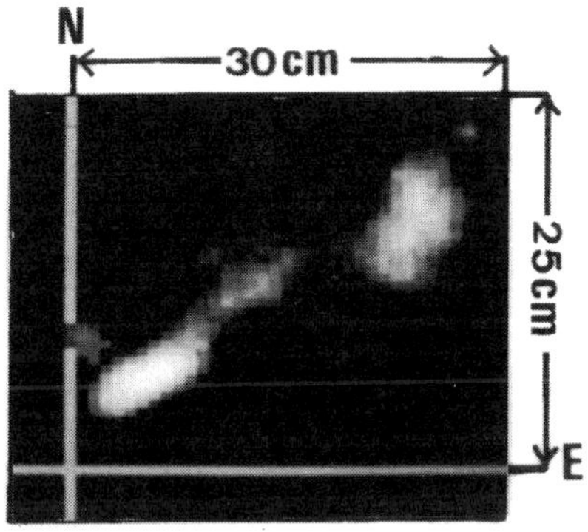

Fig.7 - Two-dimensional representation of an experimental spatio-temporal cross-correlation function obtained with the star Capella. The integration time is 5,5 min, the exposure time of the atmospheric speckles is 0.5 ms and the time lag is 6 ms. Black pixels refer to the lowest correlation value whereas the white pixels refer to the highest. The orientation and the scale are indicated in the figure. For example, the upper right correlation peak is 0.36 m apart from the origin, corresponding to a 60 m/s wind speed with a northeasterly orientation.

Several peaks are visible, each one corresponding to a different layer. More recently, Caccia [12-13] demonstrated that not only the wind speed can be assessed but also the optical turbulence strength $C_N{}^2$ (h) δh and the speed dispersion σ_V(h) within each turbulent slab. Simultaneous classical meteorological soundings gave similar profiles as those deduced from the above mentioned optical experiment.

Different optical methods [14...17] have been used simultaneously during many astronomical site testing campaigns and performing different techniques (Shearing interferometer, image motion monitor, Scidar), which all lead to a very good overall agreement. Thus, it seems that, at least for optical-based instrumentation, the whole set of physical hypotheses above described are *a posteriori* confirmed.

Discrepancy between Scidar and *in situ* turbulence profiles

Fig.8 shows the comparison between Scidar optical turbulence profiles and *in situ* soundings given by micro-thermal probes attached to free-flight balloons. Those results are extracted from an astronomical site testing campaign detailed by Vernin [15]. It is clear from Fig.8 that one needs a large factor if one wants to superimpose both profiles. But the case of Fig.8 is not unusual and after having processed many balloon flights a mean factor of about 2.8 has been necessary to account for weaker *in situ* $C_N{}^2$ profiles. Many precautions were taken to insure the right calibration of both experiments. The Scidar technique is self-calibrated throughout expressions like (7) which refers to relative flux fluctuations. The *in situ* technique gives $C_T{}^2$

measurements directly related to the temperature coefficient of resistivity corresponding to the metal used.

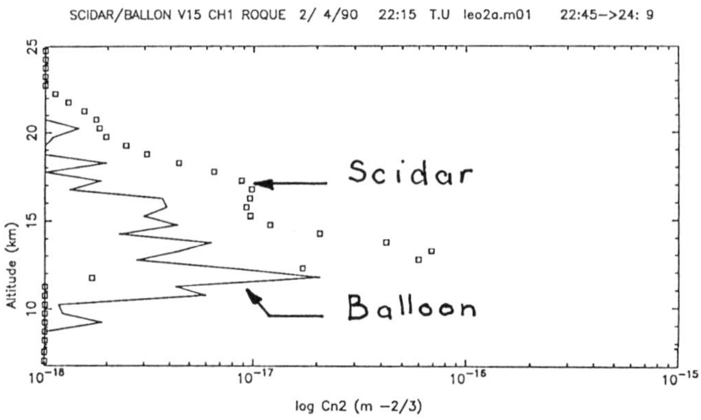

Fig.8 - Vertical C_N^2 (h) profiles simultaneously assessed with the Scidar (Optical) and balloon (*in situ*) technique.

Anisotropy of the thermal temperature field

Several authors [18,19,20] published C_T^2 profiles based on the measurements of the temperature structure function, generally deduced from a couple of sensors one meter apart placed on a horizontal arm carried by a free flight balloon. In order to test the various hypotheses made to describe the statistical properties of the thermal field, we recently duplicated a similar but more versatile instrumentation depicted by Vernin [15]. Now it is possible to assess up to four different estimations of the temperature structure function in the same gondola. This allows us to test different geometrical configurations at the same time, such as various couples of probes situated on horizontal and vertical arms or distributed at different separations. Fig. 9 shows a vertical profile of the coefficient A(h) called anisotropy ratio :

$$A(h) = \frac{C_T^2(h)\,\text{vertical}}{C_T^2(h)\,\text{horizontal}} \qquad (12)$$

computed from two couples of sensors, placed 1 m apart on a vertical and horizontal arm respectively. One would expect A(h) to be equal to 1 corresponding to an isotropic flow assumed for a scale of 1 meter (1m is well inside the inner and outer scales l_0 and L_0).

But it is obvious that the ratio is larger than unity for most of the altitudes which leads us to admit that the thermal field is no longer isotropic even for such small length scales.

This observation comes from our gondola during a joint Radar / Scidar / balloon campaign held at the balloon launching site of the Centre National d'Etudes Spatiales (CNES), in March 1990. Following Dalaudier [21] this anisotropic behaviour might be interpreted as the frequent occurence of "sheets" in the atmospheric turbulence field. In order to verify that there is no possible defect in our apparatus, we performed another flight with two couples of probes situated on the same horizontal arm, but a few centimeters apart. In this case, the A(h) ratio is shown in Fig. 10.

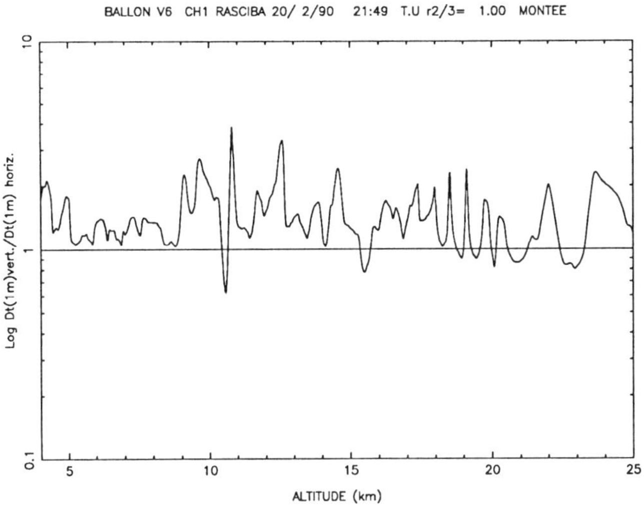

Fig 9 - Vertical profile of the anisotropy ratio deduced from structure functions measured with two pairs of temperature sensors, 1 m. apart, disposed on a vertical and a horizontal arm.

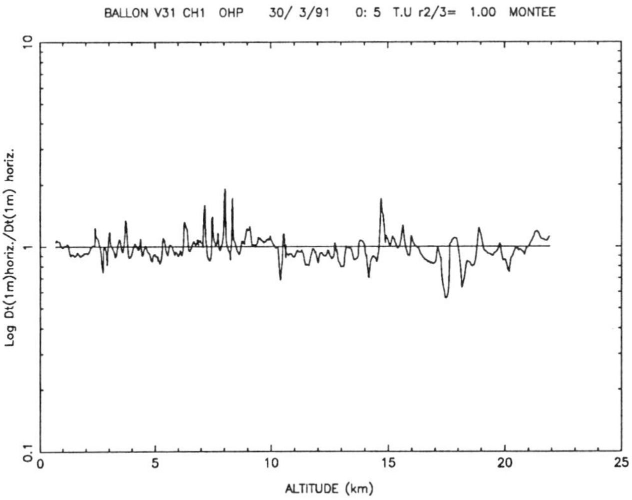

Fig 10 - Same as Fig.9 , but both arms are horizontally disposed.

Now, A(h) fluctuates around 1 and its integrated value :

$$< A > = \frac{1}{h_2 - h_1} \int_{h_1}^{h_2} A(h)\, dh \tag{13}$$

is < A > = 0.98, very close to 1. The same computation for couples on crossed arms corresponding to Fig. 9 gives $< A > \simeq 1.5$.

Finally, Fig. 11 shows "Kolmogorov" ratio profile :

$$K\left(h, \frac{r_1}{r_2}\right) = \frac{D_T(r_1\,,\ h)}{D_T(r_2\,,\ h)} = \left(\frac{r_1}{r_2}\right)^{\frac{2}{3}} \tag{14}$$

according to the well-known "two-thirds" law. In the corresponding instrumental set up r_1 and r_2 are respectively 1m and 0.3 m for which a "Kolmogorov" ratio equal to 2.23 is expected.

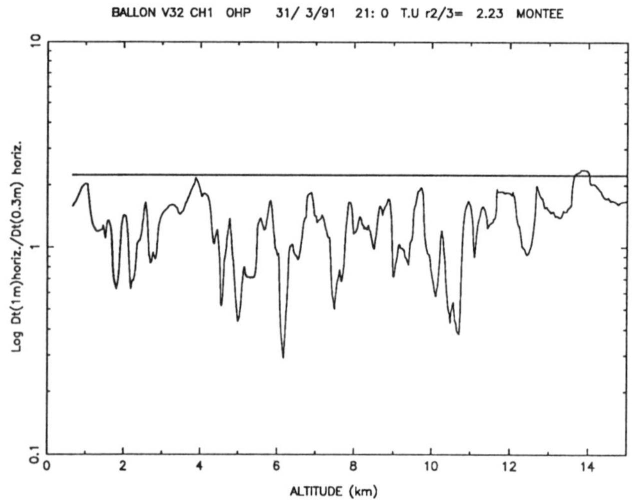

Fig.11 - Ratio of structure functions measured with two different separations (1m and 0.3 m), on a horizontal arm.

The measured ratio is almost everywhere less than 2.23 leading to the conclusion that the temperature structure function increases, with respect to the separation r, less rapidly than the expected "two-thirds" law.

Conclusion

It is demonstrated that atmospheric turbulence profiles are obtainable with scintillation-based optical techniques assuming a small set of generally admitted hypotheses. Among them one can cite that optical turbulent layers are approximated by thin phase screens, within which the small perturbation approximation can be made as well as a Kolmogorov statistical description. Turbulence profiles are retrieved using the concept of multidimensional analysis of the scintillation, combining different pairs of variables such as spatio-angular ($\mathbf{r}$, θ), spatio-temporal ($\mathbf{r}$, t) or two-color ($\mathbf{r}$, λ) analysis. Comparison between several optical-based instruments, such as shearing interferometer, Scidar and Seeing monitor, giving turbulence profiles or integrated values leads to a coherent description of light propagation phenomena. Thus, it is generally admitted that one has to rely on those few assumptions.

But recent experiments involving comparison between *in situ* and Scidar techniques or *in situ* sensing alone give rise to puzzling results which cannot be interpreted in the frame of a Kolmogorovian turbulence. Strong stable temperature stratification within horizontal sheets and structure functions departing from the "two-thirds" law, even at one-meter scale lengths, are many clues which open the question of the passive behaviour of the temperature field.

References

1 -V.I. Tatarski, *Wave propagation in a turbulent medium*, Dover (1961).

2 -F. Roddier, *Progress in Optics*, **19**, Wolf (1981).

3 - A.N. Kolmogorov, "Local structure of turbulence in incompressible fluids with very high Reynolds number", DAN SSSR, **30**, 229 (1941).

4 - A.M. Obukov, "Structure of the temperature field in turbulent flows", Izv. Akad. Nauk SSSR, **13**, 58 (1949).

5 - A.M. Yaglom, DAN SSSR, **69**, 743 (1949).

6 - J. Vernin et M. Azouit, "Traitement d'image adapté au speckle atmosphérique. II. Analyse multidimensionnelle appliquée au diagnostic à distance de la turbulence", J. Optics (Paris), **14**, 131 (1983).

7 - J. Vernin and F. Roddier, "Experimental determination of two-dimensional spatio-temporal power spectra of stellar light scintillation. Evidence for a multilayer structure of the air turbulence in the upper troposhere, J. Opt. Soc. Am., **63**, 270 (1973).

8 - J. Vernin and F. Roddier, "Détection au sol de la turbulence stratosphérique par intercorrélation spatioangulaire de la scintillation stellaire", C.R. Acad. Sc. Paris, **280**, B-463 (1975).

9 - R.B. Frieden, "Restoring with maximum likelihood and maximum entropy", J. Opt. Soc. Am., **62**, 511 (1972).

10 - J.L. Caccia, J. Vernin and M. Azouit, "Structure function $C_N{}^2$ profiling by two-color stellar scintillation with atmospheric dispersion", Appl. Opt., **27**, 2229 (1988).

11 -A. Rocca, F. Roddier and J. Vernin, "Detection of atmospheric turbulent layers by spatio-temporal and spatio-angular correlation measurements of stellar light scintillation", J. Opt. Soc. Am., **64**, 1000 (1974).

12 - J.L. Caccia, M. Azouit and J. Vernin, "Wind and $C_N{}^2$ profiling by simple-star scintillation analysis", Appl. Opt., **26**, 1288 (1987).

13 - J.L. Caccia and J. Vernin, "Wind fluctuation measurements in the buoyancy range by stellar scintillation analysis", J. Geophys. Res., **95**, 13683 (1990).

14 - C. Roddier and J. Vernin, "Relative contribution of upper and lower atmosphere to integrated refractive index profiles", Appl. Opt., **16**, 2252 (1977).

15 - J. Vernin and C. Muñoz-Tuñon, "Optical seeing at La Palma observatory. I : General guidelines and preliminary results at the Nordic Optical Telescope", Astron. Astrophys., **257**, 811 (1992).

16 - J. Vernin and M. Azouit, "Free atmosphere turbulence and seeing Scidar measurements during LASSCA", VLT Report, **55**, 58 (1987).

17 - F. Roddier, L. Cowie, J.E. Graves, A. Songaila, D. McKenna, J. Vernin, M. Azouit, J.L. Caccia, E.J. Limburg, C. Roddier, D.A. Salmon, S. Beland, D.J. Cowley, S. Hill, "Seeing at Mauna Kea : a joint UH - UN - NOAO - CFHT Study", SPIE, **1236**, Symp. on Advanced Technology Optical Telescope IV, L. Barr (1990).

18 - C.E. Coulman, "Vertical profiles of small-scale temperature structure in the atmosphere", Bound. Layer Met., **4**, 169 (1973).

19 - J.L. Bufton, "Correlation of microthermal turbulence data with meteorological soundings in the troposphere", J. Atmos. Sci., **30**, 83 (1973).

20 - R. Barletti, G. Ceppatelli, E. Moroder, L. Paternò and A. Righini, "A vertical profile of turbulence in the atlantic air mass measured by balloon-borne radiosondes", J. Geophys. Res., **79**, 4545 (1974).

21 - F. Dalaudier, C. Sidi, M. Crochet and J. Vernin, "Direct evidences of "sheets" in the atmospheric temperature field", submitted to J. Atmos. Sci..

Interplanetary Scintillation Imaging of Disturbances in the Solar Wind

A. Hewish

Mullard Radio Astronomy Observatory
Cavendish Laboratory
Madingley Road
Cambridge CB3 0HE
ENGLAND

The scintillation of celestial radio sources caused by turbulence in the solar wind is strongly correlated with the mean plasma density along the line of sight. Routine daily observations of about 1000 sources over the sky can be analysed to provide images which correspond to the two-dimensional projection of large scale structures having higher or lower density than the long term average value. Comparison of the images with computer-simulated models enables three-dimensional structures to be derived for large scale features such as major interplanetary shocks or corotating high speed streams. The method is briefly outlined and illustrated with examples taken during periods of high solar activity.

1. Introduction

It was discovered in 1964 that celestial radio sources having angular sizes of less than about 2 arcsec exhibit scintillation caused by turbulent density variations in the solar wind[1]. Early observations of this phenomenon were used to estimate the speed of the solar wind by measuring the drift motion of the speckle pattern across the ground[2,3]. A correlation of the degree of scintillation with the mean plasma density observed simultaneously by spacecraft was soon established[4] and this raised the possibility of imaging large scale structures in the solar wind by monitoring the variations of scintillation across a grid of sources suitably distributed over the sky. A schematic diagram illustrating the method is shown in Fig. 1. The importance of this technique for remote-sensing lies in the fact that spacecraft situated far beyond the magnetosphere have not yet been, and may never be, sufficiently numerous to obtain a global picture of disturbances in the solar wind. In addition, it is only recently that any spacecraft has ventured far outside the plane of the ecliptic. Remote-sensing of conditions above the solar poles by means of scintillation has therefore been of particular value and has shown that the polar solar wind is generally much faster than that emitted from the equatorial regions[5,6,7].

Interplanetary scintillation imaging imposes quite stringent requirements concerning the type of radio telescope that is suitable for this task. The typical bandwidth of scintillation for lines of sight at large angles to the Sun-Earth line is about 0.5 Hz so that each source must be observed for about 200s in order to measure the intensity of scintillation with an accuracy of ten per cent. Reasonable imaging over the whole sky requires a grid of ~1000 sources corresponding to a total observing time of 56 hours. A dedicated multiple beam antenna is thus necessary for daily observations. The size of the antenna is determined by the angular resolving power and sensitivity needed to observe at least 1000 sources.

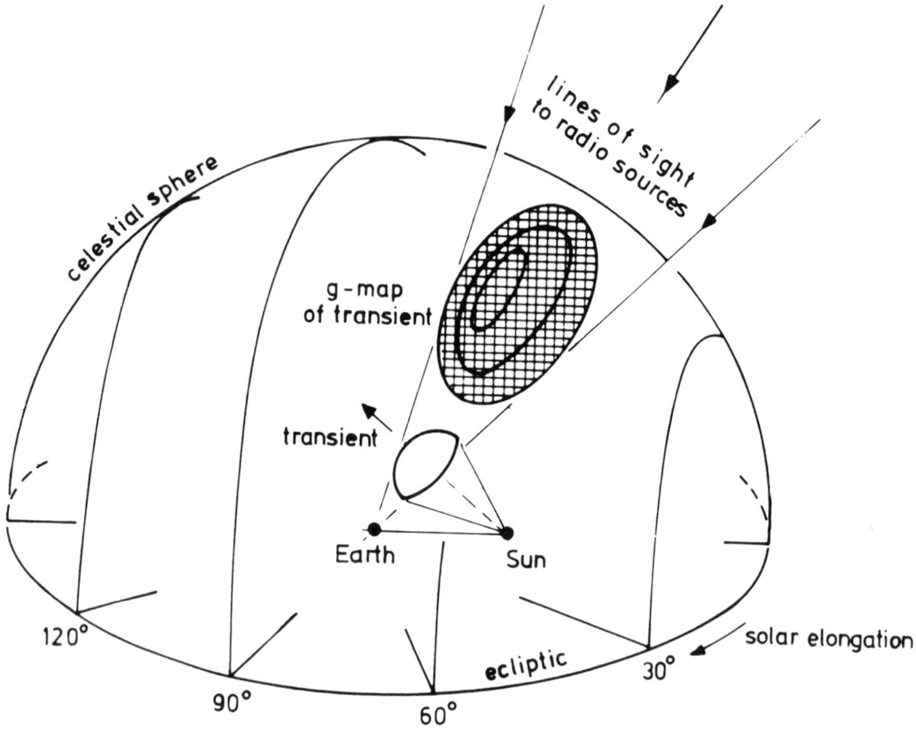

Fig 1. Schematic diagram of the IPS-imaging method.

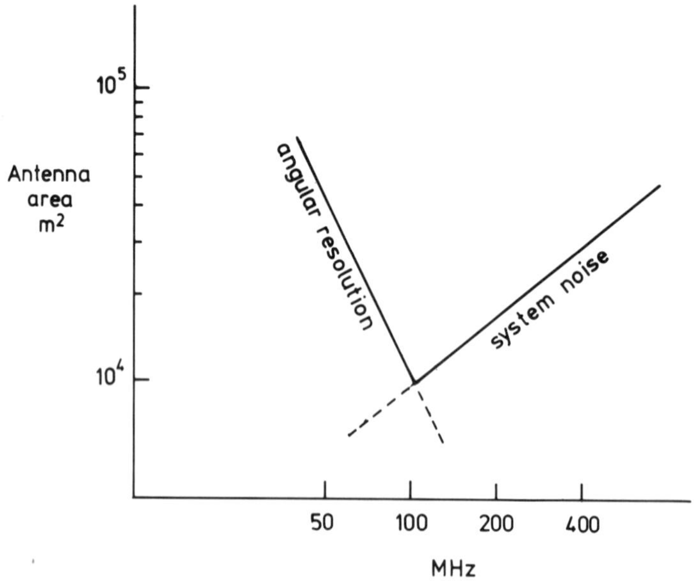

Fig 2. Size of antenna required for imaging with a grid of ~1000 sources.

The area required varies with the radio frequency as shown in Fig. 2. Large antennas having an area exceeding $10^4 m^2$ are essential and the minimum area occurs at a frequency near 100 MHz. The only dedicated antenna is the 3.6 hectare array at Cambridge[8] which operates at a frequency of 81.5 MHz and observes sources at meridian transit. 16 beams in declination are recorded simultaneously on a routine daily basis and up to 2,500 sources have been employed for imaging. Other non-dedicated instruments with imaging capability exist at Puschino in the former Soviet Union and at Ootacamund in India[9,10], while a dedicated array is under construction at Ahmedabad[11].

2. The imaging method

Typical daily observations showing the rms fluctuation of intensity on one radio source are illustrated in Fig. 3. Large day to day variations are imposed upon a systematic increase as the line of sight approaches the Sun until a maximum is reached near an elongation of 30°. This trend follows the variation of mean plasma density N with radial distance r such that $N \alpha\ r^{-2}$ appropriate to a steady radial outflow. The maximum at 30° corresponds to the saturation of scintillation and at smaller elongations scintillation decreases due to smearing by the angular size of the radio source and the finite receiver bandwidth. Angular-size smearing normally dominates so that each source has its characteristic curve of average scintillation versus solar elongation.

The first stage in mapping is to determine the average elongation curve for each source which requires daily observations for about one year. Observations on any particular day are then used to compute the ratio of the observed rms scintillation to the average value for all the sources. These ratios, known as "g-factors", are then combined to produce a map showing the relative enhancement or reduction of scintillation over the sky. Such "g-maps" are related to the two-dimensional projection of large-scale density structures in the interplanetary medium, but their detailed interpretation requires careful consideration.

Typical g-maps obtained from observations of ~1000 sources with the Cambridge 3.6 hectare array during a period of 10 days in 1980 are shown in Fig. 4. The grey scale ranges from g=1.5 (lighter) to g=0.67 (darker) and the whole-sky maps are plotted in ecliptic coordinates with the Sun at the centre. The curved lower boundary of each map marks the limit of the survey at declination -20°. A roughly spherical disturbance causing enhanced scintillation is clearly seen leaving the Sun on 28-29 September and this is followed by a zone of reduced scintillation on 3-5 October.

3. Interpreting the images

To interpret the maps in terms of the plasma density and three-dimensional structure of the disturbances it is first necessary to determine the relation between scintillation and the mean plasma density. The effect of the Fresnel filter in diffraction theory means that intensity scintillation responds only to density variations on a scale less than a few hundred km, essentially the high wavenumber tail of the spectrum of density fluctuations in the solar wind. Much effort has been devoted to

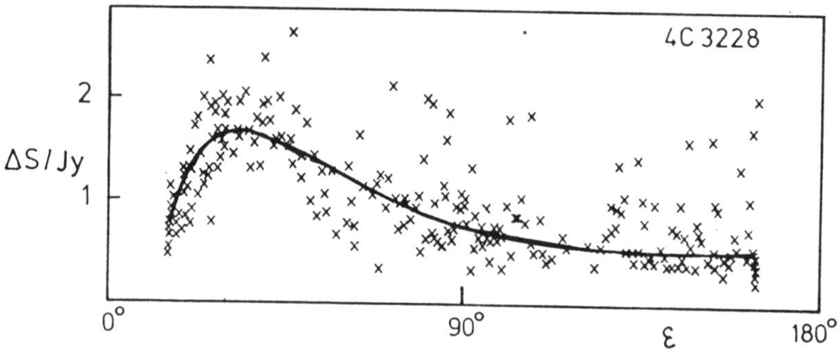

Fig. 3. The rms scintillating flux density ΔS of a radio source as a function of solar elongation E.

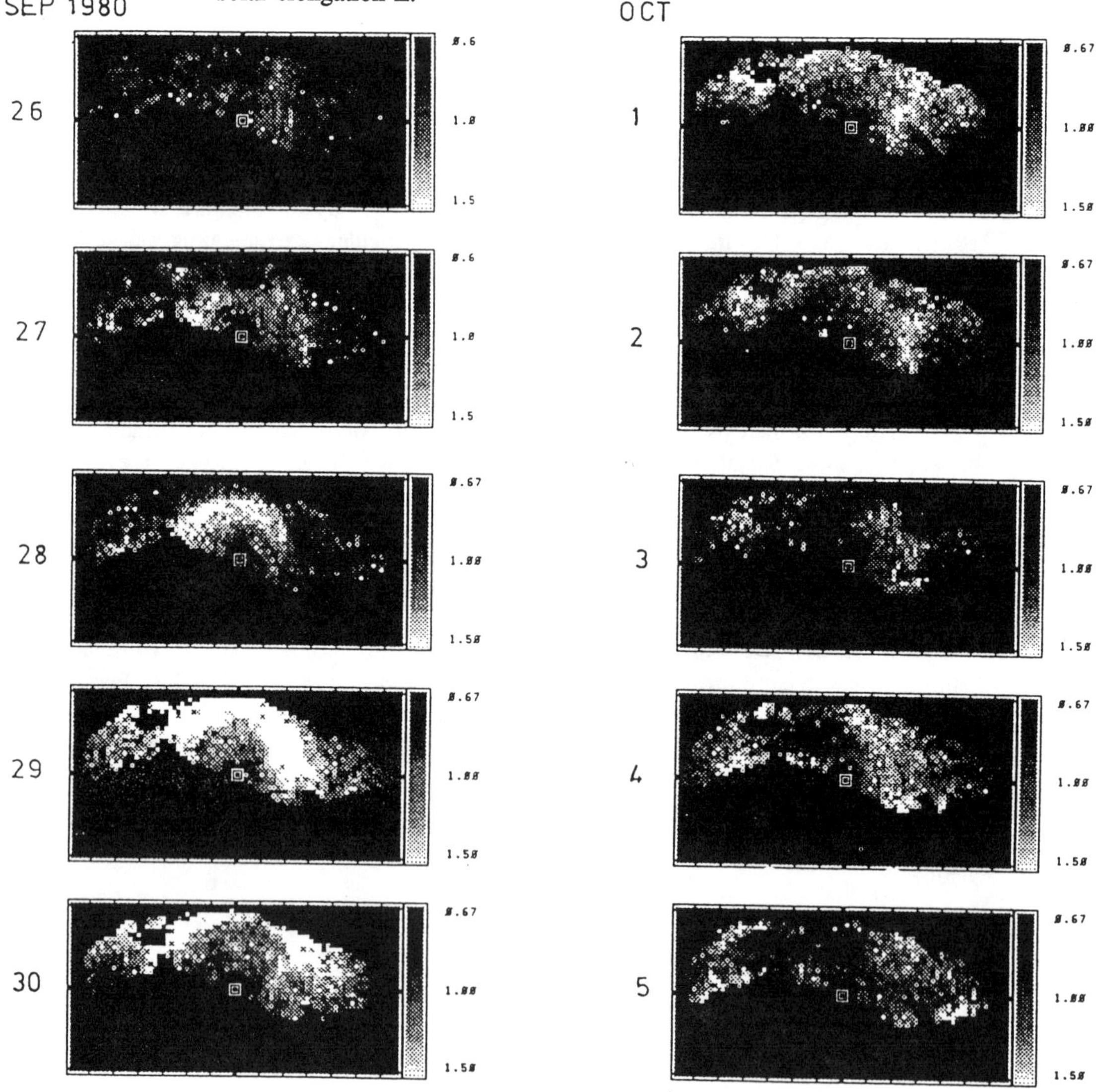

Fig. 4. Scintillation images (g-maps) for 26 September to 5 October 1980 showing a major transient leaving the Sun.

determining the form of the spectrum[12] but in the region of interest for scintillation computations it has been found that a gaussian spectrum having a characteristic scale which depends upon radial distance from the Sun is a good approximation[13]. Adopting this model gives a weighting function for scintillation along the line of sight at 81.5 MHz as shown in Fig. 5. Values of the weighting function are normalised to unity at the point of maximum weight on each line of sight, the locus of which is indicated by the heavy line in Fig. 5. For elongations near 90° it is evident that scintillation is dominated by a path through the solar wind which extends for a length of about 1 AU (the Sun-Earth distance) from the Earth. By choosing many occasions when spacecraft data indicated that the plasma density was roughly uniform over this path it was found that the variation of scintillation relative to the long term average obeyed the relation $g \alpha \sqrt{N}$[14]. For 1978-80 the quantitative relation was $g^2 = N/9$ cm^{-3} and this held over the range $3<N<40$ cm^{-3} to an accuracy of about ±20%.

For elongations smaller than 30° scintillation becomes saturated at 81.5 MHz and the speckle pattern also suffers from source-size smearing. In this region no simple relation exists between scintillation and plasma density. Observations at higher frequencies could, in principle, reduce the size of this limiting cone but larger antennas would then be necessary for comparable mapping as explained in Section 1.

When g-maps are confused by the superposition of more than one disturbance along the dominant portions of the lines of sight their interpretation in terms of three dimensional structures is clearly ambiguous. Often this is not the case and major disturbances are seen in isolation, as exemplified in Fig. 4. On such occasions approximations to actual structures can be obtained by numerical simulation. Disturbances of this kind include major density enhancements following strong interplanetary shocks, and long-lived, fast solar wind streams of low density which rotate with the Sun and interact with slower flows of higher plasma density to produce a corotating density structure in the form of spiral arms. Such disturbances are shown schematically in Fig. 6. They would be detectable within the zone indicated in Fig. 6a for density perturbations a factor of 2 above or below the mean, which are typical for moderately strong disturbances. Because radially directed transients like that in Fig. 6b cross the zone of detection in 2 or 3 days, as compared to 10 or 12 days for corotating features as in Fig. 6c, the day to day changes in the g-maps usually enable a distinction to be made between these typical structures in the solar wind. Quite simple models have been found to simulate observations of each kind of structure rather well[15,16], and even in more complex situations the additional information obtained from scintillation imaging has been of great value[17].

Systematic monitoring of the inner heliosphere over the range 0.5 to 1.5 AU has been carried out with the Cambridge 3.6 hectare array over the periods August 1978 - September 1979, February 1980 - March 1981 and from March 1990 onwards. Summaries of the results obtained for the first two periods have been given[18,19]; in the current observations data-analysis is carried out in near real-time in order to evaluate the use of scintillation mapping for predicting the arrival at the Earth of disturbances likely to cause major geomagnetic storms.

March 1991 was a period of high solar activity during which an outstanding

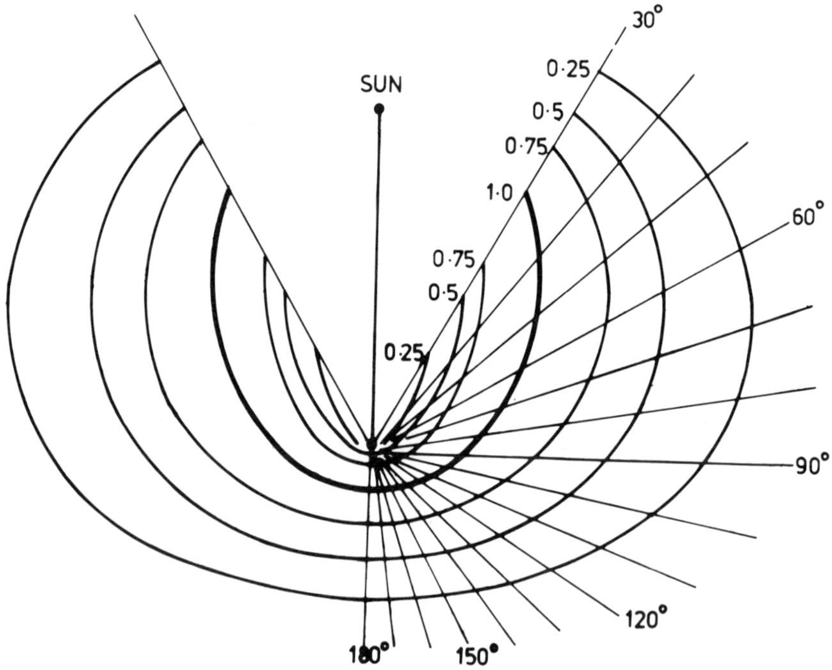

Fig. 5. The normalised weighting function for scintillation along the line of sight.

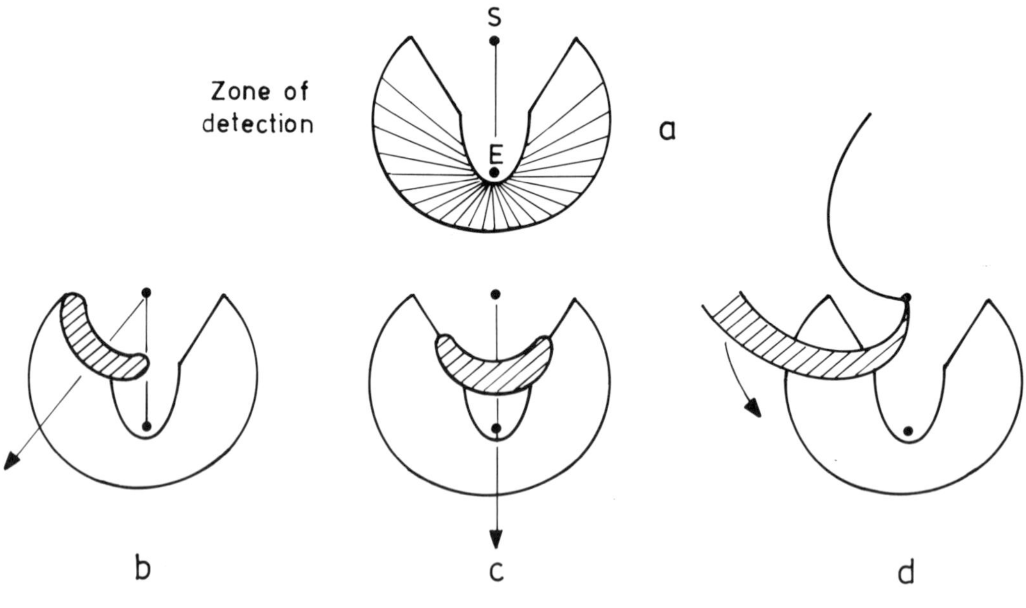

Fig. 6. Schematic diagram showing the zone through which typical disturbances can be detected.

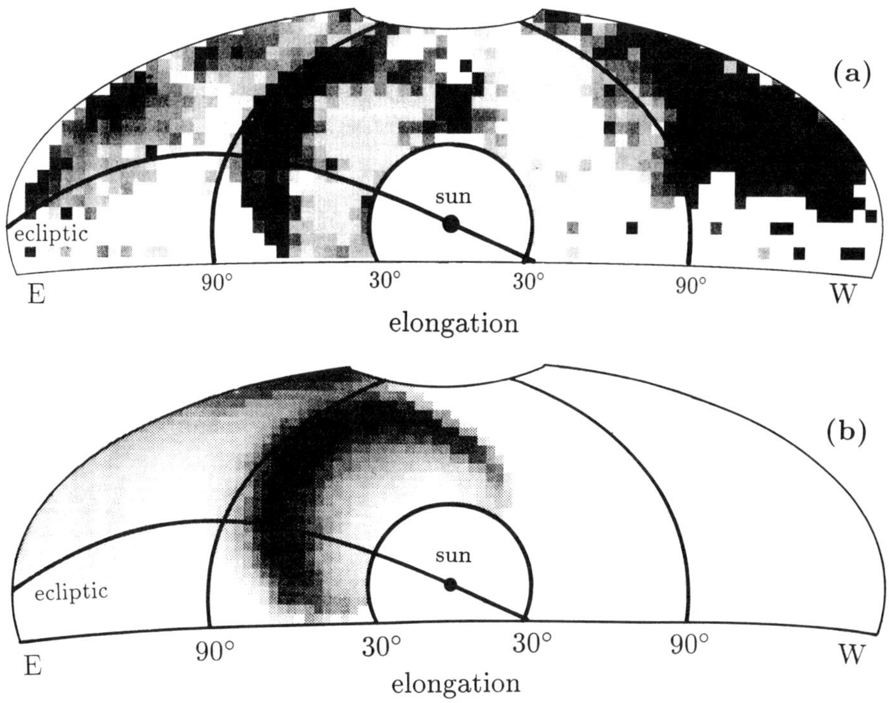

Fig. 7. (a) The scintillation image of an exceptionally fast interplanetary shock observed on 23 March 1991. It was travelling eastwards of the Sun-Earth line and reached Ulysses on 25 March. (b) A simulated image for a model assuming enhanced plasma density in a spherical shell of thickness 0.15 AU subtending an angle of ±40° at the Sun.

disturbance swept over the Earth at more than 1000 km/s. This event was well observed by the spacecraft Ulysses soon after it was launched on its path to Jupiter and has been the focus of considerable interest in view of the intensity of energetic particles which were associated with it. An image of this exceptional transient on the day before it reached the Earth is shown in Fig. 7a. Because of its high speed it had moved beyond the zone of detection on the following day so that only one image could be obtained. Ulysses was located near the ecliptic plane at this time, somewhat below the centre of the shock. A simulation (Fig. 7b) assuming a curved shell of high density having a radial thickness of 0.15 AU and an angular extend of ±40° gives a good fit to the observations for a density enhancement of roughly 6 to 9 times the mean density.

Fig. 7 also illustrates some of the practical problems of scintillation imaging. The dark area to the West of the Sun at elongations greater than 90° is caused by ionospheric scintillation. During periods of high solar activity the temporal spectrum of ionospheric scintillation can extend upwards into the interplanetary scintillation band, thus contaminating the image. This defect is normally confined to elongations exceeding 90°, corresponding to nocturnal observations. Another problem is radio emission from the Sun itself which can generate severe interference during periods of intense flare activity. The dark patch in the image directly above the Sun is due to solar interference.

4. Conclusion

Interplanetary scintillation imaging has shown itself to be a useful means of mapping the structure of disturbances leaving the Sun out to about 1.5 AU. This provides a global view of such events which complements the wealth of detailed information obtained from in-situ observations by spacecraft. The latter cannot normally determine the angular extent and location of major transients which is essential to understanding their origin at the Sun.

An interesting recent development is the possibility of mapping the large scale pattern of solar wind speed, in addition to the structure of plasma density. Routine observations of the solar wind speed using scintillation have so far been carried out by observing a small number of source at multiple stations located about 100 km apart. It has been shown that reasonable estimates of the speed can be made at a single site from suitably calibrated observations of scintillation spectra[20]. This raises the possibility of mapping the solar wind speed over the sky by an analogous process to scintillation imaging. Some preliminary results are encouraging[21] and the extension of this method to a larger source grid would provide a most valuable addition to scintillation imaging.

References

1. A. Hewish, P.F. Scott and D. Wills, "Interplanetary scintillation of small diameter radio sources," *Nature*, **203**, 1214-1217, (1964).

2. P.A. Dennison and A. Hewish, "The solar wind outside the plane of the ecliptic," *Nature*, **213**, 343-346, (1967).

3. W.A. Coles, and S. Maagoe, "Solar wind velocity from IPS observations," *J. Geophys. Res.*, **77**, 5622-5624, (1972).

4. Z. Houminer and A. Hewish, "Correlation of interplanetary scintillation and spacecraft plasma density measurements," *Planet. Space Sci.*, **22**, 1041-1043, (1974).

5. A. Hewish and P.A. Dennison, "Measurements of the solar wind and the small-scale structure of the interplanetary medium," *J. Geophys. Res.*, **72**, 1977-1978, (1967).

6. M. Kojima and T. Kakinuma, "Solar cycle evolution of the solar wind speed structure between 1973 and 1985 observed with the interplanetary scintillation method," *J. Geophys. Res.*, **92**, 7269-7279, (1987).

7. W.A. Coles and B.J. Rickett, "IPS observations of the solar wind speed out of the ecliptic," *J. Geophys. Res.*, **81**, 4797-4799, (1976).

8. A. Purvis, S.J. Tappin, W.G. Rees, A. Hewish and P.J. Duffett-Smith, "The Cambridge IPS Survey at 81.5 MHz," *Mon. Not. R. astr. Soc.*, **229**, 589-619, (1987).

9. V.T. Vlasov, "Transients in the solar wind," *Astron. Zh.*, **56**, 96-105, (1979).

10. V. Balasubramanian, P.K. Manoharan, S. Ananthakrishnan, T.R. Detman, M. Dryer and H. Leinbach, "Solar wind velocity and normalised scintillation index from single station IPS Observations," *Paper T22* (This meeting)

11. S.K. Alurkar, "Proposed plan for 'g-maps' using IPS telescope at Thaltej," Proceedings of the Indo-US Workshop on interplanetary scintillations scintillations and solar activity, Physical Research Laboratory, Amedabad, India, Feb 1-5, (1988).

12. W.A. Coles, "Solar wind scintillation," *Paper S10* (this meeting).

13. A.C.S. Readhead, M.C. Kemp and A. Hewish, "The spectrum of small-scale density fluctuations in the solar wind," *Mon. Not. R. astr. Soc.*, **185**, 207-225, (1978).

14. S.J. Tappin, "Interplanetary scintillation and plasma density," *Planet Space Sci.*, **34**, 93-101, (1986).

15. S.J. Tappin, A. Hewish and G.R. Gaper, "Tracking a major interplanetary disturbance," *Planet Space Sci.*, **31**, 1171-1176, (1983).

16. S.J. Tappin, A. Hewish and G.R. Gapper, "Tracking a high-latitude corotating stream for more than half a solar rotation," *Planet Space Sci.*, **32**, 1273-1281, (1984).

17. K.W. Behannon, L.F. Burlaga and A. Hewish, "Structure and evolution of compound streams at ≤ 1 AU," *J. Geophys. Res.* , **96**, 21213-21225, (1991).

18. A. Hewish and S. Bravo, "The sources of large scale heliospheric disturbances," *Solar Phys.* **106**, 185-200, (1986).

19. S.J. Tappin, "Transient disturbances in the solar wind," *Ph.D. Thesis, Cambridge UK,* (1984).

20. P.K. Manoharan and S. Ananthakrishnan, "Determination of solar-wind velocities using single-station measurements of interplanetary scintillation," *Mon. Not. R. astr. Soc.,* **244**, 691-695, (1990).

Section 4

Theory of Wave Scintillation and Its Applications

THEORY OF WAVE SCINTILLATIONS

V. I. Shishov
P. N. Lebedev Physical Institute
Astro Space Center
Russian Academy of Science
53, Leninskii Prospect
Moscow 117924, Russia

Abstract

Solutions of the fourth moment equation are qualitatively analyzed. The dependence of the solution on the inhomogeneities spectrum type is emphasized. There are three types of inhomogeneities spectra (the Kolmogorovian type, the Gaussian type, and the spectrum for which there exists the first moment and the third moment is infinite) that produce qualitatively different types of scintillations.

The solutions describing coherent plane and spherical wave propagation in the cases of randomly homogeneous medium and of the phase screen are compared. They represent different models of medium distribution between the source and the observer.

The intensity fluctuations of the two-frequency and two-directional waves and of the partial coherence irradiance are discussed. It is shown that the typical scale of diffraction pattern and frequency correlation radius depend weakly on the angular size of a source in the extended medium model, while the dependence of these parameters on the size of the source is stronger for the phase screen model.

The fluctuations of a pulse propagating in a random medium are considered. The initial narrow coherent pulse is transformed into a train of subpulses with an envelope, which is described by the expression for the average pulse shape.

In conclusion, the statistical characteristics of the frequency fluctuations of a monochromatic wave propagating through a moving turbulent medium are presented.

1. Introduction

There are many theoretical results for scintillation statistics that have been presented in the last 25 years using the field fourth moment equation. A short review of these results are given in this paper. The dependence of scintillation characteristics on turbulent spectrum type, distribution of the medium between the source and the observer and the properties of the initial source are considered. Statistical homogeneity in the plane normal to the propagation direction is assumed.

2. Plane wave scintillations

The problem of plane wave propagation through a random medium is the simplest. Let us present the wave field in form

$$U = u \exp(-ikz), \tag{1}$$

where k is the wavenumber and z is the propagation direction.

The intensity fluctuations are described by the fourth moment of u,

$$M = \left\langle u(\mathbf{r}_1; z)\, u^*(\mathbf{r}_1'; z)\, u(\mathbf{r}_2; z)\, u^*(\mathbf{r}_2'; z) \right\rangle, \tag{2}$$

where $\mathbf{r}_{1,2}$, $\mathbf{r}_{1,2}'$ are the transverse coordinates, and angular brackets denote ensemble average over the medium fluctuations.

The equation for M has the form[1-4]

$$\frac{\partial M}{\partial z} - \frac{i}{k} \nabla_\rho \nabla_{\rho'} M + fM = 0,$$
$$f = D(\rho) + D(\rho') - D(\rho + \rho')/2 - D(\rho - \rho')/2, \tag{3}$$

when the points of observation are defined by the apexes of a parallelogram, i.e.,

$$\rho = \mathbf{r}_1 - \mathbf{r}_1' = \mathbf{r}_2' - \mathbf{r}_2$$
$$\rho' = \mathbf{r}_1 - \mathbf{r}_2' = \mathbf{r}_2 - \mathbf{r}_1'. \tag{4}$$

The initial condition is

$$M|_{z=0} = I_0^2. \tag{5}$$

The function $D(\rho)$ is determined by the spectrum of refraction index fluctuations Φ_n,

$$D(\rho) = 4\pi k^2 \int d^2\kappa\, \Phi_n(\kappa, \kappa_z = 0) \left[1 - \cos(\kappa \cdot \rho)\right]. \tag{6}$$

$D(\rho)$ is the gradient of phase structure function in the linear geometric-optics approximation.

If we introduce the spectral function

$$\tilde{M}(\kappa,\rho',z) = (1/2\pi)^2 \int d^2\rho \;\exp(-i\kappa\cdot\rho)\; M(\rho,\rho',z), \tag{7}$$

then Eq. (3) becomes

$$\frac{\partial\tilde{M}}{\partial z} - \frac{\kappa}{k}\nabla_\rho \tilde{M} + D(\rho')\tilde{M} = \int d^2\kappa_1 \;\Psi(\kappa-\kappa_1;\rho')\;\tilde{M}(\kappa_1;\rho';z),$$
$$\Psi(\kappa) = 4\pi k^2\Phi_n(\kappa)\left[1-\cos(\kappa\cdot\rho)\right]. \tag{8}$$

The initial condition is transformed to

$$\tilde{M}\big|_{z=0} = I_0^2\,\delta(\kappa). \tag{9}$$

The solutions of the Eqs. (4) and (8) are qualitatively different when $D(\rho)$ has or does not have finite second and fourth derivative at zero lag. Three types of turbulence spectrum correspond to these versions of the function $D(\rho)$.

The first type is the Kolmogorov-type spectrum

$$\Phi_n(\kappa) = C|\kappa|^{-\mu}, \quad 3<\mu<4. \tag{10}$$

$D(\rho)$ does not have a finite second derivative at $\rho = 0$ in this case.

The second type is the steep power law spectrum,

$$\Phi_n(\kappa) = \begin{cases} C_1|\kappa|^{-\mu_1} & |\kappa| < 1/L_0 \\ C_2|\kappa|^{-\mu_2} & |\kappa| > 1/L_0. \end{cases} \tag{11}$$

At zero lag, $D(\rho)$ has a finite second derivative but does not have a finite fourth derivative. The solutions of Eqs. (4) and (8) exist formally when $L_0\rightarrow\infty$, but the parabolic approximation that is used for obtaining these equations is violated because of large-angle scattering.

The third type of spectrum is the Gaussian spectrum

$$\Phi_n(\kappa) = \frac{\left\langle(\Delta n)^2\right\rangle}{\pi\sqrt{\pi}}\;\alpha^2\exp\left[-(\kappa\alpha)^2\right], \tag{12}$$

for which the corresponding $D(\rho)$ has a finite fourth derivative at zero lag.

2.1 The phase screen

The phase screen is a thin layer of turbulent medium that modulates only the phase of wave. The diffraction effects in free space beyond the layer lead to the intensity fluctuations. This model is investigated the most in the theory of wave scintillations.[5-15]

If diffraction effects are ignored, we can obtain from Eq. (4) that

$$\begin{aligned} M\big|_{z=0} &= I_0^2 \exp[-F(\rho,\rho')]\,, \\ F(\rho,\rho') &= H(\rho) + H(\rho') - H(\rho+\rho')/2 - H(\rho-\rho')/2\,, \end{aligned} \tag{13}$$

where $H(\rho)$ is the phase structure function (in the linear approximation)

$$H(\rho) = \int_{-\Delta z_0}^{0} dz' D(z',\rho) \tag{14}$$

and Δz_0 is the layer thickness.

The Fourier transform of M in free space is given by the expression

$$\tilde{M}(\kappa,\rho',z) = \int d^2\rho \exp\left[-i\kappa\cdot\rho - F\left(\rho,\rho' - \frac{\kappa z}{k}\right)\right]. \tag{15}$$

The solution of Eq. (8) in free space corresponds to the substitution $\rho' \to \rho'\kappa z/k$. We shall investigate expression (15) for the different types of spectrum Φ_n. When $\kappa z/k$ is small, we can expand the exponent $\exp[-F - H(\kappa z/k)]$ in a power series and obtain the low-frequency series expansion for $\tilde{M}$, i.e.,

$$\left.\begin{aligned} \tilde{M} &= \tilde{M}_0 + \tilde{M}_1 + \ldots, \\ \tilde{M}_0 &= I_0^2\,\delta(\kappa)\,, \\ \tilde{M}_1 &= 2 I_0^2 \sin^2\left(\frac{\kappa^2 z}{2k}\right)\Phi_s(\kappa)\exp\left[-H(\kappa z/k)\right], \\ \Phi_s(\kappa) &= 4\pi k^2\,\Phi_n(\kappa)\,\Delta z_0\,. \end{aligned}\right\} \tag{16}$$

$\tilde{M}_1$ corresponds to the Born and Rytov approximation when $H(\kappa z/k) << 1$:

$$\tilde{M}_{1B} = 2 I_0^2 \sin(\kappa^2 z/2k)\ \Phi_s(\kappa). \tag{17}$$

For small z, (17) describes the scintillation spectrum for the important spatial frequencies. In this case, we have a weak scintillation regime. The weak scintillation condition is determined by the condition

$$m^2 = \frac{(I - \langle I \rangle)^2}{\langle I \rangle^2} \approx m_B^2 = \frac{1}{I_0^2} \int d^2\kappa\ \tilde{M}_{1B}(\kappa) << 1, \tag{18}$$

where $I = uu^*$ is the instantaneous intensity and m_B^2 is the scintillation index in the Born approximation.

For large $\kappa z/k$, $F - H(\rho) \to 0$, and we can expand $\exp[-F + H(\rho)]$ in a high-frequency power series expansion for $\tilde{M}$, i.e.,

$$\begin{aligned} \tilde{M} &= \tilde{M}^{(0)} + \tilde{M}^{(1)} + \ldots \\ \tilde{M}^{(0)} &= \left(\frac{1}{2\pi}\right)^2 I_0^2 \int d^2\rho\ \exp\left[-i\kappa\cdot\rho - H(\rho)\right]. \end{aligned} \tag{19}$$

For large z, $\tilde{M}^{(0)}$ is the dominant term and corresponds to the Gaussian approximation for the field statistics, and the scintillation index saturates at a value equal to 1.

To analyze the transition region between weak and saturated regimes, we must take into account the type of turbulence spectrum.

2.1.1 The Kolmogorovian-type spectrum

In this case, we have two regimes of scintillation: weak and saturated.[11] For the weak scintillation regime, the normalized variance and the scale of intensity fluctuations are equal to

$$\begin{aligned} m^2 \approx m_B^2 &= K(\mu)\ C\Delta z_0\ k^2 (z/k)^{\frac{\mu-2}{2}} \\ \mathrm{B} &\approx \mathrm{B}_\Phi \sqrt{z/k}, \end{aligned} \tag{20}$$

where $K(\mu)$ is the numerical coefficient.

For $m_B^2 > 1$, the scintillation spectrum has two components: high frequency (or diffraction) and low frequency (or refraction). The diffraction component is determined by wave diffraction on inhomogeneities with scale about ρ_0 (the transverse coherence scale) and is described by $\tilde{M}^{(0)}$. The low-frequency component is determined by the irradiance focusing on inhomogeneities with scale of the order of

$$R_s = z/k\rho_0 . \tag{21}$$

This component is described by $\tilde{M}_1$.

The scintillation index is equal to

$$m^2 \approx 1 + N(\mu)(m_B^2)^{-2(4-\mu)/(\mu-2)}, \tag{22}$$

where $N(\mu)$ is the numerical coefficient.

2.1.2 The steep power law spectrum

The peculiar properties of this spectrum are shown when $H(\rho = L_0) >> 1$. In this case, the weak scintillations are similar to those in the case of the Kolmogorov spectrum, but for $m_B^2 > 1$ there is the quasi-saturated regime.[10,12-15] In this regime, the main effect is determined by the focusing of irradiance by the inhomogeneities with scales

$$\beta_{lf} \approx \beta_\Phi (C \, \Delta z_0 z^2)^{1/(6-\mu)} . \tag{23}$$

The value of β_{lf} is much less than R_s. The scintillation spectrum has two components: a low-frequency component with spacial scale β_{lf}, and a high-frequency component with scale β_{hf},

$$\beta_{hf} \approx \frac{z}{k\beta_{lf}} . \tag{24}$$

The scintillation index in the quasi-saturated regime is constant

$$\begin{aligned} m^2 &= 1 + \Delta m^2 \\ \Delta m^2 &\approx 2 \sqrt{\mu - 4/6 - \mu} . \end{aligned} \tag{25}$$

There is the peculiarity when $\mu \to 6$. In this case, we must take into account the outer scale L_0.

2.1.3 The Gaussian spectrum

The main peculiarity of this spectrum is a strong focusing regime,[8,9] that occurs when

$$S_0^2 = \frac{1}{2} H|_{\rho \to \infty} \gg 1 . \tag{26}$$

The strong focusing regime divides the weak and saturated regimes. At distances

$$z_{foc} \simeq k\alpha^2 / \sqrt{2}\, S_0 , \tag{27}$$

we have a maximum of the scintillation index

$$m_{foc}^2 \approx ln\,(S_0^2) . \tag{28}$$

Near z_{foc}, the scintillation spectrum has a power law form

$$\left.\begin{aligned} \tilde{M}\,(\kappa;\rho'=0;z) &\simeq (1/|\kappa|^2)\,\phi\,(z), \qquad \frac{1}{\alpha} \le |\kappa| \le \frac{k\alpha}{z} \\ \phi\,(z) &\simeq \frac{I_0}{2\pi}\,(z_{foc}/z)^2 \exp\left[-(z_{foc}/z)^2\right] \end{aligned}\right\} . \tag{29}$$

2.2 The statistically homogeneous medium

The solution of Eq. (8) in the form of the iteration series is analogous to the low-frequency expansion for the phase screen case.[16] The zero term of this series is

$$\tilde{M}_0 = I_0^2\,\delta(\kappa)\,\exp\left[-H(\rho)\right]. \tag{30}$$

Substituting M_0 into the integrand in (8), we obtain the first term

$$\begin{aligned} \tilde{M}_1(\kappa,\rho'=0,z) &= 8\pi k^2 I_0^2 \int_0^z dz_1 \sin^2\left[\frac{\kappa^2 (z-z_1)}{k}\right] \Phi_n(\kappa) \\ &\cdot \exp\left[-\int_0^{z_1} dz'\, D\left[\frac{\kappa\,(z-z')}{k}\right] - \int_{z_1}^{z} dz'\, D\left[\frac{\kappa\,(z-z')}{k}\right]\right] . \end{aligned} \tag{31}$$

For small $|\kappa| z/k$, the first term is dominant. When

$$H(\kappa z/k) = \int_0^z dz' \, D(\kappa z'/k) << 1 , \tag{32}$$

$\tilde{M}_1$ has the simple form,

$$\tilde{M}_{1B}(\kappa, \rho' = 0, z) = 8\pi k^2 I_0^2 \int_0^z dz_1 \, \sin^2 \left[\frac{\kappa^2 (z - z_1)}{2k} \right] \Phi_n(z_1, \kappa) . \tag{33}$$

This expression corresponds to the Born and Rytov approximations and describes the important region of the scintillation spectrum for small z. The condition for using (33) is determined by (18). For the Kolmogorov-type spectrum, inequalities (32) and (18) are similar, but for steep power law and Gaussian spectra inequality (18) is weaker than (32).

The terms with large n provide the important contribution in the summation for large $|\kappa| \, z/k$. A complicated analysis leads to the next asymptotic expression for M,[16-22]

$$\tilde{M}(\kappa, \rho' = 0, z) \approx \tilde{M}^{(0)}(\kappa, \rho' = 0, z) = \left(\frac{1}{2\pi} \right)^2 I_0^2 \int d^2\rho \, \exp\left[-i\kappa \cdot \rho - H(\rho) \right] . \tag{34}$$

We can obtain expression (34) using the symmetry of $M(\rho, \rho')$ with respect to the transposition of ρ and ρ'.

The high-frequency iteration series may be constructed using $\tilde{M}^{(0)}$ as a zero term. $M^{(0)}$ is the main term in this expanding for large z.

The properties of scintillations in the region between the weak and saturated regimes are similar to those in the phase screen model. In the case of Kolmogorov spectrum, expressions (20)-(22) are suitable for the extended medium if Δz_o is reduced to z. In the case of the steep power law spectrum, the situation is similar.

There is some difference between the phase screen and extended medium solutions for the Gaussian spectrum. The scintillation properties depend on the parameter

$$S_{0,F}^2 = 1/2 \; H(\rho = \infty) \big|_{z = ka^2} . \tag{35}$$

The strong focusing regime exists when $S_{0,F}^2 >> 1$. We can use the expressions (27)-(29), but the function $\phi(z)$ has another form.[23]

$$\phi(z) \approx \begin{cases} \dfrac{I_0^2}{A(z/z_{foc})} \exp\left[-(z_{foc}/z)^3\right], & z \le z_{foc} \\ (3/4\pi) \exp\left[-B(z/z_{foc})\right], & z > z_{foc} \end{cases} \tag{36}$$

$$A \approx 16/3^{1/4}\left(6 - 3\sqrt{3}\right), \qquad B = \left[48/\left(2\sqrt{3} - 3\right)\right]^{1/3}.$$

3. The spherical wave scintillation

In the case of the spherical wave propagating through a phase screen, we can use results that were obtained for the plane wave when distance z is substituted by

$$z \to (1/z_1 + 1/z_2), \tag{37}$$

where z_1 is the distance from the source to the phase screen and z_2 is the distance from the phase screen to the observer. The substitution

$$\rho \to \frac{z_1}{z_1 + z_2}\rho, \qquad \rho' \to \frac{z_1}{z_1 + z_2}\rho' \tag{38}$$

is also required.

We do not have a simple a generalization for the statistically homogeneous medium. It is convenient to use the spherical coordinate system in this case. Let us represent the wave field in the form

$$U = \frac{1}{R} u_{sph} \exp(-ikR). \tag{39}$$

The equation for the fourth moment[24] of the field U is given by

$$\frac{\partial M_{sph}}{\partial R} + \frac{1}{kR^2} \nabla_\eta \nabla_{\eta'} M_{sph} + f(R\eta, R\eta') M_{sph} = 0, \tag{40}$$

with initial condition

$$m_{sph}\big|_{R=0} = I_0^2. \tag{41}$$

Here, the angle variables η and η' are similar to the variables ρ and ρ'. The methods of analysis for the plane wave case can be applied to the spherical wave case.

The weak scintillation spectrum is determined by the expression

$$\tilde{M}_{sph} = \tilde{M}_{sph,B}(\kappa,\rho' = 0,z) = \\ = 8\pi k^2 I_0^2 \int_0^R dR_1 \left(\frac{R}{R_1}\right)^2 \sin^2\left[\frac{|\kappa|^2 R}{2k}\left(\frac{R}{R_1} - 1\right)\right] \Phi_n\left(\kappa\frac{R}{R_1}\right). \quad (42)$$

The high-frequency component of the saturated scintillation spectrum is given by

$$\tilde{M}_{sph} = \tilde{M}_{sph}^{(0)} = \frac{I_0^2}{(2\pi)^2} \int d^2\rho' \exp\left[-H_{sph}(\rho) - i\kappa\cdot\rho\right], \\ H_{sph}(\rho) = \int_0^R dR' \, D\left(\rho\frac{R'}{R}\right). \quad (43)$$

We considered three cases of medium distribution between the source and the observer. The behavior of the solutions are similar for these three cases. Maybe there are some peculiarities when the parameters of the medium have a power law dependence on the distances from the observer or the source.

4. The scintillations of a finite angular incoherent source

In this case, we have two independent statistics for an incoherent source and for medium fluctuations, respectively. Let us consider, for example, a radio astronomical observation, when a receiver measures the short-integration mutual coherence function $\overline{u_1 u_2^*}$ or the intensity $\overline{|u_1|^2}$ averaged only over the source statistics. It is possible, if a receiver integration time is shorter than the medium temporal scale, and longer than the source coherence time. Then, fluctuations of this signal caused by medium variations are described by the fourth moment written in the following form:[25]

$$M = \left\langle \overline{u_1 u_2^*} \cdot \overline{u_3 u_4^*} \right\rangle. \quad (44)$$

The upper bars denote ensemble average over the source statistics.

For the plane wave propagating through the phase screen, we have the clearest picture. Two plane waves with directions of propagation θ_1 and θ_2 have the same intensity patterns displaced by $z(\theta_1 - \theta_2)$,

$$h\left(\rho_1;\theta_1;z\right) = h\left(\rho_2=\rho_1-(\theta_1-\theta_2)z;\theta_2;z\right). \tag{45}$$

For a source with a brightness distribution $J_o(\theta)$, we have

$$I(\rho;z) = \int d^2\theta \ J_o(\theta)\ h\left(\rho-\theta z;0,z\right). \tag{46}$$

Using (46), we obtain the simple expression for the spatial spectrum of the intensity fluctuation,[7,26]

$$\tilde{M}(\kappa,z) = \tilde{M}_0(\kappa,z)\left|B_{u,0}(\kappa z/k)\right|^2, \tag{47}$$

where M_0 is the plane wave spectrum and $B_{u,0}$ is the initial field coherence function.

For the extended medium, we can use Eq. (8), but the initial condition has the form[27]

$$\tilde{M}(\kappa,\rho',z)\Big|_{z=0} = \left|B_{u,0}(\rho')\right|^2. \tag{48}$$

The weak scintillation spectrum is determined by

$$\tilde{M}_{1,B}(\kappa;\rho'=0;z) = \\ = 8\pi k^2\int dz'\sin^2\left[\frac{\kappa^2(z-z')}{2k}\right]\Phi_n(z',\kappa)\left|B_{u,0}\left(\kappa(z-z')/k\right)\right|^2. \tag{49}$$

In the extended medium case, the source size influence is weaker than for the phase screen case, but the spectrum depends upon the source angular size.

For the saturated regime, the main component of the intensity correlation function has the form

$$B_I^{(0)}(\rho,z) = \int d^2\theta\ J_1(\theta)\ \exp\left[-\int_0^z dz'\,D(\rho-\theta z')\right], \\ J_1 = \int d^2\theta_1\ J_0(\theta-\theta_1)\ J_0(\theta_1). \tag{50}$$

The exponential factor in the integrand (50) is the correlation coefficient of the intensity of two plane waves with the different propagation directions. For the phase screen case, B_I depends very much on the angular shape of the J_0, while for the extended random medium the dependence is not pronounced. The low-frequency component becomes the main component for very large source size. It is determined by (49) when the scattering angle is less than the source angle size.

In the case of the source placed at finite distance, the generalization of Eq. (40) is needed for the correlation of the spherical waves with displaced centers. Let us represent the second intensity moment in the form

$$M_I(\rho,\rho';z) = \frac{1}{z^4}\int d^2\rho_0\, J_1(\rho_0)\, M_h(\rho,\rho',\rho_0;z), \tag{51}$$

where M_h is the fourth moment of the spherical waves amplitude with different centers. J_1 is determined by initial distribution of the source plane

$$J_1(\rho_0) = \int d^2r_0\, I\left(r_0 + \frac{1}{2}\rho_0\right) I\left(r_0 - \frac{1}{2}\rho_0\right). \tag{52}$$

The moment M_h is described by the equation in the form of (40) if R is replaced by z and

$$\eta = (\rho - \rho_0)/z, \quad \eta' = \rho'/z. \tag{53}$$

Then, f has the following form:

$$\begin{aligned} f &= D(\rho_0 + z\eta) + D(z\eta') - \frac{1}{2}D\left[\rho_0 + z(\eta+\eta')\right] \\ &\quad - \frac{1}{2}D\left[\rho_0 + z(\eta - \eta')\right]. \end{aligned} \tag{54}$$

The methods for solving this equation are similar to those discussed above. In the weak scintillation case, the spatial intensity fluctuation spectrum is [25,28]

$$\begin{aligned} &\tilde{M}(\kappa,\rho=0,z) \\ &= (8\pi k^2/z^2)\int \frac{dz_1}{z_1^2}\sin^2\left[\frac{\kappa z^2}{k}\left(\frac{z}{z_1}-1\right)\right]\Phi_n\left(\kappa\frac{z}{z_1}\right)\tilde{J}_1\left[\kappa\left(\frac{z}{z_1}-1\right)\right], \end{aligned} \tag{55}$$

where $\tilde{J}_1$ is the Fourier transform of J_1.

In the saturated regime, the main diffraction component has the form[25,28]

$$B_I^{(0)}(\rho) = \int d^2\rho_0 \, J_0 \exp\left[-\int_0^z dz_1 \, D\left(\rho_0 \frac{z - z_1}{z} + \rho \frac{z_1}{z}\right)\right]. \tag{56}$$

5. Two frequency correlations of the intensity fluctuations

In the case of plane waves with frequency ω_1 and ω_2, the problem of the intensity correlation is reduced to calculation of the following moment:[29-31]

$$M = \left\langle u(\boldsymbol{r},\omega_1) \; u^*(\boldsymbol{r}+\rho',\omega_1) \; u(\boldsymbol{r}+\rho,\omega_2) \; u^*(\boldsymbol{r}+\rho'',\omega_2)\right\rangle . \tag{57}$$

This moment is determined by the equation

$$\frac{\partial M}{\partial z} + \frac{i}{2}\left(\frac{1}{k_1} - \frac{1}{k_2}\right)\Delta_\rho M - i\nabla_{\rho_1}\left(\frac{1}{k}\nabla_{\rho'} + \frac{1}{k}\nabla_{\rho''}\right)M + fM = 0,$$

$$f = \frac{1}{2}\left[D_{11}(\rho') + D_{12}(\rho) + D_{22}(\rho'') + D_{12}(\rho + \rho' + \rho'') - D_{12}(\rho - \rho') - D_{12}(\rho - \rho'')\right],$$

$$D_{lm}(\rho) = 4\pi k_l k_m \int d^2\kappa \left[1 - \cos(\kappa\cdot\rho)\right] \Phi_{n,l,m}(\kappa),$$

$$k_{l,m} = \omega_{l,m}/c, \tag{58}$$

where c is the velocity of light.

The initial condition is

$$M\big|_{z=0} = I_0^2. \tag{59}$$

Transforming Eq. (58) to the spectral representation, we can obtain a solution in the form of an iterated series. The first term of this series determines the spectrum of weak scintillation

$$\tilde{M}_1\left(\kappa,\rho' = 0,\rho'' = 0,\omega_1,\omega_2\right) =$$

$$= 8\pi\, I_0^2\, k_1\, k_2 \int_0^z dz' \sin^2\left[\frac{\kappa^2 (z - z')}{2k_1}\right] \sin\left[\frac{\kappa^2 (z - z')}{2k_2}\right] \Phi_{n,1,2}\left(\kappa,z'\right). \tag{60}$$

This expression corresponds to the Born and Rytov approximations. In this case, the frequency decorrelation scale is approximately the average of the two frequencies.

The peculiarity of saturated scintillation regime consists in the fine frequency structure of the intensity fluctuations which is connected with the diffraction component of scintillations. For a single-scale medium, the Gaussian approximation for the field statistics can be extended on the multi-frequency case. Then, the diffraction component of saturated scintillation can be determined from the square modulus of the field coherence function $B_u(\omega_1,\omega_2)$. However, for a multi-scale, turbulent medium, such an equation should be corrected. The point is that B_u contains an exponential depending on $D(\infty)$. At the same time, the function f in Eq. (58) does not contain it. The corrected equation takes the form

$$B_I(\omega_1,\omega_2) = \left|B_u(\omega_1,\omega_2)\right|^2 \exp\left\{-\int dz' \left[D_{11}(\infty) + D_{22}(\infty) - 2D_{12}(\infty)\right]\right\}. \tag{61}$$

The exponential factor presented above excludes the contribution of the large inhomogeneities because they do not affect the intensity decorrelation on different frequencies.

There are solutions for the next form of $D(\rho)$,

$$D(\rho) = (\kappa\theta_0\rho)^2. \tag{62}$$

When $\Delta\omega = \omega_1 - \omega_2 << \omega = (\omega_1 + \omega_2)/2$, for the phase screen case[29] we have

$$R_I(\Delta\omega) = \frac{B_I(\Delta\omega)}{I_0^2} = \frac{1}{1 + (\Delta\omega/\Delta\omega_0)},$$
$$\Delta\omega_0 = \frac{2c}{z\theta^2}. \tag{63}$$

For the extended statistically homogeneous medium,[30,31]

$$R_I(\Delta\omega) = \frac{2}{\cosh x - \cos x}$$
$$x = \sqrt{|\Delta\omega/\Delta\omega_0|}. \tag{64}$$

For the Kolmogorov-type spectrum, the asymptotical expression is[32]

$$R_I(\Delta\omega) \approx 1 - |\Delta\omega/\Delta\omega_0|^{(\mu-2)/2}, \quad |\Delta\omega| << \Delta\omega_0 . \tag{65}$$

The frequency correlation depends upon the inhomogeneities spectrum type and the distribution of turbulence along the wave propagation direction.

6. Pulse scintillation

Let us represent the initial pulse in the form

$$U_0(t) = \int d\omega \; u_0(\omega) \; \exp\left[-i\left(\omega - \omega_0\right)t\right] \tag{66}$$

and assume that the pulse width τ_0 is greater than $1/\omega_0$,

$$\tau_0 \omega_0 >> 1 . \tag{67}$$

After propagation through a turbulent medium, we have

$$U(t) = \int d\omega \; u_1(\omega) \; u_0(\omega) \; \exp\left[-i\left(\omega - \omega_0\right)t\right].$$

The statistical moments of the modulating factor $u_1(\omega)$, which is simply a monochromatic wave amplitude, are discussed above.

In the case of the weak scintillations, the modulation factor $u_1(\omega)$ has a wide bandwidth and the pulse is displaced in time without a change of its structure,[33]

$$\begin{aligned} I(t) &= I_0(t - t_0) \\ t_0 &\approx \int dz \; n(z)/c . \end{aligned} \tag{68}$$

In the case of saturated scintillations, as was noted before, the modulation factor u_1 has a fine frequency structure. It can be shown using the normal distribution for u_1, that the scattered pulse consists of the train of subpulses. The width of each subpulse is approximately the initial width τ_0 and the envelope of chain has the width $\approx 1/\Delta\omega_0$, where $\Delta\omega_0$ is the frequency decorrelation scale.

7. Spectral line fluctuation in a moving turbulent medium

The interaction of the irradiance with a moving turbulent medium leads to the temporal intensity modulation. Usually, we have the transformation of the spatial pattern into a temporal one using the velocity. But for an initial narrowband spectral line, we have a new effect, the fluctuation of the line frequency and the line structure.[33]

The initial coherent irradiance

$$U_0 = u_0 \exp\left(-i\omega_0 t\right) \tag{69}$$

propagating through a turbulent medium becomes

$$U(\boldsymbol{r},z,t) = u_1(\boldsymbol{r},z,t)\, U_0 . \tag{70}$$

The average intensity distribution of the spectrum is

$$\begin{aligned} \left\langle I(\omega)\right\rangle &= |U_0|^2 \int d\tau\, B_{u,1}(\tau)\, \exp\left[-i\,(\omega-\omega_1)\,\tau\right], \\ B_{u,1} &= \left\langle u_1(t)\; u_1(t+\tau)\right\rangle . \end{aligned} \tag{71}$$

In the case of the uniform medium moving relatively to the line of sight with velocity V, we have

$$B_{u,1} = \exp\left[-(1/2)\; H\left(\rho = Vt\right)\right]. \tag{72}$$

The average line width is $\Delta\omega_0 \approx V/\rho_0$, where ρ_0 is the field coherence scale.

The phase modulation is a main factor for weak scintillations. The specific realizations may be represented as a displacement of the spectral line. In the case of saturated fluctuations, the irradiance has a noiselike form.

8. References

1. V. I. Shishov, "Theory of wave propagation in random media," *Radiophys. Quantum Electron.* **11**(6), 500-505 (1968).

2. M. J. Beran and T. L. Ho, "Propagation of the fourth-order coherence function in a random medium," *J. Opt. Soc. Am.* **59**(9), 1134-1138 (1969).

3. V. I. Tatarskii, "Light propagation in a medium with random refractive index inhomogeneities in the Markov random process approximation," *Sov. Phys. JETP* **29**(6), 1133-1138 (1969).

4. L. A. Chernov, "Equations for the statistical moments of the field in a randomly inhomogeneous medium," *Sov. Phys. Acoust.* **15**(4), 511-517 (1969).

5. V. V. Pisareva, "The diffraction of radio waves by random nonuniformities, and intensity variations of solar and cosmic radio emission," *Sov. Astron.* **2**(1), 97-111 (1958).

6. R. P. Mercier, "Diffraction by a screen causing large random phase fluctuations," *Proc. Cambridge Phyl. Soc.* **58**, 382-400 (1962).

7. E. E. Salpeter, "Interplanetary scintillations. I. Theory," *Astrophys. J.* **147**(2) 433-446 (1967).

8. R. Buckley, "Diffraction by a random phase screen with very large r.m.s. phase deviation," *Austral. J. Phys.* **24**, 351-396 (1971).

9. V. I. Shishov, "Diffraction of waves by a strongly refracting random phase screen," *Radiophys. Quantum Electron.* **14**(1), 70-75 (1971).

10. V. I. Shishov, "Dependence of the form of the scintillation spectrum on the form of the spectrum of refractive-index inhomogeneities," *Radiophys. Quantum Electron.* **17**(11), 1287-1292 (1974).

11. K. S. Gochelashvili and V. I. Shishov, "Saturation of laser irradiance fluctuation beyond a turbulent layer," *Opt. Quant. Electron.* **7**, 524-536 (1975).

12. E. Jakeman and J. G. McWhirter, "Correlation function dependence of the scintillation behind a deep random phase screen," *J. Phys. A. Math. Gen.* **10**, 1599-1643 (1977).

13. B. J. Uscinski, H. G. Booker, and M. Marians, "Intensity fluctuations due to a deep phase screen with power-law spectrum," *Proc. R. Soc. Lond.* A **374**, 503-530 (1981).

14. C. L. Rino and J. Owen, "Numerical simulation of intensity scintillation using the power-law phase screen model," *Radio Sci.* **19**(3), 891 (1984).

15. J. Goodman and R. Narayan, "Slow pulsar scintillation and the spectrum of interstellar electron density fluctuations," *Month. Not. R. Astr. Soc.* **214**, 519-537 (1985).

16. K. S. Gochelashvili and V. I. Shishov, "Saturated fluctuations in the laser radiation intensity in a turbulent medium," *Sov. Phys. JETP* **39**(4), 605-609 (1974).

17. R. L. Fante, "Electric field spectrum and intensity covariance of a wave in a random medium," *Radio Sci.* **10**(1), 77-85 (1975).

18. C. Macaskill, "An improved solution to the fourth moment equation for intensity fluctuations," *Proc. R. Soc. Lond.* A **386**(1791), 461-474 (1983).

19. S. Frankenthal, M. J. Beran, and A. M. Whitman, "Caustic correction using coherence theory," *J. Opt. Soc. Am.* **71**(2), 348-358 (1982).

20. B. J. Uscinski, "Analytical solution of the fourth-moment equation and interpretation as a set of phase screens," *J. Opt. Soc. Am.* A **2**(12), 2077-2091 (1985).

21. A. S. Gurvich, B. C. Elepov, V. V. Pokolsov, K. K. Sabelfeld, and V. I. Tatarskii, "Spatial structure of strong fluctuations of light intensity in a turbulent medium," *Radiophys. Quantum Electron.* **22**(2), 135-141 (1979).

22. J. M. Martin and S. M. Flatté, "Intensity images and statistics from numerical simulation of wave propagation in 3-D random media," *Appl. Opt.* **27**(11), 2111-2126 (1988).

23. V. I. Shishov, "Strong scintillation of a plane wave in a medium with Gaussian spectrum of inhomogeneities," *Radiophys. Quantum Electron.* **31**(8), 667-673 (1988).

24. K. S. Gochelashvili, V. G. Pevgov, and V. I. Shishov, "Saturation of fluctuations of the intensity of laser radiation at large distances in a turbulent atmosphere (Fraunhofer zone of transmitter)," *Sov. J. Quant. Electron.* **4**(5), 632-637 (1974).

25. R. G. Frehlich, "Space-time fourth moment of waves propagating in random media," *Radio Sci.* **22**(4), 481-490 (1987).

26. L. T. Little and A. Hewish, "Interplanetary scintillation and its relation to the angular structure of radio sources," *Month. Not. R. Astr. Soc.* **134**, 221-237 (1966).

27. V. I. Shishov and T. D. Shishova, "Influence of source sizes on the spectra of interplanetary scintillations. Theory," *Sov. Astron.* **22**(2), 235-239 (1978).

28. V. I. Shishov, "Effect of source size on scintillation characteristic," *Radiophys. Quantum Electron.* **33**(4), 325-332 (1990).

29. V. A. Alimov and L. M. Erukhimov, "The influence of the receiver on the magnitude of the frequency correlation of fluctuations in received radio signals," *Radiophys. Quantum Electron.* **10**(5), 340-344 (1967).

30. V. I. Shishov, "Frequency correlation of scintillations," *Radiophys. Quantum Electron.* **16**(3), 319-327 (1973).

31. L. C. Lee, "Strong scintillations in astrophysics. IV. Cross-correlation between different frequencies and finite bandwidth effects," *Astrophys. J.*, **206**, 744-752 (1976).

32. V. E. Ostashev and V. I. Shishov, "Pulse shape of a pulsar scattered in the interstellar medium," *Radiophys. Quantum Electron.* **20**(6), 581-585 (1977).

33. V. I. Shishov, "Frequency fluctuation of a wave propagating in a moving, randomly inhomogeneous medium," *Radiophys. Quantum Electron.* **19**(10), 1056-1059 (1976).

PULSE STATISTICS IN RANDOM MEDIA

C. H. Liu

National Central University, Chung-Li, Taiwan, ROC

Abstract

The study of the effects of turbulence and random scatterers on the propagation of pulsed signals has for many years received attention of researchers from different fields, such as communication, remote sensing, material science, biophysics, etc. In this paper theoretical treatment of the problem will be reviewed. Some applications will be discussed.

Early efforts to study pulse propagation in random media have been limited to weakly turbulent cases in which only the effects of single scattering were taken into account. Later the problem was formulated in terms of the two-frequency mutual coherence function. This approach has the potential to include multiple scattering effects on the pulses. However, the mutual coherence function is governed by a transport equation which can only be solved numerically when turbulence is strong. Further numerical computations are needed even after the mutual coherence function is obtained. The advantages and difficulties involved in this approach will be discussed. Next, the formulation in terms of temporal moments will be introduced. The technique is based on the idea of describing a transient signal by its temporal moments. A procedure will be developed to solve each temporal moment analytically even under multiple scattering conditions. Thus, in principle, the propagation of pulsed signals through turbulent media can be studied analytically using this procedure.

We shall discuss several aspects of the pulse statistics using this technique. First, the reconstruction of the pulse from the temporal moments will be discussed, emphasizing the effects of multiple scattering. These procedures will all be applied to examine the detailed statistics of the pulse such as the arrival time of the pulse, the spreading and wandering of the pulse, etc. Because of the analytical approach, it becomes easier for us to understand the physics of the phenomena from the solution.

The extension of the treatment to cases of pulsed beam waves will be discussed. Examples will be presented. Recent work involving pulse propagation in turbulent media with inhomogeneous background will be discussed with ionospheric applications in mind.

1. Introduction

The problem of propagation of pulsed signals in random media has been studied by many investigators[1,2,3,4,5]. This interest is sustained by applications in areas such as high data rate communication systems, precise ranging and navigation systems, remote sensing using radars and lidars as well as astrophysical applications. Early efforts have been limited to weakly turbulent media in which only the effects of single scattering were taken into account. However, for many applications, multiple scattering effects may become dominant causing saturation in the fluctuations of pulse intensity and drastically modifying the behavior of the pulse. Therefore, recent efforts in this field have been concentrating on the study of the effects of multiple scattering. Several approaches have been developed in treating the problem[6,7]. One of the basic ideas in developing these approaches is based on the observation that for almost all practical cases, forward scattering plays the principal role in the propagation phenomena. This means that the Parabolic Equation Approximation can be used to study the various aspects of the problem. Under this approximation, a set of transport equations can be derived for the higher order moments of the field[6,8]. Some of the important developments in the investigation of pulse propagation are based on these equations.

Initially, the pulse problem was treated using Born's approximation[1,2]. Then an approach based on the two-frequency mutual coherence function was introduced[3,4,9]. This approach has the potential to take into account the effects of multiple scattering on the pulses under fairly general conditions. The mutual coherence function is governed by a transport equation, as mentioned above, derived from the parabolic equation. This equation has been solved under different approximations corresponding to the different regimes of the strength of turbulence. The solutions can then be used to study the pulse problem. Because of the nature of the approximations, the features that can be investigated are often limited. Therefore, a different approach has been developed in which the idea of temporal moments was introduced[10,11]. This approach is unique in that the pulse propagation problem can be studied analytically even when the multiple scattering effects are important.

The idea of moments is not new. It is extensively used in probability theory[12]. Temporal moments have been applied to characterize stochastic transients[13]. Each moment has its unique physical meaning. The zeroth moment relates to the total energy of the pulse; the first moment is related to the mean arrival time; the second moment related to the pulse width, the third moment related to the skewness of the pulse, etc. Therefore, many of the important statistical properties can be studied when the various temporal moments are known. In section 2 the temporal moments will be introduced and the basis for the technique will be formulated. These moments are used to study the pulse arrival time in section 3 and the pulse width in section 4. In section 5 the mean pulse intensity is investigated. The case of a pulse beam wave is discussed in section 6. Some concluding remarks are included in section 7.

2. Temporal moments of stochastic signals

Let us consider a monochromatic wave propagating in the z direction in a random medium. The wave can be expressed as $u(z, \vec{\rho}, \omega) \exp[j(\omega t - kz)]$ where u is the complex amplitude of the wave, $k = \omega/v$ is the wave number with v the propagation speed, and $\vec{\rho}$ is the transverse coordinate. Any pulsed signal can be built from this basic wave. For example, a pulse with spectrum $f(\omega)$ incident at $z = 0$ can be expressed as follows at z in the medium

$$p(z, \vec{\rho}, t) = \int_{-\infty}^{+\infty} f(\omega)u(z, \vec{\rho}, \omega) \exp[j\omega(t - z/v)]d\omega = a(z, \vec{\rho}, t) \exp[j\omega_c(t - z/v)] \tag{1}$$

where ω_c is the carrier frequency and $a(z, \vec{\rho}, t)$ is the complex envelope of the pulse.

The complex amplitude u is a random function, therefore $a(z, \vec{\rho}, t)$ is a stochastic signal. To study this signal in detail we define the nth temporal moment of the signal p(t) by the expression

$$M^{(n)}(z, \vec{\rho}) = \int_{-\infty}^{+\infty} t^n p^2(z, \vec{\rho}, t)dt = \int_{-\infty}^{+\infty} t^n a^*(z, \vec{\rho}, t)\, a(z\, \vec{\rho}, t)dt \tag{2}$$

Since $a(z, \vec{\rho}, t)$ is a stochastic function, $M^{(n)}$ is a random quantity. Its average is given by

$$< M^{(n)} (z, \vec{\rho}) > = \int_{-\infty}^{+\infty} t^n <a^*(z, \vec{\rho}, t)\, a(z, \vec{\rho}, t)> dt \tag{2a}$$

where the brackets indicate ensemble averaging.

Substituting (1) into (2) and making use of the relation

$$\int_{-\infty}^{+\infty} t^n \exp[j(\omega_1 - \omega_2)t]\, dt = 2\pi\, (-j)^n\, \delta^{(n)}\, (\omega_1 - \omega_2) \tag{3}$$

where $\delta^{(n)} (\omega_1 - \omega_2)$ is the nth-order Dirac delta function, we obtain

$$< M^{(n)} (z, \vec{\rho}) > = 2(j)^n \int_{-\infty}^{+\infty} f^*(\omega_2) \exp(j\omega_2 z/v) \frac{\partial^n}{\partial \omega_1^n}$$
$$\bullet\, [f(\omega_1)\, \Gamma \exp(-j\omega_1, z/v)] \,\Big|_{\omega_1 = \omega_2} d\omega_2 \tag{4}$$

where the property of the nth-order Dirac delta function has been used, and Γ is the one-position, two-frequency mutual coherence function defined by

$$\Gamma(z, \vec{\rho}, \omega_1, \omega_2) = <u(z, \vec{\rho}, \omega_1)\, u^*(z, \vec{\rho}, \omega_2)> \tag{5}$$

The form of equation (4) suggests an expansion of Γ in the series of a parameter δ defined by $\delta = (\omega_1 - \omega_2)/\omega_2$ such that

$$\Gamma(z, \vec{\rho}, \omega_1, \omega_2) = \Gamma_0 + \Gamma_1\, \delta + \Gamma_2\, \delta^2 + \ldots \tag{6}$$

where

$$\Gamma_0 = \Gamma(z, \vec{\rho}, \omega_1, \omega_1) \text{ and } \Gamma_n = (-1)^n(\omega_2^n/n!)\, \partial^n\Gamma/\partial\omega_1^n \,\Big|_{\omega_1 = \omega_2}, \quad n = 1, 2, \ldots, \tag{7}$$

It is easy to show from equations (4), (6), and (7) that the nth temporal moment $M^{(n)}$ depends only on $\Gamma_0, \Gamma_1, \ldots \Gamma_n$ but not on Γ_{n+1} or beyond. This is an important property which will become clear in the following developments.

The next step is to solve for the mutual coherence function Γ, which now amounts to solving for the coefficients Γ_n. The starting point is the parabolic equation. Under fairly general conditions, it has been shown that the complex amplitude function u satisfies the equation[6]

$$\nabla_T^2 u - 2jk\partial u/\partial z + k^2 Q(\vec{\rho}, z)\, u = 0 \tag{8}$$

where $\nabla_T^2 u = \partial^2/\partial x^2 + \partial^2/\partial y^2$ is the transverse Laplacian and $Q = \Delta\varepsilon/\varepsilon$ is the fluctuating part of the dielectric permittivity. Starting from (8) it is possible to derive the equation for the two-position, two-frequency mutual coherence function $\Gamma(z, \vec{\rho}_1, \vec{\rho}_2, \omega_1, \omega_2)$[8]

$$\frac{\partial \Gamma}{\partial z} + \frac{j}{2k_1k_2}(k_2\nabla_1^2 - k_1\nabla_2^2)\,\Gamma + \frac{1}{8}[(k_1^2 + k_2^2)\,A(0) - 2k_1k_2A(\xi)]\,\Gamma = 0 \tag{9}$$

where $\nabla_i^2 = \partial^2/\partial x_i^2 + \partial^2/\partial y_i^2$, $\vec{\rho}_i = (x_i, y_i)$, $i = 1, 2$; $\vec{\xi} = \vec{\rho}_2 - \vec{\rho}_1$ and the two-dimensional correlation function defined by

$$A(\vec{\xi}) = \int_{-\infty}^{+\infty} B_\varepsilon(z, \vec{\xi})\,dz \tag{10}$$

In (10) $B_\varepsilon(z, \vec{\xi})$ is the correlation function for the random function Q, the fluctuating part of the permittivity. Here, to avoid introducing too many new functions, we have used the symbol Γ for the two-position mutual coherence function. Similarly, the expansion in (6) is also used for the two-position case. The substitution of this expansion into (9) yields an hierarchy of equations when terms of equal power in δ are collected. The first two are given below:

$$\frac{\partial \Gamma_0}{\partial z} + \frac{j}{2k_2}(\nabla_1^2 - \nabla_2^2)\,\Gamma_0 + \frac{k_2^2}{4}[A(0) - A(\xi)]\,\Gamma_0 = 0$$

$$\frac{\partial \Gamma_1}{\partial z} + \frac{j}{2k_2}(\nabla_1^2 - \nabla_2^2)\,\Gamma_1 + \frac{k_2^2}{4}[A(0) - A(\xi)]\,\Gamma_1 \tag{11}$$

$$= \frac{j}{2k_2}\nabla_1^2\Gamma_0 - \frac{k_2^2}{4}[A(0) - A(\xi)]\,\Gamma_0$$

The rest of the hierarchy equations all have similar form with the same differential operator on the left-hand side for the coefficient Γ_n, while on the right-hand side only terms involving Γ_0 to Γ_{n-1} appear. This set of equations can then be solved sequentially given the "initial" condition at $z = 0$. This hierarchy forms the basis of our discussion on the temporal moments for the transient wave propagating in the random medium. As discussed earlier, under the assumption of forward scattering, equation (11) is exact including all the important multiple scattering effects.

Let us consider a plane wave Gaussian pulse at carrier frequency ω_c incident at $z = 0$ in the random medium. The envelope is described by

$$a(t) = \exp(-t^2/T_0^2) \tag{12}$$

From equations (1) and (12) we obtain $f(\omega)$ which can then be used in (4) to compute the temporal moments once the mutual coherence function Γ is found. For plane wave incident, such that $\Gamma = 1$ at $z = 0$, the terms $(\nabla_1^2 - \nabla_2^2)\,\Gamma_n$ in (11) cancel and the set of equations can be solved analytically. In the derivation of these solutions, it is convenient to expand the correlation function $A(\xi)$ in a series

$$A(\xi) = A_0 + A_2\xi^2 + A_4\xi^4 + A_6\xi^6 + \ldots. \tag{13}$$

Here we have assumed isotropic turbulence in the x - y plane

The solutions $\Gamma_0, \Gamma_1, \Gamma_2$ .... can be obtained from (11) in a straightforward manner. By setting $\xi = 0$ in these solutions, we obtain the coefficients needed to calculate the various mean temporal moments from (4). The first few mean moments are given as follows:

$$<M^{(0)}> = \sqrt{2\pi} \quad T_0/2$$

$$<M^{(1)}> = (z/v)\,(1 - A_2\, z/4)\, <M^{(0)}>$$

$$\begin{aligned} <M^{(2)}> &= [(T_0^2/4) + (z/v)^2 - z^3A_2/2v^2 + zA_0/4v^2 + A_2^2z^4/16v^2 \\ &\quad + 4z^3A_4/3\omega_c^2]\, <M^{(0)}> \end{aligned} \qquad (14)$$

$$\begin{aligned} <M^{(3)}> &= [z^3/v^3 + 3(z/v)\,(1 - A_2z/4)\, T_0^2/4 + 3z^2A_0/4v^3 + 3z^5A_2^2/16v^2 \\ &\quad + 4A_4z^4/v\omega_c^2 - 12A_2A_4z^5/5v\omega_c^2]\, <M^{(0)}> \end{aligned}$$

Higher order moments can be obtained in a similar manner.

These moments are exact. They apply to the case of plane wave incident. Equation (11) can also be solved for the beam wave case which will be discussed in a later section.

3. Pulse statistics -- time of arrival

For many applications it is important to have information about the statistical behavior of the arrival time and the pulse width. Let us first consider the time of arrival of the pulse.

To discuss the arrival time it is important first to introduce the idea of a "time position" of the pulse. In the presence of noise and pulse broadening it is natural to define the time position of the pulse by its "time centroid." This, of course, is just the first temporal moment of the pulse. For the incident pulse given by (12), the initial time position of the pulse is at $t = 0$. For an observer sitting at position z, the arrival time is given by $t_a = M^{(1)} / <M^{(0)}>$. From (14) we obtain the mean arrival time:

$$<t_a> = (z/v)\,(1 - A_2z/4) \qquad (15)$$

This expression is exact under the forward scattering assumption. For a random medium with power-law irregularity spectrum as described in the Appendix, A_2 is seen to be negative and proportional to the strength of the turbulence. Therefore the additional time delay is proportional to how much scattering the signal has undergone in the medium, which is measured by the product of the distance the pulse has traveled and the strength of the turbulence. Physically this additional mean time delay comes about because of the additional distances the scattered waves have to travel during the multiple scattering process.

The arrival time is a random variable as discussed earlier. Therefore it is of interest to investigate the higher order statistics associated with t_a other than just the mean value. To study those statistics we introduce the ith statistical moments for the nth temporal moments.

$$<M^{(n)}(i)> = \int_{-\infty}^{+\infty} \cdots \int_{-\infty}^{+\infty} <p^2(t_1)p^2(t_2)\ldots p^2(t_i)> (t_1, t_2 \ldots t_i)^n \, dt_1, dt_2 \ldots dt_n \quad i = 1,2,3\ldots \tag{16}$$

For the arrival time statistics, we need $<M^{(n)}(i)>$'s. Combining (1) and (16) we have

$$<M^{(1)}(2)> = (2\pi)2 \int_{-\infty}^{+\infty} \int f^*(\omega_2)\, f^*(\omega_4) \left[\frac{z^2}{v^2} + j\frac{z}{v}\frac{\partial}{\partial\omega_1} - j\frac{z}{v}\frac{\partial}{\partial\omega_3} - \frac{\partial^2}{\partial\omega_1\partial\omega_3}\right] \cdot [f(\omega_1)\, f(\omega_3)\, \Gamma(4)]\Big|_{\omega_1 = \omega_2,\, \omega_3 = \omega_4} d\omega_2 d\omega_4 \tag{17}$$

where the fourth position, four-frequency mutual coherence function $\Gamma(4)$ is defined by

$$\Gamma(4) = <u(z, \vec{\rho}_1, \omega_1)\, u^*(z, \vec{\rho}_2, \omega_2)\, u(z, \vec{\rho}_3, \omega_3)\, u^*(z, \vec{\rho}_4, \omega_4)> \tag{18}$$

Similar expressions can be derived for the higher order statistical moments in which higher order multifrequency, multiposition mutual coherence function $\Gamma(2m)$ will appear.

The statistics of the arrival time which is related to the first temporal moment $M^{(1)}$, can be investigated by computing the statistical moments $<M^{(1)}(i)>$. The crucial quantities are the various higher order coherence functions. The general solutions for these functions under multiple scattering conditions have yet to be worked out[8,14]. Liu and Yeh[14], using the asymptotic expressions for the higher order mutual coherence functions under the extreme case of very strong scattering derived by the path-integral technique[7], studied the statistics of the pulse arrival time. They found that in the limit of extremely strong turbulence such that the fluctuations are in the fully saturated regime, the fluctuations about the mean arrival time are small as compared with the excess time delay. This suggests that in this regime the excess time delay of the pulse is caused mainly by pulse spreading rather than pulse wandering[15].

Additional work needs to be carried out to study the statistics of the pulse arrival time for the general case of partially saturated regime.

4. Pulse statistics -- pulse width

The mean width of a pulse can be computed from the second temporal moment. When the square of the mean arrival time is subtracted out from the normalized second moment, the resulting expression is the mean square pulse width. From (14) and (15) we obtain

$$<B_w^2> = <M^{(2)}>/<M^{(0)}> - <t_a>^2 = T_0^2/4 + zA_0/4v^2 + 4z^3A_4/3\omega_c^2 \tag{19}$$

The first term is the original pulse width. The other two terms are the broadening due to the multiple scattering in the random medium. The expression in (19) is exact under the forward scattering assumption. Two possible mechanisms are responsible for the temporal broadening of pulses. The ensemble averaged pulse may be broadened owing to the fact that each realization of the ensemble is spread in time as a consequence of scattering. It may also come about because of the wandering of the pulse arrival time. Although we have the exact expression of the mean square pulse broadening as given in (19), to distinguish the contributions from the two mechanisms requires additional information. Liu and Yeh[15] introduced an ensemble averaged correlation function for the intensity

$$C(t) = \langle \int I(t+\tau)\, I(\tau) d\tau \rangle = \int \int f(\omega_1) f^*(\omega_2) f(\omega_3) f^*(\omega_4)\, \Gamma(4)$$
$$\bullet \exp j[(\omega_3 - \omega_4)t + (\omega_1 - \omega_2 + \omega_3 - \omega_4)\tau - (k_1 - k_2 + k_3 - k_4)z]$$
$$\bullet\, d\tau d\omega_1 d\omega_2 d\omega_3 d\omega_4 \qquad (20)$$

where $\Gamma(4)$ is the fourth order mutual coherence function defined in (18). Since C(t) is a correlation function it is not affected by the wandering of the pulse arrival time. The width of the function C(t) will provide the information about the spreading of the pulse. To study C(t) one needs to solve for the fourth order coherence function $\Gamma(4)$ which is governed by a moment equation[15]:

$$\frac{\partial \Gamma(4)}{\partial z} = \frac{-j}{2}\left(\frac{\nabla_1^2}{k_1} - \frac{\nabla_2^2}{k_2} + \frac{\nabla_3^2}{k_3} - \frac{\nabla_4^2}{k_4}\right)\Gamma(4) + F\Gamma(4) \qquad (21)$$

where

$$F = -[(k_1^2 + k_2^2 + k_3^2 + k_4^2)\, A(0)$$
$$+ 2k_1k_3A(\vec{\rho}_1 - \vec{\rho}_3) + 2k_2k_4A(\vec{\rho}_2 - \vec{\rho}_4)$$
$$- 2k_1k_2A(\vec{\rho}_1 - \vec{\rho}_2) - 2k_1k_4A(\vec{\rho}_1 - \vec{\rho}_4)$$
$$- 2k_2k_3A(\vec{\rho}_2 - \vec{\rho}_3) - 2k_3k_4A(\vec{\rho}_3 - \vec{\rho}_4)]\,/\,8 \qquad (22)$$

This equation was solved for two limiting cases by Liu and Yeh[15]. One is the case of weak turbulence so that the Rytov approximation can be applied. The other is the fully saturated regime where asymptotic solution can be constructed[7]. It turns out that for the weak turbulence case the pulse spreading is small. It increases with the distance the pulse has traveled in the medium. Yet by far the most important contribution to pulse broadening is due to the wandering of the pulse[15]. For the asymptotic saturation regime, however, the situation is quite different. Pulse spreading dominates the broadening process. It is thought that the coherence among the spectral components of the signal is severely destroyed under this condition[15]. This is consistent with what has been discussed in section 3. There it was found that in the fully saturated regime the excess delay in pulse arrival time is mainly due to pulse spreading.

5. The intensity of the pulse

Obviously the most crucial statistics of the pulse is the mean pulse shape, or the mean intensity of the pulse. Indeed, this has been the main concern in many investigations of pulse propagation problems (see Liu and Yeh[11] and references therein). Results have been obtained under different approximations corresponding to the various regimes of turbulence strength. In this section we concentrate on the investigation based on the temporal moment technique. Since we have already shown that the temporal moments can be obtained exactly and analytically, the mean pulse intensity derived from this technique is also exact.

Let us consider the mean intensity of the pulse $I(t) = \langle |a(t)|^2 \rangle$. We introduce a polynomial $I_N(t)$

$$I_N(t) = \sum_{n=0}^{N} b_n w(t)\phi_n(t) \tag{23}$$

where $w(t)$ is a non-negative weighting function and $\phi_n(t)$ are polynomials given by

$$\phi_n(t) = \sum_{k=0}^{n} \alpha_{nk} t^k \qquad \alpha_{nn} \neq 0 \tag{24}$$

These functions satisfy the orthogonal relation

$$\int_{-\infty}^{+\infty} w(t)\phi_n(t)\phi_m(t)dt = 0 \qquad m \neq n \tag{25}$$

Depending on the weighting functions, different orthogonal polynomials have been introduced[16]. The most commonly used ones are the Legendre, Laguerre, Chebyshev, and Hermite polynomials.

We want to find an $I_N(t)$ such that it will approximate the mean intensity $I(t)$. The criterion for the approximation is to minimize the weighted mean square error defined by

$$\varepsilon = \int_{-\infty}^{+\infty} \{[I(t) - I_N(t)]^2 / w(t)\}dt \tag{26}$$

by choosing the set of coefficients b_n in (23). Substituting (23) into (26) and going through the standard procedure of minimizing ε with respect to b_n, we obtain

$$b_n = \int_{-\infty}^{+\infty} \phi_n(t)\, I(t)dt \Big/ \int_{-\infty}^{+\infty} w(t)\phi_n^2(t)dt \tag{27}$$

Using the definition of temporal moments (2) and (24), we can rewrite b_n as

$$b_n = \sum_{k=0}^{n} \alpha_{nk} \langle M^{(k)} \rangle \Big/ \int_{-\infty}^{+\infty} w(t)\phi_n^2(t)dt \tag{28}$$

Thus, the expansion coefficients b_n for the approximate mean pulse intensity $I_N(t)$ are related to the mean temporal moments. Knowing the first N temporal moments, we can construct the polynomial expansion $I_N(t)$ as an approximation for the average pulse intensity in the least square sense. For a given N, the error in the approximation can be estimated[16].

The next question is how to choose the weighting function and the orthogonal polynomials. The starting point is the incident pulse. A non-negative function $Y(t, C_0, C_1, ...C_r)$ with r + 1 parameters is chosen. The parameters are determined by requiring

$$\int_{-\infty}^{+\infty} t^n \, Y(t;C_0,C_1,...C_r)dt = M^{(n)}, \qquad n = 0, 1, 2, ...r \tag{29}$$

The weighting function w(t) is set to be equal to Y. This weighting function is close to the intensity function of the pulse such that the first r + 1 moments are exact. The polynomial expansion of the pulse intensity becomes:

$$I_N(t) = Y(t) + \sum_{n=r+1}^{N} b_n Y(t)\phi_n(t) \tag{30}$$

Thus Y(t) can be considered as the initial estimate of the mean pulse intensity. In general it is convenient to take Y as the similar functional form as the incident signal, but with the effects of scattering taken into account through the determination of the r + 1 parameters. The additional terms in (30) are corrections to the initial estimation such that the improved approximation has the same moment through order N as the average intensity I(t).

To see how the procedure works we consider an example with

$$a(t) = \exp(-t^2/T_0^2) \tag{31}$$

A two parameter weighting function of the form

$$w(t) = B \exp(-\lambda^2 t^2) \tag{32}$$

is chosen where B and λ are determined through equation (29).

$$\lambda = [\overline{M}^{(0)} / \overline{M}^{(2)}]^{1/2}, \qquad B = \lambda \overline{M}^{(0)} / \sqrt{\pi} \tag{33}$$

where the moments $\overline{M}^{(0)}$, $\overline{M}^{(2)}$, etc. are related to the original moments by

$$\overline{M}^{(n)} = \int_{-\infty}^{+\infty} (t-<t_a>)^n \, I(t)dt = \sum_{k=0}^{n} \frac{n!}{k!(n-k)!} (-<t_a>)^{n-k} M^{(k)} \quad n = 0, 1, 2, ... \tag{34}$$

These are time shifted moments with the average intensity centered at the mean arrival time $<t_a>$.

With the weighting function given in (32), the corresponding orthogonal polynomials are the Hermite polynomials given by[17]

$$\phi_n(t) = \sum_{k=0}^{k} (-1)^k \frac{n!}{k!(n-2k)!} (2\lambda t)^{n-2k} \tag{35}$$

With equations (34) and (35), b_n can be computed from (28). The polynomial expansion of the mean pulse intensity is then obtained from (30). The first few terms are given by

$$I_N(t) = B[1 - 2(\overline{M}^{(3)} / (\overline{M}^{(0)})\, \lambda^4 t(1 - 2\lambda^2 t^2/3) + ...] \exp(-\lambda^2 t^2) \qquad (36)$$

where the parameter λ^2 is given explicitly by

$$\lambda^2 = (1/2)[T_0^2/4 + (A_0/4v^2)z + (4A_4/3\omega_c^2)z^3]^{-1} \qquad (37)$$

This example illustrates how the mean pulse shape can be determined through the temporal moments technique. As mentioned earlier, the moments obtained in this procedure are exact, even in the case of multiple scattering under strong turbulence conditions. Since both the solution for the moments and the polynomial expansion are carried out in sequential order, it is very convenient in practice to apply this procedure. As indicated above, for a given number of terms in the expansion, the error can be estimated. If more accuracy is desired, additional terms in the expansion can be computed without changing any of the terms already calculated.

6. Propagation of pulsed beam waves

Up to this point the discussion has been concentrated on plane waves. In many applications the pulsed signals cannot be considered as plane waves. Rather they are really beam waves. In principle the temporal moment technique can also be applied to the beam wave case. In fact equation (4) still applies. The only difference is that the mutual coherence function Γ has to be solved for the beam wave from equation (9). For this case, in addition to the new variable

$$\vec{\xi} = \vec{\rho}_2 - \vec{\rho}_1, \text{ we define } \vec{\rho} = \tfrac{1}{2}(\vec{\rho}_1 + \vec{\rho}_2).$$

For a Gaussian beam wave of the form $u_1 = \exp[-\rho_1^2/2a^2 - jk_1\rho_1^2/2F]$. The initial mutual coherence function is given by

$$\Gamma(z{=}0) = \exp[-\frac{\rho^2+\xi^2/4}{a^2} - \frac{jk_2\,\vec{\rho}\cdot\vec{\xi}}{F} + j\frac{\delta k_2(\rho^2+\xi^2/4-\vec{\rho}\cdot\vec{\xi})}{2F}\ ] \qquad (38)$$

To solve equation (9), it is convenient to introduce the new function $\overline{W}$ by

$$\Gamma = \overline{W} \exp -[k_2^2(2 - 2\delta + \delta^2)A(0)z/8] \qquad (39)$$

and expand $\overline{W}$ in power series of δ.

$$\overline{W} = \overline{W}_0 + \delta\,\overline{W}_1 + \delta^2\,\overline{W}_2 + ... \qquad (40)$$

Substituting (40) into (9) and equating terms of same order in δ, we obtain an hierarchy of equations[18]

$$L\,\overline{W}_0 = 0$$

$$L\,\overline{W}_1 = L_1\,\overline{W}_0 - k_2^2 A(\vec{\xi}\,)\,\overline{W}_0 / 4 \qquad (41)$$

$$L\overline{W}_2 = L_1(\overline{W}_0 + \overline{W}_1) - k_2^2 A(\vec{\xi})\overline{W}_1 / 4$$

where the operators L and L_1 are given by

$$\begin{aligned} L &= \frac{\partial}{\partial z} + (j/k_2)\, \nabla_\rho \cdot \nabla_\xi - k_2^2 A(\vec{\xi}) / 4 \\ L_1 &= -(j/2k_2)[(1/2)\, \nabla_\rho^2 + \nabla_\xi^2 + \nabla_\rho \cdot \nabla_\xi] \end{aligned} \tag{42}$$

The initial conditions at z = 0 obtained from equations (38) and (39) are given by

$$\begin{aligned} \overline{W}_0^0(\vec{\rho}, \vec{\xi}, \omega_2) &= \exp[\,-(\rho^2 + \xi^2/4) / a^2 - j\, k_2\, \vec{\rho} \cdot \vec{\xi} / F] \\ \overline{W}_1^0(\vec{\rho}, \vec{\xi}, \omega_2) &= j(k_2/2F)(\rho^2 + \xi^2/4 + \vec{\rho} \cdot \vec{\xi}\,)\overline{W}_0(0) \\ \overline{W}_2^0\vec{\rho}, \vec{\xi}, \omega_2) &= -\,(k_2^2/8F^2)(\rho^2 + \xi^2/4 + \vec{\rho} \cdot \vec{\xi}\,)^2\overline{W}_0(0) \end{aligned} \tag{43}$$

To solve (41) we define the Fourier transforms for $\overline{W}_i$'s in $\vec{\rho}$:

$$\hat{\overline{W}}_i\,(z, \vec{p}, \vec{\xi}, \omega_2) = \int \overline{W}_i\,(z, \vec{\rho}, \vec{\xi}, \omega_2)\, \exp\,(-j\vec{p} \cdot \vec{\rho}\,)\, d\vec{p} \tag{44}$$

Then (41) with initial conditions (43) can be solved to yield

$$\begin{aligned} \hat{\overline{W}}_0\,(z, \vec{p}, \vec{\xi}, \omega_2) &= \hat{\overline{W}}_0^0(\vec{p}, \vec{\xi} + \vec{p}\, z / k_2, \omega_2)\, \exp\,[\chi(z)] \\ \hat{\overline{W}}_1\,(z, \vec{p}, \vec{\xi}, \omega_2) &= \hat{\overline{W}}_1^0(\vec{p}, \vec{\xi} + \vec{p}\, z / k_2, \omega_2)\, \exp\,[\chi(z)] \\ &\quad + \int_0^z \exp[\chi(z) - \chi(z')]\, \hat{F}_i(z', \vec{p}, \vec{\xi} + \frac{\vec{p}\,(z-z')}{k_2})\, dz' \qquad i = 1, 2, ... \end{aligned} \tag{45}$$

where

$$\chi(z) = \exp\,\{(k_2^2/4) \int_0^z A[\vec{\xi} + \vec{p}\ (z - z') / k_2]dz'\}$$

$$\hat{F}_1(z, \vec{p}, \vec{\xi}\,) = -(j/2k_2)\, [\nabla_\xi^2 + j\vec{p}\ \nabla_\xi - p^2/2 - jk_2^3 A(\xi)/2]\, \hat{\overline{W}}_0\,(z, \vec{p}, \vec{\xi}, \omega_2) \tag{46}$$

$$\begin{aligned} \hat{F}_2(z, \vec{p}, \vec{\xi}\,) &= -(j/2k_2)\, [\nabla_\xi^2 + j\vec{p}\ \nabla_\xi - p^2/2\,]\, [\hat{\overline{W}}_0 + \hat{\overline{W}}_1] \\ &\quad - (k_2^2/4)A(\vec{\xi}\,)\, \hat{\overline{W}}_1\,(z, \vec{p}, \vec{\xi}, \omega_2) \end{aligned}$$

The $\hat{\overline{W}}_i^0$)'s are the transforms of the initial conditions. Higher order terms can be solved in a similar manner. The solutions to $\hat{\overline{W}}_i$'s can then be obtained by carrying out the inverse transform. These $\overline{W}_i$'s, when substituted back into equation (39) yield the coefficients $\Gamma_0, \Gamma_1, \Gamma_2$, ...etc. for the beam wave case. The temporal moments can then be computed in the same manner as described in section 2. The statistical parameters such as mean arrival time, mean pulse width, etc. can also be computed. As an example, Liu and Yeh[18] investigated the case where the receiving aperture has uniform weighting and is much larger than the beam dimension. For example, the mean arrival time has been found to be[18]

$$\langle t_a \rangle = (z/v)\,(1 + a^2/2F^2 + 1\,/\,2k_c^2 a^2 - \nabla_\rho^2 A_0/z/10) \quad (47)$$

The mean square pulse width has also been computed by Liu and Yeh[18].

The same procedure can also be applied to the case of discrete random scatterers so that the propagation of pulsed beam waves through cloud, rain or fog can be investigated[18].

7. Conclusion

In this paper the problem of pulse propagation is reviewed. We have concentrated on the temporal moment technique. The technique is based on the parabolic equation approximation for wave propagation in turbulent media which is valid under the condition of forward scattering. A transport equation for the mutual coherence function is derived from the parabolic equation. The transport equation however is difficult to solve when the turbulence is strong. Instead of solving the equation for Γ directly, the technique of temporal moments seeks to solve for the individual moment separately. Recognizing that the nth moment $M^{(n)}$ depends only on the first n expansion coefficients, $\Gamma_0, \Gamma_1, \ldots \Gamma_n$, a heirarchy of equations for the Γn's is derived from the original transport equation. What makes the temporal moment technique powerful is the fact that this set of equations can be solved exactly in sequential manner even for the case of strong turbulence when multiple scattering is important. Once the Γn's are obtained the temporal moments can be solved exactly. With these moments pulse statistics such as the mean arrival time, pulse width, etc., can be investigated. In particular, the mean pulse intensity, or the mean pulse shape, can be constructed to any desired accuracy by using the temporal moments to find the coefficients of a polynomial expansion. The case of pulsed beam wave can also be studied.

The technique has been applied to study transient signals propagating through a perturbed ionosphere[4,19,20,21,22] as well as through rain, clouds, and fog[18]. Experimental work is needed to verify the theory as well as the models characterizing the turbulence or the discrete scattering media.

Recently applications of HF radar to probe the ionosphere have proved to be quite fruitful. In this application radio pulses are sent to the ionosphere to be scattered back by the irregularities in the F region. Since at HF bandwidth, the inhomogeneous background of the ionosphere becomes important. Therefore the problem of how pulsed signals propagating through turbulent media with inhomogeneous background needs to be investigated. As a starting point, single scattering, Born approximation should be studied first. For this purpose one needs the Green's function for the inhomogeneous medium. Recently an exact analytical solution for the three-dimensional Green's function in a linearly inhomogeneous medium has been obtained[23]. The Green's function can be used to study the scattering of waves in the most important region in the medium where total reflection occurs. How pulses are affected by the irregularities in the region will determine to a great extent the statistics of the pulse. Once we understand the problem for the single scattering case, the effects of multiple scattering should be investigated applying the temporal moment technique if possible.

Appendix

For the turbulence the most general spectrum is that introduced by Shkarofsky[24]:

$$\Phi_\varepsilon = \frac{(\kappa_0 l_0)^{(p-3)/2}\, l_0^3}{(2\pi)^{3/2} K_{(p-3)/2}(\kappa_0 l_0)} \bullet \frac{K_{p/2}(l_0\sqrt{\kappa^2+\kappa_0^2})}{(l_0\sqrt{\kappa_2+\kappa_0^2})^{p/2}} \bullet \sigma_n^2 \tag{A.1}$$

where $L_0 = 1/\kappa_0$ is the outer scale, l_0 is the inner scale, K is the Hankel function of imaginary argument and σ_n^2 is the strength of the turbulence. For this spectrum, the expansion coefficients for $A(\xi)$ are[11]

$$A_0 = \sqrt{2\pi l_0/\kappa_0}\; \sigma_n^2 K_{(p-2)/2}(\kappa_0 l_0)/K_{(p-3)/2}(\kappa_0 l_0)$$

$$A_2 = -\sqrt{\pi\kappa_0/2l_0}\; \sigma_n^2 K_{(p-4)/2}(\kappa_0 l_0)/K_{(p-3)/2}(\kappa_0 l_0) \tag{A.2}$$

$$A_4 = \sqrt{\pi\kappa_0^3/2^5 l_0^3}\; \sigma_n^2 K_{(p-6)/2}(\kappa_0 l_0)/K_{(p-3)/2}(\kappa_0 l_2)$$

References

1. D. Mintzer, "Wave propagation in a randomly inhomogeneous medium," *J. Acoust. Soc.Am.*, **25**, 922-927 (1953).
2. H. H. Su, and M. A. Plonus, "Optical pulse propagation in a turbulent medium," *J. Opt. Soc. Am.*, **61**, 256-260 (1971).
3. L. M. Erukhimov, I. G. A. Zarnitsyna, and P. I. Kirsh, "Selective properties and shape of pulse signals that has passed through a statistically inhomogeneous layer of arbitrary thickness," *Radiophys. Quantum Electron.*, **16**, 436-551 (1973).
4. C. H. Liu, A. W. Wernik, and K. C. Yeh, "Propagation of pulse trains through a random medium," *IEEE Trans. Antennas Propagat.*, **AP-22**, 624-627 (1974).
5. V. I. Shishov, "Effects of refraction on scintillation characteristics and average pulse shape of pulsors," *Sov. Astron*, **17**, 598-602 (1974) [*Astron. Zh.* **59**, 941-950, 1973].
6. V. I. Tatarskii, *The Effects of Turbulent Atmosphere on Wave Propagation*, (U.S. Dept. of Commerce, National Technical Information Service, Springfield, VA, 1971).
7. R. Dashen, "Path integrals for waves in random media," *J. Math. Phys.* **20**, 894-920, 1979.

8. L. C. Lee, "Wave propagation in a random medium: A complete set of the moment equations with different wave numbers," *J. Math. Phys.*, **15**, 1431-1435 (1974).
9. S. T. Hong, and A. Ishimaru, "Two-frequency mutual coherence function, coherence bandwidth, and coherence time of millimeter and optical waves in ran, fog and turbulence," *Radio Sci.*, **11**, 551-559 (1976).
10. K. C. Yeh, and C. H. Liu, "An investigation of temporal moments of stochastic waves," *Radio Sci.*, **12**, 671-680 (1977).
11. C. H. Liu, and K. C. Yeh, "Pulse propagation in random media," *IEEE Trans. Antennas Propagat.*, **AP-26**, 561-566 (1978).
12. R. von Mises, *Mathematical Theory of Probability and Statistics*, Academic, New York, 1964.
13. W. D. Mark, "Characterization of stochastic transients and transmission media: the method of power-moments spectra," *J. Sound and Vibration*, **22**, 249-295 (1972).
14. C. H. Liu, and K. C. Yeh, "Statistics of pulse arrival time in turbulent media," *J. Opt. Soc. Am.*, **70**, 168-172 (1980).
15. C. H. Liu, and K. C. Yeh, "Pulse spreading and wandering in random media," *Radio Sci.*, **14**, 925-931 (1979).
16. F. B. Hildebrand, *Introduction to Numerical Analysis*, 2nd Ed., McGraw-Hill, New York, 1974.
17. K. Rektorys, *Survey of Applicable Mathematics*, Cambridge, MA: MIT Press, 1969.
18. C. H. Liu, and K. C. Yeh, "Propagation of pulsed beam waves through turbulence, cloud, rain or fog," *J. Opt. Soc. Am.*, **67**, 1261-1266 (1977).
19. Y. W. Kiang, and C. H. Liu, "Propagation of beam waves through a turbulent layer," *Radio Sci.*, **17**, 1211-1219 (1982).
20. Y. W. Kiang, and C. H. Liu, Multiple phase screen simulation of HF wave propagation in the turbulent stratified ionosphere, *Radio Sci.*, **20**, 652-668 (1985).
21. W. F. Wagen, and K. C. Yeh, "A numerical study of waves reflected from a turbulent ionosphere," *Radio Sci.*, **21**, 583-604 (1986).
22. J. C. Lin, and Y. W. Kiang, "HF beam wave propagation in a stratified random ionosphere," *Radio Sci.*, **23**, 685-692 (1988).
23. Y. L. Li, C. H. Liu, and S. J. Franke, "Three dimensional Green's function for wave propagation in a linearly inhomogeneous medium -- the exact analytic solution," *J. Acoust. Soc. Am.*, **87**, 2285-2291 (1990).
24. I. P. Shkarofsky, "Generalized turbulence space-correlation and wave-number spectrum-function pairs," *Can. J. Phys.*, **46**, 2133-2153 (1968).

Review of the Extended Huygens-Fresnel Principle

Ronald L. Fante
The MITRE Corporation
Bedford, MA 01730

Abstract

The utility of the extended Huygens-Fresnel principle in calculating the moments of the field in a random medium is discussed.

Introduction

The Huygens-Fresnel Principle has proven quite useful[1] in the analysis of optical boundary value problems in vacuum. It is, therefore, desirable to extend it to the case when the vacuum is replaced by a random medium. This can be done in the limit when the fluctuation n_1 in the index of refraction of the random medium is very small in comparison with unity, and the scale size l_0 of the smallest turbulent eddy is large in comparison with the wavelength λ. When these conditions are satisfied it can be shown[2,3,4] that the electric field $E(\underline{r}_0)$ at a point $\underline{r}_0 = (x_0, y_0, z_0)$ is related to the electric field $E_s(\underline{\rho})$ at a point $\underline{r} = (\underline{\rho}, 0) = (x, y, 0)$ on the plane Λ shown in Fig. 1 as

$$E(\underline{r}_0) = -2ik \iint_{\Lambda} d^2\rho E_s(\underline{\rho}) \exp[\varphi(\underline{r}_0, \underline{r})](\hat{z} \cdot \hat{r}_0) \quad , \tag{1}$$

where $k = 2\pi/\lambda$, $\hat{z}$ is a unit vector along the z axis, $\hat{r}_0$ is a unit vector along $\underline{r}_0$, and φ satisfies the nonlinear differential equation

$$\nabla^2\varphi + (\nabla\varphi)^2 + k^2(1 + 2n_1) = 0 \quad . \tag{2}$$

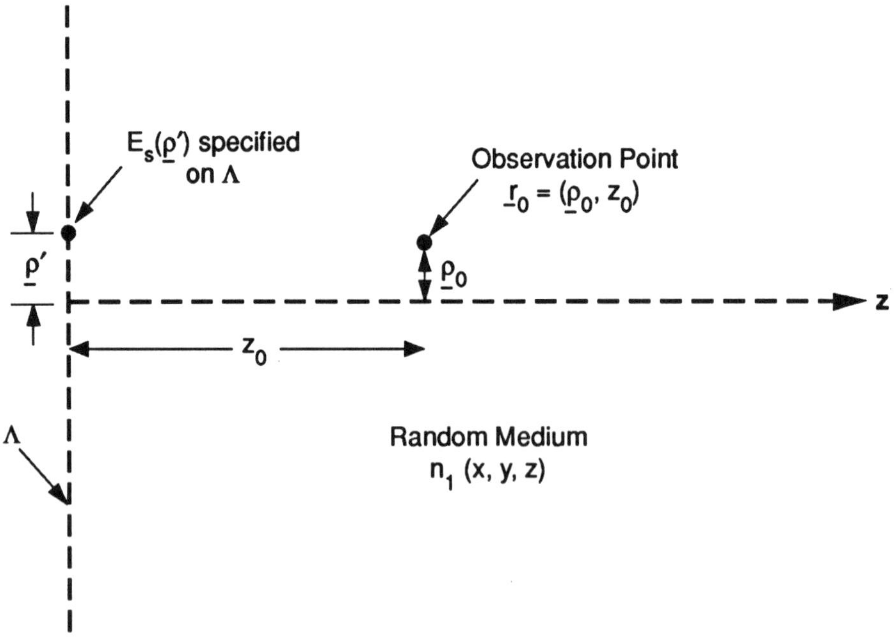

Fig. 1. Geometry assumed in derivation of equation (1).

If φ is written as a perturbation series $\varphi = \varphi_0 + \varphi_1 + \varphi_2 + \ldots$, then in the limit when

$$k^{7/6}\left\langle n_1^2\right\rangle z_0^{11/6} L_0^{-2/3} << 1 \quad , \tag{3}$$

where L_0 is the largest scale size of the turbulent eddies, Eq. (1) may be approximated by

$$E(\underline{r}_0) = -2ik \iint_\Lambda d^2\rho G(\underline{r}_0, r) E_s(\underline{\rho})(\hat{z}\cdot\hat{r}_0)\exp\left[\varphi_1(\underline{r}_0, \underline{r})\right] \quad , \tag{4}$$

where (Rytov approximation) φ_1 satisfies

$$\nabla^2\varphi_1 + 2\nabla\varphi_1\cdot\nabla\varphi_0 + 2k^2 n_1 = 0 \quad , \tag{5}$$

and

$$G(\underline{r}_0, \underline{r}) = \exp\left[\varphi_0(\underline{r}_0, \underline{r})\right] = \frac{\exp\left[ik\left|\underline{r}_0 - \underline{r}\right|\right]}{4\pi\left|\underline{r}_0 - \underline{r}\right|} \quad . \tag{6}$$

Finally, if $\rho << z_0$ and $\rho_0 << z_0$, where $\rho_0{}^2 = x_0{}^2 + y_0{}^2$, we may use the paraxial approximation in Eq. (4) to obtain

$$E(\underline{r}_0, k) = \iint_\Lambda d^2\rho\, E_s(\underline{\rho}) g_0\left(\underline{\rho}_0 - \underline{\rho}, k\right)\exp\left[\varphi_1\left(\underline{\rho}_0, \underline{\rho}, z_0\right)\right] \quad , \tag{7}$$

where

$$g_0\left(\underline{\rho}_0 - \underline{\rho}, k\right) = -\frac{ik}{2\pi z_0}\exp\left\{ik\left[z_0 + \frac{1}{2z_0}\left(\underline{\rho}_0 - \underline{\rho}\right)^2\right]\right\} \quad . \tag{8}$$

Eq. (7) appears to have been first used by Feizulin & Kravtsov[3], shortly thereafter by Lutomirski and Yura[4], and since then by a multitude of others. It will serve as the basis for the analysis in the next section.

Although φ has been discussed only for turbulence, if aerosols are present, their effect can be included by writing $\varphi(\underline{\rho}_0, \rho) = \varphi_t(\underline{\rho}_0, \underline{\rho}) + \varphi_a(\underline{\rho}_0, \underline{\rho})$, where φ_t is the additional (beyond that in vacuum) complex phase shift for a spherical wave propagating in turbulence from $(\underline{\rho}, 0)$ to $(\underline{\rho}_0, z_0)$, and φ_a is the corresponding complex phase for aerosols.

Another effect that is readily included is transmitter jitter. Thus, suppose that the transmitter plane Λ is tilted from the vertical by some small angle (θ_x, θ_y). Then at the transverse distance $\underline{\rho}$, the phase difference between the actual transmitter plane Λ and the vertical (i.e., the z = 0 plane) is approximately $k(\rho_x\theta_x + \rho_y\theta_y) = k\underline{\rho}\cdot\underline{\theta}$. Consequently, the field in the vertical plane z = 0 is given by

$$E_s(\underline{\rho})\exp(ik\underline{\rho}\cdot\underline{\theta}) \quad ,$$

where $E_S(\underline{\rho})$ is the distribution in the tilted plane Λ. This effect can be included in equation (7) by replacing φ by $\varphi = \varphi_t + \varphi_a + ik\underline{\rho}\cdot\underline{\theta}$.

Generalized second moment

Because <expφ>, where < > denotes an expectation, is quite small in an extended random medium, there is usually little interest in the first moment of the field, so the lowest moment that is of interest is the second. By using Eq. (7) it is readily shown that the generalized second moment of the field is

$$\begin{aligned}\left\langle E\left(\underline{\rho}_0, z_0, k\right)E^*\left(\underline{\rho}_1, z_1, k'\right)\right\rangle &= \int_{-\infty}^{\infty}\cdots\int d^2\rho\, d^2\rho' \left\langle E_s\left(\underline{\rho}, k\right)E_s^*\left(\underline{\rho}', k'\right)\right\rangle \\ &\cdot g_0\left(\underline{\rho}_0 - \underline{\rho}, k\right) g_1^*\left(\underline{\rho}_1 - \underline{\rho}', k'\right) \\ &\cdot \left\langle \exp\left[\varphi\left(\underline{\rho}_0, \underline{\rho}, z_0, k\right) + \varphi^*\left(\underline{\rho}_1, \underline{\rho}', z_1, k'\right)\right]\right\rangle \quad ,\end{aligned} \tag{9}$$

where it has been assumed that the source fluctuations are statistically independent of the medium fluctuations. The mutual coherence function evaluated at two different wave numbers is required in the study of transient propagation,[5,6] and for this reason the generalized result has been presented.

It is interesting to note that the use of Eq. (9) allows one to separate the turbulence effects, the free space propagation, and the source coherence structure.

In many cases of practical interest the field observations are made in the same plane. Thus, if we set $z_0 = z_1$ and then assume that the transmitter jitter, aerosols, and turbulence are all statistically independent, we get

$$\begin{aligned}\left\langle E\left(\underline{\rho}_0, k\right)E^*\left(\underline{\rho}_1, k'\right)\right\rangle &= \int_{-\infty}^{\infty}\cdots\int d^2\rho\, d^2\rho'\, g_0\left(\underline{\rho}_0 - \underline{\rho}, k\right) \\ &\cdot g_0^*\left(\underline{\rho}_1 - \underline{\rho}', k'\right) M_t M_a M_j \Gamma_s\left(\underline{\rho}, \underline{\rho}', k, k'\right) \quad ,\end{aligned} \tag{10}$$

where

$$\Gamma_s\left(\underline{\rho},\underline{\rho}',k,k'\right)=\left\langle E_s\left(\underline{\rho},k\right)E_s^*\left(\underline{\rho}',k'\right)\right\rangle \quad , \tag{11}$$

$$M_t=\left\langle \exp\left[\varphi_t\left(\underline{\rho}_0,\underline{\rho},k\right)+\varphi_t^*\left(\underline{\rho}_1,\underline{\rho}',k'\right)\right]\right\rangle \quad , \tag{12}$$

$$M_a=\left\langle \exp\left[\varphi_a\left(\underline{\rho}_0,\underline{\rho},k\right)+\varphi_a^*\left(\underline{\rho}_1,\underline{\rho}',k'\right)\right]\right\rangle \quad , \tag{13}$$

$$M_j=\left\langle \exp\left[i\theta\cdot\left(k\underline{\rho}-k'\underline{\rho}'\right)\right]\right\rangle \quad . \tag{14}$$

In the limit when the turbulence is weak and the aerosols are diffuse, we can assume that φ_t and φ_a are approximately Gaussian. If we also assume that $\underline{\theta}$ is a zero mean, Gaussian random variable, we find

$$M_t=\exp\left\{-\frac{1}{2}D_t\left(\underline{\rho}_0-\underline{\rho}_1,\underline{\rho}-\underline{\rho}',k,k'\right)\right\} \quad , \tag{15}$$

$$M_a=\exp\left\{-\alpha z_0-\frac{1}{2}D_a\left(\underline{\rho}_0-\underline{\rho}_1,\underline{\rho}-\underline{\rho}',k,k'\right)\right\} \quad , \tag{16}$$

$$M_j=\exp\left\{-\frac{1}{2}\sigma^2\left(k\underline{\rho}-k'\underline{\rho}'\right)^2\right\} \quad , \tag{17}$$

where α is the attenuation coefficient for the aerosols and $\sigma^2=\langle\theta^2\rangle$. The spherical wave structure functions D_t and D_a for the turbulence and aerosols, respectively, are presented in references 2 and 7.

It is interesting to examine Eq. (10) in the limit when the source E_s is spatially incoherent. In this case the mutual coherence function Γ_s can be approximated by

$$\Gamma_s\left(\underline{\rho},\underline{\rho}'\right)=\frac{\lambda^2}{\pi}I_0\left(\underline{\rho}\right)\delta\left(\underline{\rho}-\underline{\rho}'\right) \quad , \tag{18}$$

where $\delta(\underline{\rho})$ is the Dirac delta function and $I_0(\underline{\rho})$ is the source irradiance distribution. If Eqs. (15) through (18) are used in Eq. (10) we find

$$\left\langle E\left(\underline{\rho}_0\right)E^*\left(\underline{\rho}_1\right)\right\rangle=\exp\left\{-\alpha z_0-\frac{1}{2}D_a\left(\underline{\rho}_0-\underline{\rho}_1,0\right)-\frac{1}{2}D_t\left(\underline{\rho}_0-\underline{\rho}_1,0\right)\right\}$$

$$\cdot\frac{1}{\pi z_0^2}\exp\left[i\frac{k}{2z_0}\left(\rho_0^2-\rho_1^2\right)\right]\iint_{-\infty}^{\infty}d^2\rho I_0\left(\underline{\rho}\right)\exp\left[i\frac{k}{z_0}\underline{\rho}\cdot\left(\underline{\rho}_1-\underline{\rho}_0\right)\right] \quad . \tag{19}$$

Eq. (19) is the generalized van Cittert-Zernike theorem. In the absence of turbulence ($D_t = 0$) and aerosols ($\alpha = D_a = 0$) it reduces to the classical van Cittert-Zernike theorem.[6]

It is also instructive to use Eq. (10) to evaluate the irradiance at some position ($\underline{\rho}_0$, z_0) when the source is perfectly coherent. Let us set $\underline{\rho}_1 = \underline{\rho}_0$ and $k' = k$ in Eq. (10), and assume a source distribution

$$E_s(\underline{\rho}) = E_0 \exp\left(\frac{-\rho^2}{w^2}\right) . \tag{20}$$

Then, if we approximate

$$M_t(0, \underline{\rho} - \underline{\rho}') = \exp\left[-\left(\frac{|\underline{\rho} - \underline{\rho}'|}{\rho_t}\right)^{5/3}\right] \tag{21a}$$

by

$$\tilde{M}_t(0, \underline{\rho} - \underline{\rho}') = \exp\left[-\left(\frac{\underline{\rho} - \underline{\rho}'}{\rho_t}\right)^2\right] , \tag{21b}$$

we find that the irradiance

$$I(\underline{\rho}_0) \equiv \left\langle E(\underline{\rho}_0, k) E_0^*(\underline{\rho}_0, k) \right\rangle$$

is given by

$$I(\underline{\rho}_0) = \frac{w^2 E_0^2}{2 R_e^2} \exp\left[-\alpha z_0 - \left(\frac{\underline{\rho}_0}{R_e}\right)^2\right] , \tag{22}$$

where ρ_t is the coherence length in turbulence defined by

$$\rho_t = \left[1.46 k^2 \int_0^{z_0} d\zeta C_n^2(\zeta) \left(1 - \frac{\zeta}{z_0}\right)^{5/3}\right]^{-3/5} , \tag{23}$$

C_n^2 is the index of refraction structure constant, and R_e is the effective radius of the beam defined as

$$R_e^2 = \frac{w^2}{2} + z_0^2\left(\theta_d^2 + \theta_t^2 + \theta_a^2 + \theta_j^2\right) \quad . \tag{24}$$

The angles $\theta_d = \lambda/4.4w$, $\theta_t = \lambda/\pi\rho_t$, $\theta_a = \lambda/\pi\rho_a$, and $\theta_J = \lambda/\pi\rho_j$ are, respectively, the beam spread due to diffraction, turbulence, aerosols, and transmitter jitter. The quantity $\rho_j = 2^{1/2}/(k\sigma)$ and ρ_a is defined in reference 2.

From Eq. (24) it is evident that the turbulence, aerosols, and jitter all cause a beam spread beyond that which is produced by classical vacuum diffraction.

Higher order moments

All higher order moments can be formaliy calculated using Eq. (7). For example, the generalized fourth moment

$$\Gamma_4 \equiv \left\langle E\left(\underline{\rho}_0, z_0, k\right) E^*\left(\underline{\rho}_1, z_1, k'\right) E\left(\underline{\rho}_2, z_2, k''\right) E^*\left(\underline{\rho}_3, z_3, k'''\right)\right\rangle$$

is given by

$$\Gamma_4 = \int_{-\infty}^{\infty}\cdots\int d^2\rho\, d^2\rho' d^2\rho'' d^2\rho''' g_0\left(\underline{\rho}_0 - \underline{\rho}, k\right) g_1^*\left(\underline{\rho}_1 - \underline{\rho}', k'\right)$$

$$\cdot\, g_2\left(\underline{\rho}_2 - \underline{\rho}'', k''\right) g_3^*\left(\underline{\rho}_3 - \underline{\rho}''', k'''\right) \Gamma_{4s} \left\langle \exp\left[\varphi\left(\underline{\rho}_0, \underline{\rho}, z_0, k\right)\right.\right.$$

$$\left.\left. + \varphi^*\left(\underline{\rho}_1, \underline{\rho}', z_1, k'\right) + \varphi\left(\underline{\rho}_2, \underline{\rho}'', z_2, k''\right) + \varphi^*\left(\underline{\rho}_3, \underline{\rho}''', z_3, k'''\right)\right]\right\rangle \quad , \tag{25}$$

where Γ_{4a} is the fourth moment of the source distribution, defined as

$$\Gamma_{4s} = \left\langle E_s\left(\underline{\rho}, k\right) E_s^*\left(\underline{\rho}', k'\right) E_s\left(\underline{\rho}'', k''\right) E_s^*\left(\underline{\rho}''', k'''\right)\right\rangle \quad . \tag{26}$$

Of course, Eq. (25) is purely formal, and its utility is limited by our ability to evaluate

$$\left\langle \exp\left[\varphi\left(\underline{\rho}_0, \underline{\rho}, z_0, k\right) + \ldots\right]\right\rangle \quad ,$$

where $\varphi(\underline{\rho}_0, \underline{\rho}, z_0, k)$ is the additional complex phase shift for a spherical wave of wave number k propagating from $(\underline{\rho}, 0)$ to $(\underline{\rho}_0, z_0)$. When the turbulence is weak, so that φ is approximately Gaussian, the required average can be expressed in terms of the structure function for the complex phase and the cross correlations $<\chi(\rho_0, \rho, k) S(\rho_1, \rho', k')>$ etc.

between the real (χ) and imaginary (S) parts of φ. The required quantities are presented in references 2 and 7, and will be omitted here.

Multiple phase screens

A useful application of the Huygens-Fresnel principle is in treating the case where an extended medium is approximated[8] by a series of phase screens, as shown in Fig. 2. A phase screen is defined such that its transmission coefficient is $T(\underline{\rho}) = \exp[i\phi(\underline{\rho})]$, where $\phi(\underline{\rho})$ may be complex. Also, its reflection coefficient is defined to be zero. We can then apply the Huygens-Fresnel principle to calculate the field $E(\underline{\rho}_{n+1})$ in the receiver plane due to a source distribution $E_s(\underline{\rho}_0)$ in the z = 0 plane. Let us show how the calculation is done, and then present the general result.

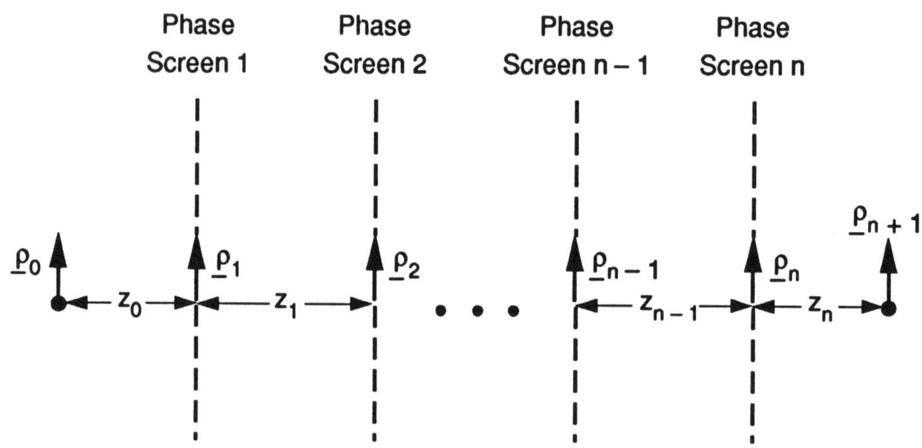

Fig. 2. Multiple phase screens.

By using Eq. (7), with $\varphi_1 = 0$, we find that the field on the left-hand side of phase screen 1 is

$$E_1(\underline{\rho}_1) = \iint_{-\infty}^{\infty} d^2\rho_0 E_s(\underline{\rho}_0) g_0(\underline{\rho}_1 - \underline{\rho}_0) \quad , \tag{27}$$

and its field on the right-hand side is

$$\exp\left[i\phi_1\left(\underline{\rho}_1\right)\right]E_1\left(\underline{\rho}_1\right) \quad .$$

If we use this field as the source and then reapply the Huygens-Fresnel principle, we see that the field on the left-hand side of phase screen 2 is

$$E_2\left(\underline{\rho}_2\right) = \int\!\!\int_{-\infty}^{\infty} d^2\rho_1 E_1\left(\underline{\rho}_1\right) \exp\left[i\phi_1\left(\underline{\rho}_1\right)\right] g_1\left(\underline{\rho}_2 - \underline{\rho}_1\right)$$

$$= \int\!\!\int_{-\infty}^{\infty} d^2\rho_0 E_s\left(\underline{\rho}_0\right) \int\!\!\int_{-\infty}^{\infty} d^2\rho_1 g_1\left(\underline{\rho}_2 - \underline{\rho}_1\right) g_0\left(\underline{\rho}_1 - \underline{\rho}_0\right) \exp\left[i\phi_1\left(\underline{\rho}_1\right)\right] \quad . \tag{28}$$

By proceeding in the above fashion, we find that the field in the receiver plane is

$$E\left(\underline{\rho}_{n+1}\right) = \int\!\!\int d^2\rho_0 E_s\left(\underline{\rho}_0\right) \int_{-\infty}^{\infty}\!\!\cdots\!\int d^2\rho_1 \ldots d^2\rho_n \exp\left[i\phi_1\left(\underline{\rho}_1\right) + \ldots + i\phi_n\left(\underline{\rho}_n\right)\right]$$

$$\cdot g_0\left(\underline{\rho}_1 - \underline{\rho}_0\right) g_1\left(\underline{\rho}_2 - \underline{\rho}_1\right) \cdots g_{n-1}\left(\underline{\rho}_n - \underline{\rho}_{n-1}\right) g_n\left(\underline{\rho}_{n+1} - \underline{\rho}_n\right) \quad . \tag{29}$$

If we now substitute for g_n from Eq. (8), we can rewrite Eq. (29) as

$$E\left(\underline{\rho}_{n+1}\right) = \int\!\!\int d^2\rho_0 E_s\left(\underline{\rho}_0\right) g_0\left(\underline{\rho}_1 - \underline{\rho}_0\right) \int\!\cdots\!\int d^{2n}R \exp\left[iS_n\left(\underline{\rho}_1, \underline{\rho}_2, \ldots\right)\right], \tag{30}$$

where

$$d^{2n}R = d^2\rho_1 d^2\rho_2 \ldots d^2\rho_n \quad , \tag{31}$$

$$S_n = \sum_{m=1}^{n} \left[kz_m + \frac{k}{2z_m}\left(\underline{\rho}_{m+1} - \underline{\rho}_m\right)^2 + \phi_m\left(\underline{\rho}_m\right) \right] \quad . \tag{32}$$

Writing Eq. (29) in the form of Eq. (30) allows us to compare it with the result obtained using path integrals (in the limit when all ray paths are linear).

It is interesting to use Eq. (30) to calculate the second moment of the field when only two phase screens (n=2) are present. Then, if the phase screens are statistically independent, we find

$$\left\langle E\left(\underline{\rho}_3\right)E^*\left(\underline{\rho}_3'\right)\right\rangle = \int\ldots\int d^2\rho_0 d^2\rho_0' \Gamma_s\left(\underline{\rho}_0,\underline{\rho}_0'\right) g_0\left(\underline{\rho}_1-\underline{\rho}_0\right) g_0^*\left(\underline{\rho}_1'-\underline{\rho}_0'\right)$$

$$\cdot\int\ldots\int d^8R \exp\left[i\tilde{S}_2\left(\underline{\rho}_1,\underline{\rho}_2,\underline{\rho}_3\right) - i\tilde{S}_2\left(\underline{\rho}_1',\underline{\rho}_2',\underline{\rho}_3'\right)\right]$$

$$\cdot M_1\left(\underline{\rho}_1,\underline{\rho}_1'\right) M_2\left(\underline{\rho}_2,\underline{\rho}_2'\right) \quad , \tag{33}$$

where

$$d^8R = d^2\rho_1 d^2\rho_2 d^2\rho_1' d^2\rho_2' \quad , \tag{34}$$

$$\Gamma_s\left(\underline{\rho}_0,\underline{\rho}_0'\right) = \left\langle E_s\left(\underline{\rho_0}\right)E_s^*\left(\underline{\rho}_0'\right)\right\rangle \quad , \tag{35}$$

$$\tilde{S}_2\left(\underline{\rho}_1,\underline{\rho}_2,\underline{\rho}_3\right) = k\sum_{m=1}^{2}\left[z_m + \frac{1}{2z_m}\left(\underline{\rho}_{m+1}-\underline{\rho}_m\right)^2\right] \quad , \tag{36}$$

$$M_m\left(\underline{\rho}_m,\underline{\rho}_m'\right) = \left\langle \exp\left[i\phi_m\left(\underline{\rho}_m\right) - i\phi_m\left(\underline{\rho}_m'\right)\right]\right\rangle \quad . \tag{37}$$

The generalization to n phase screens is trivial, and simply involves replacing d^8R by $d^{4n}R$, $\tilde{S}_2$ by $\tilde{S}_n$ and M_1M_2 by $M_1M_2 \ldots M_n$.

Phase compensation

Consider one final example. In the weak turbulence limit we can rewrite Eq. (7) as

$$E\left(\underline{\rho}_0,z_0\right) = \iint d^2\rho\, E_s(\underline{\rho}) g_0\left(\underline{\rho}_0-\underline{\rho}\right)\exp\left[iS\left(\underline{\rho}_0,\underline{\rho},z_0\right)\right] \quad , \tag{38}$$

where $\exp[\varphi] = \exp[\chi + iS]$ has been approximated by exp(iS), which is acceptable when the condition in Eq. (3) is satisfied. Now assume that the phase distribution $E_s(\rho_0)$ is obtained by placing a cooperative point source at $(\underline{\rho}_0 = 0, z_0)$ and then measuring in the z = 0 plane. Once $S(\underline{\rho},0)$ is known, then $S(0,\underline{\rho})$ is also known, by virtue of reciprocity. Thus, the transmitter phase can be modified by inserting a phase $S_c(\underline{\rho}) = S(0,\underline{\rho})$ to correct for the effects of the random medium. That is, in Eq. (38) we replace $E_s(\underline{\rho})$ by

$$E_d(\underline{\rho})\exp\left[-iS(0,\underline{\rho})\right] \quad ,$$

where $E_d(\underline{\rho})$ is the source field distribution that produces some desired field response $\hat{E}_d(\underline{\rho}_0, z_0)$ in the $z = z_0$ plane. Therefore, we now get

$$E(\underline{\rho}_0, z_0) = \iint d^2\rho \, E_d(\underline{\rho}) g_0(\underline{\rho}_0 - \underline{\rho}) \exp\left[iS(\underline{\rho}_0, \underline{\rho}, z_0) - iS(0, \underline{\rho}, z_0) \right] \quad . \tag{39}$$

From Eq. (39) we see that the image at $\rho_0 = 0$ is perfectly compensated, but there is not necessarily compensation for $\rho_0 \neq 0$. In order to see over what range of values compensation occurs, let us assume S is a Gaussian random variable, and then take the expectation of Eq. (39). We find

$$\left\langle E(\underline{\rho}_0, z_0) \right\rangle = \exp\left[-\frac{1}{2}\left(\frac{\rho_0}{\rho_I}\right)^{5/3} \right] \hat{E}_d(\underline{\rho}_0, z_0) \quad , \tag{40}$$

where $\hat{E}_d(\underline{\rho}_0, z_0)$ is the desired distribution in the $z = z_0$ plane as given by

$$\hat{E}_d(\underline{\rho}_0, z_0) = \iint d^2\rho \, E_d(\underline{\rho}) g_0(\underline{\rho}_0 - \underline{\rho}) \quad , \tag{41}$$

and ρ_I is defined as

$$\rho_I = 1.46 k^2 \int_0^{z_0} dz \left(\frac{z}{z_0}\right)^{5/3} C_n^{\,2}(z) \quad . \tag{42}$$

From Eq. (39) it is evident that (on the average) we can obtain a perfectly coherent image over transverse distances $|\underline{\rho}_0| < \rho_I$, whereas for $|\underline{\rho}_0| > \rho_I$ the coherent image is strongly attenuated. Thus, ρ_I is a measure of the isoplanatic distance.

Summary

The extended Huygens-Fresnel principle is quite useful because it allows us to separate source effects, vacuum propagation, and random medium effects. However, it is primarily a "bookkeeping" procedure and obtaining useful results ultimately depends on whether we can evaluate the moments

$$\left\langle \exp\left[\varphi(\rho_0, \rho, z_0, k) + \varphi^*(\rho_1, \rho', z_1, k') + \ldots \right] \right\rangle$$

of the spherical-wave, complex phase φ. The required results are available[2,7] for weakly random media, but not for the general, extended random medium.

References

1. M. Born and E. Wolf, *Principles of Optics*, Pergamon Press, New York (1964).

2. R. L. Fante, "Wave propagation in random media: a systems approach," in *Prog. in Optics*, (E. Wolf, Ed), Vol. 22, Elsevier Science Publishers (1985).

3. Z. Feizulin and Y. Kravtsov, "Broadening of a laser beam in a turbulent medium," *Radiophys. & Quant. Elect. (USSR)* **10**, 33-35 (1967).

4. R. Lutomirski and H. Yura, "Propagation of a finite optical beam in an inhomogeneous medium," *Appl. Opt.* **10**, 1652-1658 (1971).

5. C. Liu and K. C. Yeh, "Pulse spreading and wandering in a random medium," *Radio Science* **14**, 925-931 (1979).

6. S. Karp, R. Gagliardi, S. Moran, and L. Stotts, *Optical Channels*, Plenum Press, New York (1988).

7. R. L. Fante, "Generalized coherence functions for propagation in a random medium," *J. Opt. Soc. Amer.* **A3**, 1326-1327 (1986).

8. D. L. Knepp, "Multiple phase-screen calculation of the temporal behavior of stochastic waves," *Proc. IEEE* **71**, 722-737 (1983).

SCINTILLATION OF REFLECTED OPTICAL WAVES IN A TURBULENT MEDIUM

V.A. Banakh, Institute of Atmospheric Optics of Russian Academy of Sciences
Tomsk, 634055, Russia
A.I. Saichev, State University of Nizhnii Novgorod
Nizhnii Novgorod, 603600, Russia

1. Introduction

The use of lasers for target location in the atmosphere and through the atmosphere has made it possible to study the stochastic nature of backscattered coherent optical waves. Due to proximity and intersection of direct and return rays at laser radiation propagation in a random inhomogeneous medium, whose characteristics, in many cases, remain constant for the period of wave propagation a possibility arises coherent superposition of the correlated random fields in a certain space angle near the axis connecting the target and the lidar.
The effect of the increase of mean intensity in the backscattered radiation was first observed when solving the problem of diffraction of a spherical wave on a mirror disk[1] and on a "point" scatterer.[2] The enhancement of mean intensity is accompanied by the increase of the intensity fluctuations,[3] and the residual spatial correlation of these fluctuations takes place.[4]
Since the receiving–transmitting telescopes of lidars form the spatially limited beams, and the targets located may have arbitrary dimensions and surface reflective properties, then the appearance of the enhancement effects and their quantitative manifestation is determined by the above factors. The influence of these factors has been studied in detail in a series of papers, generalized in the monograph.[5] In this case the particular problem is to study the effects of enhancement in the image plane of a telescope.[6,7]

2. Formulation and Methods for Solving the Problems on Reflected Wave Propagation in the Turbulent Atmosphere

Let $u(z', \rho')$ be the field of laser radiation being propagated along the axis OZ′; the expression for this field, using the Green function of the propagation medium $G(z', z_0; \rho', t)$ can be represented as

$$u(z', \rho') = \int dt\, u_0(t)\, G(z', z_0; \rho', t)\,, \tag{2.1}$$

where $u_0(\mathbf{t}) = u(z', \rho')/_{z'=z_0}$ is the initial field, $G(z_0, z_0; \rho', \mathbf{t}) = \delta(\rho' - \mathbf{t})$, $\delta(\rho)$ is the Dirac delta function, ρ', $\mathbf{t}$ is the two-dimensional vectors.
Let the reflection occur in the plane $z' = z$. We introduce the function $V(\rho', \mathbf{r})$, characterizing a local reflection coefficient. Then the field on the reflecting surface is given by[8]

$$u_0^R(\mathbf{r}) = \int d\rho' u(z, \rho')\, V(\rho', \mathbf{r}) \tag{2.2}$$

and using the integral expression (2.1), for the field of a reflected wave in the plane $z' < z$ we have:

$$u^R(z', \rho) = \int d\mathbf{r}\, u_0^R(\mathbf{r})\, G^R(z', z; \rho, \mathbf{r})\,. \tag{2.3}$$

The Green functions of the forward G and backward G^R directions of propagation satisfy the conjugate equations of parabolic type[8] and are connected, due to this fact, with the reciprocal relation:

$$G^R(z', z; \rho, \mathbf{r}) = G(z, z'; \mathbf{r}, \rho)\,. \tag{2.4}$$

This enables one to reduce the reflected wave problem to the problem of direct propagation.[8,5] As a result, from Eqs. (2.3), (2.4), (2.2) and (2.1) for the field of a reflected wave we have:

$$u^R(z', \rho) = \int d\mathbf{t}\, d\rho' d\mathbf{r}\, u_0(\mathbf{t})\, G(z', z_0; \rho', \mathbf{t})\, G(z, z'; \mathbf{r}, \rho)\, V(\mathbf{r}, \rho')\,. \tag{2.5}$$

Equation (2.5) is a basis for analyzing the products of the reflected field of order $2n$

$$U_{2n}^R(z', \rho_{\underline{2n}}) = u^R(z', \rho_1)\, u^{R*}(z', \rho_2) \ldots u^R(z', \rho_{2n-1})\, u^{R*}(z', \rho_{2n}) \tag{2.5'}$$

and the mutual correlation functions of the corresponding order

$$\Gamma_{2n}^R(z', \rho_{\underline{2n}}) = \langle U_{2n}^R(z', \rho_{\underline{2n}}) \rangle\,, \tag{2.6}$$

where $\underline{2n}$ denotes variation of the index from 1 to $2n$. Averaging in (2.6) is carried out on the assumption of statistical independence of fluctuations of the medium parameters and characteristics of an initial field and a reflection coefficient which can also be random. As a result, for $z' = z_0$ we have

$$\Gamma^{R}_{2n}(z_0, \underline{\rho}_{2n}) = \int\int\int < U_{2n}(\underline{\mathbf{t}}_{2n}) > < V_{2n}(\underline{\rho}'_{2n}, \underline{\mathbf{r}}_{2n}) >$$

$$\times < \tilde{G}_{2n}(z, z_0; \underline{\rho}'_{2n}, \underline{\mathbf{t}}_{2n}; \underline{\mathbf{r}}_{2n}, \underline{\rho}_{2n}) > d\underline{\rho}'_{2n}\, d\underline{\mathbf{t}}_{2n}\, d\underline{\mathbf{r}}_{2n} , \tag{2.7}$$

where

$$\tilde{G}_{2n} = \prod_{j=1}^{n} \tilde{G}(z, z_0; \rho'_{2j-1}, \mathbf{t}_{2j-1}; \mathbf{r}_{2j-1}, \rho_{2j-1}) \tilde{G}^{*}(z, z_0; \rho'_{2j}, \mathbf{t}_{2j}; \mathbf{r}_{2j}, \rho_{2j}) \tag{2.8}$$

$\tilde{G} = G(z, z_0; \rho', \mathbf{t})\, G(z, z_0; \mathbf{r}, \rho)$ is the "lidar Green function" (LGF), which was first introduced in,[9]

$$U_{2n}(\underline{\mathbf{t}}_{2n}) = \prod_{j=1}^{n} u_0(\mathbf{t}_{2j-1})\, u_0^{*}(\mathbf{t}_{2j}) \tag{2.9}$$

$$V_{2n}(\underline{\rho}'_{2n}, \underline{\mathbf{r}}_{2n}) = \prod_{j=1}^{n} V(\rho'_{2j-1}, \mathbf{r}_{2j-1})\, V^{*}(\rho_{2j}, \mathbf{r}_{2j}) . \tag{2.10}$$

In the case of a coherent laser source the initial distribution of the field $u_0(\mathbf{t})$ is usually given as a Gaussian beam with an effective radius of output aperture a, phase front curvature F and an amplitude u_0 in its center. Statistical moments of initial field of partially coherent laser beams are assigned in a more complicated form. In particular, for the function of mutual spatial coherence of the second order $< U_2 >$ we have:[10]

$$< U_2(\underline{\mathbf{t}}_2) > = u_0^2 \exp\left\{ -\frac{t_1^2 + t_2^2}{2a^2} - i\frac{k}{2F}(t_1^2 - t_2^2) - \frac{(\mathbf{t}_1 - \mathbf{t}_2)^2}{4a_k^2} \right\} \tag{2.11}$$

where a_k is the effective radius of spatial coherence of irradiance, $k = 2\pi/\lambda$ is the wave number, and the fourth moment $< U_4 >$ on the assumption of strong phase fluctuations of the source field must be given as:

$$< U_4(\underline{\mathbf{t}}_4) > = < U_2(\underline{\mathbf{t}}_2) > < U_2(\mathbf{t}_3, \mathbf{t}_4) > , \tag{2.12}$$

if the coherence time of the source τ_s is less than the time of averaging of the receiver τ_r, or as

$$< U_4(\underline{t}_4) > = < U_2(\underline{t}_2) > < U_2(t_3, t_4) > + < U_2(t_1, t_4) > < U_2(t_2, t_3) > , \quad (2.13)$$

if the inverse condition $\tau_s \gtrsim \tau_r$ is fulfilled.[11,12]

The local reflection coefficient in the case of mirror targets is assigned as[5]

$$V(\rho', \mathbf{r}) = A(\mathbf{r})\, \delta(\rho \pm \mathbf{r}) , \quad (2.14)$$

where $A(\mathbf{r}) = V_0 \exp\{-r^2/2a_r^2\}$, a_r is the effective radius of the reflector, V_0 is the amplitude factor, and the sign - corresponds to the plane mirror, and the sign + corresponds to the retroreflector.

In the case of a diffuse surface with a random reflection coefficient for $< V_2 >$ we have:[12,13]

$$< V_2(\underline{\rho}'_2, \underline{\mathbf{r}}_2) > = \frac{4\pi}{k^2} A(\mathbf{r}_1) A^*(\mathbf{r}_2)\, \delta(\mathbf{r}_1 - \mathbf{r}_2)\, \delta(\rho'_1 - \mathbf{r}_1)\, \delta(\rho'_2 - \mathbf{r}_2) , \quad (2.15)$$

and the fourth moment $< V_4 >$ depending on the relation between the time of the receiver averaging τ_r and the time of correlation of surface roughness fluctuations τ_k is represented as[14,12]

$$< V_4(\underline{\rho}'_4, \underline{\mathbf{r}}_4) > = < V_2(\underline{\rho}'_2, \underline{\mathbf{r}}_2) > < V_2(\rho'_{3,4}, \mathbf{r}_{3,4}) > , \; \tau_k << \tau_r , \quad (2.16)$$

$$< V_4(\underline{\rho}'_4, \underline{\mathbf{r}}_4) > = < V_2(\underline{\rho}'_2, \underline{\mathbf{r}}_2) > < V_2(\rho'_{3,4}, \mathbf{r}_{3,4}) >$$

$$+ < V_2(\rho'_{1,4}, \mathbf{r}_{1,4}) > < V_2(\rho'_{2,3}, \mathbf{r}_{2,3}) > , \; \tau_k \gtrsim \tau_r . \quad (2.17)$$

Thus, for calculating the reflected wave mutual coherence functions $\Gamma^R_{2n}(z_0, \underline{\rho}_{2n})$ it is necessary to know the statistical moments of the lidar Green function of corresponding order. Equations for those[15,16,5] follow from the parabolic equation for the lidar Green function:[9]

$$2ik \frac{\partial \widetilde{G}}{\partial z'} + (\Delta\rho' + \Delta\rho)\widetilde{G} + k^2(\varepsilon_1(z', \rho') + \varepsilon_1(z', \rho))\, \widetilde{G} = 0 \quad (2.18)$$

with the boundary condition $\widetilde{G}(z_0, z_0; \rho', \mathbf{t}; \mathbf{r}, \rho) = \delta(\rho' - \mathbf{t})\, \delta(\mathbf{r} - \rho)$. Therefore, the methods for solving the lidar problems on the basis of the functions of mutual coherence (2.7) are connected mainly with those or another approachs when solving equations for statistical moments of LGF. The above methods, namely, the perturbation method[17] or the asymptotic technique,[15,16] are described in the monographs.[5,12] The use of approximate methods in this case[18-20] is possible, but this should be done carefully since the use, for example, of the methods[19,20] in the reflected wave problems can lead to incorrect results due to violation of the reciprocity principle between direct and return waves.

3. Relationships Between the Statistical Intensity Moments of Direct and Reflected Waves

General relationship between the even moments of the incident U_{2n} and reflected U_{2n}^R fields[21,16] may be written, using Eqs. (2.1), (2.5′), (2.8)−(2.10) for the case of backscattering of a spherical wave $(u(\mathbf{t}) = 2\pi\delta(k\mathbf{t}))$ by a point reflector: $(A(\mathbf{r}) = 2\pi V_0 \delta(k\mathbf{r}))$

$$U_{2n}^R(z_0, 0) = |u_0|^{-2n} |V_0|^{2n} [U_{2n}(z, 0)]^2 . \tag{3.1}$$

Equation (3.1) yields the relationship between the statistical intensity moments of incident and reflected from a point targets spherical waves:

$$< (I^R(z_0, 0))^n > = |u_0|^{-2n} |V_0|^{2n} < I^{2n}(z, 0) > . \tag{3.2}$$

Similarly, it may be shown that the mean intensity of a spherical wave reflected from an infinite mirror $(A(\mathbf{r}) \equiv V_0)$ is given by expression

$$< I^R(z_0, \mathbf{R}) > = \frac{|u_0|^2 |V_0|^2}{4k^2 L^2} [1 + B_{I,S}(z, \mathbf{R})] , \tag{3.3}$$

where $L = z - z_0$, $B_{I,S}(z, \mathbf{R})$ − is the normalized to the square of the mean intensity correlation function of a spherical wave intensity along direct path. The same expression correct to the constant coefficient takes place when the spherical wave is scattered by the point reflector.[2]

4. Basic Reflected Wave Phenomena Due to Correlation of Counter Waves

4.1 Enhancement of Wave Backscattering

From Eq. (3.3) as well as from similar formula for the point target[2] it follows that in the strictly backward direction ($\mathbf{R} = 0$) the mean intensity of a reflected spherical wave increases as compared with the propagation in a homogeneous medium by the value, determined by the variance of the spherical wave intensity $\sigma_{I,S}^2 = B_{I,S}(0)$ and this effect occurs around the strictly backward direction inside the cone of angle $\theta \sim r_I/L$, where r_I is the scale of spatial correlation of intensity fluctuations.

In a general case the value of enhancement depends essentially on the reflector size and the divergence of an incident laser beam.[22,5,12]

For quantitative estimate of the mean intensity enhancement we introduce the factor $N(\vec{R}) = \langle I^R(z_0, \mathbf{R})\rangle / \langle I^R(z_0, \mathbf{R})\rangle_{inc}$, where $\langle I^R(z_0, \mathbf{R})\rangle_{inc}$ corresponds to the mean intensity of a reflected wave, neglecting the correlation of the incident and backward waves. For spatially unlimited plane ($\Omega = ka^2/L >> 1$) and spherical ($\Omega << 1$) waves $\langle I^R\rangle_{inc}$ coincides with the intensity in a homogeneous medium, and in this case $N(\mathbf{R})$ characterizes an absolute amplification of mean intensity.

Figure 1 gives the results[22,12] of the calculation of the normalized value $\tilde{N} = (N(0) - 1)/\beta_0^2$ in the regime of weak fluctuations ($\beta_0^2 < 1$) at arbitrary values of the parameters Ω and $\Omega_r = ka_r^2/L$ for the plane mirror, retroreflector and target with diffuse random surface. Here the parameter $\beta_0^2 = 1.23\, C_n^2 k^{7/6} L^{11/6}$ characterizes the strength of optical turbulence, C_n^2 – structural characteristics of ε_1.

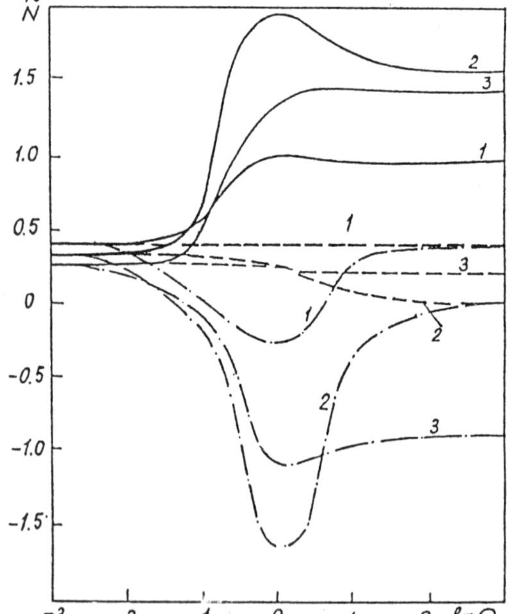

Fig. 1.

——— retroreflector

–·–·–·– plane mirror

– – – – diffuse target

1 - $\Omega = 10^{-3}$

2 - $\Omega = 1$

3 - $\Omega = 10^3$

One can see the correlation of direct and backward waves does not lead to amplification but to the opposite effect, i.e., reduction of mean intensity of reflected radiation, if a space–limited beam with a finite diffraction size $\Omega_{eff} \sim 1(\Omega_{eff}^{-1} = \Omega^{-1} + \Omega_r^{-1})$ is formed at the mirror. In the case of the incidence of a plane wave to an infinite size mirror ($\Omega_r >> 1$) the amplification effect is absent, as is seen from the figure. In contrast to the plane mirror at the values of $\Omega_{eff} \sim 1$

the mean intensity of radiation reflected from the retroreflector is not decreased but becomes maximum one. When a spherical wave is scattered by a large corner reflector, the amplification effect is not determined by the correlation function of the spherical wave intensity, as it is in the case of the mirror, but by the plane wave intensity correlation function, and $N(0) = 1 + \beta_0^2$.[23] The increase of mean intensity occurs also, when an infinite retroreflector intercepts a plane wave. In the case the amplification factor in the direction $\mathbf{R} = 0$ is defined by the value $N(0) = 1 + 1{,}56\,\beta_0^2$.

Fig. 1 presents the results for Gaussian model of $A(\mathbf{r})$ when diffraction on edges of reflector is not taken into consideration. The example of the calculation of $N(0)$, when edge diffraction is taken into account, is presented in Fig. 2. As it is seen, the real picture is more complicated than that shown in Fig. 1. The amplitude of enhancement nonmonotonously varies until $\Omega_r < 50$.

The amplification of backscattering has been experimentally confirmed for the point reflector in works,[24,25,26] for the infinite plane mirror in work,[27] substantiation of the reduction of the reflected wave mean intensity, when $\Omega_{eff} \sim 1$ (Fig. 1, Fig. 2), was obtained in.[28]

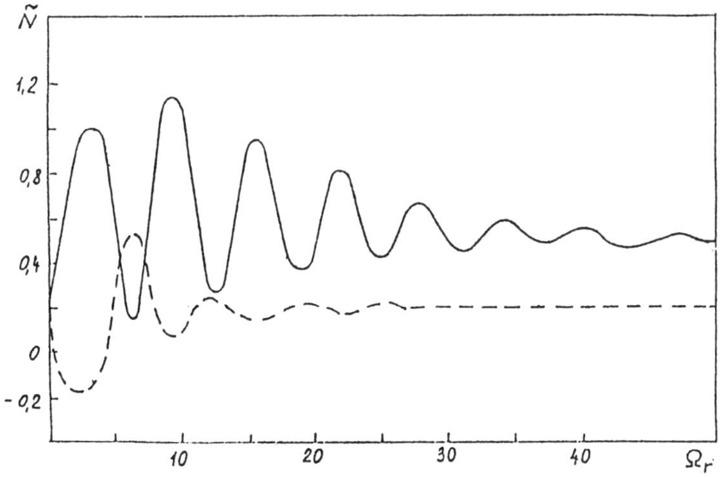

Fig. 2. $\beta_0^2 = 0.5$; $\Omega = 10^{-3}$ ——— retroreflector – – – – plane mirror

Let us consider the amplification of the mean intensity, in the regime of strong fluctuations, which is realized at the values of the parameter β_0^2 exceeding unit. In this case the regimes of a plane, a spherical waves and a spatially–limited beam are characterized by the conditions $\Omega >> \beta_0^{12/5}$, $\Omega << \beta_0^{-12/5}$ and $\beta_0^{-12/5} << \Omega << \beta_0^{12/5}$ respectively. Similar conditions characterize the "infinite" reflector, the "point" scatterer and the target of finite size: $\Omega_r >> \beta_0^{12/5}$,

$\Omega_r << \beta_0^{-12/5}, \beta_0^{-12/5} << \Omega_r << \beta_0^{12/5}$.

The mutual coherence function of the reflected wave can be represented as:

$$\Gamma_2^R(z_0, \underline{\rho_2}) = \Gamma_2^{(1)}(z_0, \underline{\rho_2}) + \Gamma_2^{(2)}(z_0, \underline{\rho_2}), \tag{4.1}$$

where $\Gamma_2^{(1)}$ describes the field coherence function at the path with a reflection when correlation between counter waves is not taken into account, $\Gamma_2^{(2)}$ is responsible for the correlation of the direct and backward waves. As follows from the analysis of corresponding asymptotic expressions for $\Gamma_2^{(1)}$ and $\Gamma_2^{(2)}$,[5,12] the amplification of the mean intensity in the regime of strong fluctuations is particularly significant in the case of scattering of a spherical wave. The both components in (4.1) are then comparable in value, and the mean intensity of the reflected spherical wave is more than 2 times as large as the intensity at the path when direct and backward waves are noncorrelated. The amplification factor $N(0)$ in this case can be represented as

$$N(0) = 2 + f(\Omega_r, \beta_0)\, \beta_0^{-4/5}, \tag{4.2}$$

where $f(\Omega_r, \beta_0)$ is the function depending on the reflector size and the turbulence parameter β_0. For the infinite mirror and the point scatterer Eq. (4.2) is of the form:

$$N(0) = 2 + 2.74\, \beta_0^{-4/5} + 0(\beta_0^{-8/5}) \tag{4.3}$$

and can be obtained from Eq. (3.3) directly if for $B_{I,S}$ its representation in the regime of strong fluctuations is used.[12]

In all other cases, when $\Omega \gtrsim \beta_0^{-12/5}$, the amplification effect is insignificant and is of the order of magnitude $\beta_0^{-4/5}$:[5,12]

$$N(\mathbf{R}) = 1 + g(\mathbf{R}, \Omega_r, \beta_0)\, \beta_0^{-4/5}$$

where $g(\mathbf{R}, \Omega_r, \beta_0)$ is a certain function.

The conditions of existence and the regions of localization of the amplification effect of backscattering at $\beta_0^2 >> 1$ in cases of a corner reflector and a diffuse surface[23,16,5,12] do not differ essentially from those stated above.

4.2 Effect of Long–Range Correlations and its Manifestation as Amplification of Mean–Reflected Intensity at the Lens Focus

Let us consider more carefully Eq. (4.1) when incident wave is plane

$\Omega >> \{D_s(\sqrt{L/k})\}^{6/5}$ $(D_s(\sqrt{L/k}) = 0.89\,\beta_0^2)$ for infinite flat mirror. In this case the two first terms in strong intensity fluctuations asymptotic expansions[5] corresponding to $\Gamma_2^{(1)}$ and $\Gamma_2^{(2)}$ can be written in the following form

$$\Gamma_{20}^{(1)} \sim \exp\left\{-\frac{8}{3}D_s(\sqrt{L/k})\rho_f^{5/3}\right\},\ \Gamma_{20}^{(2)} \sim \left(1+\frac{16}{3}\left[\frac{4}{3}D_s(\sqrt{L/k})\right]^{12/5}\right)^{-1}$$

$$\times \exp\left\{-\frac{2(4D_s(\sqrt{L/k})\,/\,3)^{6/5}\rho_f^2}{1+\frac{16}{3}\left[\frac{4}{3}D_s(\sqrt{L/k})\right]^{12/5}}\right\}, \tag{4.4}$$

where $\rho_f = |\rho_1 - \rho_2| \,/\, \sqrt{L/k}$. Similar expressions we shall have, also, in the case of plane wave for targets of other size.[5] It is clear from Eq. (4.4) that the term $\Gamma_{20}^{(2)}$ is of the order of $(D_s(\sqrt{L/k})^{-12/5}$, and is significantly smaller than the term $\Gamma_{20}^{(1)}$, so that it would seem that it does not make any considerable contribution in the $\Gamma_2^R(z_0, \underline{\rho}_2)$. But this is not the case. We have to take this term into account because it describes the so called long-range correlation effect.[6,33]

The point is that the terms $\Gamma_2^{(1)}$and $\Gamma_2^{(2)}$ in Eq. (4.1) when $\Omega >> \{D_s(\sqrt{L/k})\}^{6/5}$ differ not only in magnitude, but also in the scale of the reduction down to the 1/e level, as it is seen from Eq. (4.4). The first term fall to this level for small separations of the points of observation $\rho_f \sim \{D_s(\sqrt{L/k})\}^{-3/5}$, the second one have the scale of their reduction much greater $\rho_f \sim \{D_s(\sqrt{L/k})\}^{3/5}$. So the small energy flux from the extended source $(\Omega >> \{D_s(\sqrt{L/k})\}^{6/5})$, defined by the term $\Gamma_2^{(2)}(z_0, \underline{\rho}_2)$, can be efficiently focused by a lens placed near the source.[6]

Indeed,[6] if one performs the focusing of the reflected plane wave by the lens at $\beta_0^2 >> 1$ whose size $2a_t$ is such large, that the condition $\Omega_t = \frac{ka_t^2}{L} >> \{D_s(\sqrt{L/k})\}^{6/5}$ is fulfilled then in the lens focus the mean intensity increases more than two fold as compared with the case when direct and return waves are noncorrelated.

Thus, if in the input of the lens the effect of enhancement of the mean intensity of a plane wave at $\beta_0^2 >> 1$ is absent (see sect. 4.1) then in the lens focus such amplification appears. And vice versa, the focusing of a reflected spherical wave by the lens of size $\Omega_t >> \{D_s(\sqrt{L/k})\}^{6/5}$ at $\beta_0^2 >> 1$ results in the fact that the twofold increase of mean intensity, occurring at the input of the lens (4.2), disappears in the image plane of this lens.[7]

For quantitative estimate of exceeding by mean intensity of a reflected wave $< I^R(l, \mathbf{R}) >$ under conditions of strong fluctuations the mean intensity value $< I^{(1)}(l, \mathbf{R}) >$, corresponding to the path without correlation of the counter waves, we introduce the factor:

$$N(l, \mathbf{R}) = < I^R(l, \mathbf{R}) > / < I^{(1)}(l, \mathbf{R}) > = 1 + < I^2(l, \mathbf{R}) > / < I^{(1)}(l, \mathbf{R}) > .$$

Figure 3 presents the results[7] of calculation of the amplification factor $N = N(F_t, 0)$ of the mean intensity in the focus ($l = F_l$) of the lens for a diffuse target.

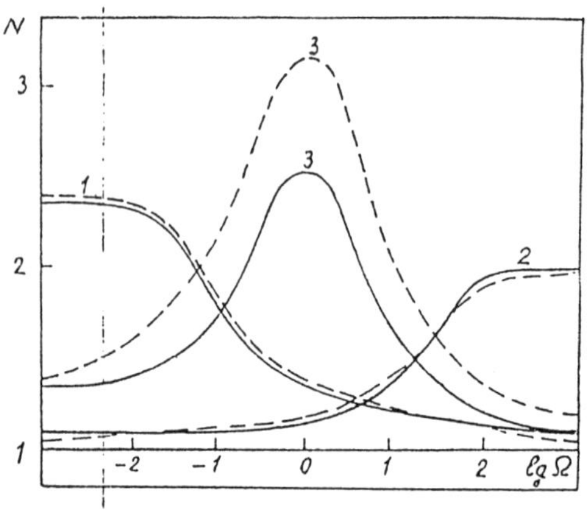

Fig. 3. $\beta_0^2 = 55$;
curves 1 to 3 correspond to
$\Omega_t = 10^{-3}, 10^3, 1$

——— $\Omega_r = 10^{-3}$

- - - - $\Omega_r = 10^3$

The figure shows that the amplification takes place only in the case of proximity of the size of the source and the receiver apertures.

This result becomes clear if we take into account[7] the circumstance that the rays are coherent with each other due to long-range correlations only in the region limited by the size of output aperture. Therefore the receiver with less, than $2a$, dimensions collects not all the coherent rays, and the use of the receiver of more than $2a$ size results in the relative decrease of the contribution of coherent component of scattered radiation $< I^{(2)} >$ as compared with the increased contribution of the incoherent component $< I^{(1)} >$.

4.3 Enhancement of an Intensity Scintillations

Correlation of the waves, which propagate in forward and backward directions through the same inhomogeneties of the medium result also in the amplification of intensity fluctuations of the reflected field.[3] Value of amplification of intensity fluctuations significantly varies depending on the parameters of a transmitting aperture, the size and the type of a reflector and the strength of the turbulence.

Figure 4 gives the result[17] of calculations of relative variance of intensity of a collimated beam, reflected from the plane mirror, on the axis of optical system at $\beta_0^2 < 1$. It is seen that the intensity fluctuations are minimal when the mirror dimensions comparable with the radius of the first Fresnel zone ($\Omega_R \simeq 1$), and they are maximal when reflecting from the "infinite" mirror.

As to the corner reflector, the expression for $\sigma_{I,R}^2(\mathbf{R})$ is the same as for a plane mirror if $\mathbf{R} = 0$. When $\mathbf{R} \neq 0$ the behavior of the variance is different from that in the case of reflection from a mirror. In particular, in contrast to an infinite plane mirror, the reflected plane wave intensity variance in the case of a retroreflector of large size ($\Omega_R >> 1$) is a function of the position of the point of observation and decreases with the increase $|\mathbf{R}|$ from $\sigma_{I,R}^2(0) = 5.23\,\beta_0^2$ to $\sigma_{I,R}^2 = 3.56\,\beta_0^2$ for $|\mathbf{R}| \gtrsim \sqrt{\lambda L}$.

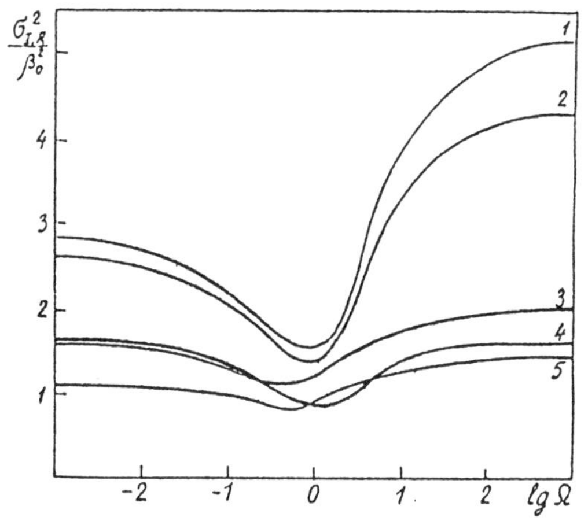

Fig. 4. Curves 1 to 5 correspond to $\Omega_r = 10^3, 10, 10^{-3}, 1, 10^{-1}$

In the regime of strong fluctuations the increase of relative variance intensity is maximal in the case of a point target $\Omega_r << \beta_0^{-12/5}$. As compared with propagation along a path without reflection when the saturation level of the relative intensity fluctuations is equal to unit for $\beta_0^2 \to \infty$, this level increases some times:[15,16]

$$\sigma_{I,R}^2 = \frac{5 + 46.6\,\beta_0^{-4/5}}{1 + 2.74\,\beta_0^{-4/5}} + 0(\beta_0^{-8/5}), \quad \Omega << \beta_0^{-12/5} \tag{4.5}$$

$$\sigma_{I,R}^2 = \frac{3 + 12.9\,\beta_0^{-4/5}}{1 + 0.82\,\beta_0^{-4/5}} + 0(\beta_0^{-8/5}), \quad \Omega >> \beta_0^{12/5} \tag{4.6}$$

When the conditions $\Omega_r << \beta_0^{-12/5}$ is violated, the effect of intensity fluctuation amplification becomes negligible.[16,5]

In case of the scattering by the diffuse surface it should be taken into account that the receiver is capable of both "follow" the fast random variations of the reflected radiation field appearing due to the diffuse surface, if the correlation time of the roughness changes $\tau_k > \tau_r$, and average them, when the relation $\tau_r >> \tau_k$ is fulfilled. Having used the corresponding models for the fourth moment of the function $V(\underline{\rho}_4, \underline{\mathbf{r}}_4)$ (2.17) for the relative variance of weak intensity fluctuations of a spherical wave reflected from the diffuse surface, at $\tau_r >> \tau_k$ we find[5,12]

$$\sigma_{I,R}^2(\mathbf{R}) = 2\sigma_{I,g}^2 + 2B_{I,g}(\mathbf{R}) \,. \tag{4.7}$$

In this case the function $B_{I,g}(\mathbf{R})$ coincides with the intensity correlation function of the incoherent source in the turbulent atmosphere[12] emitting in backward direction from the reflector plane, $\sigma_{I,g}^2 = B_{I,g}(0)$.

From Eq. (4.7) it follows that in the directions close to a backward one ($|\mathbf{R}| \lesssim \sqrt{\lambda L}$) the amplification of intensity fluctuations occurs: $\sigma_{I,R}^2(0) = 4\sigma_{I,g}^2$ and $\sigma_{I,R}^2 = 2\sigma_{I,g}^2$, if $|\mathbf{R}| >> \sqrt{\lambda L}$. For the point reflector the relative fluctuations of the reflected spherical wave are maximal. The increase of the size of the diffuse target results in their decrease. In particular, from (4.7) we have

$$\sigma_{I,R}^2(0) = 1.6\,\beta_0^2 + 0(\beta_0^4)\,,\ \Omega_r << 1$$

$$\sigma_{I,R}^2(0) = 0.68\,\beta_0^2 + 0(\beta_0^4)\,,\ \Omega_r = 1$$

$$\sigma_{I,R}^2 \sim \Omega_r^{-7/6}\,\beta_0^2\,, \qquad \Omega_r >> 1, \tag{4.8}$$

whence it appears that in the limit $\Omega_r \to \infty$ the intensity of the reflected radiation does not fluctuate.

The analogous results take place also in the case of scattering by the diffuse surface of a plane wave.[5,12]

In the regime of strong fluctuations when scattering spherical and plane wave by the diffuse target of small transverse size $\Omega_r << \beta_0^{-12/5}$ the relative variance is described, as in the case of the mirror, by the formulas (4.5) and (4.6). For a diffuse target of finite size we have[5,12]

$$\sigma_{I,R}^2 = 5.48\,\beta_0^{-4/5} + 0(\beta_0^{-8/5})\,, \quad \Omega << 1\,, \tag{4.9}$$

$$\sigma_{I,R}^2 = 2.6\,\beta_0^{-4/5} + 0(\beta_0^{-8/5}), \quad \Omega >> 1, \tag{4.10}$$

From (4.9) and (4.10) it follows that with the increase of the turbulence $\beta_0 \to \infty$, the relative variance of intensity, reflected by a surface with reflection coefficient fluctuations averaged over τ_r tends to zero. This means that a diffuse surface illuminated by a coherent light becomes a source of incoherent spherical waves, whose intensities add at the point of reception. In this case,[30,12] the variance of strong intensity fluctuations has no the saturation level different from zero.
With the increase of size of the diffuse surface ($\Omega_r >> \beta_0^2$) the averaging of strong fluctuations of reflected radiation takes place in accordance with the same law (4.8) as in the case of weak intensity fluctuations.
As the analysis has shown,[16,5,12] in conditions of $\tau_k \geq \tau_r$, when receiver could follow the signal fluctuations, appearing due to phase distortions contributed by the surface, the value of the variance of intensity is defined by:

$$\sigma_{I,R}^2(R) = 1 + 2\sigma_{I,RS}^2(R), \tag{4.11}$$

where $\sigma_{I,RS}^2$ coincides with the variance of the reflected radiation intensity at $\tau_k << \tau_r$. Thus, the significant increase of the variance due to the amplification effect of the intensity fluctuations in accordance with the formulas (4.11), (4.5), (4.6) takes place at scattering by the "point" random reflector, whose transverse size satisfy the condition $\Omega_r << \beta_0^{-12/5}$. The saturation level of $\sigma_{I,R}^2$ in the case of a spherical wave scattered by such reflector becomes equal to eleven, and in the case of a plane wave the saturation level equals seven. The increase of size of diffuse surface results in averaging of intensity fluctuations of the reflected radiation and when fulfilling the condition $\Omega_r \gtrsim \beta_0^{-12/5}$ the relative variance of intensity slightly exceeds unit at any values of the turbulence parameter β_0^2.[16,5,12] In a series of particular cases the theoretical results concerning the intensity fluctuations of reflected radiation, given in this section, are confirmed by the experimental results.[25,31,32]

5. Reflected Wave Intensity Correlation and Effect of Residual Correlation Persistence

The spatial structure of intensity scintillations in the reflected radiation differs from the case of direct propagation. As it has been first shown in[4] for weak fluctuations, in contrast to direct propagation the spatial intensity correlation of a spherical wave scattered by point reflector does not fall to zero but persists on certain non zero level even for an infinite separation between the points of observation ρ_1 and ρ_2. In general case this level depends on type of the incident wave, type of the reflector and strength of the atmospheric turbulence.

For a point reflector the level of weak intensity correlation coefficient at $R = (\rho_1 + \rho_2)/2 = 0$, $|R| = |\rho_1 - \rho_2| \to \infty$ (symmetric disposition of observation points) is defined by formula[17]

$$b_{I,R} = \sigma_{I,b}^2 / (\sigma_{I,Sph}^2 + \sigma_{I,b}^2) \tag{5.1}$$

where $\sigma_{I,b}^2$ and $\sigma_{I,Sph}^2$ are the relative intensity variances of the incident beam and the spherical wave respectively. An increase of the size of the reflecting mirror reduces the residual correlation down to zero because of the averaging of intensity fluctuations in the incident radiation. In contrast to reflection by the infinite mirror the spatial structure of fluctuations in the intensity reflected by an infinite retroreflector is statistically inhomogeneous. The correlation falls to zero for an asymmetric disposition ($\rho_1 = 0$, $|\rho_2| \to \infty$) of the points of observation as in the case of mirror, but axially symmetric disposition of these points produces a reduction in the correlation function down only to a level determined by plane wave intensity variance on direct path of double length $\sigma_{I,pl}^2(2L) = 3.56\,\beta_0^2$. The behaviour of a reflected wave intensity correlation coefficient for weak fluctuation is shown in Fig. 5.

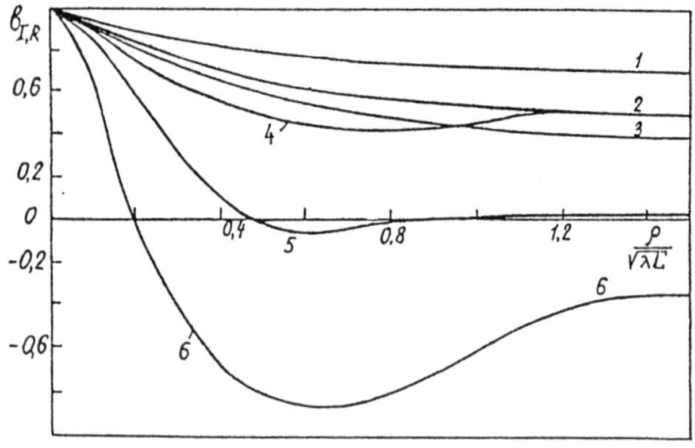

Fig. 5. Curves 1,2,3 – $\Omega_r = 10^{-3}$; curves 1 to 3 correspond to $\Omega = 10^3$, 10^{-3} and 0.316; curve 5 – $\Omega = \Omega_r = 10^3$; curve 6 – $\Omega = 10^3$, $\Omega_r = 1$; curve 4 – retroreflector with $\Omega_r = 10^3$, $\Omega = 10^3$, symmetric disposition of observation points.

The effect of residual correlation persistence takes place also in the case of strong intensity fluctuations when size of target is small. If a plane wave is incident on a point reflector, the spatial structure of correlation is homogeneous and the level of residual correlation is equal to $b_{I,R} = 1/3$ independently of the disposition of the points of observation. For spherical wave this level is different for different dispositions of the points of observation: $b_{I,R} = 1/3$, when $\mathbf{R} = 0$, $|\rho| \to \infty$ and $b_{I,R} = 2/\sqrt{15}$, when $\rho_1 = 0$, $|\rho_2| \to \infty$ (Fig. 6).

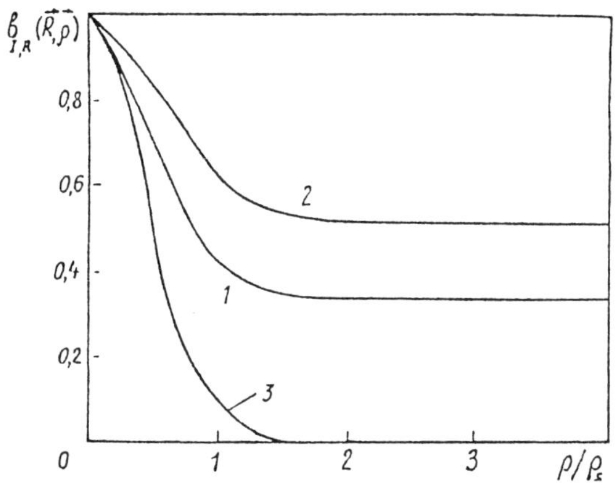

Fig. 6. Curve 1 corresponds to incident plane wave or spherical one (symmetric disposition of observation points) and point reflector, curve 2 corresponds to incident spherical wave, point reflector, asymmetric disposition; curve 3 corresponds to spherical wave and infinite mirror.

So, just as in the case of weak fluctuations the strong intensity fluctuations of a spherical wave reflected by a point target is not statistically homogeneous.

6. Manifestation of Reflected Wave Phenomena in Target Location and in Remote Sensing

When estimating the efficiency of operation of the lidars in the atmosphere one should take into account both "geometric" parameters of the receiver–transmitter, and the influence of the turbulence on laser radiation propagation. From the point 4.2 it follows that the conditions of existence of the effects due to the correlation of counter waves, behind the focusing lens, can vary significantly from those taking place in the input plane. Therefore it is important to know the value of photocurrent fluctuations in the receiving system of a lidar and in what way the enhancement fluctuations effect is manifested in this case.

As it is known[34] the complex amplitude of the field on output of a photodetector is determined by equation:

$$I = \int d\mathbf{r}\, m(\mathbf{r})\, |u_l^R(\mathbf{r})|^2 , \tag{6.1}$$

where $u_l^R(\mathbf{r})$ is the complex amplitude of the field of the reflected wave after its focusing by lens, $m(\mathbf{r})$ is the amplitude transmission coefficient of the diaphragm in the plane of photodetector. Relative variance of the photodetector output signal σ_J^2 is expressed through the second $< u_{\underline{2}}^R >$ and the fourth $< u_{\underline{4}}^R >$ moments of the reflected wave field (2.6) and it may be written in the form

$$\sigma_J^2 = \sigma_{Jd}^2 + \sigma_{Jr}^2 + \sigma_{Jc}^2 \tag{6.2}$$

where σ_{Jd}^2 corresponds to direct path, σ_{Jr}^2 corresponds to return one and σ_{Jc}^2 corresponds to correlation of the counter waves. Fig. 7[35] demonstrates the behaviour of the σ_J^2 under conditions of the weak turbulent fluctuations in the case of mirror reflector, when Fresnel number of the effective radius of the lens Ω_t is $\Omega_t = 10^3$.

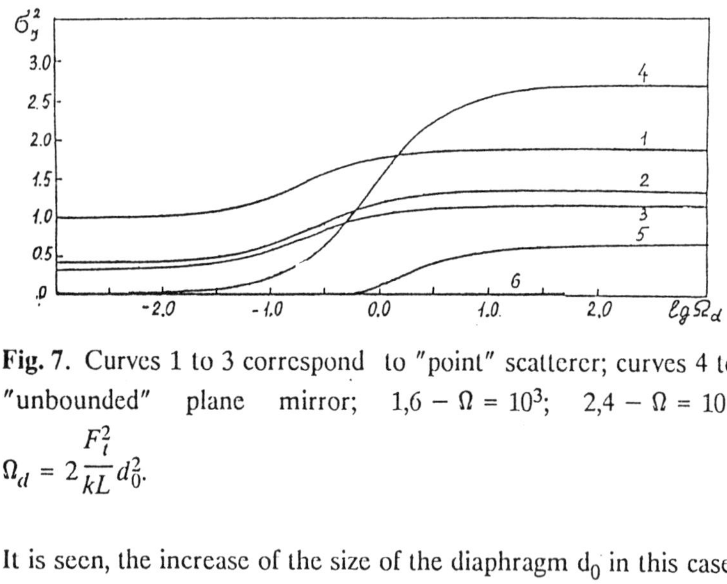

Fig. 7. Curves 1 to 3 correspond to "point" scatterer; curves 4 to 6 corresponds to "unbounded" plane mirror; 1,6 – $\Omega = 10^3$; 2,4 – $\Omega = 10^{-3}$; 3,5 – $\Omega = 1$; $\Omega_d = 2\frac{F_t^2}{kL}d_0^2$.

It is seen, the increase of the size of the diaphragm d_0 in this case leads to decrease of photocurrent fluctuations. Owing to a residual correlation effect (previous section) the photocurrent fluctuations of any wave scattered by point reflector don't fall to zero level even when $d_0 \to \infty$. This level is defined by term σ_{Jd}^2 in (6.2) and it is equal to intensity fluctuations variance of the incident wave. This result was used for development of the lidar method of determination of C_n^2 in atmosphere by means of large receiving aperture.[36] However, when $\Omega_t \lesssim 10^2$ the condition $\sigma_J^2 \equiv \sigma_{Jd}^2$ is violated and photocurrent fluctuations at $d_0 \to \infty$ contain also a component stipulated by both wave fluctuations on return path and correlation between direct and backward waves.[35]

7. Conclusion

Thus, correlation between counter waves in double passage through a turbulent atmosphere leads to appearance of a set of the specific enhancement effects which do not take place on path in a turbulent atmosphere without a reflection. The existence conditions and quantitative manifestation of these effects significantly depend on the turbulence strength, target size, reflecting surface properties and angular divergence of the wave beams illuminating the reflectors. All these factors must be taken into account when estimating the influence of the enhancement effects on effectiveness of the target location and lidar system operation in a turbulent atmosphere.

References

1. M.S. Belen'kii and V.L. Mironov, *Kvant. Electron.*, No. 5(11), 38–45 (1972).
2. A.G. Vinogradov, Yu.A. Kravtsov, and V.I. Tatarskii, *Izv. Vyssh. Uchebn. Zaved. Radiofiz.* **16**, No. 7, 1064–1070 (1973).
3. A.G. Vinogradov and Yu.A. Kravtsov, In: *VI All–Union Symposium on Diffraction and Propagation of Waves*, Paper abstracts, Moscow–Erevan **1**, 294 (1973).
4. M.S. Belen'kii and V.L. Mironov, *Kvant. Electron.* **1**, No. 10, 2253–2262 (1974).
5. V.A. Banakh and V.L. Mironov, *Lidar in a Turbulent Atmosphere,* Boston and London, Artech House (1987).
6. A.B. Krupnik and A.I. Saichev, *Izv. Vyssh. Uchebn. Zaved. Radiofiz.* **24**, No. 10, 1234–1239 (1981).
7. V.A. Banakh, *Izv. Vyssh. Uchebn. Zaved. Radiofiz.* **29**, No. 12, 1507–1509 (1986).
8. V.I. Gel'fgat, *Akust. Zh.* **22**, No. 1, 123–124 (1976).
9. V.P. Aksenov and V.L. Mironov, *J.Opt. Soc. Am.* **69**, No. 11, 1609–1614 (1979).
10. A.I. Kon and V.I. Tatarskii, *Izv. Vyssh. Uchebn. Zaved. Radiofiz.* **15**, 1547–1553 (1972).
11. V.A. Banakh, I.N. Smalikho, and B.N. Chen, *Opt. Spektrosk.* **61**, 582–586 (1986).
12. V.E. Zuev, V.A. Banakh, and V.V. Pokasov, *Optika Turbulentnoi Atmosfery*, Leningrad, Gigrometeoizdat (1988).
13. V.M. Orlov, I.V. Samokhvalov, G.M. Krekov et. al., *Signals and Noises in Laser Ranging*, M., Radio i Svyaz (1985) Translated in *Journal of Soviet Laser Research* **8**, No. 4, 281–428 (1987).
14. V.P. Aksenov, V.A. Banakh, V.L. Mironov et. al., *Opt. Spektrosk.* **61**, 839–844 (1986).
15. V.P. Aksenov, V.A. Banakh, V.L. Mironov, In: *II All–Union Conference on Atmospheric Optics,* Paper abstracts, Tomsk **2**, 120–130 (1980).
16. V.P. Aksenov, V.A. Banakh, V.L. Mironov, *J.Opt. Soc. Am.* **A.1**, No. 3, 263–274 (1984).
17. V.A. Banakh and O.V. Tikhomirova, *Opt. Spektrosk.* **56**, 857–863 (1984).

18. V.A. Banakh and V.L. Mironov, *Optics Letters* **1**, No. 5, 172–174 (1977).

19. V.P. Aksenov and V.L. Mironov, *Optics Letters* **3**, No. 5, 184–186 (1978).

20. S. Frankenthal, A.M. Whitman, and H.J. Beran, *J.Opt. Soc. Am.* **A.1**, 585–590 (1984).

21. V.P. Aksenov, V.A. Banakh, V.L. Mironov, In: *Paper Abstracts of IV All–Union Conference on Physical Principles of Information Transfer Using Laser Radiation*, Kiev, 149 (1976).

22. V.A. Banakh, V.M. Buldakov, and I.N. Smalikho, In: *Waves and Diffraction–85*, XI All–Union Symposium on Diffraction and Propagation of Waves, Tbilisi **1**, 410–413 (1985).

23. V.P. Aksenov and V.A. Banakh, In: *VI All-Union Symposium on Laser Radiation Propagation in the Atmosphere*, Paper abstracts, Tomsk **3**, 67–70 (1981).

24. A.S. Gurvich and S.S. Kashkarov, *Izv. Vyssh. Uchebn. Zaved. Radiofiz.* **20**, No. 5, 794–796 (1977).

25. S.S. Kashkarov, T.N. Nesterova, and A.S. Smirnov, *Izv. Vyssh. Uchebn. Zaved. Radiofiz.* **27**, No. 10, 1272–1278 (1984).

26. S.S. Kashkarov, *Izv. Vyssh. Uchebn. Zaved. Radiofiz.* **26**, No. 1, 44–48 (1983).

27. A.S. Gurvich, A.P. Ivanov, S.S. Kashkarov, G.Ya. Patrushev, and A.P. Rostov, In: *XI All-Union Symposium on Laser Radiation Propagation in the Atmosphere and Water Media,* Paper Abstracts, Tomsk, 6 (1991).

28. B.S. Agrovskii, A.N. Bogaturov, A.S. Gurvich, S.V. Kireev, and V.A. Myakinin, *J.Opt. Soc. Am.* **A.8**, No. 7, 1142–1147 (1991).

29. V.P. Aksenov and V.L. Mironov, *Izv. Vyssh. Uchebn. Zaved. Radiofiz.* **22**, No. 2, 141–149 (1979).

30. V.A. Banakh, V.M. Buldakov, and V.L. Mironov, *Opt. Spektrosk.* **57**, 1054–1059 (1983).

31. P.A. Pincus, M.E. Fossey, J.F. Holmes, and J.R. Kerr, *J.Opt. Soc. Am.* **68**, No. 6, 760–762(1978).

32. G.Ya. Patrushev, A.I. Petrov, and V.V. Pokasov, *Izv. Vyssh. Uchebn. Zaved. Radiofiz.* **26**, No. 7, 823–831 (1983).

33. Yu.A. Kravtsov and A.I. Saichev, *Uspekhi Fiz. Nauk* **137**, 501–527 (1982).

34. V.V. Protopopov and N.D. Ustinov, *Lasers Heterodying*, M.,Nauka (1985).

35. V.A. Banakh, O.V. Tikhomirova, and A.P. Shelekhov, *Atmospheric Optics* (to be published).

36. M.S. Belen'kii, V.L. Mironov, P.I. Netreba, V.V. Pokasov, and A.P. Shelekhov, *Izv. Vyssh. Uchebn. Zaved. Fiz*, No. 12, 103–105 (1986).

Random gravitational lenses

L.M.Erukhimov P.I.Shpiro

Scientific Research Radiophysics Institute
Nizhny Novgorod, Russia

Abstract

We analyse new application of scintillation: the propagation of electromagnetic waves through an interstellar space characterized by random mass distribution. The random refractive index fluctuations are related to the random gravitational potential. We first consider the statistics of the geometrical-optics phase fluctuations, angular scattering, frequency correlation and pulse broadenning which are related to the statistics of the fluctuating random potential. The total cross section of scattering is also evaluated in the context of the Born approximation. The gravitational effects are compared with the same effects of strong magnetic fields.

1.Introduction.

The purpose of this paper is to discuss the application of some methods of wave propagation in random media to wave scintillations and scattering induced by random gravitational fields (or strong magnetic fields).

The gravity can lead to electromagnetic waves refraction and change pulse signal time-delays [1,2,3].

The gravitational lense leads to focusing effects which could be important not only in the black holes problem and invisible matter observations but also for non-linear effects in the envelope plasma surrounding double objects: pulsar and black hole.Apparently, the micro-scale black holes scatter electromagnetic waves. The irregularity of strong magnetic fields inducing vacuum polarization could lead to vacuum polarization, lenses and scatters.

But the majority of effects mentioned above is uncommon. These effects are rather problematically to observe. And it is the reason the critical views presented below.

2.A layer with random gravitating masses.

According to nonrelativistic approximation $U \ll c^2$, U is the potential and c is light velocity, gravitational fields affect the propagation of electromagnetic waves as through a medium with the effective refraction index $n = 1 + 2U/c^2$ being optically more dense near gravitating mass [2,3]:

$$U(\mathbf{r}) = G \int \frac{\mu(\mathbf{r}')d\mathbf{r}'}{|\mathbf{r} - \mathbf{r}'|}. \quad (1)$$

Here $\mu(\mathbf{r})$ is the mass density and G is the gravitational constant. Because of random fluctuations of mass density the effective refraction index also varies randomly in addition to regular variation with coordinate. Therefore,the medium is characterized by $n(\mathbf{r}) =< n(\mathbf{r}) > +\tilde{n}(\mathbf{r})$, which consists of an average part $< n(\mathbf{r}) >$ and a fluctuating part

$$\tilde{n}(\mathbf{r}) = \frac{2G}{c^2} \int \frac{\tilde{\mu}(\mathbf{r}')d\mathbf{r}'}{|\mathbf{r} - \mathbf{r}'|}, \tag{2}$$

where $\tilde{\mu}(\mathbf{r})$ is an irregular part of $\mu(\mathbf{r})$.

Let us consider a case when gravitating masses are located in a layer of finite depth $2L$ in the ray propagation direction Z and infinite one- transverse it.Then (2) can be written in the form

$$\tilde{n}(\mathbf{r}) = \frac{2G}{c^2} \int_{-\infty}^{\infty} dz' \int_{-\infty}^{\infty} d\rho'_\perp \frac{\tilde{\mu}(z', \rho'_\perp)}{|\mathbf{r} - \mathbf{r}'|} M(z'), \tag{3}$$

where, for simplicity, we choose $M(z) = e^{-z^2/L^2}$, $\mathbf{r} = \mathbf{r}(z, \rho_\perp)$.
Using the known presentation
$\frac{1}{|\mathbf{r}-\mathbf{r}'|} = \frac{1}{2\pi^2} \int \frac{e^{i\mathbf{K}(\mathbf{r}-\mathbf{r}')}}{K^2} d\mathbf{K}$, $\mathbf{K} = \mathbf{K}(K_z, \mathbf{K}_\perp)$, $K^2 = K_z^2 + K_\perp^2$,
one can obtain a correlation function of random refractive index[4]

$$\Gamma_n(\mathbf{r}_1, \mathbf{r}_2) = \frac{G^2}{(\pi c)^4} \int_{-\infty}^{\infty} \int dz' dz'' \int_{-\infty}^{\infty} \int d\rho'_\perp d\rho''_\perp d\mathbf{K}_1 d\mathbf{K}_2 \times$$
$$\times \frac{M(z')M(z'')}{K_1^2 K_2^2} \Gamma_\mu(\mathbf{r_1}', \mathbf{r_2}'') \exp\left[i\mathbf{K}_1(\mathbf{r}_1 - \mathbf{r}') - i\mathbf{K}_2(\mathbf{r}_2 - \mathbf{r}'')\right] \tag{4}$$

Apparently, in considered case the correlation function of mass density fluctuations $\Gamma_\mu(\mathbf{r_1}', \mathbf{r_2}'') =< \tilde{\mu}(\mathbf{r}')\tilde{\mu}(\mathbf{r}'') >$ only depends on the difference of variables $\mathbf{r} = \mathbf{r}' - \mathbf{r}''$ and can be expressed as

$$\Gamma_\mu(\mathbf{r}) = (2\pi)^{-3/2} \int \Phi_\mu(\mathbf{K}) e^{i\mathbf{K}\mathbf{r}} d\mathbf{K}, \tag{5}$$

where $\Phi_\mu(\mathbf{K})$ is the three - dimensional Fourier spectrum of mass density fluctuations.

Inserting (5) into (4) yields

$$\Gamma_n(z_1, z_2, \rho) = \frac{2\pi G^2 L}{c^4} \int\int d\mathbf{K}_\perp dK_{1z} dK_{2z} \exp[i(K_{1z}z_1 - K_{2z}z_2)] \times$$
$$\times e^{i\mathbf{K}_\perp \rho} \frac{\Phi_\mu((K_{1z} + K_{2z})/2, \mathbf{K}_\perp)}{(K_\perp^2 + K_{1z}^2)(K_\perp^2 + K_{2z}^2)} \exp\left[-\frac{(K_{1z} - K_{2z})^2 L^2}{8}\right]. \tag{6}$$

We have a wave that has travelled a path of length Z in a turbulent medium. The random component of the phase is defined by the relation[5]:

$$\tilde{S}(z, \rho_\perp) = k_0 \int_0^Z \tilde{n}(\zeta, \rho_\perp) d\zeta, \tag{7}$$

$k_0 = \omega/c, \omega$ is the wave frequency.

By using (6) and (7) the expression for the phase structure function $D_S(z,\rho_1,\rho_2) = <[\tilde{S}(z,\rho_1) - \tilde{S}(z,\rho_2)]^2>$ can be readily obtained:

$$D_S(z,\rho) = \frac{4\pi k_0^2 G^2 L}{c^4} \int_0^Z d\zeta \int_{\zeta/2}^{Z-\zeta/2} d\bar{z} \int dK_z d\Delta K_z d\mathbf{K}_\perp e^{i\Delta K_z \bar{z}} e^{iK_z\zeta} \times$$
$$\times [1 - \cos(\mathbf{K}_\perp \rho)] \frac{\Phi\mu(K_z, \mathbf{K}_\perp)}{K_\perp^2 + K_z^2} e^{-\frac{(\Delta K_z)^2 L^2}{8}}. \quad (8)$$

Here $\bar{z} = (\zeta_1 + \zeta_2)/2$, $\zeta = \zeta_1 - \zeta_2$, $K_z = (K_{1z} + K_{2z})/2, \Delta K_z = K_{1z} - K_{2z}$.

In case of isotropic fluctuations $\tilde{\mu}$ for $Z \gg L$ (8) becomes

$$D_S(\rho) = \frac{(2\pi)^3 k_0^2 G^2 L}{c^4} \int dK_\perp K_\perp^{-3} [1 - J_0(K_\perp \rho)] \Phi_\mu(K_\perp), \quad (9)$$

$J_0(x)$ is the Bessel function and $\rho = |\rho|$.

In a paper by Erukhimov and Shpiro [4],to avoid the integral divergence on small scales $\Phi_\mu(\mathbf{K})$ has been chosen in the form

$$\Phi_\mu(\mathbf{K}) = C_\mu^2 K^{-p} \exp(-\frac{K^2}{K_{max}^2} - \frac{K_{min}^2}{K^2}), \quad (10)$$

where K_{min}, K_{max} are the characteristic scales of the spectrum cut off. The structural constant C_μ^2 is defined by the condition $<\tilde{\mu}^2> = \int d\mathbf{K}\Phi_\mu(\mathbf{K})$ as

$$C_\mu^2 = \begin{cases} \frac{<\tilde{\mu}^2>}{2\pi\Gamma(\frac{3-p}{2})K_{max}^{-p+3}}, & p < 3 \\ \frac{<\tilde{\mu}^2>}{2\pi\Gamma(\frac{3-p}{2})K_{min}^{-p+3}}, & p > 3 \end{cases} \quad (11)$$

Then, using the known relation

$$1 - J_0(x) = \sum_{n=1}^{\infty} (-1)^{n+1} \left(\frac{x}{2}\right)^{2k} (n!)^{-2}$$

one can obtain

$$D_S(\rho) = 2S_0^2 \sum_{n=1}^{\infty} \frac{(-1)^{n+1}}{(n!)^2} \left(\frac{K_{max}K_{min}}{4}\right)^n \frac{K_{n-1-p/2}(2\frac{K_{min}}{K_{max}})}{K_{-1-p/2}(2\frac{K_{min}}{K_{max}})} \rho^{2n}, \quad (12)$$

$K_\nu(x)$ is the modified Bessel function, the phase variance $S_0^2 \equiv <\tilde{S}^2>$ is given by

$$S_0^2 = \frac{4\pi^3 k_0^2 G^2}{c^4} L C_\mu^2 (K_{min}K_{max})^{-\frac{p}{2}-1} K_{-\frac{p}{2}-1}\left(2\frac{K_{min}}{K_{max}}\right) \quad (13)$$

Taking into account $K_\nu(x) = \frac{1}{2}\Gamma(|\nu|)(\frac{2}{x})^{|\nu|}$ for values $x = 2\frac{K_{min}}{K_{max}} \ll 1, \nu \neq 0$, we have

$$S_0^2 = \begin{cases} \pi^2 <\tilde{\mu}^2> \frac{k_0^2 G^2 L}{c^4} \frac{\Gamma(\frac{2+p}{2})}{\Gamma(\frac{3-p}{2})} \frac{K_{min}^{-p-2}}{K_{max}^{-p+3}}, p < 3 \\ \pi^2 <\tilde{\mu}^2> \frac{k_0^2 G^2 L}{c^4} \frac{\Gamma(\frac{2+p}{2})}{\Gamma(\frac{3-p}{2})} K_{min}^{-5}, p > 3 \end{cases} \quad (14)$$

lense

$$\Gamma_\omega\left(z_p, \mathbf{t}_p = \frac{\rho_p}{L}\right) = \frac{\chi_0^2(1-\delta^2)}{4\pi^2} I_0 e^{i2k_0\delta z_p} e^{i\chi_0\delta t_p^2} \times$$
$$\times \int d\mathbf{t}_1 d\mathbf{t}_2 \exp\left\{ i\chi_0 \left[\frac{t_1^2 - t_2^2}{2} + \frac{\delta(t_1^2 + t_2^2)}{2} - \ln\frac{|\mathbf{t}_1|}{|\mathbf{t}_2|} - \right.\right.$$
$$\left.\left. -\delta \ln(|\mathbf{t}_1||\mathbf{t}_2|) - \mathbf{t}_p(\mathbf{t}_1 - \mathbf{t}_2) - \delta \mathbf{t}_p(\mathbf{t}_1 + \mathbf{t}_2) \right]\right\} e^{-\frac{1}{2}D_S^\omega(\mathbf{t}_1, \mathbf{t}_2)}; \tag{25}$$

$$q(z_p, t_p) = \chi_0^2 \int dt_1 dt_2 t_1 t_2 \exp\left\{ i\chi_0 \left[\frac{t_1^2 - t_2^2}{2} - \ln\frac{t_1}{t_2} \right] - \right.$$
$$\left. -i\frac{\chi_S^2}{8} \ln^2 \frac{t_1}{t_2} \right\} J_0(\chi_0 t_p t_1) J_0(\chi_0 t_p t_2). \tag{26}$$

Here $L = \sqrt{2 < r_g > z_p}$, $\chi_0 = 2k_0 < r_g >$, $\chi_S = 2k_0\sqrt{< \tilde{r}_g^2 >}$, $t_{1,2} = |\mathbf{t}_{1,2}|$,

$$D_S^\omega = \frac{\chi_S^2}{4}\left[\ln\frac{t_2}{t_1} + \delta \ln\frac{4}{t_1 t_2}\right]^2. \tag{27}$$

In the case of sufficiently weak scattering $\chi_S^2 \ll \chi_0$, $\chi_0 \gg 1$, using (26), we go to the known expression [6,7]

$$q(z_p, t_p) \approx \pi\chi_0 J_0^2(\chi_0 t_p). \tag{28}$$

An alternative form of this result for $\chi_S \ll \chi_0 \ll \chi_S^2$ (or $\theta_S^2 \ll \theta_g\theta_d$, where $\theta_S^2 =< \tilde{r}_g^2 > /L^2$, the diffraction angle on the aperture $\theta_d = 1/k_0 L$ and the ray deviation angle in undisturbed gravity field $\theta_g = 2 < r_g > /L$) becomes

$$q(z_p, t_p) \approx \chi_S^{-1}\chi_0^{3/2} J_0^2(\chi_0 t_p). \tag{29}$$

4.Scattering by gravitation field small-scale inhomogeneities. The Born approximation.

We start from the wave equation[6]

$$\triangle\, \mathbf{E} + k_0^2 n_g^2(\mathbf{r})\mathbf{E} + \nabla(\mathbf{E} \nabla \ln n_g) + [\nabla \ln n_g, rot\mathbf{E}] = 0, \tag{30}$$

in which the effective dielectric and magnetic permittivity tensors $\varepsilon_{\alpha\beta} = \mu_{\alpha\beta} = n_g(\mathbf{r})\delta_{\alpha\beta}$ ($\delta_{\alpha\beta}$ -Kronecker symbols) and $n_g = 1 + \frac{r_g}{r}$ consists of an average $< n_g >$ and a chaotic $\tilde{n}$ components.

In order to analyse this kind of scattering we can neglect two last terms in (30) which describe the polarization effects and, accordingly, we may deal with the Helmholtz equation

$$\triangle\, u + k_0^2(1 + 2\frac{< r_g >}{r})u = -2k_0^2 \tilde{n} u, \tag{31}$$

whose the Born solution has the form

$$u(\mathbf{r}) = -2k_0^2 \int d\mathbf{r}' G(\mathbf{r}, \mathbf{r}')\tilde{n}(\mathbf{r}')u_0(\mathbf{r}'), \tag{32}$$

$u_0(\mathbf{r})$ is an initial field and $G(\mathbf{r}, \mathbf{r}')$ is the Green function.

where $D = 4Z/kl^2$, $l \sim K_{min}^{-1}$. From (13) and (18) the estimation for frequency coherence bandwidth Ω_{fc} follows:$\Omega_{fc} = c/z\theta_S^2 \sim 10^7 c^{-1}$. Here $z = 10^{25} m$, the effective scattering angle $\theta_S \sim S_0\lambda/2\pi l \sim 10^{-12}, S_0 \sim 10^6$. Apparently, the pulse spreading duration can be obtain as $\tau \sim \Omega_{fc}^{-1}$.

3.Rough gravitational lense.

Let us consider a case when a gravity potential is created by spherical symmetric body of finite volume V and the mass density has a random component $\tilde{\mu}$ inside it.

Then, the correlation function Γ_μ depended both on difference $\mathbf{r} = \mathbf{r}_1 - \mathbf{r}_2$ and mean coordinate $\mathbf{R} = (\mathbf{r}_1 + \mathbf{r}_2)/2$ can be expressed as

$$\Gamma_\mu(\mathbf{R}, \mathbf{r}) =< \tilde{\mu}^2 > e^{-\frac{R^2}{R_0^2}} \gamma_\mu(\mathbf{r}), \tag{19}$$

where $< \tilde{\mu}^2 >$ is the strength of density fluctuations and R_0 is the focusing volume radius.

On the analogy with regular lense gravitational radius [1]:$r_g = 2Gm/c^2$, m is total mass of gravitating volume, we can introduce "average" and "random" gravitation radii[4]

$$< r_g >= \frac{2G}{c^2} \int < \mu > d\mathbf{r}, \tag{20}$$

$$\tilde{r}_g = \frac{2G}{c^2} \int \tilde{\mu}(\mathbf{r}) d\mathbf{r} \tag{21}$$

and define "the strength" of lense random gravitation radius as

$$< \tilde{r}_g^2 >= \frac{4G^2}{c^4} \int \int d\mathbf{R}\, d\mathbf{r} \Gamma\mu(\mathbf{R}, \mathbf{r}) \tag{22}$$

The structural function of the phase can be obtain as

$$D_S(z, \rho_{\perp 1}, \rho_{\perp 2}) = k_0^2 < \tilde{r}_g^2 > \ln^2 \frac{|\rho_{\perp 1}|}{|\rho_{\perp 2}|} \tag{23}$$

at distances $|\mathbf{r}_1|, |\mathbf{r}_2| \gg R_0$, where $R_0 \gg l$ and l is the characteristic scale of the function $\gamma_\mu(\mathbf{r})$. Therefore, the effective scattering angle of "rough" gravitational lense $\theta_S^2 = \frac{1}{2k_0^2} \Delta_{\perp \rho_\perp} D_S(\mathbf{R}_\perp, \rho_\perp)$ only depends on the mean transverse coordinate $\mathbf{R}_\perp = (\rho_{\perp 1} + \rho_{\perp 2})/2 : \theta_S^2 =< \tilde{r}_g^2 > /R_\perp^2$.

By Huygens principle the field $E_\omega(z_p, \rho_p)$ at the observation point $P(z_p, \rho_p)$ can be defined as [6]

$$E_\omega(z_p, \rho_p) = \frac{ik}{2\pi} \int E_{o\omega} e^{i\tilde{S}_\omega} \frac{e^{i\mathbf{k}\mathbf{R}}}{R} df, \tag{24}$$

where R is the distance from the element df of the wave surface up to the observation point, $E_{0\omega}$ is the field on wave surface and $\tilde{S}_\omega$ is a random phase change on the lense.

Then, for constant intensity of incident wave I_0 we obtain the two -frequency mutual coherence function and amplification coefficient of "rough" gravitational

To find the wave field in observation point $P(z,\rho_1)$ being sufficiently far distant from the layer of randomly gravitating masses we use the approximation of the phase screen which is placed in the plane $z=0$[6,7]. In this approximation

$$E_\omega(z,\rho_1)=\frac{ik}{2\pi z}\int E_{o\omega}e^{\frac{ik(\rho-\rho_1)^2}{2z}}e^{i\tilde{S}_\omega(\rho)}d\rho, \tag{15}$$

$E_{o\omega}$ is the incident wave amplitude, $\tilde{S}_\omega(\rho)$ is the random phase change in the screen. The two- frequency mutual coherence function $\Gamma_\omega(z,\rho_1,\rho_2) = < E_{\omega_1}(z,\rho_1)E^*_{\omega_2}(z,\rho_2) >$ can be expressed as

$$\Gamma_\omega(z,\rho_1,\rho_2)=\frac{k_1k_2}{4\pi^2z^2}\int\int d\rho' d\rho'' E_{o\omega_1}E^*_{o\omega_2}e^{\frac{i}{2z}[k_1(\rho'-\rho_1)^2-k_2(\rho''-\rho_2)^2]}\times$$
$$\times\left\langle \exp\left(i\left[\tilde{S}_{\omega_1}(\rho')-\tilde{S}_{\omega_2}(\rho'')\right]\right)\right\rangle \tag{16}$$

Since the contributions to the phase fluctuations come from many inhomogeneities in the gravitating masses layer the central limit theorem predicts that $\tilde{S}_\omega(\rho)$ has a Gaussian distribution and

$$< \exp\{i[\tilde{S}_{\omega_1}(\rho')-\tilde{S}_{\omega_2}(\rho'')]\} >= \exp\left\{-\frac{1}{2}\left\langle\left[\tilde{S}_{\omega_1}(\rho')-\tilde{S}_{\omega_2}(\rho'')\right]^2\right\rangle\right\}=$$
$$=\exp\left[-\frac{1}{2}D_S^\omega(\rho',\rho'')\right],$$

where $D_S^\omega=(1-\delta)^2D_S+4\delta^2S_0^2$, $\delta=(\omega_1-\omega_2)/2\omega_0$, $\omega_0=(\omega_1+\omega_2)/2$.

Using $\mathrm{R}=(\rho'+\rho'')/2$, $\rho=\rho'-\rho''$, $\mathrm{R}_0=(\rho_1+\rho_2)/2$, $\rho_0=(\rho_1-\rho_2)/2$ we arrive at

$$\Gamma_\omega(z,\mathrm{R}_0,\rho_0)=\frac{k_0^2(1-\delta^2)}{4\pi^2z^2}e^{i\frac{k_0}{z}}\left[\mathrm{R}_0\rho_0+\delta\left(R_0^2+\frac{\rho_0^2}{4}\right)\right]\times$$
$$\times\int\Gamma_{0\omega}e^{i\frac{k_0}{z}}\left[\mathrm{R}\rho-\mathrm{R}_0\rho-\mathrm{R}\rho_0+\delta\left(R^2+\frac{\rho^2}{4}-2\mathrm{R}\mathrm{R}_0-\frac{\rho\rho_0}{2}\right)\right]\times$$
$$\times\exp\left[-\frac{1}{2}D_s^\omega(\rho)\right]d\mathrm{R}d\rho \tag{17}$$

Substituting $\delta = 0$ in (17) one can show that the correlation function $\Gamma_2(z,\rho_1,\rho_2) =< E(z,\rho_1)E^*(z,\rho_2) >$ and the intensity $I =< E(z,\rho)E^*(z,\rho) >$ keep constant in passing behind the inhomogeneous layer: $\Gamma_2(z,\rho_1,\rho_2) = I_0\exp\left[-\frac{1}{2}D_S(\rho_1-\rho_2)\right]$, $I = I_0$ in the considered case of a plane wave incidence.

From here the evident result follows: an infinitely elongated layer of randomly gravitating masses does not induce the focusing of electromagnetic waves.

In the case of strong phase fluctuations ($S_0^2 \gg 1$) using the first term of the series (12) we have

$$\Gamma_\omega(z,\rho)=e^{-2\delta^2S_0^2}\frac{1+i\delta DS_0^2}{1+\delta^2D^2S_0^4}\exp\left[-\frac{S_0^2(1-\delta^2)\rho^2}{(1+\delta^2D^2S_0^4)l^2}\right]\times$$
$$\times\exp\left[-i\frac{\delta DS_0^2(1-\delta^2)}{1+\delta^2D^2S_0^4}\frac{\rho^2}{l^2}\right] \tag{18}$$

It seems natural that

$$G(\mathbf{r},\mathbf{r}') \approx G_0(\mathbf{r},\mathbf{r}') = -(\frac{1}{4\pi})\frac{\exp(ik|\mathbf{r}-\mathbf{r}'|)}{|\mathbf{r}-\mathbf{r}'|} \tag{33}$$

if the observation point $P(\mathbf{r})$ moves away from the scattering region and $\tilde{n} \ll 1$. In this case an average intensity $< I(\mathbf{r}) >$ may be written in traditional form[5]

$$< I(r) >= \frac{\pi k_0^4 I_0}{2r^2} \int d\mathbf{R} M(\mathbf{R}) \Phi_\varepsilon(\mathbf{q},\mathbf{R}). \tag{34}$$

In (34) an initial plane wave intensity I_0 is constant and the cut-off function $M(\mathbf{R})$ characterizes a scattering volume.

Correspondingly, the scattering cross section $\sigma(\mathbf{q})$ depends on the spectrum Φ_ε as $\sigma(\mathbf{q}) = (\pi k_0^4/2)\Phi_\varepsilon(\mathbf{q})$ and the total (integrated over the whole solid angle) scattering cross section of unit volume $\sigma_t = \oint \sigma(\mathbf{q}) d\Omega, d\Omega = \sin\theta \sin\varphi$.

The spectral density of fluctuations $\tilde{\varepsilon} = 2\tilde{n}$

$$\Phi_\varepsilon(\mathbf{q},\mathbf{R}) = \frac{4}{(2\pi)^{3/2}} \int d\rho \Gamma_n(\mathbf{R},\rho) \exp(-i\mathbf{q}\rho) \tag{35}$$

where Γ_n can be expressed through the correlation function $\Gamma_n(\mathbf{r}_1 = \mathbf{R}+\frac{\rho}{2}, \mathbf{r}_2 = \mathbf{R}-\frac{\rho}{2})$ which is defined by (4).

Here the scattering angle $q(\mathbf{R}) = \mathbf{k}_S(\mathbf{R}) - \mathbf{k}_i(\mathbf{R})$, $q = |\mathbf{k}_S - \mathbf{k}_i| = 2k_0 \sin\frac{\theta_S}{2}$, θ_S is the angle between incident ($\mathbf{k}_i$) and scattered ($\mathbf{k}_S$) wave vectors.

Using (2) and (35), it is possible to arrive at the following expressions

$$\Phi_\varepsilon(\mathbf{q},\mathbf{R}) = (8\pi G/c^2)^2 \Phi_\mu(q_z, \mathbf{q}_\perp)/(q_z^2 + q_\perp^2) \tag{36}$$

and

$$\sigma_t = (32\pi^3 k_0^4 G^2/c^4) \int_0^{2\pi} d\varphi \int_{\theta_0}^{\pi} d\theta \sin\theta (2k_0 \sin\frac{\theta}{2})^{-4} \Phi_\mu(2k_0 \sin\frac{\theta}{2}) \tag{37}$$

In (37) in order to analyse the scattering at large angles the lower limit θ_0 is chosen by condition $2k_0 \sin\frac{\theta_0}{2} = K_0$ where K_0 is the corresponding cut-off scale of $\Phi_\mu(\mathbf{K})$. Apparently, $\tilde{\mu} = (c^2/2G) r_g \tilde{N}$,where r_g is the gravitational radius and $\tilde{N}$ is the random variation of the point-like scatterer density, then

$$\Phi_\mu(\mathbf{k}) = (c^2/2G)^2 r_g^2 \Phi_N(\mathbf{k}) \tag{38}$$

To evaluate the cross section σ_t we introduce the spectrum $\Phi_N(\mathbf{k})$ in the form

$$\Phi_N(\mathbf{K}) = C_N^2 K^{-p} \exp(-\frac{K^2}{K_{max}^2} - \frac{K_{min}^2}{K^2}), \tag{39}$$

with outer $L_0 \sim K_{min}^{-1}$ and inner $l_0 \sim K_{max}^{-1}$ turbulence scales.

Making a change of variable $t = \sin\frac{\theta}{2}$ in equation (37) we have

$$\sigma_t = 32\pi^4 k_0^3 r_g^2 C_N^2 \int_{K_0/2k_0}^{1} dt (2k_0 t)^{-3-p} \exp\left[-(2k_0 l_0 t)^2 - \frac{1}{(2k_0 L_0 t)^2}\right]. \tag{40}$$

It should be noted that always $l_0 K_0 \ll 1$, $K_0 \ll 2k_0$ and $K_{min}/K_{max} = l_0/L_0 \ll 1$ and then further calculations yield the following results:

$$\sigma_t \approx 8\pi^4 \Gamma(p/2+1) k_o^2 r_g^2 C_N^2 L_0^{p+2} \tag{41}$$

if $l_0 K_0 \ll 2k_0 l_0 \ll 1$ and

$$\sigma_t \approx 8\pi^4 \Gamma(p/2+1) k_o^2 r_g^2 C_N^2 L_0^{3p/2+3} l_0^{-p/2-1} \tag{42}$$

if $l_0 K_0 \ll 1 \ll 2k_0 l_0$.

Also we define C_N^2 on analogy (11).

Finally, we have for $l_0 \ll \lambda$

$$\sigma_t \approx 4\pi^3 \frac{\Gamma(p/2+1)}{\Gamma((3-p)/2)} k_o^2 r_g^2 < \tilde{N}^2 > \begin{cases} (L_0/l_0)^p L_0^2 l_0^3, \; p < 3 \\ L_0^5, \; p > 3 \end{cases} \tag{43}$$

and, alternatively, for $l_0 \gg \lambda$

$$\sigma_t \approx 4\pi^3 \frac{\Gamma(p/2+1)}{\Gamma((3-p)/2)} k_o^2 r_g^2 < \tilde{N}^2 > \begin{cases} (L_0/l_0)^{3p/2} L_0^3 l_0^2, \; p < 3 \\ (L_0/l_0)^{p/2} L_0^6 l_0^{-1}, \; p > 3 \end{cases} \tag{44}$$

Unfortunately, the analysis of large-angle scattering (Eq.44) can be nowadays considered something like the play upon formulas or the imagination because,as it is easy to show,outside of black holes ($< n_G > -1 \ll 1$) the scattering signals are significant ($\sigma_t R \geq 1$, where R is the thickness of the layer) only for scattering angles $\theta_S < L_0/R$.It is the result of inside the gravitational objects regular refraction which is extremely large even when the ray departures from the axis of symmetry (where $n_G \simeq 1$) exceed L_0. It leads to interesting effects of ray concentration near this axis in the thick gravitational transparent medium.Moreover, these rays must pass through the medium along the axis of the symmetry because of the significant refraction for scattered rays.

5.Vacuum polarization lenses in strong magnetic fields.

Induced by strong magnetic field $\mathbf{B}_0$ the vacuum polarization also feeds the refraction index n. If, for instance, the wave propagating in the vacuum is linearly polarized and its electric field vector $\mathbf{E}$ is in the $(\mathbf{kB}_0)$ -plane, the correction to n takes the form[8]

$$\triangle n_{B_\perp} \equiv n_{B_\perp} - 1 \approx c_\perp \alpha (\frac{\hbar\omega_B}{m_e c^2})^2 \sin^2\theta \tag{45}$$

Here the coefficient $c_\perp = 7/90, \alpha$ is the fine structure constant, ω_B is the electron cyclotron frequency, m is the electron rest mass ($\hbar\omega_B \ll m_e c^2$) and θ is the angle between k and B_0. For $\mathrm{Ek} = \mathrm{EB}_0 = 0$ the refraction index $n_{B_\parallel}$ differs from $n_{B_\perp}$ only in coefficient $c_\perp$ which in this case equals $2/45$.

Thus, in contrast to gravitational correction to n, the magnetic field correction depends on wave polarization and the difference between $n_{B_\perp}$ and $n_{B_\parallel}$ may yield some polarization effects. In this sense the vacuum polarization phenomenon compares favourably with gravitational one where polarization effect is induced by the gyration and the correction to n_G is proportional $(r_g/r)(v/c)$.

But usually the gravitational correction Δn_G is much greater Δn_B. One can obtain that $\Delta n_B \geq \Delta n_G$ only if

$$[M/M_\odot] \leq 3.5 \cdot 10^2 (B_0/B_{00})^2 (R/R_\odot), \tag{46}$$

where $M_\odot$ and $R_\odot$ is the mass and radius of the Sun, and

$$B_{00} = \frac{m_e^2 c^3}{e\hbar} \approx 4 \cdot 10^{13} Gs. \tag{47}$$

It means even if the pulsar field equals 10^{13} at $R_\odot \approx 10^6$ on the right side we have only $5.6 \cdot 10^{-2}$. But if the small structure of $\mathbf{B}_0$ (with length-scale $l \ll R$) could exist the scattering angle $\theta_{SB} \sim < n_B^2 >^{1/2} \sqrt{R/l}$ could be pronounced.

It is interesting to note that the electromagnetic wave depolarization effect could help to evaluate the upper limit of B_0 near the source of emission.

In the case of rarefield medium the magnetic correction could be much greater gravitational one if there were the source of current which can induced the strong field $\mathbf{B}_0$ ($B_0 \sim 4\pi e N L$, where we put $v \sim c$ and denote L as a system dimension). This situation could take place in the black hole environments and near the Galactic nuclei.

6.Conclusions.

The theory of electromagnetic wave scattering in the random gravity field media based on the real and vague approximations has been discussed. The aim of this discussion is to make an emphasis on the interesting field of application of the wave propagation standard theory to somewhat new problem for the professionals of waves in random media. We hope that some well known effects can play unexpected role.In particular, the gravity focusing of a strong pulsar radiation by a massive partner in the double star system could bring additional effects in the radio emissions by means of nonlinear effects in the plasma envelopes.The useful phenomena for gravitational lense parameter diagnostics can be observed if the plasma irregularity lense effects are also take place.In the case of drifting plasma lenses we can expect the temporal variations of various parts of gravity focused sources as a result of the plasma irregularities pass through the various rays which come to an observer.

We can also expect the wave conversion in the pulsar plasma due to the vacuum polarization irregularities in the strong magnetic fields. It will be interesting to find the objects where the vacuum polarization effect could exceed the gravity one. Also,the interesting effects can be found in the case of the gravity waveguides in the transparent gravity systems. Maybe in the future we'll be able to look at wave scattering effects due to "micro black holes" not only as at unreal and unobserved inventions.

Certainly, the theory of the random gravitational lenses needs the development to evaluate some effects which take also place in usual plasma media.Particularly, one of them could show the effects of caustic influences on the wave scattering in the thick layer with gravity irregularities.

References

1.Zel'dovich Ya.B.,"Observations in a universe, which is homogeneous the mean", *Astronomich.Zh.* 41(1), 19-24 (1964).

2. Fok V.A.,"The theory of space, time and gravity", Moscow: Fizmatgiz (1961).

3.Landau L.D. and E.M.Lifshits,"Field theory", Moscow: Nauka (1967).

4.Erukhimov L.M. and P.I.Shpiro, "On electromagnetic wave propagation in random gravitation fields",*Preprint NIRFI,* 184, Gorky (1984).

5.Rytov S.M.,Yu.A.Kravtsov and V.I.Tatarskii, "Principles of statistical radiophysics", Berlin: Springer (1989).

6.Bliokh P.V. and A.A.Minakov, "Gravitational lenses" Kiev: Dumka (1989)
7.Bliokh P.V. and A.A.Minakov, "Diffraction of light and lens effect of the stellar gravitational field",*Astrophys. and Space Sci.* 34(2), L7-L9 (1975).
8.Berestetskii V.B.,E.M.Lifshits and L.P.Pitaevskii, "Quantum electrodynamics", Moscow: Nauka (1989).

Section 5

New Theoretical Approaches in Wave Propagation in Random Media

Multi-Phase-screen Analysis

B.J. Uscinski

Department of Applied Mathematics and Theoretical Physics
Silver Street, Cambridge CB3 9EW, U.K.

1. Introduction

The randomly modulated phase changing screen is an important building block in the study of wave propagation in random media. It has given rise to new and useful concepts in the field and helped in providing the first solutions to some of the more important problems involving multiple scattering.

The concept of an angular spectrum of plane waves, which has proved so useful in describing wave fields in random media, arose directly from the introduction of the single shallow phase screen as approximating a thin irregular layer. Sets of many such screens were subsequently used to describe an extended volume of random medium. This model gave the first analytical expressions for the directional spectrum and complex coherence of a multiply-scattered wave field. Subequently a similar model provided the first expressions for the time spread function of a multiply-scattered pulse.

Finally, the multiple-phase screen model played a central role in interpreting the results of the first analytical solutions of the fourth-moment equation and has also provided a connection between solutions of the moment equations and the results of the path-integral methods. This note aims to place the random phase screen in historical perspective and examine the physical significance of modelling an extended random medium as a set of phase screens.

2. The physical picture and early work

The single phase screen

One of the earliest problems that provided an incentive to study wave propagation in random media was the ''fading" of radio signals reflected from the earth's ionosphere. These irregular fluctuations in signal amplitude were due to the modulations imposed by random structures in the two main ionospheric layers, the E and F layers. Signals from distant radio stars, discovered only a few years previously, also exhibited the same random variations, and it was in order to explain these that the concept of the randomly modulated phase screen was introduced (Fig. 1).

The ionospheric layers caused very little absorption of radio waves at frequencies higher than about 10MHz, the main effect being to produce an irregular time delay over the wave-front. Compared with the distance to the source the layers were of negligible thickness, and so were modelled by a plane that modulated the phase of the wave in a random manner over the wave-front. This device was called a ''random phase screen".

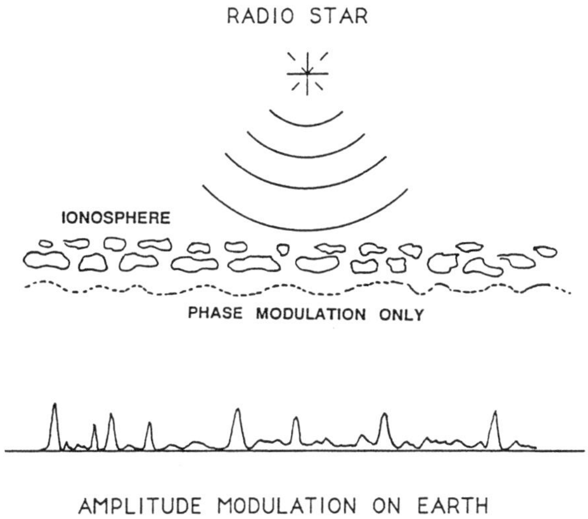

Fig. 1. Ionosphere produces phase fluctuations in the signal from a radio star. These become amplitude fluctuations on the Earth.

The r.m.s. phase fluctuation imposed by the ionosphere ϕ_o was usually a good deal less than a radian at the megahertz frequencies and so the term Shallow Phase Screen was adopted for $\phi_0 \ll 1$. Much of the early work was concerned with the shallow phase screen. Radiophysicists Booker and Clemmow[1] showed that the field emerging from the layer or ''screen" could be represented as a set of plane waves with normals directed over a narrow spread of angles. The spectrum giving the power in the plane wave components as a function of the angles of the normals became known as the ''angular spectrum" (Fig. 2).

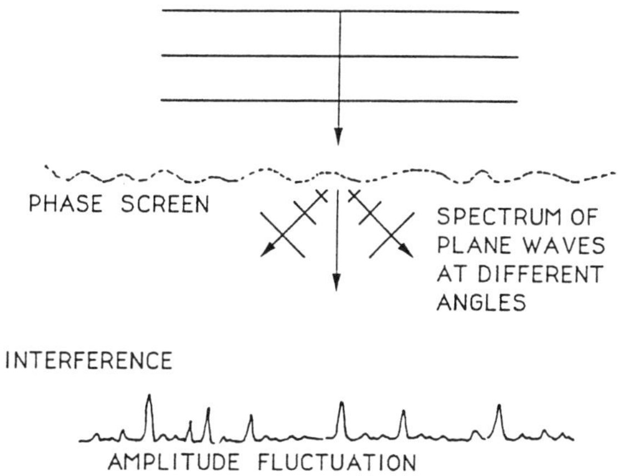

Fig. 2. Field modulated by the phase screen represented as a set or spectrum of plane waves travelling at different angles.

Booker et al[2] showed that the Fourier Transform of the angular spectrum gave the spatial autocorrelation function of complex amplitude of the wave-field beyond the screen. This is the complex second moment and has since come to be known as the "complex coherence" of the random wave-field. Booker et al[2] also noted that this complex correlation remained unaltered as it propagated away from the screen provided that a plane wave was incident on the screen. This meant that observations on the ground could give the field correlation at the screen and hence the autocorrelation of phase variations in the ionospheric layer.

The shallow phase screen continued to be used extensively by workers in the fields of ionospheric wave propagation and radio-astronomy. It provided a theory to describe not only the complex coherence of the field beyond an ionospheric layer, but also the gradual build-up of intensity fluctuations as the radio wave propagated from the layer to the ground (Hewish[3]).

Theoretical interest in the phase screen spread to include screens with very large phase variances $\phi_0^2 \gg 1$. Such screens provided a realistic model for ionospheric layers in cases when the phase modulation was large. Such screens came to be known as "deep phase screens" and theory describing wave propagation beyond them presented and continues to present many difficult and challenging problems. Mercier[4] in 1962 was the first to determine the characteristics of the intensity fluctuations beyond a deep phase screen. This question is still an important topic of research even today.

Multiple scattering

Some of the earliest treatments of multiple wave scattering in a random medium were carried out using the phase-screen concept.

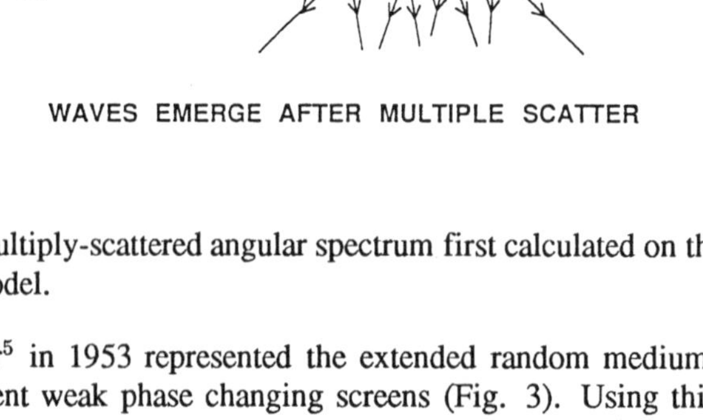

Fig. 3. Multiply-scattered angular spectrum first calculated on the basis of a multi-phase-screen model.

Fejer[5] in 1953 represented the extended random medium as a set of statistically independent weak phase changing screens (Fig. 3). Using this model, he was able to derive the first analytical expressions for the angular spectrum and spatial autocorrelation function of a multiply scattered wave-field. Williamson[6] in 1972 used a similar approach to derive the first analytical expressions for a pulse that has been broadened by multiple scattering in a random medium (Fig. 4).

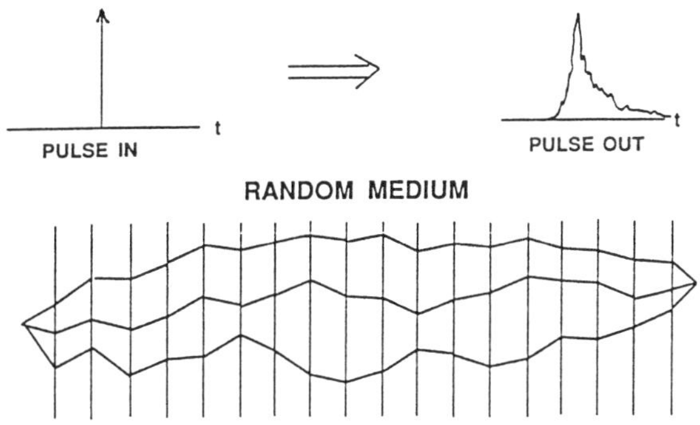

Fig. 4. Pulse broadening by scatter first calculated on the basis of a multi-phase-screen model.

Both of these results, which were important firsts in their field, were later confirmed by several different methods. Although the model of an extended random medium as a set of statistically independent phase screens was only given a firm theoretical basis many years later, it is clear that this device played a key role in the development of both single and multiple scattering theory.

3. The moment equations

The multi-phase-screen model for an extended random medium can be justified by considering the parabolic form of the wave equation

$$\frac{\partial u}{\partial z} = \frac{-i}{2k}\nabla_T^2 u - ikn(\mathbf{r}, z)u \quad . \tag{1}$$

The right hand side of (1) shows that the propagating wave-field is modified by two effects, distance in space and scattering. Here $\mathbf{r} = (x, y)$.

Distance effect

If we retain only the first term on the right hand side of (1) the resulting equation describes the forward spreading of the field from a point source in free space. Its solution is the Green's function for such a source situated at (x', y', z')

$$u = \frac{1}{|z - z'|} \exp\left\{-ik\frac{(x - x')^2 + (y - y')^2}{2|z - z'|}\right\} \quad . \tag{2}$$

Scattering effect

If only the second term on the right hand side is retained the resulting equation describes the effect of the random refractive index of the medium. Its solution

$$u = \exp\left\{-ik\int^z n(\mathbf{r}, z')dz'\right\} \tag{3}$$

is a phase modulation equal to the integrated optical path in the propagation direction.

The early idea of the multiple phase-screen model makes use of the assumption that these two effects could be applied separately and in succession. The medium is split into a set of parallel slabs. The phase modulation (3) due to the slab is assigned to a screen situated at the centre of the slab. The field, phase modulated by the screen, is propagated using (2), to the next screen and so on. The procedure would seem reasonable provided that the slabs are made thin enough. In this case it may be necessary to take correlation between the screens into account. Algorithms for integrating (1) numerically operate in this way. However if the screens are statistically independent, as in the early approach, then each elementary slab must be no thinner than the correlation length of irregularities in the medium.

The equivalent phase screen

Such independent elementary screens can be constructed provided that two conditions are satisfied.

The first condition is that the phase modulation (scattering) caused by a single irregularity of scale size L be small

$$k\mu L \ll 1 \ . \tag{4}$$

Here μ^2 is the variance of $n(\mathbf{r}, z)$,

$$\mu^2 = < n^2(r, z) > \ , \tag{5}$$

and we can write

$$n(r, z) = \mu n_1(r, z) \ , \tag{6}$$

where n_1 now has zero mean and variance unity. Condition (4) means that multiple scattering within an elementary slab can be neglected and that the equivalent phase screen causes only single scatter. The second condition is that the slab thickness Δz be greater than the scale size L

$$L < \Delta z \ . \tag{7}$$

This means that the screens, which are formed by projection of the contents of each slab, are statistically independent of each other. The screen formed by projection produces the following phase modulation (Fig. 5)

$$\phi(\mathbf{r}) = -k\mu \int_z^{z+\Delta z} n_1(\mathbf{r}, z')dz' \tag{8}$$

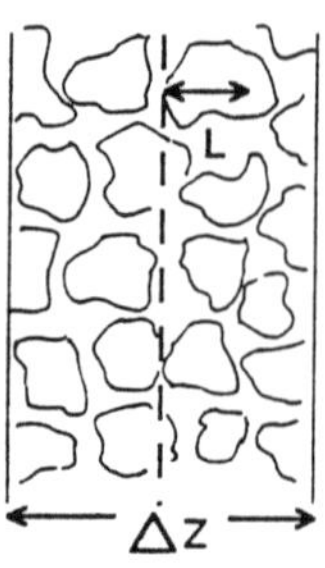

Fig. 5. Slab of thickness $\Delta z > L$ projects on to a central plane to give an equivalent phase screen with phase $\phi(\mathbf{r})$, (8).

The Transverse or Projected Autocorrelation Function

The equivalent phase screen now has the autocorrelation function

$$< \phi_1\phi_2 >= k^2\mu^2 \int_z^{z+\Delta z} \int_z^{z+\Delta z} < n_1(\mathbf{r}_1, z')n_1(\mathbf{r}_2, z'') > dz'dz'' \quad , \tag{9}$$

$$z' - z'' = \zeta \quad , \quad z' + z'' = 2Z \quad . \tag{10}$$

This involves the three-dimensional autocorrelation function of the medium integrated in the direction of propagation. The integral with respect to the difference variable ζ can be taken with infinite limits since the autocorrelation falls to zero over the scale size L, which is less than the slab thickness Δz, and (9) becomes

$$< \phi_1\phi_2 >= k^2\mu^2\Delta z\sigma(r_1 - \mathbf{r}_2) \quad , \tag{11}$$

$$\sigma(\mathbf{r}_1 - \mathbf{r}_2) = \int_{-\infty}^{\infty} \rho(r_1 - \mathbf{r}_2, \zeta)d\zeta \quad . \tag{12}$$

Here ρ is the three-dimensional autocorrelation function of the medium and σ, the so-called transverse or projected autocorrelation function, gives the phase autocorrelation of the equivalent screen. We note that σ has a different form from that of ρ.

The random medium as a set of phase screens

By dividing into slabs and projecting we can represent an extended medium as a set of phase screens (Fig. 6). The screens are statistically independent since the slab thickness is greater than the correlation length. Thus the screens are sometimes said to have a ‘‘Delta-function" correlation in the direction of propagation. Of course the medium is actually correlated in this direction but such correlation has now been taken into account by the integration (12).

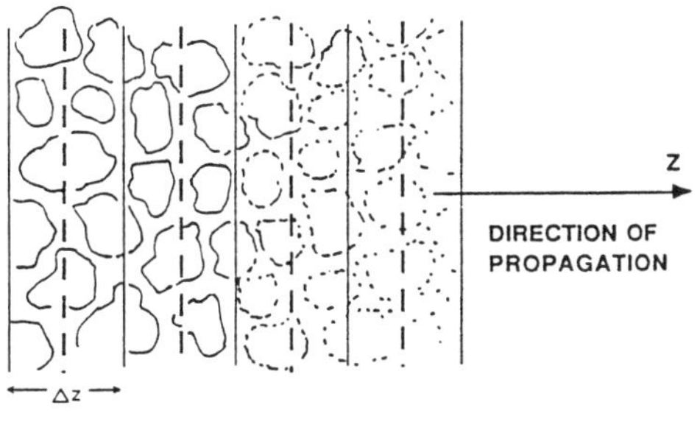

Fig. 6. Extended random medium becomes a set of independent (delta-correlated) phase screens.

Derivation of the moment equations

The parabolic-moment equations can easily be derived using the multi-screen model. We illustrate the method using a concrete example of the second moment

$$m_2(\xi, \eta, z) =< u(x+\xi, y+\eta, z)u^*(x, y, z) > \quad . \tag{13}$$

Let $u_i(z)$, the field with transverse coordinates (x_i, y_i), be incident on a slab of thickness Δz and let $u_i(z+\Delta z)$ be that which emerges.

The Scattering Effect

From (3)

$$u_1 u_2^*(z+\Delta z) = u_1 u_2^*(z) \times$$
$$\exp\left\{-ik \int_z^{z+\Delta z} [n(\mathbf{r}_1, z')dz' - n(\mathbf{r}_2, z'')dz'']\right\} \quad . \tag{14}$$

Expand the left hand side of (14) in a Taylor series in Δz, expand the exponential on the right hand side, take the ensemble averaging of both sides and retain the leading non-vanishing term to get

$$\frac{\partial m_2}{\partial z}\Delta z = -[< \phi^2 > - < \phi_1 \phi_2 >]m_2 \quad , \tag{15}$$

remembering that u and ϕ are statistically independent. Use of (11) in (15) gives, for the contribution due to scattering

$$\frac{\partial m_2}{\partial z} = -k^2 \mu^2 [\sigma(0) - \sigma(\mathbf{r}_1 - \mathbf{r}_2)]m_2 \quad . \tag{16}$$

The Distance Effect

Differentiating $u_1 u_2^*$ with respect to z we have

$$\frac{\partial u_1 u_2^*}{\partial z} = u_2^* \frac{\partial u_1}{\partial z} + u_1 \frac{\partial u_2^*}{\partial z} \quad . \tag{17}$$

Use of the first term on the right hand side of the parabolic wave equation (1) for $\frac{\partial u_i}{\partial z}$ gives for the distance effect contribution

$$\frac{\partial m_2}{\partial z} = \frac{-i}{2k}(\nabla_{T1}^2 - \nabla_{T2}^2)m_2 \quad . \tag{18}$$

Addition of the two effects (16), (18) gives the parabolic equation for the second moment

$$\frac{\partial m_2}{\partial z} = \frac{-i}{2k}(\nabla_{T1}^2 - \nabla_{T2}^2)m_2 - k^2\mu^2[\sigma(0) - \sigma(\mathbf{r}_1 - \mathbf{r}_2)]m_2 \quad . \tag{19}$$

All the other parabolic moment equations can be obtained in the same way.

IV Solution of the Moment Equations

The various moments of a field in a random medium can be found directly from the multi-screen model. They coincide with the solutions of the moment equations.

(a) The first moment $< u >$

The field of a unit amplitude plane wave is phase modulated on passing through a single screen

$$u\,(\text{single}) = \exp\{i\phi\} \quad . \tag{20}$$

Taking the ensemble average gives

$$< u\,(\text{single}) >= \exp\{-\frac{1}{2} < \phi^2 >\} \quad . \tag{21}$$

The result (21) does not necessarily assume that the medium has Gaussian statistics, but simply that the phase change introduced by an elementary layer is small. The process is simply repeated for each successive screen since they are statistically independent, to give for N screens

$$\begin{aligned} < u > &= \exp\{-N\frac{1}{2} < \phi^2 >\} \\ &= \exp\{-\frac{1}{2}k^2\mu^2\sigma(0)N\Delta z\} \\ &= \exp\{-\frac{1}{2}k^2\mu^2\sigma(0)z\} \quad . \end{aligned} \tag{22}$$

This is exactly the solution obtained by solving the first moment equation for an incident plane wave.

(b) The second moment

The second moment is obtained in exactly the same way. Multiplying the fields passing through a single screen and taking the ensemble average gives

$$\begin{aligned} < u_1 u_2^*\,(\text{single}) > &=< \exp\{i[\phi(\mathbf{r}_1) - \phi(\mathbf{r}_2)]\} > \\ &= \exp\{-[< \phi^2 > - < \phi_1\phi_2 >]\} \quad . \end{aligned} \tag{23}$$

Multiplying the results for N independent screens gives

$$\begin{aligned} < u_1 u_2^* > &= \exp\{-N[< \phi^2 > - < \phi_1\phi_2 >]\} \\ &= \exp\{-k^2\mu^2[\sigma(0) - \sigma(\mathbf{r}_1 - \mathbf{r}_2)]N\Delta z\} \\ &= \exp\{-k^2\mu^2 z[\sigma(0) - \sigma(\mathbf{r}_1 - \mathbf{r}_2)]\} \quad , \end{aligned} \tag{24}$$

which coincides with the solution of the second-moment equation (19) for an incident plane wave.

(c) The Fourth Moment

Construction of the fourth-moment

$$m_4 =< u_1 u_2^* u_3 u_4^* > \tag{25}$$

is slightly more complicated. It is easier to consider $M_4(\nu_a, \nu_b, z)$, the Fourier Transform of $m_4(\xi_a, \xi_b, z)$. Here ξ_a, ξ_b are two separations in the transverse x direction. The y direction is omitted. For a single screen the expression for M_4 is well known

$$M_4\,(\text{single}) = \frac{1}{(2\pi)^2}\int\int \exp\{-[1-g(\xi_a,\xi_b)]\}\exp\{-i(\nu_a\xi_a+\nu_b\xi_b)+i\nu_a\nu_b\Delta z\}d\xi_a d\xi_b \qquad (26)$$

where

$$g(\xi_a, \xi_b) = [\sigma(\xi_a) + \sigma(\xi_b) - \frac{1}{2}\sigma(\xi_a + \xi_b) - \frac{1}{2}\sigma(\xi_a - \xi_b)]/\sigma(0) \ . \qquad (27)$$

The result (26) for one screen is convolved with the next screen, and so on for N screens to give[7]

$$M_4(\nu_a, \nu_b, z) = \frac{1}{(2\pi)^{N+1}}\int\int$$

$$\exp\left\{-\sum_{j=1}^{N}\{[1 - g(\xi_{a_j}, \xi_b + Q_j)] - i\xi_{a_j}(\nu_{a_j} - \nu_{a_j-1}) - i\xi_b\nu_b\}\right\}$$

$$d\xi_b;\, d\xi_{a_1}\ldots d\xi_{a_N} d\nu_a \ldots d\nu_{a_{N-1}} \ , \qquad (28)$$

$$Q_j = (\nu_a + \nu_{a_{N-1}} + \ldots + \nu_{a_j})\Delta z \ . \qquad (29)$$

Expression (28) can also be obtained by integrating the fourth moment equation by the method of successive scatters. This result is almost an exact solution for the fourth moment. It has the disadvantage that it is a multiple convolution and requires some work to evaluate. However, various analytical forms now exist that represent (28) to a high degree of accuracy and have been verified by comparison with numerical solutions of the fourth-moment equation.

(d) Path integrals and multi-screen model

The solution of the parabolic wave equation can be written as a path integral

$$u = \frac{-i}{2k}\int\ldots\int \exp\left\{-ik\int^{z}\left(\frac{1}{2}[\frac{dr(z')}{dz'}]^2 + n(r(z'), z')\right)dz'\right\} d\,(\text{paths}) \ . \qquad (30)$$

Inspection shows the *free space distance effect* as the first term in the exponential, and the *phase modulation* as the second term. The path integral has a well known interpretation as expressing all the micro-paths from one transverse plane to the next in the random medium (Fig. 7). The phase-screen model is exactly parallel to this picture, except that it has combined the effect of a number of these planes that are adjacent and correlated to form a *single-phase screen*. In the phase-screen model the adjacent screens are statistically independent.

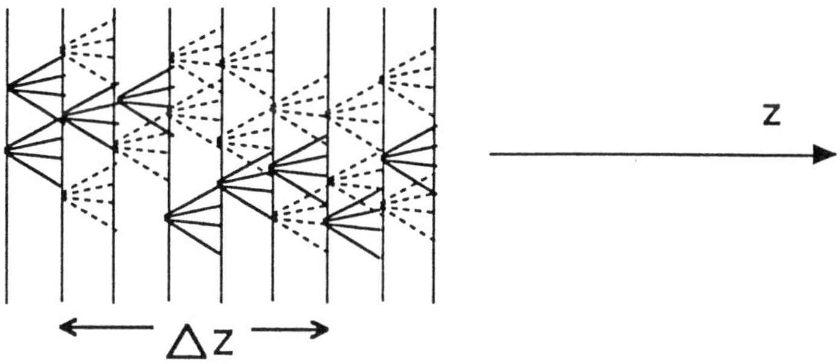

Fig. 7. the well known picture of micropaths in path-integrals. The screen model combines the paths in Δz.

V Numerical simulations and the multi-screen model

The principle of carrying out numerical simulations of a wave propagating in a random medium is well known[8]. The wave front is given a phase modulation corresponding to a thin layer of random medium. Its complex amplitude is then altered according to the law of free-space propagation within the layer thickness. The procedure is repeated from layer to layer. This is the familiar "split-step" algorithm, and has been discussed in some detail in section 3.

The present multi-screen model ensures the most efficient implementation of the method since all correlation in a layer is integrated beforehand to give a single screen. The model also assigns the correct transverse or projected autocorrelation to the screen.

The method has been extended to deal with point sources by writing the wave equation in circular coordinates and deriving the appropriate projected a.c.f. for circular screens. An example of such a point-source simulation is present here[9] (Fig. 8).

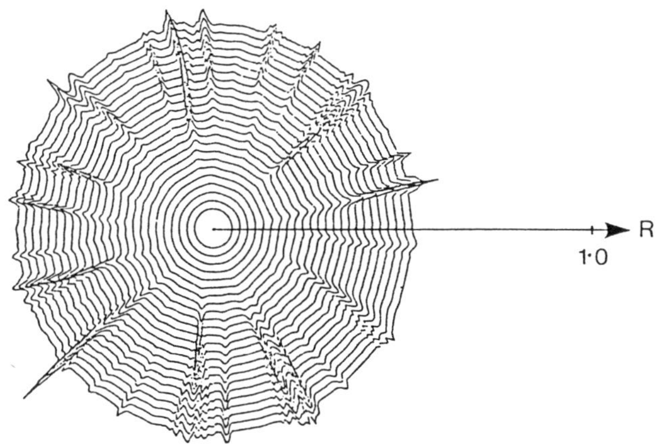

Fig. 8. Multi-phase-screen method extended to deal with a point source in circular coordinates.

The scintillation index obtained by point source simulations is compared with the index obtained by solving the fourth-moment equation (Fig. 9).

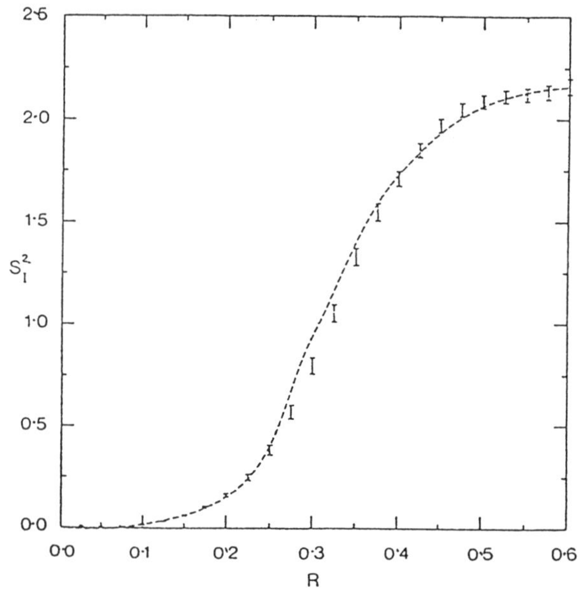

Fig. 9. Fourth moment obtained by numerical simulations compared with solution of the fourth-moment equation.

A notable feature of the random wave field is the existence of long high-intensity ribbons which were not predicted by the available average quantities such as the moments.

Summary

The multi-phase-screen model was one of the earliest to be used in multiple-scattering theory. In its present form it rationalises much of the early work and explains its success. The multi-phase-screen model gives us a clear physical picture of the processes involved in random wave propagation. It provides a simple way of deriving both the moment equations and their solutions.

Numerical simulations of wave scattering in a random medium are most efficiently carried out by applying the present multi-phase-screen model.

Finally, this model has given us many firsts in scattering theory -- it may well still turn out some winners.

References

1. H.G. Booker & P.C. Clemmow, *Proc. Inst. Elect. Engrs. Pt. III*, **97**, 11 (1950).

2. H.G. Booker, J.A. Ratcliffe & D.H. Shinn, *Phil. Trans. Roy. Soc.* A **242**, 579 (1950).

3. A. Hewish, *Proc. Roy. Soc.* A **209**, 81 (1951).

4. R.P. Mercier, *Proc. Camb. Phil. Soc.* A **58**, 382 (1962).

5. J.A. Fejer, *Proc. Roy. Soc.* A **220**, 455 (1953).

6. I.P. Williamson, *Mon. Not. Roy. Astronom. Soc.* **157**, 55 (1972).

7. B.J. Uscinski, *J. Opt. Soc. Am.* **A, 2,** No. 12, 2077 (1985).

8. C. Macaskill & T.E. Ewart, *IMA J. Appl. Math.* **33**, 1 (1984).

9. B.J. Uscinski, *Mod. Opt.* **36** No. 12, 1631 (1989).

Applications of the Two-scale Embedding Technique

M.J. Beran and S. Frankenthal, Tel Aviv University
R. Mazar, University of the Negev
A.M. Whitman, Villanova University

Abstract

The two-scale embedding technique is a systematic approach to the solution of the moment equations which describe the propagation of radiation in a random medium. This technique combines Fourier transforms, two-scale embedding and asymptotics. The moment equations are transformed into a hierarchy of first-order partial differential equations which can be solved iteratively. These lead to approximate solutions which produce a given coherence moment in terms of its prescribed distribution function in a source plane or aperture. Rays and spreading ray tubes emerge as characteristics of the first-order partial differential equation. The scattering enters as an integral of the appropriate scattering coefficients of the moment equations over a tube surrounding the main ray. The results converge to known solutions in the perturbation and saturation limits. The two-scale technique also permits the incorporation of caustic corrections and has been extended to situations where the main direction of propagation follows a curved ray.

In this article we summarize the two-scale embedding approach and point out that the iterative procedure may be constructed so that the zero-order approximation is the first term of a convergent series. Since the subsequent terms in the series are difficult to calculate, we show how approximate expressions for these terms may be derived. We pay principal attention to finding the scintillation index and correlation of intensities for quasi-monochromatic light propagation in a turbulent atmosphere. We discuss the difference between initial plane-wave and point-source boundary conditions and emphasize the importance of including a turbulent inner scale.

We shall discuss the solutions of the two-frequency, two-point moment equation. Here an exact solution is available for a quadratic scatterer and the results will be compared to the two-scale approach. Finally, we conclude with a formulation of the two-scale method on the field rather than on the ensemble-averaged moments.

1. Introduction and Moment Equations

The basic equation governing the quasi-monochromatic field $u(\mathbf{r},z)$, mean frequency ν, in the parabolic approximation is

$$2ik\partial u/\partial z + \nabla_T^2 u + k^2\mu(\mathbf{x})u = 0 \qquad (1)$$

where k is the mean wavenumber and $\mu(\mathbf{r},z)$ is the fluctuating part of the index-of-refraction $n(\mathbf{r},z)$. The transverse coordinates are denoted by the vector $\mathbf{r}$ (x,y) and the mean propagation direction by z. The subscript T on the Laplacian indicates that the Laplacian is with respect to the transverse coordinates.

To treat the problem from a statistical point of view, the coherence functions
$\langle\Gamma(\mathbf{r}_1,\mathbf{r}_2,z)\rangle=\langle u(\mathbf{r}_1,z)u^*(\mathbf{r}_2,z)\rangle$

$\langle\Gamma_4(\mathbf{r}_1,\mathbf{r}_2,\mathbf{r}_3,\mathbf{r}_4,z)\rangle=$
$\langle u(\mathbf{r}_1,z)u^*(\mathbf{r}_2,z)u^*(\mathbf{r}_3,z)u(\mathbf{r}_4,z)\rangle$

are defined. The brackets indicate an ensemble average over a sequence of realizations of the turbulent field. We assume that the statistics of the turbulent field are homogeneous and isotropic. The intensity $\langle I(\mathbf{r},z)\rangle = \langle\Gamma(\mathbf{r},\mathbf{r},z)\rangle$. The mean square intensity $\langle I^2(\mathbf{r},z)\rangle = \langle\Gamma_4(\mathbf{r},\mathbf{r},\mathbf{r},\mathbf{r},z)\rangle$ and the scintillation index $\sigma_I^2 = [\langle I^2\rangle/\langle I\rangle^2 - 1]$. The correlation of intensities
$\langle R_I(\mathbf{r}_1,\mathbf{r}_2,z)\rangle = \langle\Gamma_4(\mathbf{r}_1,\mathbf{r}_1,\mathbf{r}_2,\mathbf{r}_2,z)\rangle$

Governing equations have been derived for the functions $\langle\Gamma\rangle$ and $\langle\Gamma_4\rangle$ [1-5]. They are

$$\partial\langle\Gamma\rangle/\partial z - (i/2k)[\nabla_{T1}^2 - \nabla_{T2}^2]\langle\Gamma\rangle + k^2[\overline{\sigma}(0) - \overline{\sigma}(|\mathbf{r}|)]\langle\Gamma\rangle = 0 \qquad (2)$$

$$\partial\langle\Gamma_4\rangle/\partial z - (i/2k)\,[\nabla^2_{T1}-\nabla^2_{T2}-\nabla^2_{T3}+\nabla^2_{T4}]\,\langle\Gamma_4\rangle - k^2\,[\overline{\sigma}(r_{12})+\overline{\sigma}(r_{13})+\overline{\sigma}(r_{24})+ \overline{\sigma}(r_{34})-\overline{\sigma}(r_{14})-\overline{\sigma}(r_{23})-2\overline{\sigma}(0)]\,\langle\Gamma_4\rangle = 0 \quad (3)$$

Here r_{ij} = ABS($\mathbf{r}_i-\mathbf{r}_j$) and

$$\overline{\sigma}(r_{ij}) = (1/4)\int_{-\infty}^{\infty}\sigma(r_{ij},r_z)\,dr_z \quad (4)$$

where $\sigma(r_{ij},r_z) = \langle\mu(\mathbf{r}_1,z_1)\mu(\mathbf{r}_2,z_2)\rangle$ and $r_z = z_1-z_2$.

The solution of Eq. (2) has been readily found[6] but the solution of Eq. (3) has proved to very difficult even for an initial plane wave. To show the cause of the difficulty we transform Eq. (3) using the following coordinates

$$\begin{aligned} \boldsymbol{R} &= [\boldsymbol{r}_1+\boldsymbol{r}_2+\boldsymbol{r}_3+\boldsymbol{r}_4]/4 \\ \boldsymbol{s} &= [\boldsymbol{r}_1-\boldsymbol{r}_2+\boldsymbol{r}_3-\boldsymbol{r}_4]/2 \\ \boldsymbol{p} &= [\boldsymbol{r}_1+\boldsymbol{r}_2-\boldsymbol{r}_3-\boldsymbol{r}_4]/2 \\ \boldsymbol{q} &= [\boldsymbol{r}_1-\boldsymbol{r}_2-\boldsymbol{r}_3+\boldsymbol{r}_4] \end{aligned} \quad (5)$$

We find

$$\partial\langle\Gamma_4\rangle/\partial z-(i/k)\,([\nabla_R\cdot\nabla_q+\nabla_s\cdot\nabla_p]\,\langle\Gamma_4\rangle-k^2\overline{h}(\boldsymbol{s},\boldsymbol{p},\boldsymbol{q})\,\langle\Gamma_4\rangle=0 \quad (6)$$

where

$$\overline{h}(\boldsymbol{s},\boldsymbol{p},\boldsymbol{q}) = [\overline{\sigma}(|\boldsymbol{s}+\boldsymbol{q}/2|)+\overline{\sigma}(|\boldsymbol{s}-\boldsymbol{q}/2|)+\overline{\sigma}(|\boldsymbol{p}+\boldsymbol{q}/2|) +\overline{\sigma}(|\boldsymbol{p}-\boldsymbol{q}/2|)-\overline{\sigma}(|\boldsymbol{p}+\boldsymbol{s}|)-\overline{\sigma}(|\boldsymbol{p}-\boldsymbol{s}|)-2\overline{\sigma}(0)] \quad (7)$$

For an initial plane wave, there is no dependence in $\langle\Gamma_4\rangle$ on **R**. This is also true if the initial wave field is statistically homogeneous. The vector **q** is then simply a parameter in Eq. (6) and, for convenience, may be set equal to 0. Equation (6) is then of the form

$$\partial\langle\Gamma_4\rangle/\partial z-(i/k)\,\nabla_s\cdot\nabla_p\langle\Gamma_4\rangle-k^2\overline{h}(\boldsymbol{s},\boldsymbol{p})\,\langle\Gamma_4\rangle=0 \quad (8)$$

Equation (8) is very similar to Eq. (2) when Eq. (2) is written in terms of sum and difference coordinates. The difference is that in Eq. (2) the scattering term depends only on the difference coordinate while in Eq. (8) the scattering term depends on both transverse coordinates **s** and **p**. In Eq. (2) it is possible to take a Fourier transform with respect to the sum coordinate and obtain a first-order partial differential equation that may be solved by the method of characteristics. On the other hand taking a Fourier transform with respect to **s** or **p** in Eq. (8) yields an integro-partial differential equation.

The fact that $\bar{h}$ depends on **s** and **p** is the source of our difficulty in solving Eq. (8) and the two-scale method that has been developed is a way of overcoming the joint dependence. The problem was to find an iteration scheme that led to a convergent series. The motivation for the technique came from trying to solve for the intensity distribution of an acoustic signal in the neighborhood of a caustic, using the coherence equations[7]. In the caustic problem there is a natural small parameter $\epsilon = (kL)^{-2/3}$, where L is a characteristic scale of the sound profile. Using the smallness of ϵ we found an equation that was identical in form to Eq. (8). It was natural to introduce a two-scale embedding in the caustic problem since the sound-speed term contained the variable **s** as a product $\epsilon\mathbf{s}$. Macaskill[8] then noticed the similarity of Eq. (8) to the caustic coherence equation and applied the method to Eq. (8).

Although the two-scale method appears cumbersome because of the number of Fourier transforms used before and after the embedding, these multiple transforms do not introduce approximations. Only in the final step of the method is an approximation introduced and it is shown that it leads to a convergent iteration procedure. In studying the caustic problem we tried to introduce the embedding method using fewer Fourier transforms, and hence different variables, but could not develop a workable iteration scheme.

In Sec. 2 we shall develop the two-scale method and show that it leads to a convergent iteration procedure. In Sec. 3 we shall discuss the lowest-order term of the iteration procedure and in Sec. 4 the next-order approximation. Results will be given for propagation of quasi-monochromatic light in a turbulent atmosphere. Initial plane-wave and point-source boundary conditions will be discussed and the importance of a turbulent inner scale be emphasized. Some discussion of a two-dimensional finite beam will also be given. In Sec. 5 the two-frequency problem will be considered and in Sec. 6 the two-scale approach will be applied to the field rather than the coherence functions.

2. Two-scale Method

To illustrate the two-scale method, we consider the solution of Eq. (8) with a plane-wave initial condition. We may obtain a similar solution for any initial boundary condition. For an initial laser beam we must, however, retain the gradient term with respect to **R** that appears in Eq. (6) and generalize the procedure. At the conclusion of the derivation for the plane-wave boundary condition we shall present solutions for other boundary conditions but not give the derivations. In addition, for simplicity of notation we shall present the details in two dimensions. The generalization to three dimensions is straightforward.

Using non-dimensional variables Eq. (8) can be made to depend upon a single parameter when the turbulence contains an inner scale (but no outer scale or a fixed outer scale). When the turbulence is described by a pure Kolmogorov structure function, the parameter may be absorbed into the non-dimensional length scales and does not appear. However, we cannot prove convergence of the iteration method for a pure power law structure function and we consider the pure Kolmogorov structure function only as a limiting case.

We use the following non-dimensional variables

$$z_1 = [1/(\epsilon^2 k l_m^2)]\, z$$
$$s_1 = [1/\epsilon l_m]\, s \qquad \textbf{(9)}$$
$$p_1 = [1/\epsilon l_m]\, p$$

where l_m is the inner scale of turbulence and

$$\epsilon = 1/[k^3 \overline{\sigma}(0)\, l_m^2]^{1/2} \qquad \textbf{(10)}$$

If we now take the Fourier transform of Eq. (8) with respect to p_1 we find, for the equation governing

$$M(s_1, \eta, z_1) = \int_{-\infty}^{\infty} \langle \Gamma_4(s_1, p_1, z_1) \rangle \exp[i p_1 \cdot \eta]\, dp_1 \qquad \textbf{(11)}$$

the result

$$\partial M/\partial z_1 - \eta\, \partial M/\partial s_1 = (1/2\pi) \int_{-\infty}^{\infty} H(\epsilon s_1, \omega)\, M(s_1, \eta - \epsilon\omega, z_1)\, d\omega \qquad \textbf{(12)}$$

where H is the Fourier transform of $h = \overline{h}/\overline{\sigma}(0)$ and we write $h(\epsilon s_1, \epsilon p_1)$.

We next embed Eq. (12) in a higher dimensional space $[s_1, s_2, \eta, \eta_2, z_1]$ where $s_2 = \epsilon s_1$ and $\eta_2 = \epsilon\eta$. This is a formal procedure and introduces no approximations. In the original work on caustics, the embedding was introduced because ϵ was a small parameter but here ϵ may have any magnitude. After embedding we find the equation

$$\partial M/\partial z_1 - \eta\, \partial M/\partial s_1 - \eta_2\, \partial M/\partial s_2 = (1/2\pi) \int_{-\infty}^{\infty} H(s_2, \omega)\, M(s_1, s_2, \eta - \epsilon\omega, \eta_2 - \epsilon^2\omega, z_1)\, d\omega \qquad \textbf{(13)}$$

We can eliminate the ϵ argument from the η argument of M and retain the same form of the equation by taking the Fourier transform of M with respect to s_1 and η in Eq. (13). The equation governing the transform

$$N(w, s_2, v, \eta_2, z_1) =$$

$$(1/2\pi) \int_{-\infty}^{\infty}\int_{-\infty}^{\infty} M(s_1, s_2, \eta, \eta_2, z_1) \exp[i(s_1 w - \eta v)]\, ds_1 d\eta \tag{14}$$

is

$$\partial N/\partial z_1 - w\partial N/\partial v - \eta_2 \partial N/\partial s_2 =$$

$$(1/2\pi) \int_{-\infty}^{\infty} \exp[-i\epsilon v\omega]\, H(s_2, \omega)\, N(w, s_2, v, \eta_2 - \epsilon^2\omega, z_1)\, d\omega \tag{15}$$

We notice that, if the ϵ^2 argument were not present in the function N appearing under the integral, that the integral could be easily evaluated. If ϵ is set equal to zero we find

$$\partial N/\partial z_1 - w\partial N/\partial v - \eta_2 \partial N/\partial s_2 - h(s_2, \epsilon v) N = 0 \tag{16}$$

The solution of Eq. (16), which we term N^o is the lowest-order approximation to Eq. (15). It was first obtained by Uscinski[9] by using a multiple-phase screen technique. We shall now show that Eq. (16) is the lowest-order term of a convergent iterative procedure. To do this we subtract the term $h(s_2, \epsilon v)N$ from both sides of Eq. (15). We find

$$\partial N/\partial z_1 - w\partial N/\partial v - \eta_2 \partial N/\partial s_2 - h(s_2, \epsilon v) N = F_N(w, s_2, v, \eta_2, z_1) \equiv$$

$$(1/2\pi) \int_{-\infty}^{\infty} \exp[-i\epsilon v\omega]\, H(s_2, \omega)$$

$$[N(w, s_2, v, \eta_2 - \epsilon^2\omega, z_1) - N(w, s_2, v, \eta_2, z_1)]\, d\omega \tag{17}$$

Here N is a function of w, s_2, v, η_2 and z_1. Using the method of characteristics, Eq. (17) can be written in the form of an integral equation

$$N(w, s_2, v, \eta_2, z_1) = N_o(w, s_2+\eta_2 z_1, v+wz_1, \eta_2, 0)\exp[P_h(w, s_2, v, \eta_2, z_1)] + \exp[P_h(w, s_2, v, \eta_2, z_1)] \int_0^{z_1} F_N[w, s_2-\eta_2(z_1'-z_1), v-w(z_1'-z_1), \eta_2, z_1'] \exp[-P_h(w, s_2, v, \eta_2, z_1')]\, dz_1' \tag{18}$$

where

$$P_h(w, s_2, v, \eta_2, z_1) = \int_0^{z_1} h[s_2-\eta_2(z_1'-z_1), \epsilon(v-w(z_1'-z_1))]\, dz_1' \tag{19}$$

Equation (18) is a Volterra equation of the second kind. Gozani[10a-d] has shown that the standard iteration procedure will lead to a convergent series provided that certain conditions are satisfied. The most important condition is that $\overline{\sigma}(0)$ is finite. This requires that the turbulence representation contain an outer scale. Of course, the series may converge even if the outer scale is equal to infinity but Gozani's proof is not valid unless the scale has a finite value. As we shall discuss below the size of the inner scale seems to have an important effect on the rate of convergence. The size of the outer scale appears to be unimportant[11].

The next-order approximation for N, N^1, is found from Eq. (18) where in the expression for F_N (see Eq. 17) N is replaced by N^o. This term is difficult to evaluate and in many previous calculations, as an approximation, the difference term in N is set equal to the first term of a Taylor series in ϵ^2 That is,

$$N(w,s_2,v,\eta_2-\epsilon^2\omega,z_1) - N(w,s_2,v,\eta_2,z_1) = -[\partial N(w,s_2,v,\eta_2,z_1)/\partial\eta_2]\epsilon^2\omega \tag{20}$$

The integral over ω in Eq. (17) may then be evaluated in terms of a derivative with respect to h.

From the function $N(w,s_2,v,\eta_2,z_1)$ we use an inverse transform with respect to w and v and find $M(s_1,s_2,\eta,\eta_2,z_1)$. Next set $s_2 = \epsilon s_1$ and $\eta_2 = \epsilon\eta$. The result is $M(s_1,\eta,z_1)$. The fourth-order coherence function Γ_4 is then found by taking a Fourier transform with respect to η. In the next section we discuss the result of approximating N by the lowest-order approximation N^o.

Finally, we mention that in some very recent work we have found that for the two-frequency quadratic scatterer that will be discussed in Sec. 5, it is useful to modify the expression we use for the zero-order approximation. In Eq. (15) we expand the function N in a Taylor series in ϵ^2. Instead of keeping only the term independent of ϵ in Eq. (16) we also retain the term proportional to ϵ^2 which contains a derivative of N with respect to η_2. The equation may still be solved by the method of characteristics but the characteristics have to be obtained numerically rather than analytically as is possible in Eq. (16). We plan to perform numerical calculations in the near future to determine if this new zero-order approximation gives results that are better than those reported in Sec. 4.

3. Lowest-order Approximation

In this section we discuss the results obtained by approximating N by N^o. The function N^o satisfies Eq. (16). The solution of Eq. (16) is the first term on the right-hand-side of Eq. (18). The solution is

$$N^{o}(w,s_2,v,\eta_2,z_1)=N_o(w,s_2+\eta_2 z_1,v+wz_1,0)\exp\left[\int_0^{z_1} h[s_2-\eta_2(z_1'-z_1),v-w(z_1'-z_1)]\,dz_1'\right] \quad (21)$$

Here $N_o(w,s_2+\eta_2 z_1,v+wz_1,\eta_2,0)$ is the function N at the beginning of the propagation. The distance of propagation is z_1.

3.1 Plane wave initial condition

For a plane wave $N_o = 2\pi\delta(w)$. The final expression for $<\Gamma_4>$ is

$$<\Gamma_4(s,p,z)>=(1/2\pi)\int_{-\infty}^{\infty}\int_{-\infty}^{\infty}\exp[i(y-p/(z/k)^{1/2})x]$$

$$\exp(k^2\overline{\sigma}(0)z\int_0^1 h[s/l_m-(z/kl_m^2)^{1/2}xt,y(z/kl_m^2)^{1/2}]\,dt)\,dxdy \quad (22)$$

The expression for $<\Gamma_4(\mathbf{s},\mathbf{p},z)>$ in three dimensions is identical in form to Eq. (22). The quantities s,p,x and y are replaced by vector quantities **s**,**p**,**x** and **y**. The two-dimensional integral over x and y is replaced by a four-dimensional integral, the $(1/2\pi)$ factor is replaced by $(1/2\pi)^2$ and the products xy and xp are replaced by dot products of **x** and **y** and **x** and **p**. The arguments within h must be treated as vector rather than scalar quantities. Finally, since the four-dimensional integral depends only on the magnitudes of the vectors **x** and **y** and the dot product between the vectors, the four-dimensional integral can be reduced to a three-dimensional integral. Details are given in Refs. 11 and 12.

Numerical solutions are available in both two and three dimensions for a Gaussian correlation function, the Kolmogorov structure function without an inner scale and the Kolmogorov structure function with an inner scale. It is not too difficult to evaluate Eq. (22) or its three-dimensional counterpart. Most solutions are for the mean square intensity (**s** = **0**, **p** = **0**) but some solutions are available when **s** or **p** is not equal to zero[13,14].

We shall mainly restrict our attention here to presenting results for the scintillation index σ_I^2 for the three-dimensional case using a Kolmogorov structure function with an inner scale. This is the case of most interest for propagation of light through a turbulent atmosphere. We shall, however, briefly discuss results for the correlation of intensities.

The inner scale may be included in a number of ways. The most correct way is to use the correlation function presented by Hill and Clifford[15]. However, this function has not been used extensively since it requires much more calculation than do simpler choices. Here, we present results obtained using a modification mentioned by Tatarski[1] (Chap. 1, Sec. 13, Eq. 44) as a convenient way to introduce an inner scale. In Figs. 1 and 2 we present a plot of the standard deviation σ_I (the square root of the variance) as a function of the distance scale $\sigma_o = 1.23k^{7/6}C_n^2z^{11/6}$. The structure constant is denoted by C_n^2. The inner scale parameter is γ_K which is defined as $\gamma_K = 1.37k^3C_n^2l_m^{11/3}$. The figures are a reproduction of Figs. 1 and 2 of Whitman and Beran[16]. The solid curve in each figure is the zero-order solution.

The zero-order solution gives the correct perturbation solution and approaches unity as σ_o approaches infinity. In between these limits, it has the correct qualitative behavior as has been observed in experiments[11]. Further, it predicts an increase in the peak value of the curve as the inner scale increases. A benchmark solution is necessary to determine the accuracy of the result.

Experimental data has too much scatter to permit a good evaluation of the accuracy of the zero-order solution. Hence most effort has focused on comparing the results with other theoretical and numerical calculations. Since the two-scale method is an approximation to the solution of the moment equation, it is desirable to compare the zero-order approximation with numerical solutions of Eq. (8). Unfortunately, there are, at present, no numerical solutions available for three dimensions, but only for two dimensions. There are, however, numerical simulations in three dimensions that have been obtained by replacing the continuous random medium by a series of phase screens[17].

Comparison of the zero-order solution to a numerical solution of Eq. (8) in two dimensions was performed by Gozani[18] for the case of a zero inner scale. The agreement between the solutions is very good for σ_I^2 and differs by only a few percent. The comparison is shown in Fig. 3 of Ref. 11. Tur[19] provided a numerical solution for a Gaussian correlation. The comparison is shown in Fig. 2a of Ref. 12. Here the agreement for σ_I^2 is not so good and the two solutions differ by almost 20% in the peak region.

For three dimensions, the comparison between the zero-order solution and the Martin and Flatte numerical simulation is shown in Figs. 1 and 2. Note that the ordinate is σ_I and not σ_I^2. We see that, while the zero-order solution is qualitatively in agreement with the numerical simulation result, there is significant disagreement in the peak region.

Solutions are also available for the correlation of intensities $\langle R_I(r,z)\rangle$ where r is the distance between two points in the transverse plane. In Fig. 3 we show a typical plot for a normalized form of $\langle R_I\rangle$, $\langle S_I\rangle$, defined as $\langle S_I(r,z)\rangle = [\langle R_I(r,z)\rangle - \langle R_I(0,z)\rangle]/\sigma_I^2$. The typical characteristic separation distance r_c is of order $(z/k)^{1/2}$. The function $\langle S_I(r,z)\rangle$ satisfies a conservation law and should go negative for a range of r. However, the calculation is not accurate enough to show the negative region. We have not found any published results with which to compare the two-scale results in three dimensions.

Comparisons for the correlation of intensities with results of Tur[10] in two dimensions are given in Ref. 12. These calculations are for a Gaussian correlation function and an asymptotic approximation was used in evaluating Eq. (22) since $\epsilon \ll 1$. The comparison of the zero-order approximation and Tur's numerical evaluation of Eq. (8) are satisfactory for most separation distances at all ranges.

It is important to point out that, although $\langle\Gamma_4(\mathbf{r},\mathbf{0},z)\rangle = \langle\Gamma_4(\mathbf{0},\mathbf{r},z)\rangle$, this relation is not true for the zero-order approximation. In all calculations that we have performed we have found that the two calculations give very similar results in the zero-order approximation. Taking $\mathbf{p} = \mathbf{r}$ does, however, give slightly better results than choosing $\mathbf{s} = \mathbf{r}$. Since the two-scale procedure is convergent, the final result after many iterations, should give identical results independent of the choice made.

Finally, we mention that calculations have been made for the scintillation index for a plane wave (light from a star) impinging on the atmosphere[20]. The structure constant C_n^2 is taken to vary with altitude and results are given as a function of the angle from the zenith.

3.2 Point source initial condition

The point source initial condition can be obtained from the plane wave case by replacing the y argument in the h function in Eq. (22). Instead of y we write y(1-t). The integrals in both two and three dimensions are not more difficult to evaluate than in the plane wave case. The results are qualitatively the same as in the plane wave case with the peak being shifted and the peak value being higher for the point source than for the plane wave. However, as we shall see, the zero-order approximation for the point source is not a good quantitative approximation of the correct solution. In Fig. 4 we show a comparison between the plane-wave and point-source solutions for the zero-order approximation using a Kolmogorov form with no inner scale[21]. In Fig. 5 we present a series of solutions for the zero-order approximation with various inner scales. In these graphs the range coordinate is

$$\zeta_K = 1.09\sigma_o^{12/11}$$

and the numbers on the curves in Fig. 5 correspond $\zeta = l_m/(z/k)^{1/2}$. The zero-order solutions may be compared to the numerical simulations of Martin and Flatte[22].

3.3 Finite beam

If the initial radiation is a finite beam we must solve Eq. (6) rather than Eq. (8) since the solution depends on **R**. The two-scale procedure may be carried through, but it leads to the calculation of four integrals even in the two-dimensional case. In order to simplify the calculation, Mazar and Bronshtein[23] have replaced a Gaussian finite beam with a directional source. They discuss the validity of the approximation and present results for various beam parameters.

4. First-order Correction

As we see from the above comparisons between the zero-order solution and numerical solutions of the moment equations, the next order correction must be included in order to obtain a very good agreement between the two types of results. Although in many practical applications the zero-order approximation may be adequate for the plane wave case with a Kolmogorov structure function, a correction is definitely required for a point source.

The correction term is obtained by using Eq. (18). Under the integral, the zero-order solution is used to evaluate the expression for F_N given in Eq. (17). Unfortunately, the resultant expression is very hard to evaluate and to our knowledge this correction has only been evaluated in two dimensions. For three dimensions the first-order correction has been evaluated approximately by replacing the difference term in Eq. (17) $N^o(w, s_2, v, \eta_2 - \epsilon^2\omega, z_1) - N^o(w, s_2, v, \eta_2, z_1)$

by the first-order term in a Taylor series. See Eq. (20). This approximation is not valid in general but it should be satisfactory when ϵ is very small.

4.1 Plane wave initial condition

In two-dimensions, adding the first order correction gives good agreement with numerical solutions of the moment equation using the approximation given in Eq. (20). The comparison with the Gaussian correlation function (ϵ=0.14) is shown in Fig. 7a of Ref. 12. For a pure Kolmogorov structure function (a zero inner scale) the comparison is given in Fig. 3 of Ref. 11. We note that when no inner scale is present the justification for using the approximation in Eq. (20) is absent but the approximation still gives a first-order correction that improves the solution. Moreover, the correction in Ref. 11 was obtained using an exponentiated form of the solution.

In three dimensions the effect of the first-order correction using the approximation in Eq. (20) is shown in Figs. 1 and 2. Here the corrected solution is compared to the numerical simulation results of Martin and Flatte[17]. For zero inner scale (Fig. 1) the correction is small (5%) and the final result is 7.5% below the peak of the Martin and Flatte solution. For γ_K=10 (ϵ=0.316) the correction is 11% and the final result is 7% below the peak of the Martin and Flatte result. We expect that as γ_K increases further the two-scale results will be closer to the numerical simulation. In addition we expect that, if we were able to implement the correction without using the approximation in Eq. (20), the corrected solution for small γ_K would give results closer to the Martin and Flatte solution. It should also be borne in mind that there is some error in the Martin and Flatte results since discretizing their grid introduces an effective inner scale.

4.2 Point source initial condition

In two dimensions, the two-scale solution, with the first-order correction obtained using Eq. (20), does not give satisfactory results unless ϵ is very small. Reeve et al.[24] present numerical solutions of the moment equation for a Gaussian correlation function and $\epsilon=0.14$. The peak value of the scintillation index obtained including the first-order correction obtained using the approximation Eq. (20) falls far below the peak value given in Fig. 4 of their paper. However, preliminary calculations indicate that the correction obtained when Eq. (20) is not used, gives results much closer to Reeve et al.

In three dimensions, the two-scale results may be compared to the numerical simulation results of Martin and Flatte[22]. The correction obtained using Eq. (20) gives poor agreement when there is no inner scale or ζ is small. For $\zeta=1$, however, preliminary calculations show a first order correction of about 30%.

5. Two-frequency Correlations

The two-frequency counterparts of the functions $\langle\Gamma\rangle$ and $\langle\Gamma_4\rangle$, denoted here $\langle\Gamma_2\rangle$ and $\langle\Gamma_{42}\rangle$, are governed by suitably modified versions of Eqs.(2) and (3), which are also amenable to treatment by the two-scale method. Interest is focussed in this case on the dependence of these functions on the wavenumber separation k_1-k_2.

The treatment of the fourth-order bichromatic coherence $\langle\Gamma_{42}\rangle$ is straightforward since it follows exactly the procedure outlined in Section 3 above. The zero-order treatment has been applied[25] to a plane wave propagating in a Gaussian medium, where the results are compared with previous treatments and shown to approach the correct asymptotic limits at small and large ranges and frequency separations.

The treatment of the second-order coherence[26] requires a preliminary transformation involving the coordinates and the wavenumbers in order to bring the equation to a form, similar to Eq.(8) above, which is amenable to treatment by the two-scale procedure. The solution of the equation can be given in terms of a Green's function, which serves to propagate the known distribution of the bichromatic coherence in the source plane (or aperture) to any generic field point. The result of the zero-order two-scale treatment has been employed[27] to compute the two-frequency coherence in a power-law medium with uniform background and with various types of excitation (plane wave, point source and a Gaussian beam). We observe that in this case (and for all power-law media) the parameter ϵ can be "scaled out" of the formulation by a suitable redefinition of the variables, or retained to order the iteration process and subsequently set equal to unity.

For a plane-wave, a single parameter $\eta \sim C_n^{[4/(3+a)]}(k_1-k_2)^{[(1+a)/(3+a)]}k^{[4(1-a)/(3+a)]}z$ serves to measure the effects of structure constant C_n, the frequency separation k_1-k_2, the average frequency k and the range z. For a pure Kolmogorov medium (a=2/3) a comparison of the plane-wave results with earlier numerical results[2] reveals a reasonable agreement for the behavior of the magnitude of the two-frequency coherence $<\Gamma_2>$, but a marked discrepancy in the asymptotic behavior of its phase at large values of η. For practical purposes, this discrepancy is of little consequence because the asymptotic magnitude of the coherence is very small.

We have also studied the two-frequency problem in the presence of a quadratic structure-function, a=1, (which approximately models a Gaussian medium). In this case Eq. (17) can be manipulated to the form

$$[\partial/\partial\eta - y\partial/\partial x + x^2]N = [-i2x\partial/\partial y + \partial^2/\partial y^2]N \tag{23}$$

Here, the right-hand member reflects the integral on the right-hand side of Eq. (17), simplified due to the quadratic character of h. (Its two terms would be multiplied by ϵ^2 and ϵ^4 , respectively, had ϵ not been "scaled out"). In the present context, this problem is of interest because (1) it possesses an exact analytical solution[28] and (2) it is eminently treatable by the two-scale method. For the two-scale method one can apply the various iteration schemes suggested in the discussion of Eqs.(17)-(20) to all orders, and thus shed light on the manner in which the exact solution is approached with repeated iterations. This advantage is somewhat mitigated, however, by the fact that the quadratic structure-function is unbounded, and violates one of the caveats of Gozani's convergence-proof.

The results reveal that repeated applications of the iteration process simply produce a power-series expansion in the parameter η of the exact solution. The various iteration schemes differ in the order at which they collect the coefficients of the various powers of η. The scheme based on the use of the Volterra equation (equivalent to Eq. 18 above) is superior to the others, in that it approaches the exact solution with fewer terms.

However, in the case at hand, the exact solution happens to have singular points at some (complex) values of η, so that its power-series (about $\eta=0$) diverges and becomes useless beyond the nearest such singularity. This is manifest in the results of successive iterations at large values of η, and also explains the aforementioned discrepancy in the asymptotic behavior of the phase. The difficulty is eliminated when the first term of the right-hand member of Eq. (23) is transposed to the left and employed in the determination of the characteristics. While the later become complex and are more difficult to compute, the results of successive iterations converge to the exact result for all values of η. As indicated in the concluding paragraph of Sec. 2, this modification can be implemented in the solution of Eq. (17) and hopefully improve the results reported in Sec. 4.

6. Use of the Two-scale Method for the Field Equations

In this section we apply the two-scale method to the field equation Eq. (1) rather than to the moment equation Eq. (8). This is done because in certain problems, such as the those involving a double passage of light through the same random medium, it is necessary to delay averaging to form the coherence functions, until after an interaction has taken place. The method we use may be applied to a random medium with a non-homogeneous deterministic background but for simplicity we consider here only the homogeneous case. We confine our attention to the two-point coherence function but the method is applicable to all even-order coherence functions[29-32].

The basic equation governing the Green's function $g(\mathbf{p},\mathbf{s},z/\mathbf{p}_o,\mathbf{s}_o,0)$ associated with the unaveraged two-point coherence function $\Gamma(\mathbf{p},\mathbf{s},,z) = u(\mathbf{p}+\mathbf{s}/2,z)u^*(\mathbf{p}-\mathbf{s}/2,z)$ is

$$\partial g/\partial z = (i/k)\nabla_p.\nabla_s g + ik[\mu(\boldsymbol{p}+\boldsymbol{s}/2,z) - \mu(\boldsymbol{p}-\boldsymbol{s}/2,z)]g \tag{24}$$

where

$$\Gamma(\boldsymbol{p},\boldsymbol{s},z) = \int_{-\infty}^{\infty}\int_{-\infty}^{\infty} g(\boldsymbol{p},\boldsymbol{s},z/\boldsymbol{p}_o,\boldsymbol{s}_o,z)\Gamma(\boldsymbol{p}_o,\boldsymbol{s}_o,z_o)\,d\boldsymbol{p}_o d\boldsymbol{s}_o$$

The procedure for solving Eq. (23) by the two scale method is the same as that described in Sec. 2 for solving Eq. (8). The result of the zero-order approximation is

$$g(\boldsymbol{p},\boldsymbol{s},z/\boldsymbol{p}_o,\boldsymbol{s}_o,z_o) = (k/2\pi)^4\int_{-\infty}^{\infty}\int_{-\infty}^{\infty} d\eta\, d\boldsymbol{v} \exp i\eta.[\boldsymbol{p}-\boldsymbol{p}_o- \boldsymbol{v}(z-z_o)]\exp[i\boldsymbol{v}.(\boldsymbol{s}-\boldsymbol{s}_o)]\exp[ik\int_{z_o}^{z} dz'[\mu(\boldsymbol{p}_o+\boldsymbol{v}(z'-z_o)+ \boldsymbol{s}/2+\eta(z'-z)/2,z')-\mu(\boldsymbol{p}_o+\boldsymbol{v}(z'-z_o)-\boldsymbol{s}/2-\eta(z'-z),z')]] \tag{25}$$

6.1 Application to double passage analysis

In location and remote sensing experiments, in the presence of random index-of-refraction fluctuations, the reflected radiation (or radiation diffracted by embedded obstacles) may traverse the same random inhomogeneities as the incident radiation. Double passage leads to new effects not observed in one-directional propagation problems[33,34].

To show how to account for the correlation between the forward and backward propagation directions, consider the problem of radiation emitted at the plane z_0 that is reflected from a plane surface located at propagation distance z. The turbulent medium lies between the planes at z_0 and z. Assume the local reflection coefficient K(x) varies with transverse position x. The coherence function Γ_{refl} of the reflected radiation passing back through the plane at z_0 is given by the expression

$$\Gamma_{refl}(\mathbf{x}_1,\mathbf{x}_2,z_o) = \int_{-\infty}^{\infty} \cdot \int_{-\infty}^{\infty} \Gamma(\mathbf{x}_{10},\mathbf{x}_{20},z_o)\, g(\mathbf{x}_{1j},\mathbf{x}_{2j},z/\mathbf{x}_{10},\mathbf{x}_{20},z_o)\, K(\mathbf{x}_{1j})\, K^*(\mathbf{x}_{2j})\, g(\mathbf{x}_1,\mathbf{x}_2,z_o/\mathbf{x}_{1j},\mathbf{x}_{2j},z)\, d\mathbf{x}_{10} d\mathbf{x}_{20} d\mathbf{x}_{1j} d\mathbf{x}_{2j} \tag{26}$$

We next ensemble average Eq. (25). The effect of the double passage through the same random fluctuations is given by the correlation <g(out)g(in)> between the two Green's functions given in the equation. If the outgoing and incoming paths do not traverse the same portion of the random medium then <g(out)g(in)>=<g(out)><g(in)>. Specific results are given by Mazar and Bronshtein[35]. To date, calculations have only been made using the zero-order approximation.

7. References

1. V.I. Tatarski, The Effects of the Turbulent Atmosphere on Wave Propagation, TT-68-50464 National Technical Information Service (1971).
2. A. Ishimaru, Propagation and Scattering in Random Media, Academic, (1978).
3. B. Uscinski, The Elements of Wave Propagation in Random Media, McGraw-Hill (1977).
4. M.J. Beran, "Propagation of a finite beam in a random medium", J. Opt. Soc. Am. **60**, 518-521 (1970).

5. M.J. Beran and T.L. Ho, "Propagation of the fourth-order coherence function in a random medium (a nonperturbative formulation)", J. Opt. Soc. Am. **59**, 1134-1138 (1969).
6. A.M. Whitman and M.J. Beran, "Beam spread of laser light propagating in a random medium", J. Opt. Soc. Am. **60**, 1595-1602 (1970).
7. S. Frankenthal, A.M. Whitman and M.J. Beran, "Caustic corrections using coherence theory", J. Acoust. Soc. Am. **71**, 348-358 (1982).
8. C. Macaskill, "Improved solution to the fourth moment equation for intensity fluctuations", Proc. Royal Soc. London Ser. **A 386**, 461-474 (1983).
9. B. Uscinski "Intensity fluctuations in a multiple scattering medium. Solution of the fourth moment equation", Proc. Royal Soc. London Ser. **A 380**, 137-169 (1982).
10a. J. Gozani, "Improvement in the two-scale solution for wave propagation in a random medium", Opt. Lett. **12**, 1-3 (1987).
10b. J. Gozani, "Applicability of the two-scale solution of a wave propagating in a random medium", J. Opt. Soc. Amer. **B5**, 721-727 (1988).
10c. J. Gozani, Multiple-Scale Solution for Wave Propagation in Random Media PhD. Thesis, Tel Aviv University, Tel Ai, Israel (1988).
10d. J. Gozani, "Two-scale expansion of wave propagation in a random medium", Computer Physics Communications **65**, 117-120 (1991).
11. A.M. Whitman and M.J. Beran, "Two-scale solution for atmospheric scintillation", J. Opt. Soc. Am. A **2**, 2133-2143 (1985).
12. S. Frankenthal, A.M. Whitman and M.J. Beran, "Two-scale solutions for intensity fluctuations in strong scattering", J. Opt. Soc. Am. A **1**, 585-597 (1984).
13. M.J. Beran and A.M. Whitman, "Effect of the atmosphere on measurements of correlations of intensity", SPIE **828**, Digital Image Recovery and Synthesis 122-126 (1987).
14. M.J. Beran and A.M. Whitman, "Non-isoplanatic effects in imaging through turbulent media", SPIE **926**, Optical, Infrared, and Millimeter Wave Propagation Engineering 306-310 (1988).
15. R.J. Hill, and S.F. Clifford, "Modified spectrum of atmospheric temperature fluctuations", J. Opt. Soc. Am. A **68**, 892-897, (1978).

16. A.M. Whitman and M.J. Beran, "First-order correction for the scintillation index and correlation of intensity function", J. Opt. Soc. Am. A **9**, 974-977 (1992).
17. J.M. Martin and S.M. Flatte, "Intensity images and statistics from numerical simulation of wave propagation in 3-D random media", Appl. Optics **27**, 2111-2126 (1988).
18. J. Gozani, "Numerical solution of the fourth-order coherence function of a plane wave propagating in a two-dimensional Kolmogorovian medium", J. Opt. Soc. Am. A **2**, 2144-2151 (1985).
19. M. Tur, "Numerical solutions of the fourth moment equation of a plane wave propagating in a random medium" , J. Opt. Soc. Am. **72**, 1683-1691 (1982).
20. M.J. Beran and A.M. Whitman, "Scintillation index calculations using an altitude-dependent structure constant", Appl. Opt. **27**, 2178-2182 (1988).
21. A.M. Whitman and M.J. Beran, "Two-scale solution for atmospheric scintillation from a point source", J. Opt. Soc. Am. A **5**, 735-737 (1988).
22. J.M. Martin and S.M. Flatte "Simulation of point-source scintillation through three-dimensional random media", J. Opt. Soc. Am. A **7**, 838-847 (1990).
23. R. Mazar and A. Bronshtein, "Finite aperture effects on intensity fluctuations in random media", Optics. Comm. **75**, 365-369 (1990).
24. D.E. Reeve, S.R. Leonard and M. Spivak, "Numerical solution from the fourth-moment equation for a point source", J. Mod. Opt. **37**, 965-975 (1990).
25. R. Mazar, J. Gozani and M. Tur, "Two-scale solution for the intensity fluctuations of two-frequency wave propagation in a random medium", J. Opt. Soc. Am. A. **2**, 2152-2160 (1985).
26. S. Frankenthal, "The mutual coherence function in a scattering channel - a two-scale solution", J. Acoust. Soc. Am. **85**, 104-113 (1989).
27. D. Cohen, "Two-scale Solution for the Mutual Coherence Function in the Presence of Turbulence", Thesis submitted towards the degree of Master of Science in Engineering Science, Tel Aviv University, September 1990.

28. I. Sreenivasiah, A. Ishimaru and S. T. Hong, "Two-frequency mutual coherence function and pulse-propagation in a random medium: an analytic solution to the plane-wave case", Radio Sci. **11**, 775-778 (1976).
29. R. Mazar and L.B. Felsen, "High-frequency coherence functions propagated along ray paths in the inhomogeneous background of a weakly random medium, I. Formulation and evaluation of the second moment", J. Acoust. Soc. Am. **81**, 925-937 (1987); "II. Higher moments", J. Acoust. Soc. Am. **82**, 593-600 (1987).
30. R. Mazar and L.B. Felsen, "Geometrical theory of diffraction of high-frequency coherence functions in a weakly random medium", Optics Letters **12**, 146-148 (1987).
31. R. Mazar and L.B. Felsen, "Stochastic Geometrical Theory of diffraction", J. Acoust. Soc. Am. **86**, 2292-2308 (1989).
32. R. Mazar, "High frequency propagators for diffraction and backscattering in random media", J. Opt. Soc. Am. A**7**, 34-46 (1990).
33. Y.A. Kravtsov and A.I. Saichev, "Effects of double passage of waves in randomly inhomogeneous media", Sov. Phys. Uspekhi **25**, 494-508 (1982).
34. V.A. Banakh and V.L. Mirinov, Locational propagation of laser radiation in a turbulent atmosphere (translated from the Russian), Artech House (1988).
35. R. Mazar and A. Bronshtein, "Double passage analysis in random media", Waves in Random Media **1**, 341-362 (1991).

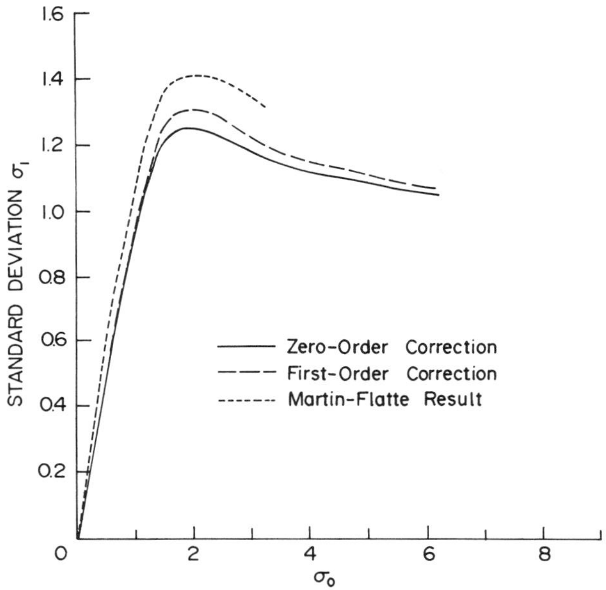

Fig. 1. Standard deviation σ_I as a function of σ_0 for $\gamma_K = 0$.

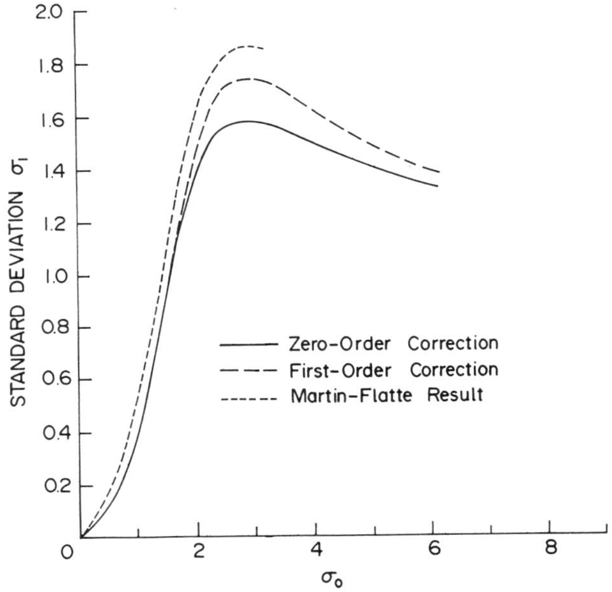

Fig. 2. Standard deviation σ_I as a function of σ_0 for $\gamma_K = 10$.

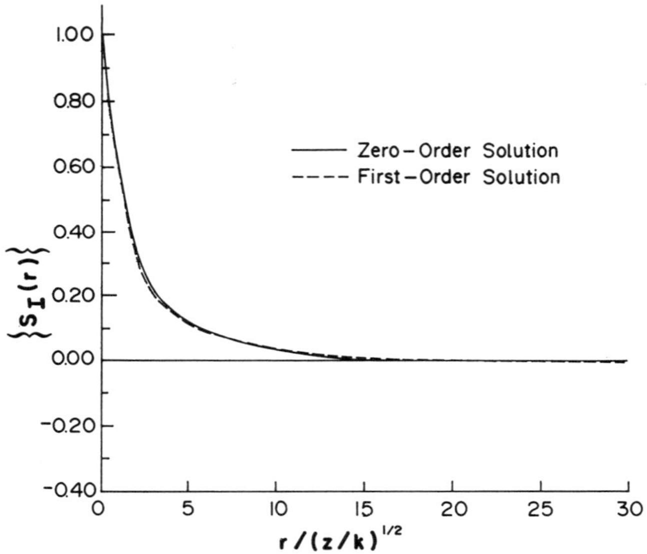

Fig. 3. $\{S_I(r)\}$ versus separation distance ($\sigma_0 = 2.06$).

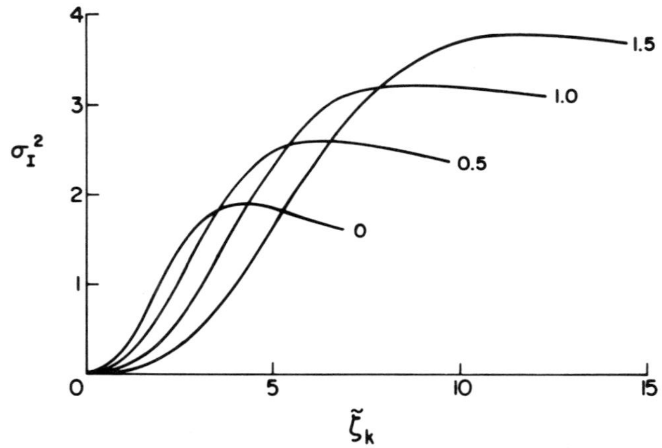

Fig. **5**. Scintillation index versus range for a point source in a Kolmogorov medium with a dissipation region. The numbers on the curves correspond to values of $\zeta^{-1/2}$. Thus each curve represents a constant value of the inner scale.

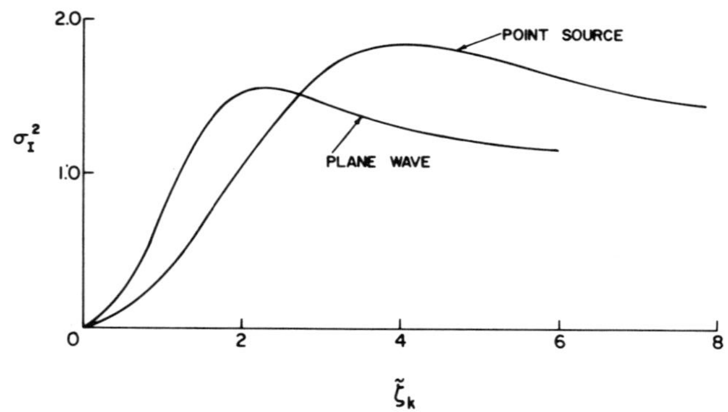

Fig. **4**. Scintillation indices as a function of dimensionless range for plane-wave and point-source initial conditions propagating through a pure Kolmogorov medium. The curve with the higher maximum corresponds to the point source.

PATH INTEGRAL APPROACH TO WAVE PROPAGATION IN RANDOM MEDIA PART I: DERIVATIONS AND VARIOUS FORMULATIONS

V.I. Tatarskii, M.I. Charnotskii, J. Gozani and V.U. Zavorotny

Cooperative Institute for Research in Environmental Sciences
University of Colorado/NOAA/Wave Propagation Laboratory
Boulder, Colorado 80309-0216

1. Introduction

The theory of wave propagation in random media with large-scale (compared with wavelength) inhomogeneities has been developed during almost half a century. It represents an extended area of study, having important applications such as atmospheric optics, ocean and atmospheric acoustics, radio meteorology, optical and radio astronomy, and plasma physics. Until the middle of the 1960's the main efforts were spent working out different approaches to developing wave propagation theory based on perturbation theory. Descriptions of those efforts can be found in books by Chernov[1] and by Tatarskii.[2] Unfortunately the results could be applied to only a narrow region of the problem, characterized by small fluctuations of a field at the receiving point. But there exist many important problems that are characterized by strong fluctuations of field intensity, in spite of the small-angle character of the scattering phenomenon. Examples are light propagation near the ground, sound propagation over long distances in the ocean, and radio wave propagation in the interstellar medium.

To develop an adequate theory for these situations, one must use methods describing multiple small-angle scattering. A parabolic equation has been used for such description. It was originally used by Leontovich,[3] Leontovich and Fock[4], and Fock[5] for the problem of radio wave propagation near the ground in a regularly inhomogeneous atmosphere. For a random medium this equation becomes a stochastic parabolic one. The application of quite different approaches (as the Markov approximation or the local method of small perturbation) to this equation has led to equations for arbitrary statistical moments of the receiving field by Chernov,[6,7] Dolin,[8] Shishov,[9] Beran and Ho,[10] Tatarskii,[11] Klyatskin and Tatarskii,[12] Klyatskin,[13] Molyneux,[14] and Brown[15]). That step represented considerable progress in the development of the theory of strong-intensity fluctuations. The main results obtained in that area can be found in numerous monographs and review papers: Barabanenkov et al.,[16] Tatarskii,[17] Klyatskin and Tatarskii,[18] Prokhorov et al.,[19] Klyatskin,[20,21] Fante,[22,23] Gurvich et al.,[24] Uscinski,[25] Ishimaru,[26] Tatarskii and Zavorotny,[27] Yakushkin,[28] Rytov et al.,[29] Kravtsov[30], and Gochelashvili and Shishov[31].

The first moment is the mean field, which is associated with the "coherent" or unscattered part of the wave field, and the second moment is the mutual coherence function, which gives the degree of correlation of the field at two points and determines the angular distribution of the scattered radiation as well as mean energy density of the radiation. The equations for the mean field and for the coherence function can be solved in a general form for arbitrary cases of initial conditions and of medium structure functions (Dolin,[8,32] Beran,[33] Klyatskin and Tatarskii,[12,52] Tatarskii,[17] Ishimaru[26]).

The next significant statistical moment is the fourth moment, because the third and all odd moments behave similarly to the mean field and thus rapidly decay with increasing distance in a random medium. The fourth moment is used to describe intensity fluctuations, by determining their variance and spatial-temporal spectrum. But the corresponding equations do not have full analytical solutions for all ranges of the parameters of the problems. Only asymptotic solutions for regions of weak and saturated intensity fluctuations were found. These asymptotic expressions for fourth statistical field moments were derived by various asymptotic methods applied directly to the moment equations (e.g., Gochelashvili and Shishov[34,35] and Yakushkin[36,37]). Significant efforts were made to obtain a numerical solution that may complement the description in the intermediate regime. The first numerical solution showing the saturation of intensity fluctuations was given by Dagkesamanskaya and Shishov.[38] But most of the numerical solutions presented until now used two-dimensional models of the medium in order to study salient effects. The full-dimensional numerical solution of the fourth-moment parabolic equation was reliably obtained by Gurvich et al.,[39] Tatarskii et al.,[40] and Martin and Flatté[41,42] as a result of numerical simulation.

Apart from the methods mentioned above, which use differential equations for the statistical moments of the field, there is an alternative method, which uses Feynman path integrals, or functional integrals. Functional integration implies integration over some measure. A special measure in the space of continuous functions, which gives the infinite-dimensional probability distribution of trajectories of a Brownian particle (Einstein[43]), was thoroughly studied by Wiener[44,45] (see, e.g., the book by Kac[46]). Then, Feynman[47] introduced the complex measure and the corresponding functional (path) integration in his formulation of nonrelativistic quantum theory. The Feynman path integral is now widely used in quantum mechanics, quantum electrodynamics, and statistical mechanics. The most well-known review papers and books of the field are written by Feynman,[48] Kac,[46] Gel'fand and Yaglom,[49] Feynman and Hibbs,[50] and Fradkin.[51]

The application of the path integral approach to wave propagation in random media was first suggested by Klyatskin and Tatarskii[52] in which the solution of both the Helmholtz equation and the parabolic equation was presented in the form of a Feynman path integral. Then the method was used for analogous purposes in works of Molyneux[53] and Chow.[54,55] Using asymptotic methods for the evaluation of Feynman integrals, Zavorotny et al.[56] and Dashen[57] derived expressions for arbitrary moments of wave intensity in the regime of saturated fluctuations. The result obtained for the fourth moment turned out to be the same as that obtained in works of Gochelashvili and Shishov[34,35] and Yakushkin[36,37] by solving the differential equation for this moment. The same asymptotic result for higher statistical moments was obtained by Yakushkin[58] using the conventional method of Green's functions to produce the solution of the moment equation. The above-mentioned asymptotic approaches were improved by using the path integral technique, in a paper by Dubovikov[59] in which the rigorous asymptotic expansion for the fourth moment was obtained.

Despite its apparent complexity, the path integral technique provides valuable physical insight and is very fruitful for producing new, useful approximations or for understanding how some known but not well-grounded ones work and what their applicability limits are. The subject is discussed by Zavorotny et al.[60] in this issue.

2. Path integral representation for the solution of a parabolic equation

As we mentioned in Section 1, for the problem considered the scale of inhomogeneities is large in comparison with the wavelength. In this case it is possible to neglect the backscattering and depolarization of waves and to describe the wave propagation by the parabolic wave equation:

$$2ik\frac{\partial u}{\partial z} + \Delta u + k^2\tilde{\epsilon}(z,\mathbf{R})u = 0. \tag{2.1}$$

In (2.1), $k = \omega(\overline{\epsilon})^{1/2}/c, \omega$ is the wave frequency, z is the initial direction of wave propagation, $\Delta = (\partial/\partial x)^2 + (\partial/\partial y)^2$, $\tilde{\epsilon} = (\epsilon - \overline{\epsilon})/\overline{\epsilon}, \epsilon$ is the dielectric permittivity, and the overbar denotes averaging over the medium fluctuations. The slowly varying complex amplitude $u(\mathbf{R}, z, \omega)$, where $\mathbf{R} = \{x, y\}$ is a radius-vector in transverse plane, is related to the electrical field $E(\mathbf{R}, z, \omega)$ as follows:

$$E(\mathbf{R}, z, \omega) = \exp[i(kz - \omega t)]u(\mathbf{R}, z, \omega).$$

It is instructive to consider first the simplest one-dimensional parabolic equation

$$\partial u/\partial t = D\partial^2 u/\partial x^2 - U(x,t)u(x,t), \qquad u(x,0) = u^0(x), \tag{2.2}$$

and to represent its solution by means of a functional integral. In (2.2) D is a complex number with a positive real part, Re $D > 0$; $U(x,t)$ is an arbitrary "good" function. [To obtain a one-dimensional analog of Eq. (2.1) we should set $D = i/2k, k = k' + ik'', k'' > 0$. The condition Re $D > 0$ is equivalent to the positiveness of the absorption coefficient k''.] This diffusion equation was studied from a probabilistic point of view by Einstein, Smoluchowski, Wiener, and Kac (see, e.g., Kac [46]). The following method of deriving the path integral representation for the solution of (2.2) can be found, for example, in the review paper by Tatarskii and Zavorotny.[27]

We integrate (2.2) from $t = t_k$ to $t = t_{k+1} = t_k + \Delta t$ and obtain

$$\begin{aligned} u(x, t_{k+1}) &= u(x, t_k) + \Delta t[D(\partial^2 u/\partial x^2) - U(x, t_k)u(x, t_k)] + o(\Delta t) \\ &= \{1 + \Delta t[D(\partial^2/\partial x^2) - U(x, t_k)] + o(\Delta t)\}u(x, t_k). \end{aligned}$$

With the same accuracy we can represent this expression in the form

$$u(x,t_{k+1}) = [1 - \Delta t U(x,t_k) + o(\Delta t)] \cdot [1 + \Delta t D(\partial^2/\partial x^2) + o(\Delta t)]u(x,t_k). \quad (2.3)$$

This expression can be represented in the operator form as follows:

$$u(x,t_{k+1}) = \exp[-\Delta t U(x,t_k) + o(\Delta t)] \exp\left(\Delta t D \frac{\partial^2}{\partial x^2}\right) u(x,t_k). \quad (2.4)$$

To verify this equality it is sufficient to expand the exponential in a Taylor series; after this we obtain (2.3) with the same accuracy.

To understand how the operator $\exp[\Delta t D(\partial^2/\partial x^2)]$ acts, we represent $u(x,t_k)$ in the form of a Fourier integral:

$$u(x,t_k) = \frac{1}{2\pi} \int\int \exp[i\kappa(x - x')]u(x',t_k)dx'd\kappa.$$

It is easy now to calculate the action of the operator $\exp\left(\Delta t D\ \partial^2/\partial x^2\right)$ because in the Fourier representation this operator takes the form $\exp(-\Delta t D\kappa^2)$. Then we can calculate the integral over κ. If Re $D > 0$, this integral converges and we obtain the simple result

$$\exp\left[\Delta t D \frac{\partial^2}{\partial x^2}\right] u(x,t_k) = \int \frac{dx'}{2\sqrt{\pi D \Delta t}} \exp\left[-\frac{(x - x')^2}{4D\Delta t}\right] u(x',t_k). \quad (2.5)$$

We then substitute this expression into (2.4) and denote $x = x_{k+1}, x' = x_k$. We can introduce the factor $\exp[-\Delta t U(x,t_k)]$ under the integral sign because it does not depend on $x' = x_k$. The final result is given by the formula

$$u(x_{k+1},t_{k+1}) = \int \frac{dx_k}{2\sqrt{\pi D \Delta t)}} \exp\left[-\frac{(x_{k+1} - x_k)^2}{4D\Delta t}\right]$$
$$\times \exp[-\Delta t U(x_{k+1},t_k) + o(\Delta t)]u(x_k,t_k). \quad (2.6)$$

Thus, we use (2.6) recurrently to obtain values of u for the next point, given that the values of u are known for the previous point. Let $x_N = X, t_N = T$ be the final point, $t_0 = 0$ be the initial point, and $u(x,0) = u^0(x)$ be known. Then we obtain for $\Delta t = t/N$

$$u(X,T) = \int \frac{dx_{N-1}}{2\sqrt{\pi D \Delta t}} \int \frac{dx_{N-2}}{2\sqrt{\pi D \Delta t}} \cdots \int \frac{dx_0}{2\sqrt{\pi D \Delta t}}$$
$$\times \exp\left\{-\frac{1}{4D\Delta t}[(X - x_{N-1})^2 + (x_{N-1} - x_{N-2})^2 + \ldots + (x_1 - x_0)^2]\right.$$
$$\left.-\Delta t[U(X,t_{N-1}) + U(x_{N-1},t_{N-2}) + \ldots + U(x_1,0)] + N\,o(\Delta t)\right\} u^0(x_0). \quad (2.7)$$

Taking the limit of (2.7) as $N \to \infty$ and $N \times o(\Delta t) \to 0$ gives the solution to (2.2). The limits of the expressions in the exponential of (2.7) are the following:

$$\lim_{N\to\infty} \frac{1}{4D\Delta t} \sum_{k=0}^{N-1} (x_{k+1} - x_k)^2 = \lim_{N\to\infty} \frac{\Delta t}{4D} \sum_{k=0}^{N-1} (\frac{x_{k+1} - x_k}{\Delta t})^2 = \frac{1}{4D} \int_0^T \left[\frac{dx(\tau)}{d\tau}\right]^2 d\tau,$$

$$\lim_{N\to\infty} \Delta t \sum_{k=1}^{N} U(x_k, t_{k-1}) = \int_0^T U\Big(x(\tau), \tau\Big) d\tau.$$

Here we have introduced the continuous function $x(\tau)$ in such a manner that $x(t_k) = x_k, x(T) = X$. Also we use the notation

$$\lim_{N\to\infty} \prod_{k=0}^{N-1} \frac{dx_k}{2\sqrt{\pi D \Delta t}} = \prod_{\tau=0}^{T} \frac{dx(\tau)}{2\sqrt{\pi D d\tau}} \equiv \mathcal{D}x(\tau). \tag{2.8}$$

Using all these notations, we can represent (2.7) in the formally closed form:

$$u(X,T) = \int^{x(T)=X} \mathcal{D}x(\tau) \exp\left\{ -\frac{1}{4D} \int_0^T [\dot{x}(\tau)]^2 d\tau - \int_0^T U[x(\tau), \tau] d\tau \right\} u^0\Big(x(0)\Big), \tag{2.9}$$

where $\dot{x}(\tau) \equiv dx(\tau)/d\tau$. The notation $x(T) = X$ in the upper limit shows that all the trajectories $x(\tau)$ have the same value X in the final point $t = T$. It easy to verify that the definition (2.8) of $\mathcal{D}x(\tau)$ is equivalent to the following normalizing condition:

$$\int^{x(T)=X} \exp\left\{ -\frac{1}{4D} \int_0^T [\dot{x}(\tau)]^2 d\tau \right\} \mathcal{D}x(\tau) = 1. \tag{2.10}$$

The right-hand side of (2.9) is known as a functional, or path, integral (see, Feynman,[47,48] Kac,[46] Gel'fand and Yaglom,[49] and Feynman and Hibbs[50]). If $D = D^*$, this functional integral is the Wiener integral (Kac [46]). If D is pure imaginary, it is the Feynman integral in a configurational (coordinate) space. For wave propagation problems, D is a complex number, having ReD << Im D; so the properties of this integral are more similar to the Feynman integral.

3. Unconditional and conditional path integrals

We call the type of path integral (2.9) the unconditional path integral because there is no special condition for the trajectories $x(\tau)$ at the initial point $\tau = 0$. Only the final point of all trajectories, $x(T) = X$, is fixed.

Green's function $G(X,T;X_0)$ corresponds to the special boundary condition, $u^0(x) = \delta(x - X_0)$ for equation (2.2). Using the general form of solution (2.9), we represent Green's function in the form of a path integral:

$$G(X,T;X_0) = \int^{x(T)=X} \mathcal{D}x(\tau)\exp\left\{-\frac{1}{4D}\int_0^T \dot{x}^2(\tau)d\tau - \int_0^T U\Big(x(\tau),\tau\Big)d\tau\right\}\delta(x_0 - X_0). \tag{3.1}$$

It is clear that all the paths really contributing something to the integral are restricted by the additional condition $x(0) = X_0$. We may include this restriction in the definition of the integral, using, for example, a discrete formulation analogous to (2.7) with obvious modifications. The result for Green's function can be written as

$$G(X,T;X_0) = \int_{x(0)=X_0}^{x(T)=X} \mathcal{D}x(\tau)\exp\left\{-\frac{1}{4D}\int_0^T [\dot{x}(\tau)]^2 d\tau - \int_0^T U\Big(x(\tau),\tau\Big)d\tau\right\}. \tag{3.2}$$

According to the well-known property of Green's function, the solution corresponding to the initial condition in (2.2) in terms of G takes the form

$$u(X,T) = \int dX_0 G(X,T;X_0)u^0(X_0). \tag{3.3}$$

If we substitute (3.2) into this formula, we obtain

$$u(X,T) =$$

$$\int u^0(X_0)dX_0 \int_{x(0)=X_0}^{x(T)=X} \mathcal{D}x(\tau)\exp\left\{-\frac{1}{4D}\int_0^T [\dot{x}(\tau)]^2 d\tau - \int_0^T U\Big(x(\tau),\tau\Big)d\tau\right\}. \tag{3.4}$$

We call this type of path integral the conditional path integral.

It should be noted that normalization (2.10) produces some normalization of the conditional path integral (see Section **4**).

4. The probabilistic interpretation

The trajectories of the path integral were considered deterministic albeit arbitrary. A different and appealing approach considers the trajectories as Brownian motion and assigns probabilistic weights to the trajectories. The resulting Wiener path integral represents an averaging over all these trajectories according to their weights (Kac[46]).

Let us consider the probability distribution

$$dP[x_0,\ldots,x_{N-1}] = \frac{dx_0}{2\sqrt{\pi D\Delta t}}\cdots\frac{dx_{N-1}}{2\sqrt{\pi D\Delta t}}\exp\left[-\frac{1}{4D\Delta t}\sum_{n=0}^{N-1}(x_{n+1}-x_n)^2\right], \quad (4.1)$$

with real D, of the random numbers $x_0,\ldots,x_{N-1}$. Here $x_N = X$ is the fixed value. This is, of course, the Gaussian probability distribution, and we can use it to represent (2.7) as a mean value of a functional

$$u(X,T) = \left\langle \exp[-\Delta t\sum_{n=0}^{N-1} U(x_n,t_{n-1})]u^0(x_0)\right\rangle \quad (4.2)$$

over x_n fluctuations, denoted by the brackets $\langle\ \rangle$. In the limit as $N\to\infty$ the continuous form of the probability density function becomes

$$\mathcal{D}P[x(\tau)] = \exp\left\{-\frac{1}{4D}\int_0^T[\dot{x}(\tau)]^2 d\tau\right\}\mathcal{D}x(\tau), \quad (4.3)$$

and it is properly normalized because of (2.10). Using (4.3), we rewrite the unconditional path integral (2.9) as a conditional mean value:

$$u(X,T) = \left\langle \exp\left\{-\int_0^T U\Big(x(\tau),\tau\Big)d\tau\right\}u^0\Big(x(0)\Big)\right\rangle_{x(T)=X}. \quad (4.4)$$

The subscript $x(T) = X$ denotes that only paths that follow the condition $x(T) = X$ are considered. The trajectory $x(\tau)$ in (4.4) is a Gaussian random process, fully characterized by its mean value $\langle x(\tau)\rangle_{x(T)=X}$ and the covariation function

$$B(\tau',\tau'') = \langle[x(\tau') - \langle x(\tau')\rangle][x(\tau'') - \langle x(\tau'')\rangle]\rangle_{x(T)=X}.$$

From (4.1) it follows that

$$\langle x(\tau)\rangle_{x(T)=X} = X, \qquad B(\tau',\tau'') = 2D\min(T-\tau',T-\tau''). \quad (4.5)$$

The variance $\sigma^2(\tau) = B(\tau,\tau) = 2D(T-\tau)$ is equal to zero at the point $\tau = T$ and to its maximum value $2D$ at the point $\tau = 0$. It corresponds to a Brownian motion beginning at the final point $\tau = T$ directed "back" (see Fig. 1).

For the case $D \neq D^*$ the probabilistic interpretation has only a formal meaning. However, a formula obtained using this interpretation will be correct even for the case of a complex D, if the result has a mathematical meaning for such a D.

In the case of the conditional path integral (3.4) it is beneficial to include the initial restriction in a continuous form of the probability density function:

$$\mathcal{D}P[x(\tau)] = N^{-1} \exp\left\{ -\frac{1}{4D} \int_0^T \dot{x}^2(\tau) d\tau \right\} \delta\Big(x(0) - X_0\Big) \mathcal{D}x(\tau). \tag{4.6}$$

Here N is a normalization coefficient given by

$$N = \int^{x(T)=X} \mathcal{D}x(\tau) \exp\left\{ -\frac{1}{4D} \int_0^T \dot{x}^2(\tau) d\tau \right\} \delta\Big(x(0) - X_0\Big). \tag{4.7}$$

It is clear that N is simply the well-known free Green function of (2.2),

$$N = G_0(X, T; X_0) = \frac{1}{2\sqrt{\pi DT}} \exp\left(-\frac{(X - X_0)^2}{4DT} \right), \tag{4.8}$$

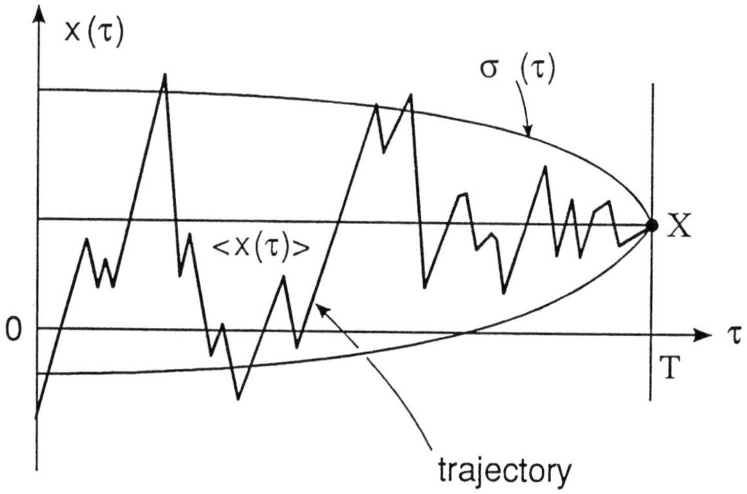

Fig. 1. Random Brownian trajectories for the unconditional path integral case starting from the point of observation X and going backward in time. The mean value $\langle x(\tau) \rangle = X$ and r.m.s. $\sigma(\tau) = \sqrt{2D(T - \tau)}$ are shown.

and we can now represent (3.4) as

$$u(X, T) = \int d^2 X_0 u^0(X_0) G_0(X, T; X_0) \left\langle \exp\left\{ -\int_0^T U(x(\tau), \tau) d\tau \right\} \right\rangle_{x(0)=X_0}^{x(T)=X}, \tag{4.9}$$

and Green's function as

$$G(X, T; X_0) =$$

$$\frac{1}{2\sqrt{\pi DT}} \exp\left[-\frac{(X - X_0)^2}{4DT} \right] \left\langle \exp\left[-\int_0^T U(x(\tau), \tau) d\tau \right] \right\rangle_{x(0)=X_0}^{x(T)=X}. \tag{4.10}$$

We denoted the averaging over these random paths by $\langle\ldots\rangle_{x(0)=X}^{x(T)=X}$. It can be shown that the new probability density function (4.3) is also a Gaussian one, but it differs from the previous one because both ends of the path are fixed. We can calculate the mean value $\langle x(\tau)\rangle$ and the correlation function $B(\tau', \tau'')$ for this random process. The result is given by the expressions

$$\langle x(\tau)\rangle_{x(0)=X}^{x(T)=X} = X_0 + \frac{\tau(X - X_0)}{T}, \tag{4.11}$$

$$B(\tau', \tau'') = 2D\min(\tau', \tau'')\left[1 - \frac{\max(\tau', \tau'')}{T}\right]. \tag{4.12}$$

The mean value for this conditional process is a linear function. The variance $\sigma^2(\tau) = B(\tau, \tau)$ is equal to $B(\tau, \tau) = 2D\tau(T - \tau)/T$ and vanishes at the ends of the paths (see Fig. 2).

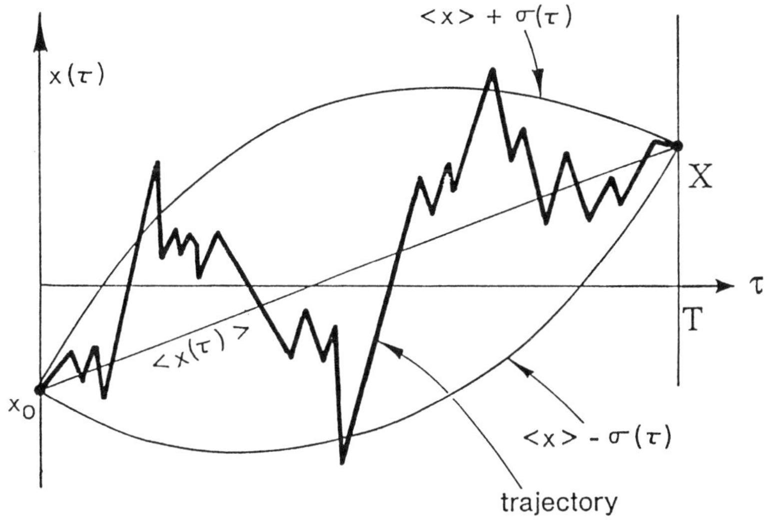

Fig. 2. Random Brownian trajectories for the conditional path integral case starting from the point of observation X and going backward in time to the point of the source X_0. The mean value $\langle x(\tau)\rangle = X_0 + (\tau/T)(X - X_0)$ and r.m.s. $\sigma(\tau) = \sqrt{2D\tau(T - \tau)/T}$ are shown.

It is possible also to include the arbitrary initial conditions in the probability density function of paths by

$$\mathcal{D}P\Big[x(\tau)\Big] = N^{-1}\exp\left[-\frac{1}{4D}\int_0^T \dot{x}^2(\tau)d\tau\right] u^0\Big(x(0)\Big)\mathcal{D}x(\tau). \tag{4.13}$$

Of course, in the framework of Wiener path-integral theory we need $u^0(x)$ to be a non-negative real function, but keeping in mind the generalization of the case of complex D and the Feynman integral, we can consider complex initial conditions as well. The normalization coefficient N here is given by

$$N = \int^{x(T)=X} \mathcal{D}x(\tau) \exp\left[-\frac{1}{4D}\int_0^T \dot{x}^2(\tau)d\tau\right] u^0\big(x(\tau)\big), \tag{4.14}$$

and comparing with (2.9) it is obviously a free solution of (2.2), which we readily can write as

$$N = u_0(X,T) = \frac{1}{2\sqrt{\pi DT}}\int dX_0 \exp\left[-\frac{(X-X_0)^2}{4DT}\right] u^0(X_0). \tag{4.15}$$

Now we can represent (2.7) as

$$u(X,T) = u_0(X,T)\left\langle \exp\left\{-\int_0^T U(x(\tau),\tau)d\tau\right\}\right\rangle_{u^0}^{x(T)=X}. \tag{4.16}$$

Here averaging is performed over paths having a random distribution of the initial coordinate $x(0)$ with the "probability density function" u^0 and a fixed final coordinate value $x(T) = X$.

If the initial distribution $u^0(X_0) = \exp[-(X_0 - a)^2/2B^2]$ is Gaussian, the probability density function (4.13) is also Gaussian with parameters

$$\langle x(\tau)\rangle = \frac{(B^2X + 2DTa) + 2D\tau(X - a)}{B^2 + 2DT}, \tag{4.17}$$

$$B(\tau_1,\tau_2) = 2D\min\{T-\tau_1, T-\tau_2\} - \frac{4D^2}{B^2 + 2DT}(T-\tau_1)(T-\tau_2). \tag{4.18}$$

For the limiting case, $B^2 \to \infty$,

$$\langle x(\tau)\rangle = X, \qquad B(\tau_1,\tau_2) = 2D\min\{T-\tau_1, T-\tau_2\}$$

corresponds to the unconditional Wiener integral. For $B^2 \to 0$,

$$\langle x(\tau)\rangle = a + (X-a)\tau/T,$$

$$B(\tau_1,\tau_2) = 2D\min(T-\tau_1, T-\tau_2) - \frac{2D}{T}(T-\tau_1)(T-\tau_2)$$

corresponds to the conditional Wiener integral.

Representation of the path integral as a mean value of some functional is an effective tool for numerical computation by the Monte-Carlo method. The first numerical calculation of some special path integral, related to solid state physics, was published by Gel'fand and Chentsov.[61]

5. Variational operator representation

Sometimes an equivalent form of the functional integral (2.9), the variational operator representation, is used (see, e.g., Fradkin,[51] Klyatskin and Tatarskii,[52]

Klyatskin,[20,21] and Zavorotny et al.[56]). We use here a graphic derivation of this form that was presented in the review by Tatarskii and Zavorotny.[27] We start from the operator form for a regular integral transformation (2.4). But now we use the right-hand side of (2.4) in a somewhat modified form:

$$u(x,t) = \exp[-\Delta t U(x,t-\Delta t)] \exp\left(\Delta t D \frac{\partial^2}{\partial x^2}\right) u(x,t-\Delta t)$$

$$= \exp\left(\Delta \tau D \frac{\partial^2}{\partial y^2}\right) \exp[-\Delta t U(x,t-\Delta t)] u(x+y,t-\Delta t)\Big|_{y=0} . \quad (5.1)$$

Let us here set $x = X, t = T$, and $y = y_{N-1}$:

$$u(X,T) =$$

$$\exp\left(\Delta\tau D \frac{\partial^2}{\partial y_{N-1}^2}\right) \exp\Big[-\Delta t U(X,T-\Delta t)\Big] u\Big(X + y_{N-1}, T-\Delta t\Big)\Big|_{y_{N-1}=0} . \quad (5.2)$$

We can apply now the same formula (5.1) with the parameters $x = X + y_{N-1}$, $t = T - \Delta t, y = y_{N-2}$ to the last factor in (5.2):

$$u(X + y_{N-1}, T-\Delta t) = \exp\left(\Delta\tau D\, \partial^2/\partial y_{N-2}^2\right)$$

$$\times \exp[-\Delta t U(X + y_{N-1}, T - 2\Delta t)] u(X + y_{N-1} + y_{N-2}, T - 2\Delta t)|_{y_{N-1}=y_{N-2}=0} . \quad (5.3)$$

We can then substitute (5.3) into (5.2). Because the differential operator in (5.3) does not act on $U(X, T-\Delta t)$, we can present the result in the form

$$u(x,t) = \exp\left[\Delta\tau D\left(\frac{\partial^2}{\partial y_{N-1}^2} + \frac{\partial^2}{\partial y_{N-2}}\right)\right]$$

$$\times \exp\{-\Delta t[U(X,T-\Delta t) + U(X + y_{N-1}, T - 2\Delta t)]\}$$

$$\times u(X + y_{N-1} + y_{N-2}, T - 2\Delta t)|_{y_{N-1}=y_{N-2}=0} .$$

Repeating this transformation n times we obtain

$$u(x,t) = \exp\left[\Delta\tau D\left(\frac{\partial^2}{\partial y_{N-1}^2} + \cdots + \frac{\partial^2}{\partial y_{N-n}^2}\right)\right]$$

$$\times \exp\{-\Delta t[U(X,T-\Delta t) + U(X + y_{N-1}, T - 2\Delta t) + \cdots$$

$$+ U(X + y_{N-1} + \cdots + y_{N-n+1}, T - n\Delta t)]\}$$

$$\times u(X + y_{N-1} + y_{N-2} + \cdots + y_{N-n}, T - n\Delta t)|_{y_{N-n}=\cdots=y_{N-1}=0} .$$

If we choose $n = N$ here and remember that $\Delta t = T/N$, the term

$$u(X + y_{N-1} + y_{N-2} + \cdots + y_{N-n}, T - n\Delta t)$$

becomes the known initial function

$$u^0(X + y_{N-1} + y_{N-2} + \ldots + y_0).$$

We obtain in this case the approximate solution to (2.2) in the form

$$u(x,t) = \exp\left[\Delta\tau D\left(\frac{\partial^2}{\partial y_{N-1}^2} + \cdots + \frac{\partial^2}{\partial y_0^2}\right)\right]$$

$$\times \exp\{-\Delta t[U(X, T-\Delta t) + U(X+y_{N-1}, T-2\Delta t) + \cdots U(X+y_{N-1}+\cdots+y_1, 0)]\}$$

$$\times u^0(X + y_{N-1} + y_{N-2} + \cdots + y_0)\big|_{y_0=y_1=\cdots=y_{N-1}=0} . \tag{5.4}$$

To perform the transition to the limit $N \to \infty$, we introduce the piecewise constant function $v(t_k) = y_k/\Delta t$. Then

$$y_k + y_{k+1} + \cdots + y_{N-1} = [v(t_k) + \cdots + v(t_{N-1})]\Delta t \to \int_t^T v(\tau)d\tau,$$

and

$$\frac{\partial}{\partial y_k} = \frac{1}{\Delta t}\frac{\partial}{\partial v_k} \to \frac{\delta}{\delta v(t_k)} .$$

Here, $\delta/\delta v(t)$ is the symbol of functional differentiation (for more details, see, e.g., Tatarskii[17] or Rytov et al. (vol. 3)[29]). Then

$$\lim_{N\to\infty} \Delta\tau D\left(\frac{\partial^2}{\partial y_{N-1}^2} + \cdots + \frac{\partial^2}{\partial y_0^2}\right) = D\int_0^T d\tau \frac{\delta^2}{\delta v^2(\tau)},$$

and the final expression for the operator representation of the path integral takes the form

$$u(x,t) = \exp\left[D\int_0^T d\tau \frac{\delta^2}{\delta v^2(\tau)}\right]$$

$$\times \exp\left[-\int_0^T d\tau U\left(X + \int_\tau^T v(\tau')d\tau', T-\tau\right)\right] u^0\left(X + \int_0^T v(\tau)d\tau\right)\Bigg|_{v\,=\,0} . \tag{5.5}$$

Note that the independent variable $v(\tau)$ in (5.5) has the physical meaning of the velocity of a Brownian particle (the velocity representation discussed in more detail by Zavorotny et al.[60]).

The form (5.5) of the solution of the parabolic wave equation (2.1) was used by Klyatskin and Tatarskii,[52] Klyatskin,[20,21] and Zavorotny et al.[56]:

$$u(\mathbf{R}, Z) = \exp\left[\frac{i}{2k}\int_0^Z d\zeta \frac{\delta^2}{\delta \mathbf{v}^2(\zeta)}\right] u^0\left(\mathbf{R} + \int_0^Z \mathbf{v}(\zeta)d\zeta\right)$$

$$\times \exp\left[\int_0^Z \tilde{\epsilon}\left(\mathbf{R} + \int_{\zeta'}^Z \mathbf{v}(\zeta)d\zeta,\ \zeta'\right) d\zeta'\right]\Bigg|_{\mathbf{v}=0} . \tag{5.6}$$

Here, $\tilde{\epsilon}$ is the permittivity fluctuation, $\mathbf{R}$ is the transverse coordinate in the plane $z = \text{Const}$.

It is also possible to obtain (5.5) using a functional Fourier transform of (2.9) (see Klyatskin and Tatarskii[52]) or using the method developed in quantum field theory (Fradkin[51]).

6. The orthogonal expansion of paths

It is often very useful to expand the path $x(\tau)$ into a series by some set of functions and replace integration over the path space by integration over the series coefficients (see, e.g., Feynman and Hibbs[50]). This approach in combination with probabilistic interpretation can create a variety of analytical approximations and numerical techniques (Tatarskii,[62,63] Sabel'feld and Tatarskii,[64] Tatarskii and Zavorotny[65]). In this section we demonstrate this idea for the one-dimensional case. Let us represent the path in the form

$$x(\tau) = X + \sum_{k=1}^{\infty} a_k \varphi_k(\tau), \tag{6.1}$$

where $\{\varphi_k(\tau)\}$ is such a set of functions that $\{\dot{\varphi}_k(\tau)\}$ is a complete orthonormal set; i.e,

$$\int_0^T \dot{\varphi}_k(\tau)\dot{\varphi}_j(\tau)d\tau = \delta_{kj}, \qquad \sum_{k=1}^{\infty} \dot{\varphi}_k(\tau_1)\dot{\varphi}_k(\tau_2) = \delta(\tau_1 - \tau_2). \tag{6.2}$$

This in fact means that we use the orthogonal expansion not for coordinate paths but for velocities. In this case we obtain the diagonal representation for the first exponential in (2.9):

$$\int_0^T \dot{x}^2(\tau)d\tau = \sum_{k,j=1}^{\infty} a_k a_j \int_0^T \dot{\varphi}_k(\tau)\dot{\varphi}_j(\tau)d\tau = \sum_{k=1}^{\infty} a_k^2. \tag{6.3}$$

To fulfill the condition $x(T) = X$, we must satisfy the condition $\varphi_k(T) = 0$; this leads to

$$\varphi_k(\tau) = -\int_\tau^T \dot{\varphi}_k(\tau')d\tau'. \tag{6.4}$$

The coefficients a_k are obtained from

$$a_k = \int_0^T \dot{x}(\tau)\dot{\varphi}_k(\tau)d\tau. \tag{6.5}$$

Probabilistic interpretation of the path integral leads us to probabilistic interpretation of the Fourier coefficients. It is clear from (6.5) that a_k are Gaussian numbers with $\langle a_k \rangle = 0$. Using (4.5) it is easy to find that

$$\langle \dot{x}(\tau_1)\dot{x}(\tau_2)\rangle = 2D\delta(\tau_1 - \tau_2). \tag{6.6}$$

Combining (6.5) and (6.6), we can obtain

$$\langle a_k a_j \rangle = 2D\delta_{kj}. \tag{6.7}$$

So, the numbers a_k are statistically independent. It is clear that averaging over all paths $x(\tau)$ is the same as averaging over all numbers a_k. Thus, (4.4) takes the form

$$u(X,T) = \left\langle \exp\left\{ -\int_0^T U\left(X + \sum_{k=1}^{\infty} a_k \varphi_k(\tau),\, \tau\right) d\tau \right\} u^0\left(X + \sum_{k=1}^{\infty} a_k \varphi_k(0)\right) \right\rangle_{\{a_k\}}. \tag{6.8}$$

Here the averaging operation over each a_k denotes an integration with the weight functions

$$P(a_k) = \frac{1}{2\sqrt{\pi D}} \exp\left(-\frac{a_k^2}{4D} \right).$$

This implies that (6.8) can be written in the form

$$u(X,T) = \int \frac{da_1}{2\sqrt{\pi D}} \int \frac{da_2}{2\sqrt{\pi D}} \cdots \exp\left[-\frac{1}{4D}\left(a_1^2 + a_2^2 + \cdots\right)\right]$$

$$\times \exp\left[-\int_0^T U\left(X + \sum_{k=1}^{\infty} a_k \varphi_k(\tau),\, \tau \right) d\tau \right] u^0\left(X + \sum_{k=1}^{\infty} a_k \varphi_k(0) \right). \tag{6.9}$$

If $D^* \neq D$, this transformation [from (4.4) to (6.9)] has no real probabilistic interpretation, but it is correct for Re$D > 0$.

The functions $\{\varphi_k(\tau)\}$, satisfying conditions (6.3), can be introduced in different ways. We can use additional information from path statistics (4.5) to choose the set that is the best (in the mean-square sense). It is the so-called Karhunen-Loève expansion of random processes (see, e.g., Papoulis[66]). That means that we can obtain the functions $\{\varphi_k(\tau)\}$ from the condition

$$\int_0^T B(\tau,\tau')\varphi_k(\tau')d\tau' = \lambda_k \varphi_k(\tau), \quad 0 \leq \tau \leq T. \tag{6.10}$$

If we substitute (4.5) into (6.10), we obtain the eigenvalue problem (Tatarskii[62])

$$\ddot{\varphi}_k + \frac{2D}{\lambda_k}\varphi_k = 0, \quad \dot{\varphi}_k(0) = 0, \varphi_k(T) = 0. \tag{6.11}$$

Its solution normalized by (6.2) is

$$\varphi_k(\tau) = \frac{\sqrt{2T}}{\pi(k - 1/2)} \cos\left[\frac{\pi(k-1/2)\tau}{T} \right],$$

and

$$\lambda_k = \frac{2DT^2}{\pi^2(k-1/2)^2}, \quad k = 1, 2, \ldots .$$

It is clear that for large k the amplitudes of $\{\varphi_k(\tau)\}$ decrease and the significance of high-order terms is lessened. This permits us to approximate (6.9) by using a finite-dimensional integral (Cameron[67]).

For the conditional path integral the expansion of $x(\tau)$ is introduced by the formula

$$x(\tau) = \langle x(\tau)\rangle + \sum_{k=1}^{\infty} a_k \varphi_k(\tau), \tag{6.13}$$

where $\langle x(\tau)\rangle$ is given by (4.11), $\varphi_k(0) = \varphi_k(T) = 0$, and again we assume $\{\dot{\varphi}_k(\tau)\}$ to be a complete orthonormal set; i.e., (6.2) is valid. Thus,

$$\dot{x}(\tau) = \frac{X - X_0}{T} + \sum_{k=1}^{\infty} a_k \dot{\varphi}_k(\tau). \tag{6.14}$$

We multiply this equation by $\dot{\varphi}_k(\tau)$ and integrate from 0 to T. The first term vanishes because $\varphi_k(0) = \varphi_k(T) = 0$, and we obtain an equation that is the same as (6.5). For $\langle a_k\rangle$ we obtain

$$\langle a_k\rangle = \int_0^T \langle \dot{x}(\tau)\rangle \dot{\varphi}_k(\tau) d\tau = \frac{X - X_0}{T} \int_0^T \dot{\varphi}_k(\tau) d\tau = 0. \tag{6.15}$$

To obtain $\langle x_k x_j\rangle$, we use the equation

$$\langle \dot{x}(\tau_1)\dot{x}(\tau_2)\rangle = 2D\delta(\tau_1 - \tau_2) - \left[\frac{2D}{T} + \left(\frac{X - X_0}{T}\right)^2\right], \tag{6.16}$$

which is the consequence of (4.12). Therefore

$$\langle a_k a_j\rangle = \int_0^T \int_0^T \langle \dot{x}(\tau_1)\dot{x}(\tau_2)\rangle \dot{\varphi}_k(\tau_1)\dot{\varphi}_j(\tau_2) d\tau_1\, d\tau_2 = 2D\delta_{k,j}\,. \tag{6.17}$$

So, the random numbers $\{a_k\}$ have the same distribution as in the previous case.

We can represent now Green's function (4.10) in a form that contains only an averaging over Gaussian random numbers a_k:

$$G(X, T; X_0) =$$

$$G_0(X, T; X_0)\left\langle \exp\left\{ - \int_0^T U\left(X_0 + (X - X_0)\tau/T + \sum_{k=1}^{\infty} a_k \varphi_k(\tau), \tau \right) d\tau \right\}\right\rangle_{\{a_k\}}$$

$$= G_0(X,T;X_0)\int \frac{da_1}{2\sqrt{\pi D}}\int \frac{da_2}{2\sqrt{\pi D}}\cdots \exp\left[-\frac{1}{4D}\left(a_1^2+a_2^2+\cdots\right)\right]$$

$$\times \exp\left[-\int_0^T U\left(X_0+(X-X_0)\tau/T+\sum_{k=1}^{\infty} a_k\varphi_k(\tau),\ \tau\right)d\tau\right]. \qquad (6.18)$$

For a specific choice of $\{\varphi_k(\tau)\}$ again we use the Karhunen-Loève expansion (6.10), which in this case results in the eigenvalue problem (Tatarskii[63]):

$$\ddot{\varphi}_k + \frac{2D}{\lambda_k}\varphi_k = 0, \qquad \varphi_k(0) = \varphi_k(T) = 0. \qquad (6.19)$$

Its solution, normalized by (6.2), is given by the formula

$$\varphi_k(\tau) = \frac{\sqrt{2T}}{\pi k}\sin\left(\frac{\pi k\tau}{T}\right), \quad \lambda_k = \frac{2DT^2}{\pi^2 k^2}. \qquad (6.20)$$

If we substitute $u^0(x) = \delta(x - X_0)$ into (6.9), we obtain another representation of $G(X,T;X_0)$ because the set of functions $\{\varphi_k(\tau)\}$ would be given by (6.12). We believe that this latter representation is not effective for numerical calculation by the Monte-Carlo method because most of the $\{a_k\}$ realizations will give zero contribution to the integral sum as a result of the δ-function. So to obtain the effective computational technique, one would have to choose the basis in accordance with the initial conditions.

In the general case (4.13), the probability density function is generally not Gaussian, but the Karhunen-Loève expansion still can be applied.

Examples of usage of the orthogonal and some more effective expansions for numerical calculations of path integrals are given in the paper by Sabel'feld and Tatarskii[64] and in the paper by Gozani et al.[68] presented in this issue.

7. Conclusions

In this introductory paper we tried to explain the path integral method to newcomers. We present here the simplest (not rigorous from a mathematical point of view) derivations of different representations of the path integral. We hope that this introduction will be enough for understanding the later material using the path integral method and for reading papers based on the path integral method.

References

1. L. A. Chernov, *Wave Propagation in a Random Medium,* McGraw-Hill, New York (1960).
2. V. I. Tatarskii, *Wave Propagation in a Turbulent Medium,* McGraw-Hill, New York (1961).
3. M. A. Leontovich, "On the new method for solving of the radio-wave propagation problem, "*Izv. AN SSSR, Ser. Fiz.* **8**, 16-22 (1944) (in Russian).
4. M. A. Leontovich and V. A. Fock, "Solution of the problem of propagation of electromagnetic waves along the earth's surface by the parabolic equation method," *Zh. Eksp. Teor. Fiz.* **16**, 557-73 (1946) (in Russian).

5. V. A. Fock, "Theory of radio-wave propagation in an inhomogeneous atmosphere for a raised source," *Bull. Acad. Sci. URSS Ser. Phys.* **14**, 70 (1950) [translation: V. A. Fock, *Electromagnetic Diffraction and Propagation Problems*, Chap. 14, Pergamon, Oxford (1965)].
6. L. A. Chernov, "The local method of computing strong fluctuations of the field in the problem of wave propagation in a medium with random inhomogeneities," *Proc. 6th All-Union Acoustical Conf.*, Nauka, Moscow, USSR (1968).
7. L. A. Chernov, "Equations for the statistical moments of the field in a randomly inhomogeneous medium," *Sov. Phys. Acoust.* **15**(4), 511-517 (1969).
8. L. S. Dolin, "Equations for the correlation functions of a wave beam in a randomly-inhomogeneous medium," *Radiophys. Quantum Electron.* **11**(6), 486-491 (1968).
9. V. I. Shishov, "Theory of wave propagation in random media," *Radiophys. Quantum Electron.* **11**(6), 500-505 (1968).
10. M. J. Beran and T. L. Ho, "Propagation of the fourth-order coherence function in a random medium (a non perturbative formulation)," *J. Opt. Soc. Am.* **59**(9), 1134-1138 (1968).
11. V. I. Tatarskii, "Light propagation in a medium with random index refraction inhomogeneities in the Markov process approximation," *Sov. Phys. JETP* **29**(6), 1133-1138 (1969).
12. V. I. Klyatskin and V. I. Tatarskii, "On the theory of the propagation of light beams in a medium having random inhomogeneities," *Radiophys. Quantum Electron.* **13**(7), 828-833 (1970).
13. V. I. Klyatskin, "Applicability of the approximation of a Markov random process in problems relating to the propagation of light in a medium with random inhomogeneities," *Sov. Phys. JETP* **30**(9), 520-523 (1970).
14. J. E.Molyneux, "Propagation of the Nth-order coherence function in a random medium: the governing equations," *J. Opt. Soc. Am.* **61**(2), 248-255 (1971).
15. W. P. Brown, Jr., "Moments equations for waves in random media," *J. Opt. Soc. Am.* **62**(8), 45-54 (1972).
16. Yu. A. Barabanenkov, Yu. A. Kravtsov, S.M. Rytov, and V. I. Tatarskii, "Status of the theory of propagation of waves in a randomly inhomogeneous medium," *Sov. Phys. Usp.* **13**(5), 551-680 (1971).
17. V. I. Tatarskii, *The Effect of the Turbulent Atmosphere on Wave Propagation*, National Technical Information Service, Springfield, Va. (1971).
18. V. I. Klyatskin and V. I. Tatarskii, "Statistical theory of light propagation in a turbulent medium (Review)," *Radiophys. Quantum Electron.* **15**(10), 1095-1112 (1972).
19. A. M. Prokhorov, F. V. Bunkin, K. S. Gochelashvili, and V. I. Shishov, "Laser irradiance propagation in turbulent media," *Proc. IEEE* **63**(5), 790-811 (1975).
20. V. I. Klyatskin, *Statistical Description of Dynamic Systems with Fluctuating Parameters,* Nauka, Moscow (1975) (in Russian).
21. V. I. Klyatskin, *Stochastic Equations and Waves in Random Inhomogeneous Media,* Nauka, Moscow (1980) (in Russian) [V. I. Klyatskin, *Ondes et equations stochastique dans les milieux aleatoirement non homogenes,* Editions de Physique, Besancon-Cedex (1985) (in French)].
22. R. L. Fante, "Electromagnetic beam propagation in turbulent media," *Proc. IEEE* **63**(12), 1669-1692 (1975).

23. R. L. Fante, "Wave propagation in random media: A systems approach," *Progress in Optics*, **22**, ed. E. Wolf, North-Holland, Amsterdam (1985).
24. A. S. Gurvich, A. I. Kon, V. L. Mironov, and S. S. Khmelevtsov, *Laser Radiation in Turbulent Atmosphere,* Nauka, Moscow (1976) (in Russian).
25. B. J. Uscinski, *The Elements of Wave Propagation in Random Media,* McGraw-Hill, New York (1977).
26. A. Ishimaru, *Wave Propagation and Scattering in Random Media,* Academic Press, New York (1978).
27. V. I. Tatarskii and V. U. Zavorotny, 1980, "Strong fluctuations in light propagation in a randomly inhomogeneous medium," *Progress in Optics*, **18**, ed. E. Wolf, North-Holland, Amsterdam (1980).
28. I. G. Yakushkin, "Intensity fluctuations during small-angle scattering of wave field," *Radiophys. Quantum Electron.* **28**(5), 365-389 (1985).
29. S. M. Rytov, Y. A. Kravtsov, and V. I. Tatarskii, *Principles of Statistical Radiophysics*, vols. 1-4, Springer, Berlin (1988).
30. Yu. A. Kravtsov, "Propagation of electromagnetic waves through a turbulent atmosphere," *Rep. Prog. Phys.* **55**(1), 39-112 (1992).
31. K. S. Gochelashvili and V. I. Shishov, *Waves in Randomly Inhomogeneous Media*, VINITI, Moscow (1981).
32. L. S. Dolin, "Beam description of weakly-inhomogeneous wave fields," *Radiophys. Quantum Electron.* **7**(3), 244-249 (1964).
33. M. J. Beran, "Propagation of the mutual coherence function through random media," *J. Opt. Soc. Am.* **56**(11), 1475-1480 (1966).
34. K. S. Gochelashvili and V. I. Shishov, "Laser beam scintillation beyond a turbulent layer," *Optica Acta* **18**(4), 313-320 (1971).
35. K. S. Gochelashvili and V. I. Shishov, "Saturated fluctuations in the laser radiation intensity in a turbulent medium," *Sov. Phys. JETP* **39**(4), 605-609 (1974).
36. I. G. Yakushkin, "Asymptotic calculation of field-intensity fluctuations in a turbulent medium for long paths," *Radiophys. Quantum Electron.* **18**(11), 1224-1229 (1975).
37. I. G. Yakushkin, "Strong intensity fluctuations in the field of a light beam in a turbulent atmosphere, *Radiophys. Quantum Electron.* **19**(3), 270-275 (1976).
38. I. M. Dagkesamanskaya and V. I. Shishov, "Strong intensity fluctuations during wave propagation in statistically inhomogeneous and isotropic media," *Radiophys. Quantum Electron.* **13**(1), 9-11 (1970).
39. A. S. Gurvich, B. S. Elepov, V. V. Pokasov, K. K. Sabel'feld, and V. I. Tatarskii, "Spatial structure of strong fluctuations of light intensity in a turbulent medium," *Radiophys. Quantum Electron.* **22**(2), 135-141 (1979).
40. V. I. Tatarskii, A. S. Gurvich, B. S. Elepov, V. V. Pokasov, and K. K. Sabel'feld, "Space structure of strong intensity fluctuations of light in a turbulent medium," *Optica Acta* **26**(5), 531-542 (1979).
41. J. M. Martin and S. M. Flatté, "Intensity images and statistics from numerical simulation of wave propagation in $3-D$ random media," *Appl. Opt.* **27**(11), 2111-2126 (1988).
42. J. M. Martin and S. M. Flatté, "Simulation of point-source scintillation through three-dimensional random media," *J. Opt. Soc. Am. A* **7**(5), 838-847 (1990).

43. A. Einstein, "On the movement of small particles suspended in a stationary liquid demanded by the molecular-kinetic theory of heat," *Ann. Phys.* **17**(8), 549-560 (1905) [translation: A. Einstein, *Investigation on the theory of the Browninan Movement,* ed. R. Furth, transl. A. D. Cowper, Dover, New York (1956)].
44. N. Wiener, "Differential space," *J. Math and Phys.* **2**, 131-174 (1923).
45. N. Wiener, "The average value of a functional," *Proc. London Math. Soc. Ser.* 2 **22**(6), 454-467 (1924).
46. M. Kac, *Probability and Related Topics in the Physical Sciences,* Interscience, New York (1959).
47. R. P. Feynman, "Space-time approach to non-relativistic quantum mechanics," *Rev. Mod. Phys.* **20**(2), 367-387 (1948).
48. R. P. Feynman, *Selected Papers on Quantum Electrodynamics,* ed. J. Schwinger, Dover, New York (1958).
49. I. M. Gel'fand and A. M. Yaglom, "Integration in functional spaces and its applications in quantum physics," *J. Math. Phys.* **1**(1), 48-69 (1960).
50. R. P. Feynman and A. R. Hibbs, *Quantum Mechanics and Path Integrals,* McGraw-Hill, New York (1965).
51. E. S. Fradkin, "Application of functional methods in quantum field theory and quantum statistics," *Nuclear Phys.* **76**, 588-624 (1966).
52. V. I. Klyatskin and V. I. Tatarskii, "The parabolic equation approximation for propagation of waves in a random medium with random inhomogeneities," *Sov. Phys. JETP* **31**(2), 335-339 (1970).
53. J. E. Molyneux, "Propagation of the Nth-order coherence function in a random medium, II. General solutions and asymptotic behaviour," *J. Opt. Soc. Am.* **61**(3), 369-377 (1971).
54. P. L. Chow, "Application of function space integrals to problems in wave propagation in random media," *J. Math. Phys.* **13**(8), 1224-1236 (1972).
55. P. L. Chow, "A functional phase-integral method and applications to the laser beam propagation in random media," *J. Stat. Phys.* **12**(2), 93-109 (1975).
56. V. U. Zavorotny, V. I. Klyatskin, and V. I. Tatarskii, "Strong fluctuations of the intensity of electromagnetic waves in randomly inhomogeneous media," *Sov. Phys. JETP* **46**(2), 252-260 (1977).
57. R. Dashen, "Path integrals for waves in random media," *J. Math. Phys.* **20**(5), 894-920 (1979).
58. I. G. Yakushkin, "Moments of field propagating in randomly inhomogeneous medium in the limit of saturated fluctuations," *Radiophys. Quantum Electron.* **21**(8), 835-840 (1978).
59. M. M. Dubovikov, "Class of non-Gaussian functional integrals," *Theor. Math. Phys.* **58**(3), 215-220 (1984).
60. V. I. Zavorotny, M. I. Charnotskii, J. Gozani, and V. I. Tatarskii, "Path integral approach to wave propagation in random media, Part II, Exact formulations and heuristic approximations," (this issue).
61. I. M. Gelfand and N. N. Chentsov, "Numerical calculation of functional integrals," *JETP* **31,** 1106-1107 (1956) (in Russian).
62. V. I. Tatarskii, "On approximate computation of the path integrals," *Monte-Carlo Method in Computational Mathematics and Mathematical Physics*, pp. 60-74, Nauka, Novosibirsk (1976).
63. V. I. Tatarskii, "On approximate computation of the conditional Wiener and

some Feynman path integrals," *Monte-Carlo Method in Computational Mathematics and Mathematical Physics*, pp. 75-90, Nauka, Novosibirsk (1976).
64. K. K. Sabel'feld and V. I. Tatarskii, "The approximate evaluation of Wiener path integrals," *Sov. Phys. Dokl.* **23**(12), 898-900 (1978).
65. V. I. Tatarskii and V. U. Zavorotny, "On the connection between the extended Huygens-Fresnel principle and the path-integral approximate computation based on orthogonal expansions," *Proc. SPIE* **642,** 276-281 (1986).
66. A. Papoulis, *Probability, Random Variables, and Stochastic Processes,* McGraw-Hill, New York (1965).
67. R. H. Cameron, "A "Simpson rule" for the numerical evaluation of Wiener's integrals in functional space," *Duke Math. J.* **18,** 111-130 (1951).
68. J. Gozani, M. I. Charnotskii, V. I. Tatarskii, and V. I. Zavorotny, "Path integral approach to wave propagation in random media, Part III, Mixed representation; orthogonal expansion," (this issue).

PATH INTEGRAL APPROACH TO WAVE PROPAGATION IN RANDOM MEDIA PART II: EXACT FORMULATIONS AND HEURISTIC APPROXIMATIONS

V. U. Zavorotny, M. I. Charnotskii, J. Gozani, and V. I. Tatarskii

Cooperative Institute for Research in Environmental Sciences
University of Colorado/NOAA/Wave Propagation Laboratory
Boulder, Colorado 80309-0216

1. Introduction

The path integral technique provides analytical and computational approach to a variety of propagation problems. By using it, we may obtain a clear representation of the complex functional of a field, which is impossible to get by any other technique. The path integral method was shown to be a powerful tool for asymptotic analysis of statistical field moments.[1,2]

At the same time some authors prefer to use heuristic analytical approximations instead of more substantiated asymptotic expressions for statistical moments, because the heuristic approximations are supposed to work with some acceptable accuracy for all distances and for all scattering strength values in a medium. These approximations yield the expression for a statistical moment in the form of an ordinary finite-fold integral that can be calculated analytically or numerically. One of them is the phase approximation of the extended Huygens-Fresnel principle (in Russian literature, this is the phase approximation of the Huygens-Kirchhoff method), which was first used in a paper by Feyzulin and Kravtsov[3] (see also Refs. 4-9). Tur and Beran[10] provided a comparison between the results of this approximation and numerical solutions for the case of a two-dimensional random medium. Other similar approaches were proposed using plane wave expansions instead of spherical wave expansions (Gochelashvili,[11,12] Aksenov and Mironov[13,14]). Mironov gives a detailed review of these methods in his book.[15] Other authors prefer to use calculated solutions of the parabolic equation, known as numerical simulations, which are not considered here.

Of course, the results produced by all these approximate methods cannot be considered satisfactory for all cases, because sometimes they give controversial or even wrong results. Therefore, their physical justification and the limits of validity should be obtained. Naturally, it is a very difficult task, because these approximations are heuristic and do not contain any small or large parameters. The path integral technique[16] helps to solve this problem. Below we examine the connection between the approximate expressions and the exact solutions written in the form of the path integral. The analysis reveals the real origin of those heuristic approximations.

2. Problem formulation and governing equations

It is known that for large-scale (in comparison with the wavelength λ) inhomogeneities it is possible to neglect the backscattering and depolarization of waves. For this case the scalar parabolic equation for the field envelope of a wave propagating in the positive direction of the z axis is valid:[17]

$$2ik\frac{\partial u}{\partial z} + \Delta u + k^2\tilde{\epsilon}(\mathbf{r}, z)u(\mathbf{r}, z) = 0. \tag{2.1}$$

In (2.1) $\mathbf{r} = (x, y)$ is the transverse, position vector, $\Delta = \partial^2/\partial x^2 + \partial^2/\partial y^2$, $k = \omega\langle\epsilon\rangle^{1/2}/c$, ω is the wave frequency, $\tilde{\epsilon}(\mathbf{r}, z) = (\epsilon - \langle\epsilon\rangle)/\langle\epsilon\rangle$, ϵ is the dielectric permittivity, and the angle brackets denote averaging over medium fluctuations. The complex envelope $u(\mathbf{r}, z)$ of the electrical field $E(\mathbf{r}, z, t)$ is introduced by

$$E(\mathbf{r}, z, t) = \exp(ikz - i\omega t)u(\mathbf{r}, z).$$

The solution of Eq.(2.1) satisfies the initial condition in the plane $z = Z_0$:

$$u(\mathbf{r}, z = Z_0) = u^0(\mathbf{r}), \tag{2.2}$$

where $u^0(\mathbf{r})$ is the field distribution in the plane $z = Z_0$. In the case of a point source placed in the point $(\mathbf{R}_0, Z_0)$, we have

$$u^0(\mathbf{r}) = \delta(\mathbf{r} - \mathbf{R}_0), \tag{2.3}$$

where $\delta(\mathbf{r})$ is a two-dimensional Dirac δ-function. We denote the field of a spherical wave at the point $(\mathbf{R}, Z)$ as $G(\mathbf{R}, Z; \mathbf{R}_0, Z_0)$. This is Green's function for (2.1). So, the field satisfying the general initial condition (2.2) can be represented as superposition of the fields created by the point sources in the initial plane:

$$u(\mathbf{R}, Z) = \int G(\mathbf{R}, Z; \mathbf{R}_0, Z_0)u^0(\mathbf{R}_0)d^2R_0, \tag{2.4}$$

which is the spherical wave expansion of the solution. It is well-known that Green's function is subject to the reciprocity principle. That is, it is invariant to interchanging of source and observation point; i.e.,

$$G(\mathbf{R}, Z; \mathbf{R}_0, Z_0) = G(\mathbf{R}_0, Z_0; \mathbf{R}, Z). \tag{2.5}$$

2.1 Plane wave expansions

Introducing the Fourier decomposition of the initial field

$$u^0(\mathbf{R}_0) = \int \hat{u}^0(\mathbf{P}_0)\exp(i\mathbf{P}_0 \cdot \mathbf{R}_0)d^2P_0, \tag{2.6}$$

which is simply a plane wave expansion of the initial conditions, we can represent (2.4) as

$$u(\mathbf{R}, Z) = \int \hat{G}(\mathbf{R}, Z; \mathbf{P}_0, Z_0)\hat{u}^0(\mathbf{P}_0)d^2P_0, \tag{2.7}$$

where

$$\hat{G}(\mathbf{R}, Z; \mathbf{P}_0, Z_0) = \int G(\mathbf{R}, Z; \mathbf{R}_0, Z_0)\exp(i\mathbf{P}_0 \cdot \mathbf{R}_0)d^2R_0. \tag{2.8}$$

According to (2.4), $\hat{G}(\mathbf{R}, Z; \mathbf{P}_0, Z_0)$ is the field of the originally plane wave with a transverse wave vector $\mathbf{P}_0$, traveling from the plane $z = Z_0$ to the plane $z = Z$. Therefore, we call (2.7) the outgoing plane wave expansion of the field.

The Fourier spectrum of the field in the observation plane is defined by

$$\hat{u}(\mathbf{P}, Z) = (2\pi)^{-2}\int u(\mathbf{R}, Z)\exp(-i\mathbf{P} \cdot \mathbf{R})d^2R. \tag{2.9}$$

Through (2.4) it can be represented in the form

$$\hat{u}(\mathbf{P}, Z) = (2\pi)^{-2}\int \tilde{G}(\mathbf{P}, Z; \mathbf{R}_0, Z_0)u^0(\mathbf{R}_0)d^2R_0, \tag{2.10}$$

where

$$\tilde{G}(\mathbf{P}, Z; \mathbf{R}_0, Z_0) = \int G(\mathbf{R}, Z; \mathbf{R}_0, Z_0)\exp(-i\mathbf{P} \cdot \mathbf{R})d^2R. \tag{2.11}$$

Because of the reciprocity principle, $\tilde{G}(\mathbf{P}, Z; \mathbf{R}_0, Z_0)$ is the field of an originally plane wave with a transverse wave vector $\mathbf{P}$, traveling from the plane $z = Z$ to the plane $z = Z_0$. We call (2.10) the incoming plane wave expansion.

Since Z_0 is arbitrary, for simplicity's sake we further choose $Z_0 = 0$. Equations (2.4), (2.7), and (2.10) permit us to reduce the solution of any problem with arbitrary initial conditions to a plane or spherical wave problem.

2.2 Heuristic field approximations

The following heuristic analytical approximations are well-known:

(a) the phase approximation of the Huygens-Kirchhoff method, which was first used by Feyzulin and Kravtsov[3] and analyzed by Banakh and Mironov;[9]
(b) the approximation that we call here the phase approximation of the plane wave expansion method, first used by Gochelashvili;[11] and
(c) the spectral phase approximation of the Huygens-Kirchhoff method, described by Aksenov and Mironov.[13]

We consider these as heuristic approximations because they were derived by replacing some part of an exact expression by an approximate one using some simple reasons, mainly motivated by obvious computational advantages. But their limits of applicability are still not properly investigated. We will show that these approximations have a direct connection with the exact field representations discussed above.

(a) The main formula of the phase approximation of the Huygens-Kirchhoff method is

$$u_s(\mathbf{R}, Z) = \frac{k}{2\pi i Z} \int d^2 R_0 \; u^0(\mathbf{R}_0) \exp\left\{ \frac{ik}{2} \left[\frac{|\mathbf{R} - \mathbf{R}_0|^2}{Z} + \int_0^Z \tilde{\epsilon}\left(\mathbf{R}_0 + \frac{z(\mathbf{R} - \mathbf{R}_0)}{Z}, z \right) dz \right] \right\}. \tag{2.12}$$

(b) The formula of the phase approximation of the plane wave expansion method is

$$u_p(\mathbf{R}, z) = \int d^2 P_0 \; \hat{u}^0(\mathbf{P}_0) \exp\left[i\mathbf{P}_0 \cdot \mathbf{R} - \frac{iP_0^2 Z}{2k} + \frac{ik}{2} \int_0^Z \tilde{\epsilon}\left(\mathbf{R} - \frac{(Z - z)\mathbf{P}_0}{k}, z \right) dz \right]. \tag{2.13}$$

(c) The formula of the spectral phase approximation of the Huygens-Kirchhoff method is

$$\hat{u}_{sp}(\mathbf{P}, Z) = \frac{1}{4\pi^2} \int d^2 R_0 \; u^0(\mathbf{R}_0) \exp\left[-i\mathbf{P} \cdot \mathbf{R}_0 - \frac{iP^2 Z}{2k} + \frac{ik}{2} \int_0^Z \tilde{\epsilon}(\mathbf{R}_0 + \frac{\mathbf{P} z}{k}, z) dz \right]. \tag{2.14}$$

3. Path integral representations of a wave in inhomogeneous media

In this section, several different wave field representations are obtained, based on some simple transformations of the functional variable. All these representations are equivalent, but they give different insight into the problem and can serve as starting points for different analytical approximations and numerical techniques.

3.1 Unconditional and conditional Feynman path integrals

The solution of (2.1) with the initial condition (2.2) can be written in the form of the unconditional Feynman path integral[18,19]

$$u(\mathbf{R}, Z) = \int^{\mathbf{s}(Z)=\mathbf{R}} \mathcal{D}^2 s(z) u^0 \left[\mathbf{s}(0)\right]$$

$$\times \exp\left\{\frac{ik}{2}\int_0^Z \left[\frac{d\mathbf{s}(z)}{dz}\right]^2 dz + \frac{ik}{2}\int_0^Z \tilde{\epsilon}[\mathbf{s}(z), z] dz\right\}. \qquad (3.1)$$

or in the form of the conditional Feynman path integral

$$u(\mathbf{R}, Z) = \int dR_0^2 u^0(\mathbf{R}_0) \int_{\mathbf{s}(0)=\mathbf{R}_0}^{\mathbf{s}(Z)=\mathbf{R}} \mathcal{D}^2 s(z)$$

$$\times \exp\left\{\frac{ik}{2}\int_0^Z \left[\frac{d\mathbf{s}(z)}{dz}\right]^2 dz + \frac{ik}{2}\int_0^Z \tilde{\epsilon}[\mathbf{s}(z), z] dz\right\}. \qquad (3.2)$$

The definition of $\mathcal{D}^2 s(z)$ and the corresponding normalization condition can be easily formulated by a generalization of the one-dimensional case considered in Ref. 20. The path restrictions included in the definition of the unconditional Feynman integral (3.1) can be replaced by an appropriate δ-function in the integrand. The same is valid for the conditional integral also. So for the general solution we obtain

$$u(\mathbf{R}, Z) = \int d^2 R_0 u^0(\mathbf{R}_0) \int \mathcal{D}^2 s(z) \delta[\mathbf{s}(0) - \mathbf{R}_0] \delta[\mathbf{s}(Z) - \mathbf{R}]$$

$$\times \exp\left\{\frac{ik}{2}\int_0^Z \left[\frac{d\mathbf{s}(z)}{dz}\right]^2 dz + \frac{ik}{2}\int_0^Z \tilde{\epsilon}[\mathbf{s}(z), z] dz\right\}. \qquad (3.3)$$

Comparing (3.3) with (2.4), we find that Green's function is

$$G(\mathbf{R}, Z; \mathbf{R}_0) = \int \mathcal{D}^2 s(z) \delta[\mathbf{s}(0) - \mathbf{R}_0] \delta[\mathbf{s}(Z) - \mathbf{R}]$$

$$\times \exp\left\{\frac{ik}{2}\int_0^Z \left[\frac{d\mathbf{s}(z)}{dz}\right]^2 dz + \frac{ik}{2}\int_0^Z \tilde{\epsilon}[\mathbf{s}(z), z] dz\right\}. \qquad (3.4)$$

3.2 Velocity representation for path integral variables

Let us introduce the variable:[21,22]

$$\mathbf{v}(z) = -\frac{d\mathbf{s}(z)}{dz}, \qquad \mathbf{s}(Z) = \mathbf{R}, \tag{3.5}$$

which is equivalent to

$$\mathbf{s}(z) = \mathbf{R} + \int_z^Z \mathbf{v}(\zeta)d\zeta. \tag{3.6}$$

Inserting (3.6) into (3.1) gives

$$u(\mathbf{R}, Z) = \int D^2 v(z) u^0 \left[\mathbf{R} + \int_0^Z \mathbf{v}(z)dz\right]$$

$$\times \exp\left\{\frac{ik}{2}\int_0^Z v^2(z)dz + \frac{ik}{2}\int_0^Z \tilde{\epsilon}\left[\mathbf{R} + \int_z^Z \mathbf{v}(\zeta)d\zeta, z\right]dz\right\}. \tag{3.7}$$

This representation is called the velocity representation because it can be related to some velocity $v(\tau) = dx(\tau)/d\tau$. This new variable seems to be more convenient for analytical transformations than the coordinate variable. Note that the path boundary condition $\mathbf{s}(Z) = \mathbf{R}$ is included in (3.6), and it is clear that by definition $\mathbf{v}(z)$ is not restricted for any z. We use here a new notation D for a differential to emphasize that

(a) the velocity functional variable $\mathbf{v}(z)$ belongs to a different functional space, and

(b) the functional integral over velocities is normalized by

$$\int D^2 v(z) \exp\left[\frac{ik}{2}\int_0^Z v^2(z)dz\right] = 1. \tag{3.7$'$}$$

Using the velocity integral for Green's function (3.4) we get

$$G(\mathbf{R}, Z; \mathbf{R}_0) = \int D^2 v(z) \delta\left[\mathbf{R} - \mathbf{R}_0 + \int_0^Z \mathbf{v}(\zeta)d\zeta\right]$$

$$\times \exp\left\{\frac{ik}{2}\int_0^Z v^2(z)dz + \frac{ik}{2}\int_0^Z \tilde{\epsilon}\left[\mathbf{R} + \int_z^Z \mathbf{v}(\zeta)d\zeta, z\right]dz\right\}. \tag{3.8}$$

Subtracting the mean slope of the trajectory $\mathbf{v}(\zeta)$ between the source and the observation point, which is just a shift of variable, i.e. $\mathbf{v}(\zeta) \to \mathbf{v}(\zeta) - (\mathbf{R} - \mathbf{R}_0)/Z$, leads to the following form:

$$G(\mathbf{R}, Z; \mathbf{R}_0) = \exp\left[\frac{ik(\mathbf{R}-\mathbf{R}_0)^2}{2Z}\right] \int D^2 v(z)\delta\left[\int_0^Z \mathbf{v}(z)dz\right] \exp\left\{\frac{ik}{2}\int_0^Z v^2(z)dz \right.$$

$$\left. + \frac{ik}{2}\int_0^Z \tilde{\epsilon}\left[\mathbf{R}_0 + \frac{z(\mathbf{R}-\mathbf{R}_0)}{Z} + \int_z^Z \mathbf{v}(\zeta)d\zeta, z\right]dz\right\}. \tag{3.9}$$

Inserting (3.9) into (2.4), we obtain the functional integral form of the spherical wave expansion:

$$u(\mathbf{R}, Z) = \int d^2 R_0 u^0(\mathbf{R}_0) \exp\left[\frac{ik(\mathbf{R}-\mathbf{R}_0)^2}{2Z}\right]$$

$$\times \int D^2 v(z)\delta\left[\int_0^Z \mathbf{v}(z)dz\right] \exp\left\{\frac{ik}{2}\int_0^Z v^2(z)dz \right.$$

$$\left. + \frac{ik}{2}\int_0^Z \tilde{\epsilon}\left[\mathbf{R}_0 + \frac{z(\mathbf{R}-\mathbf{R}_0)}{Z} + \int_z^Z \mathbf{v}(\zeta)d\zeta, z\right]dz\right\}. \tag{3.10}$$

3.3 The plane wave expansion for a field in the form of the path integral

We note that our transformations from (3.1) to (3.3) allowed us to represent end restrictions by means of δ-functions. We use this simple method to express the boundary conditions of path variables in an implicit way. Applying the Fourier expansion of the δ-function in (3.8), we obtain

$$G(\mathbf{R}, Z; \mathbf{R}_0) = \int \frac{d^2 p}{4\pi^2} \int D^2 v(\zeta) \exp\left\{i\mathbf{p}\cdot\left[\int_0^Z \mathbf{v}(\zeta)d\zeta + \mathbf{R} - \mathbf{R}_0\right]\right\}$$

$$\times \exp\left\{\frac{ik}{2}\int_0^Z v^2(z)dz + \frac{ik}{2}\int_0^Z \tilde{\epsilon}\left[\mathbf{R} + \int_z^Z \mathbf{v}(\zeta)d\zeta, z\right]dz\right\}. \tag{3.11}$$

If we now use the Fourier transform of the initial condition (2.6), insert (2.6) and (3.11) into (2.4), and shift the velocity variable $\mathbf{v}(\zeta) \to \mathbf{v}(\zeta) - \mathbf{P}_0/k$, we are able to integrate over $\mathbf{R}_0$. This integration gives $4\pi^2\delta(\mathbf{P}_0 - \mathbf{p})$, and the final result becomes

$$u(\mathbf{R}, Z) = \int d^2 P_0 \hat{u}^0(\mathbf{P}_0) \exp\left(i\mathbf{P}_0 \cdot \mathbf{R} - \frac{iP_0^2 Z}{2k}\right) \int D^2 v(z) \exp\left\{\frac{ik}{2}\int_0^Z v^2(z)dz\right.$$

$$\left. + \frac{ik}{2}\int_0^Z \tilde{\epsilon}\left[\mathbf{R} - \frac{\mathbf{P}_0(Z-z)}{k} + \int_z^Z \mathbf{v}(\zeta)d\zeta, z\right]dz\right\}, \qquad (3.12)$$

which is the functional integral form of the outgoing plane wave expansion (2.7).

Now we obtain another representation of the field. Beginning from (3.10) we can replace

$$\int_z^Z \mathbf{v}(\zeta)d\zeta \text{ by } \quad -\int_0^z \mathbf{v}(\zeta)d\zeta \qquad (3.13)$$

in an argument of $\tilde{\epsilon}$, using the equation $\int_z^Z(\zeta)d\zeta = 0$ stemming from the δ-function in (3.10). Again, as in the case of (3.11), we use Fourier expansion of the δ-function and shift a variable, but then, we perform a Fourier transform on the final field (3.10), making it possible to integrate over $\mathbf{R}$. The final result,

$$\hat{u}(\mathbf{P}, Z) = (2\pi)^{-4}\int d^2 R_0 u^0(\mathbf{R}_0) \exp\left(-i\mathbf{P}\cdot\mathbf{R}_0 - \frac{iP^2 Z}{2k}\right)$$

$$\times \int D^2 v(z) \exp\left\{\frac{ik}{2}\int_0^Z v^2(z)dz\right.$$

$$\left. + \frac{ik}{2}\int_0^Z \tilde{\epsilon}[\mathbf{R}_0 + \mathbf{P}z/k - \int_0^z \mathbf{v}(\zeta)d\zeta, z]dz\right\}, \qquad (3.14)$$

is the functional form of the incoming plane wave expansion, which is seen by comparing (3.14) with (2.10).

3.4 The connection between heuristic approximations and path integral representations

Now we can easily connect the exact representation of fields (3.10), (3.12), and (3.14) and the following heuristic analytical approximations: (a) (2.12), the phase approximation of the Huygens-Kirchhoff method; (b) (2.13), the phase approximation of the plane wave expansion method; and (c) (2.14), the spectral phase approximation of the Huygens-Kirchhoff method.

It is seen that (2.12) can be derived from the spherical wave expansion (3.10) if one neglects the dependence over $\mathbf{v}(\zeta)$ in the argument in $\tilde{\epsilon}$ and integrates over $\mathbf{v}(z)$, using the formula stemming from (3.7′)

$$\int D^2 v(z)\delta\Big[\int_0^Z \mathbf{v}(z)dz\Big]\exp\left[\frac{ik}{2}\int_0^Z v^2(z)dz\right]=\frac{k}{2\pi iZ}. \tag{3.15}$$

In the same way (2.13) can be obtained from outgoing plane wave expansion (3.12), again omitting the $\mathbf{v}(\zeta)$ dependence in the argument of $\tilde{\epsilon}$. Considering the incoming plane wave expansion (3.14) and omitting the $\mathbf{v}(\zeta)$ dependence in the argument of $\tilde{\epsilon}$, we obtain the first term of the spectral phase approximation of the Huygens-Kirchhoff method.

All these approximations can be called "straight-line approximations" because they can be obtained by neglecting all the deviations of the paths from the straight line. In terms of the probabilistic interpretation of paths, this means replacing the averaging of a functional over paths by a substitution of the mean path value instead of the path realization. In terms of orthogonal expansions of paths, these approximations are equivalent to neglecting all the terms of expansion but zero-order terms.

Approximations (2.12)-(2.14) were produced heuristically by neglecting the amplitude fluctuations in corresponding spherical and plane waves and assuming a simple geometrical optics formula for the phase fluctuations. These approximations ignore the cumulative effect of the perturbed paths in the contribution to the field by inhomogeneities. The connection between the phase approximation of the Huygens-Kirchhoff method and the functional integral representation was discussed in Refs. 1 and 23. It should be noted that, although the spherical wave expansion, the outgoing plane wave expansion, and the incoming plane wave expansion are simply different forms of the same exact solution of (2.1), approximations (2.12)-(2.14) are quite different if $\tilde{\epsilon}\neq 0$.

It is obvious that we can obtain a series expansion of the exact path representations of fields (3.10), (3.12), and (3.14) using (2.12)-(2.14) as first terms. For example we can rewrite (3.10) as

$$\begin{aligned}
u(\mathbf{R},Z) &= \int d^2R_0 u^0(\mathbf{R}_0)\exp\left[\frac{ik(\mathbf{R}-\mathbf{R}_0)^2}{2Z}\right]\int D^2v(z)\delta\left[\int_0^Z \mathbf{v}(z)dz\right]\\
&\times\exp\left\{\frac{ik}{2}\int_0^Z v^2(z)dz+\frac{ik}{2}\int_0^Z \tilde{\epsilon}\left[\mathbf{R}_0+\frac{z(\mathbf{R}-\mathbf{R}_0)}{Z},z\right]dz\right\}\\
&\times\exp\left\{\frac{ik}{2}\int_0^Z\left[\tilde{\epsilon}\left(\mathbf{R}_0+\frac{z(\mathbf{R}-\mathbf{R}_0)}{Z}+\int_z^Z \mathbf{v}(\zeta)d\zeta,z\right)\right.\right.\\
&\left.\left.-\tilde{\epsilon}\left(\mathbf{R}_0+\frac{z(\mathbf{R}-\mathbf{R}_0)}{Z},z\right)\right]dz\right\}.
\end{aligned} \tag{3.16}$$

Now we can expand the last exponential in (3.16) in a series and perform the $\mathbf{v}(z)$ integration to obtain the series in which the first term coincides with (2.12).

4. Path integral representations of moments

Equations for statistical moments of the field are of the same parabolic type as the instantaneous field equation (2.1). So the path integral representation of a solution is available, similar to the representations described in Section 3. But here we use the solution of (2.1) in the form of the path integral that provides an explicit expression for $u(\mathbf{r}, \mathbf{z})$. Then we construct the statistical moments Γ_{nm}, using the Gaussian δ-correlated model of $\tilde{\epsilon}$ fluctuations.

4.1 The path integral representation for the second moment

The general scheme of the computation of the statistical moments is demonstrated by the second moment:

$$\Gamma_2(\mathbf{r}_1, \mathbf{r}_1', Z) \equiv \langle u(\mathbf{r}_1, Z) u^*(\mathbf{r}_1', Z) \rangle$$

$$= \mathbf{LL}' \Big\langle \exp\Big\{ \frac{ik}{2} (\varphi[\mathbf{s}_1(z)] - \varphi[\mathbf{s}_1'(z)]) \Big\} \Big\rangle, \tag{4.1}$$

where $\mathbf{L}$ symbolyzes the path integration, and

$$\varphi[\mathbf{s}(z)] = \int_0^Z \tilde{\epsilon}\left[\mathbf{s}(z) \equiv \mathbf{r} + \int_z^Z \mathbf{v}(\zeta) d\zeta, z \right] dz.$$

For Gaussian statistics of $\varphi(\mathbf{s}) - \varphi(\mathbf{s}')$, (4.1) takes the form

$$\Gamma_2 = \mathbf{LL}' \exp\Big\{ -\frac{k^2}{8} \int_0^Z dz \int_0^Z dz' \big[B(\mathbf{s}_1(z), \mathbf{s}_1(z'); z, z') + B(\mathbf{s}_1'(z), \mathbf{s}_1'(z'); z, z')$$

$$- B(\mathbf{s}_1(z), \mathbf{s}_1'(z'); z, z') + B(\mathbf{s}_1'(z), \mathbf{s}_1(z'); z, z') \big] \Big\}, \tag{4.2}$$

where

$$B(\mathbf{r}, \mathbf{r}'; z, z') = \langle \tilde{\epsilon}(\mathbf{r}, z) \tilde{\epsilon}(\mathbf{r}', z') \rangle.$$

The next step that helps us simplify (4.2) is the Markov approximation for the longitudinal correlation of $\tilde{\epsilon}$. This means that the correlation function $B(\mathbf{r}, \mathbf{r}'; z, z')$ is approximated by

$$B(\mathbf{r}, \mathbf{r}'; z, z') = \delta(z - z') A(\mathbf{r} - \mathbf{r}', z).$$

Let us consider a general case of the spherical wave expansion. For the second moment we introduce new variables in (4.1):

$$\mathbf{r}_0 = \mathbf{r}_1 - \mathbf{r}_1'; \qquad \mathbf{b}(z) = \mathbf{v}_1(z) - \mathbf{v}_1'(z);$$

$$\mathbf{R} = (\mathbf{r}_1 + \mathbf{r}_1')/2; \qquad \mathbf{V}(z) = [\mathbf{v}_1(z) + \mathbf{v}_1'(z)]/2.$$

Then using the above-mentioned scheme for the case of the spherical wave expansion, we obtain the following formula:

$$\Gamma_2(\mathbf{r}_0, \mathbf{R}, Z)$$

$$= \int d^2r_0'd^2R' \; \Gamma_2^0(\mathbf{r}_0', \mathbf{R}') \exp\left\{\frac{ik}{Z}[(\mathbf{R}-\mathbf{R}')\cdot(\mathbf{r}_0-\mathbf{r}_0')]\right\} \int D^2V(z)D^2b(z)$$

$$\times \delta\Big[\int_0^Z \mathbf{V}(z)dz\Big]\delta\Big[\int_0^Z \mathbf{b}(z)dz\Big]\exp\Big[ik\int_0^Z \mathbf{V}(z)\cdot\mathbf{b}(z)\,dz\Big]$$

$$\times \exp\left\{-\frac{\pi k^2}{4}\int_0^Z H\left[\mathbf{r}_0' + \frac{z(\mathbf{r}_0-\mathbf{r}_0')}{Z} + \int_z^Z \mathbf{b}(\zeta)d\zeta, z\right]dz\right\}, \tag{4.3}$$

where $H(\mathbf{r}, z) = \frac{1}{\pi}[A(0,z) - A(\mathbf{r}, z)]$. Here the dependence over $\mathbf{V}(z)$ is also in the δ-function argument, but we can use the Fourier expansion of the δ-function to rewrite the integral over $\mathbf{V}(z)$ as

$$\Psi[\mathbf{b}(\cdot)] = (2\pi)^{-2}\int d^2p\int D^2V(z)\exp\left\{ik\int_0^Z \mathbf{V}(z)\cdot[\mathbf{b}(z)-\mathbf{p}/k]\,dz\right\}. \tag{4.4}$$

Then we apply the formula[24]

$$\int D^2V(z)\exp\left[ik\int_0^Z \mathbf{V}(z)\cdot\mathbf{b}(z)\,dz\right] = \delta_\infty[\mathbf{b}(z)] \tag{4.5}$$

to (4.4), where $\delta_\infty(\cdot)$ is a δ-function functional, i.e.,

$$\int D^2b(z)\;\delta_\infty[\mathbf{b}(z)-\mathbf{b}_0(z)]\;\Phi[\mathbf{b}(z)] = \Phi[\mathbf{b}_0(z)].$$

Here $\Phi[\mathbf{b}(z)]$ is a continuous functional. As a result (4.4) takes the form:

$$\Psi[\mathbf{b}(\cdot)] = (2\pi)^{-2}\int d^2p\delta_\infty[\mathbf{b}(\cdot)-\mathbf{p}/k]. \tag{4.6}$$

After substituting this formula into (4.3) and integrating over $\mathbf{b}(z)$ and $\mathbf{p}$, we obtain

$$\Gamma_2(\mathbf{r}_0, \mathbf{R}, Z) = \frac{k^2}{4\pi^2Z^2}\int d^2R'd^2r_0' \; \Gamma_2^0(\mathbf{r}_0', \mathbf{R}')\exp\left\{\frac{ik}{2Z}[(\mathbf{R}-\mathbf{R}')\cdot(\mathbf{r}_0-\mathbf{r}_0')]\right\}$$

$$\times \exp\left\{-\frac{\pi k^2}{4}\int_0^Z H\left[\mathbf{r}_0' + \frac{z(\mathbf{r}_0-\mathbf{r}_0')}{Z}, z\right]dz\right\}. \tag{4.7}$$

This analytical formula for Γ_2 is obtained here using the path integral technique as it was first derived by Klyatskin and Tatarskii.[25] The same expression can be obtained as a solution of the corresponding equation for the second moment.[26]

4.2 The fourth moment: spherical wave and plane wave expansions

The definition for the fourth moment is

$$\Gamma_4(\mathbf{R}_1, \mathbf{R}_1', \mathbf{R}_2, \mathbf{R}_2'; Z) \equiv < u(\mathbf{R}_1, Z) u^*(\mathbf{R}_1', Z) u(\mathbf{R}_2, Z) u^*(\mathbf{R}_2', Z) > . \qquad (4.8)$$

Further we use new variables:[17]

$$\mathbf{r}_1 = (\mathbf{R}_1 - \mathbf{R}_2 - \mathbf{R}_1' + \mathbf{R}_2')/2; \qquad \mathbf{R} = (\mathbf{R}_1 + \mathbf{R}_2 + \mathbf{R}_1' + \mathbf{R}_2')/4;$$

$$\mathbf{r}_2 = (\mathbf{R}_1 - \mathbf{R}_2 + \mathbf{R}_1' - \mathbf{R}_2')/2; \qquad \mathbf{r} = \mathbf{R}_1 + \mathbf{R}_2 - \mathbf{R}_1' - \mathbf{R}_2'.$$

The same type of transformation we use for integration variables in regular and path integrals. Substituting (3.10) into (4.8) and using Gaussian statistics of $\varphi_1 - \varphi_1' + \varphi_2 - \varphi_2'$ and the Markov approximation for the longitudinal correlation of $\tilde{\epsilon}$, we obtain the spherical wave expansion for the fourth moment

$$\Gamma_4(\mathbf{r}_1, \mathbf{r}_2, \mathbf{r}, \mathbf{R}; Z)$$

$$= \int d^2 r_{10} \int d^2 r_{20} \int d^2 R_0 \int d^2 r_0 \; \Gamma_4^0(\mathbf{r}_{10}, \mathbf{r}_{20}, \mathbf{R}_0, \mathbf{r}_0)$$

$$\times \exp\left\{ \frac{ik}{Z} [(\mathbf{R} - \mathbf{R}_0) \cdot (\mathbf{r} - \mathbf{r}_0) + (\mathbf{r}_1 - \mathbf{r}_{10}) \cdot (\mathbf{r}_2 - \mathbf{r}_{20})] \right\}$$

$$\times \int D^2 v_1 D^2 v_2 D^2 V D^2 v \delta \left[\int_0^Z \mathbf{V}(z) dz \right] \delta \left[\int_0^Z \mathbf{v}(z) dz \right]$$

$$\times \delta \left[\int_0^Z \mathbf{v}_1(z) dz \right] \delta \left[\int_0^Z \mathbf{v}_2(z) dz \right] \exp\left\{ \frac{ik}{2} \int_0^Z [\mathbf{V}(z) \cdot \mathbf{v}(z) + \mathbf{v}_1(z) \cdot \mathbf{v}_2(z)] dz \right.$$

$$- \frac{\pi k^2}{4} \int_0^Z F_4 \Big[\mathbf{r}_{10} + \frac{z(\mathbf{r}_1 - \mathbf{r}_{10})}{Z} + \int_z^Z \mathbf{v}_1(\zeta) d\zeta, \;\; \mathbf{r}_{20} + \frac{z(\mathbf{r}_2 - \mathbf{r}_{20})}{Z}$$

$$\left. + \int_z^Z \mathbf{v}_2(\zeta) d\zeta, \;\; \mathbf{r}_0 + \frac{z(\mathbf{r} - \mathbf{r}_0)}{Z} + \int_z^Z \mathbf{v}(\zeta) d\zeta; z \Big] \right\}, \qquad (4.9)$$

where

$$F_4(\mathbf{r}_1, \mathbf{r}_2, \mathbf{r}; z) = H(\mathbf{r}_1 + \mathbf{r}/2, z) + H(\mathbf{r}_1 - \mathbf{r}/2, z) + H(\mathbf{r}_2 + \mathbf{r}/2, z)$$

$$+ H(\mathbf{r}_2 - \mathbf{r}/2, z) - H(\mathbf{r}_1 + \mathbf{r}_2, z) - H(\mathbf{r}_1 - \mathbf{r}_2, z). \qquad (4.10)$$

After integration over $\mathbf{V}(z)$, using (4.5) and (4.6), we obtain

$$\Gamma_4(\mathbf{r}_1, \mathbf{r}_2, \mathbf{r}, \mathbf{R}; Z)$$

$$= \int d^2R_0 d^2r_{10} d^2r_{20} d^2r_0 \; \Gamma_4^0(\mathbf{r}_{10}, \mathbf{r}_{20}, \mathbf{r}_0, \mathbf{R}_0)$$

$$\times \exp\left\{\frac{ik}{Z}[(\mathbf{R}-\mathbf{R}_0)\cdot(\mathbf{r}-\mathbf{r}_0) + (\mathbf{r}_1-\mathbf{r}_{10})\cdot(\mathbf{r}_2-\mathbf{r}_{20})]\right\}$$

$$\times \int D^2v_1 D^2v_2 \delta\left[\int_0^Z \mathbf{v}_1(z)dz\right]\delta\left[\int_0^Z \mathbf{v}_2(z)dz\right]\exp\left[\frac{ik}{2}\int_0^Z \mathbf{v}_1(z)\cdot\mathbf{v}_2(z)\, dz\right]$$

$$\times \exp\left\{-\frac{\pi k^2}{4}\int_0^Z F_4\left[\mathbf{r}_{10} + \frac{z(\mathbf{r}_1-\mathbf{r}_{10})}{Z} + \int_z^Z \mathbf{v}_1(\zeta)d\zeta,\ \mathbf{r}_{20} + \frac{z(\mathbf{r}_2-\mathbf{r}_{20})}{Z}\right.\right.$$

$$\left.\left. + \int_z^Z \mathbf{v}_2(\zeta)d\zeta,\ \mathbf{r}_0 + \frac{z(\mathbf{r}-\mathbf{r}_0)}{Z};\ z\right]\right\}. \tag{4.11}$$

Note that only two functional variables are present in (4.11) and the result depends on $\mathbf{R}$ and $\mathbf{r}$ as simple (not functional) variables. Since the main problem with (4.11) is connected with functional variables, for simplicity's sake we further limit ourselves by the spatially averaged fourth moment and set $\mathbf{r} = 0$:

$$\Gamma_4(\mathbf{r}_1, \mathbf{r}_2; Z)$$

$$= \int d^2R \; \Gamma_4(\mathbf{r}_1, \mathbf{r}_2, 0, \mathbf{R}; Z)$$

$$= \int d^2r_{10} d^2r_{20} \; \Gamma_4^0(\mathbf{r}_{10}, \mathbf{r}_{20}) \exp\left\{\frac{ik}{Z}[(\mathbf{r}_1-\mathbf{r}_{10})\cdot(\mathbf{r}_2 - \mathbf{r}_{20})]\right\}$$

$$\times \int D^2v_1 D^2v_2 \delta\left[\int_0^Z \mathbf{v}_1(z)dz\right]\delta\left[\int_0^Z \mathbf{v}_2(z)dz\right]\exp\left[\frac{ik}{2}\int_0^Z \mathbf{v}_1(z)\cdot\mathbf{v}_2(z)dz\right]$$

$$\times \exp\left\{\int_0^Z F_4\left[\mathbf{r}_{10} + \frac{z(\mathbf{r}_1-\mathbf{r}_{10})}{Z} + \int_z^Z \mathbf{v}_1(\zeta)d\zeta,\right.\right.$$

$$\left.\left.\mathbf{r}_{20} + \frac{z(\mathbf{r}_2-\mathbf{r}_{20})}{Z} + \int_z^Z \mathbf{v}_2(\zeta)d\zeta;\ z\right]\right\}, \tag{4.12}$$

where the scattering function is

$$F_4(\mathbf{r}_1, \mathbf{r}_2; z) = 2H(\mathbf{r}_1, z) + 2H(\mathbf{r}_1, z) - H(\mathbf{r}_1 + \mathbf{r}_2, z) - H(\mathbf{r}_1 - \mathbf{r}_2, z).$$

Note that the presence of the variables $\mathbf{v}_1$ and $\mathbf{v}_2$ in the scattering function F makes it impossible to calculate (4.12) analytically. In the case of the second moment, only one of the two functional variables was present in H. This made it possible to calculate the functional integral analytically, using (4.5) and (4.6).

Here we do not consider approaches[1,2] based on asymptotical computation of (4.12). We address instead another idea, which uses the principle of simplification of the corresponding path integral under some reliable assumptions. This allows us to reduce the initial path integral to some ordinary finite-fold integral, which can be calculated analytically or numerically. We might start directly from (4.11) or (4.12), but other new approximate formulae exist for the statistical field moments using various expansions described above. According to them formula (4.11) represents the fourth moment of the field in the spherical wave expansion (3.10). To compose the statistical moment of the wave, one can produce two different forms by multiplying and averaging forms of the field, presented by the outgoing plane wave expansion (3.12) and by the incoming plane wave expansion (3.14). Several different formulae can be derived using mixed representations of the product of the fields. For our convenience, we prefer to use the velocity representations for the variables of Green's function, but coordinate representations can be used as well. It is possible also to derive the forms corresponding to the outgoing plane wave expansion and to the incoming plane wave expansion directly from the spherical wave expansion (4.11).

4.3 The outgoing plane wave expansion

As in the case of the fields (2.7) and (3.12) dealt with before, we can exclude the initial boundary conditions on both paths. The first step is Fourier representation of the δ-functions in (4.12), written as

$$\delta\Big[\int_0^Z \mathbf{v}_{1,2}(z)dz\Big] = \frac{1}{4\pi^2}\int d^2p_{2,1}\exp\Big[i\mathbf{p}_{2,1}\cdot\int_0^Z \mathbf{v}_{1,2}(z)dz\Big]. \tag{4.13}$$

Then we shift the functional variables

$$\mathbf{v}_j(\zeta) \to \mathbf{v}_j(\zeta) - \mathbf{p}_j/k, \qquad j = 1,2, \tag{4.14}$$

and change the variables

$$\mathbf{p}_i = \mathbf{q}_i - (\mathbf{r}_i - \mathbf{r}_{i0})k/z. \tag{4.15}$$

As a result we can rewrite (4.12) as

$$\Gamma_4(\mathbf{r}_1,\mathbf{r}_2;Z) = \frac{1}{(2\pi)^4}\int d^2r_{10}d^2r_{20}\Gamma_4^0(\mathbf{r}_{10},\mathbf{r}_{20})\,G(\mathbf{r}_1,\mathbf{r}_2;Z;\mathbf{r}_{10},\mathbf{r}_{20}), \tag{4.16}$$

where

$$G = \int d^2q_1 d^2q_2 \exp\left[-i\frac{Z}{k}\mathbf{q}_1 \cdot \mathbf{q}_2 + i\mathbf{q}_1 \cdot (\mathbf{r}_2 - \mathbf{r}_{20}) + i\mathbf{q}_2 \cdot (\mathbf{r}_1 - \mathbf{r}_{10})\right] \int D^2v_1 D^2v_2$$

$$\times \exp\left\{ik \int_0^Z \mathbf{v}_1(z) \cdot \mathbf{v}_2(z)\, dz - \frac{\pi k^2}{4} \int_0^Z F_4\left[\mathbf{r}_1 - \mathbf{q}_1\frac{(Z-z)}{k}\right.\right.$$

$$\left.\left. + \int_z^Z \mathbf{v}_1(\zeta)d\zeta,\ \mathbf{r}_2 - \mathbf{q}_2\frac{(Z-z)}{k} + \int_z^Z \mathbf{v}_2(\zeta)d\zeta;\ z\right]\right\}. \tag{4.17}$$

This Green function representation of the problem differs from the representation in (4.12). It is more convenient to write (4.17) in terms of the complete Fourier transform of the moment at the initial plane, defined by

$$\tilde{\Gamma}_0(\mathbf{q}_2, \mathbf{q}_1) = \frac{1}{(2\pi)^4} \int d^2r_{10} d^2r_{20} \Gamma_4^0(\mathbf{r}_{10}, \mathbf{r}_{20}) \exp\left(-i\mathbf{q}_2 \cdot \mathbf{r}_{10} - i\mathbf{q}_1 \cdot \mathbf{r}_{20}\right). \tag{4.18}$$

Then we can integrate over $\mathbf{r}_{10}$ and $\mathbf{r}_{20}$. This produces a simplified form of Γ_4:

$$\Gamma_4(\mathbf{r}_1, \mathbf{r}_2; Z)$$

$$= \int d^2q_1 d^2q_2 \tilde{\Gamma}_4^0(\mathbf{q}_2, \mathbf{q}_1) \exp\left[-i\frac{Z}{k}\mathbf{q}_1 \cdot \mathbf{q}_2 + i\mathbf{q}_1 \cdot \mathbf{r}_2 + i\mathbf{q}_2 \cdot \mathbf{r}_1\right]$$

$$\times \int D^2v_1 D^2v_2 \exp\left\{ik \int_0^Z \mathbf{v}_1(z) \cdot \mathbf{v}_2(z)dz - \int_0^Z F_4\left[\mathbf{r}_1 - \mathbf{q}_1\frac{(Z-z)}{k}\right.\right.$$

$$\left.\left. + \int_z^Z \mathbf{v}_1(\zeta)d\zeta,\ \mathbf{r}_2 - \mathbf{q}_2\frac{(Z-z)}{k} + \int_z^Z \mathbf{v}_2(\zeta)d\zeta;\ z\right]dz\right\}. \tag{4.19}$$

Comparing (4.19) with (4.12) we can see that (4.19) contains no end restrictions on the functional variables, and comparing it with (3.12) we can see that the form of the arguments of $\tilde{\epsilon}$ in (3.12) and of $F_4(\cdot)$ in (4.19) are similar. In fact, (4.19) can be derived directly as a fourth moment of (3.12). So, we can treat (4.19) as a fourth moment in the form of the outgoing plane wave expansion.

4.4 The incoming plane wave expansion

We can exclude the final boundary condition on both paths in the same way as we did to derive (3.14). First, we make the replacement in (4.12)

$$\int_z^Z \mathbf{v}_i(\zeta)d\zeta \to -\int_0^z \mathbf{v}_i(\zeta)d\zeta, \qquad i = 1, 2,$$

which is possible because of the δ-functions in (4.12). Then we repeat the transformations from (4.13) through (4.19). The final result is more clearly expressed for the spectrum $\tilde{\Gamma}_4(\mathbf{q}_1, \mathbf{q}_2; Z)$ instead of $\Gamma_4(\mathbf{r}_1, \mathbf{r}_2; Z)$. The relationship between these two functions is analogous to (4.18). Following these steps, we finally have

$$\tilde{\Gamma}_4(\mathbf{q}_2, \mathbf{q}_1; Z)$$

$$= \frac{1}{(2\pi)^4} \int d^2r_{10} d^2r_{20} \Gamma_4^0(\mathbf{r}_{10}, \mathbf{r}_{20}) \exp\left(-i\frac{Z}{k}\mathbf{q}_1 \cdot \mathbf{q}_2 - i\mathbf{q}_1 \cdot \mathbf{r}_{20} - i\mathbf{q}_2 \cdot \mathbf{r}_{10}\right)$$

$$\times \int D^2v_1 D^2v_2 \exp\left\{ ik \int_0^Z \mathbf{v}_1(z) \cdot \mathbf{v}_2(z) dz - \frac{\pi k^2}{4} \int_0^Z F_4\left[\mathbf{r}_{10} + \frac{z}{k}\mathbf{q}_1\right.\right.$$

$$\left.\left. - \int_0^z \mathbf{v}_1(\zeta) d\zeta,\ \mathbf{r}_{20} + \frac{z}{k}\mathbf{q}_2 - \int_0^z \mathbf{v}_2(\zeta) d\zeta;\ z\right] dz \right\}. \qquad (4.20)$$

Again, comparing (4.20) with (3.14), we can see that the form of the arguments of $\tilde{\epsilon}$ in (3.14) and of $F_4(\cdot)$ in (4.20) are similar. In fact (4.20) can be derived directly as a fourth moment of (3.14). So, (4.20) is a fourth moment in the form of the incoming plane wave expansion.

4.5 The connection between exact fourth moments and corresponding heuristic approximations

Now, if we return to the spherical wave expansion for the fourth moment (4.12) and omit one of the functional arguments of the function $F_4(\cdot)$, then the path integration is trivial, and we obtain the formula of the phase approximation of the Huygens-Kirchhoff method for the fourth moment:

$$\Gamma_s(\mathbf{r}_1, \mathbf{r}_2; Z)$$

$$= \frac{k^2}{4\pi^2 Z^2} \int d^2r_{10} d^2r_{20}\ \Gamma_4^0(\mathbf{r}_{10}, \mathbf{r}_{20}) \exp\left[i\frac{k}{Z}(\mathbf{r}_1 - \mathbf{r}_{10}) \cdot (\mathbf{r}_2 - \mathbf{r}_{20}) \right]$$

$$\times \exp\left\{ -\frac{\pi k^2}{4} \int_0^Z F_4\left[\mathbf{r}_{10} + \frac{z}{Z}(\mathbf{r}_1 - \mathbf{r}_{10}), \mathbf{r}_{20} + \frac{z}{Z}(\mathbf{r}_2 - \mathbf{r}_{20}); z\right] dz \right\} \qquad (4.21)$$

which can be obtained also as the fourth moment of (2.12). This approximation was widely used for a variety of problems,[4-9,15] although it is clear that it has some drawbacks. First, it is not a difficult task to show that (4.21) gives zero scintillation index value for the point source without regard to intensity of medium fluctuations. Second, (4.21) does not match the energy conservation principle, having here the form

$$\int \Gamma_4(0, \mathbf{r}_2; Z) d^2r_2 = \ const. \qquad (4.22)$$

So, approximation (4.21) is known to be invalid for calculations of scintillation for small sources, large incoherent sources, and intensity spatial covariance for large separations. At the same time (2.12) gives the exact result for the second moment and qualitatively correct asymptotic results to the plane wave scintillation index.[15]

The same procedure of omitting the functional arguments applied to the outgoing plane wave expansion (2.13) gives us the formula of the phase approximation of the plane wave expansion for the fourth moment,

$$\Gamma_p(\mathbf{r}_1, \mathbf{r}_2; Z)$$

$$= \int d^2q_1 d^2q_2 \tilde{\Gamma}_4^0(\mathbf{q}_2, \mathbf{q}_1) \exp\left(-i\frac{Z}{k}\mathbf{q}_1 \cdot \mathbf{q}_2 + i\mathbf{q}_1 \cdot \mathbf{r}_2 + i\mathbf{q}_2 \cdot \mathbf{r}_1\right)$$

$$\times \exp\left\{-\int_0^Z F_4\left[\mathbf{r}_1 - \mathbf{q}_1\frac{(Z-z)}{k},\ \mathbf{r}_2 - \mathbf{q}_2\frac{(Z-z)}{k};\ z\right]dz\right\}. \quad (4.23)$$

which coincides with the fourth moment of (2.13). This approximation does not describe the plane wave scintillation index correctly, but gives qualitatively correct asymptotic results for the point source scintillation index.[15] It can be shown that $\Gamma_p(\mathbf{r}_1, \mathbf{r}_2; Z)$ does not match the energy conservation principle (4.22), or satisfy the reciprocity principle.

Neglecting both of the $\mathbf{v}(z)$ variables in the incoming plane wave expansion (4.20) gives us

$$\tilde{\Gamma}_{sp}(\mathbf{q}_2, \mathbf{q}_1; Z)$$

$$= \frac{1}{(2\pi)^4}\int d^2r_{10} d^2r_{20} \Gamma_4^0(\mathbf{r}_{10}, \mathbf{r}_{20}) \exp\left(-i\frac{z}{k}\mathbf{q}_1 \cdot \mathbf{q}_2 - i\mathbf{q}_1 \cdot \mathbf{r}_{20} - i\mathbf{q}_2 \cdot \mathbf{r}_{10}\right)$$

$$\times \exp\left[-\frac{\pi k^2}{4}\int_0^Z F_4\left(\mathbf{r}_{10} + \frac{z}{k}\mathbf{q}_1, \mathbf{r}_{20} + \frac{z}{k}\mathbf{q}_2; z\right)dz\right]. \quad (4.24)$$

which is the same as the fourth moment of (2.14), i.e., the first term of the spectral phase approximation of the Huygens-Kirchhoff method. This gives qualitatively correct asymptotic results for the point source and plane wave scintillation index,[13,14] but it needs two more integrations than (4.21) to obtain the second moment of intensity. It can be shown that $\tilde{\Gamma}_{sp}(\mathbf{q}_1, \mathbf{q}_2; Z)$ does not match the energy conservation principle (4.22), or satisfy the reciprocity principle. The last three formulae written in terms of $\tilde{\Gamma}_4$ can be found in the book by Mironov.[15]

We argue that the outcome of applying these approximations depends strongly on the problem to which they are applied. One can find many examples when the result was unsuccessful. For instance, the scintillation index of a spherical wave in the phase approximation of the Huygens-Kirchhoff method, and for a plane wave in the phase approximation of the plane wave expansion, which erroneously is zero for an arbitrary level of inhomogeneity. This means that all these approximations describe unsatisfactorily the intensity fluctuation phenomena, and need corrections and improvements.[9,14,23]

5. Conclusions

In this paper we considered the theory of wave propagation in a continuous large-scale random medium based on the path integral technique. Attention was mainly devoted to the connection between the path integral technique and different analytical approximations for solutions of the statistical moment equations, such as the phase approximation of the Huygens-Kirchhoff method and other heuristic approaches that are based on various plane wave expansions. A treatment of path integrals considered here helped us to clarify the nature of these approximations and indicate possible improvements.

If we look at the advantages of the path integral technique in wave propagation studies more generally, we note that this technique provides a powerful approach to a variety of propagation problems. One can expect further intensive exploitation of the path integral method, possibly for creating numerical techniques, and for expanding the field of an application, for example, for a random medium combined with the regular refractivity background, wave guides, resonators, or rough surfaces.

References

1. V. U. Zavorotnyi, V. I. Klyatskin, and V. I. Tatarskii, "Strong fluctuations of the intensity of electromagnetic waves in randomly inhomogeneous media," *Sov. Phys. JETP* **46**(2), 252-260 (1977).

2. R. Dashen, "Path integrals for waves in random media," *J. Math. Phys.* **20**(5), 894-920 (1979).

3. Z. I. Feyzulin and Yu. A. Kravtsov, "Broadening of a laser beam in a turbulent atmosphere," *Radiophys. Quantum Electron.* **10**(1), 33-35 (1967).

4. R. F. Lutomirskii and H. T. Yura, "Propagation of a finite optical beam in an inhomogeneous medium," *Appl. Opt.* **10**(7), 1652-1658, (1971).

5. H. T. Yura, "First and second moment of an optical wave propagating in a random medium: Equivalence of the solution of the Dyson and Bethe-Salpeter equation to that obtained by the Huygens-Fresnel principle," *J. Opt. Soc. Am.* **69**(7), 1292-1304, (1972).

6. M. H. Lee, J. F. Holmes, and J. R. Kerr, "Statistics of speckle propagation through the turbulent atmosphere," *J. Opt. Soc. Am.* **66**(11), 1164-1172 (1976).

7. R. L. Fante, "Electromagnetic beam propagation in turbulent media," *Proc. IEEE* **63**(12), 1669-1692 (1975).

8. R. L. Fante, "Wave propagation in random media: A systems approach," *Progress in Optics*, **22**, ed. E. Wolf, North-Holland, Amsterdam (1985).

9. V. A. Banakh and V. L. Mironov, "Phase approximation of the Huygens-Kirchhoff method in problems of laser beam propagation in the turbulent atmosphere," *Opt. Lett.* **1**(5), 172-174 (1977).

10. M. Tur and M. Beran, "Wave propagation in random media: A comparison of two theories," *J. Opt. Soc. Am.* **73**(10), 1343-1349 (1983).

11. K. S. Gochelashvili, "Saturation of the fluctuations of focused radiation in a turbulent medium," *Radiophys. Quantum Electron.* **14**(4), 470-473 (1971).

12. K. S. Gochelashvili, "Propagation of focused laser radiation in a turbulent medium," *Sov. J. Quant. Electron.* **4**(4), 465-470 (1974).

13. V. P. Aksenov and V. L. Mironov, "Spectral expansion method in problems of laser-beam propagation in the turbulent atmosphere," *Opt. Lett.* **3**(5), 184-186 (1978).

14. V. P. Aksenov and V. L. Mironov, "Method of spectral expansions in problems of optical wave propagation in turbulent media," *Radiophys. Quantum Electron.* **22**(5), 414-421 (1979).

15. V. L. Mironov, *Propagation of a Laser Beam in a Turbulent Atmosphere*, Nauka, Novosibirsk (1981) (in Russian).

16. V. I. Tatarskii and V. U. Zavorotnyi, "Strong fluctuations in light propagation in a randomly inhomogeneous media," *Progress in Optics*, **18**, ed. E. Wolf, North-Holland, Amsterdam (1980).

17. S. M. Rytov, Y. A. Kravtsov, and V. I. Tatarskii, *Principles of Statistical Radiophysics*, vols. 1-4, Springer, Berlin (1988).

18. R. P. Feynman and A. R. Hibbs, *Quantum Mechanics and Path Integrals*, McGraw-Hill, New York (1965).

19. P. L. Chow, "Application of function space integrals to problems in wave propagation in random media," *J. Math. Phys.* **13**(8), 1224-1236 (1972).

20. V. I. Tatarskii, M. I. Charnotskii, J. Gozani, and V. U. Zavorotny, "Path integral approach to wave propagation in random media. Part I: Derivations and various formulations," (this issue).

21. V. I. Klyatskin, *Statistical Description of Dynamic Systems with Fluctuating Parameters*, Nauka, Moscow (1975) (in Russian).

22. V. I. Klyatskin, *Stochastic Equations and Waves in Random Inhomogeneous Media*, Nauka, Moscow (1980) (in Russian) [V. I. Klyatskin, *Ondes et Equations Stochastique dans les Milieux Aleatoirement non Homogènes*, Editions de Physique, Besançon Cedex (1985) (in French)].

23. V. I. Tatarskii and V. U. Zavorotnyi, "On the connection between the extended Huygens-Fresnel principle and the path-integral approximate computation based on orthogonal expansions," *Proc. SPIE* **642**, 276-281 (1986).

24. E. A. Novikov, "The solutions of some variational differential equations," *Usp. Mat. Nauk SSSR* **16**(2), 135-141 (1961) (in Russian).

25. V. I. Klyatskin and V. I. Tatarskii, "The parabolic equation approximation for propagation of waves in a medium with random inhomogeneities," *Sov. Phys. JETP* **31**(2), 335-339 (1970).

26. V. I. Klyatskin and V. I. Tatarskii, "On the theory of the propagation of light beams in a medium having inhomogeneities," *Radiophys. Quantum Electron.* **13**(7), 828-833 (1970).

PATH INTEGRAL APPROACH TO WAVE PROPAGATION IN RANDOM MEDIA PART III: MIXED REPRESENTATION; ORTHOGONAL EXPANSION

J. Gozani, M. I. Charnotskii, V. I. Tatarskii, and V. U. Zavorotny

Cooperative Institute for Research in Environmental Sciences
University of Colorado/ NOAA/ Wave Propagation Laboratory
Boulder, Colorado 80309-0216

1. Introduction

A path integral formulation is clearly described in the book by Feynman and Hibbs.[1] A detailed discussion in our field of research can be found in a book by Klyatskin,[2] in the paper by Dashen,[3] and in a review by Tatarskii and Zavorotnyi.[4] The main use of the path integral until the mid-1980's was for physical interpretation and for asymptotic results. Simultaneously, other analytical techniques were developed. The extended Huygens-Fresnel principle[5], and the incoming[6] and the outgoing[7] plane wave expansions were widely used for practical purposes. However these heuristic techniques were suspected of being crude.

In 1982 Uscinski[8] derived an approximate solution for intensity correlation function, which gave good qualitative results. This solution is known as Uscinski's fundamental solution. It so happened that in 1983, the same approximate solution was obtained by Macaskill using a different technique.[9] He adopted the two-scale expansion method originally developed in 1982 by Frankenthal, Whitman, and Beran for acoustic propagation in the ocean.[10,11] This solution was called the two-scale zero-order solution. Some researchers were hopeful that this was more than a mere coincidence, especially when results of a numerical solution for a two-dimensional Kolmogorov medium agreed with the results by the two-scale expansion, with a first-order correction term.[11-13]

The two-scale expansion was admittedly an asymptotic method;[11,13] however, even in its regime of applicability results were unsatisfactory, and the method was criticized as being crude. At the end of the 1980's, Martin and Flatté[14] introduced results of a simulation that gave a scintillation peak much higher than the two-scale expansion zero-order and first-order approximations. Their simulations were in agreement with measurements obtained by Coles and Frehlich.[15] The resulting controversy attracted more attention in the 1990's, because additional two-scale computations and simulations disagreed.[16-18]

In 1985, another version of the two-scale expansion was developed by Gozani.[19] His method is not asymptotic, and introduced a solution in terms of a uniformly convergent iterated series. Therefore though the iteration is bound to converge, many terms of the series may be required to produce a result that is agreeable with the results of simulations. In their 1993 review[20] of the two-scale embedding technique, Beran *et al.* agreed that their two-scale formulation is not an asymptotic expansion. The controversy mentioned above is the motivation of our paper.

The present paper is the third in a series by the same authors, in which a path integral approach is adopted to study propagation in random media. In this section we define the problem and the path integral formulation. Then, in the second section, we introduce the mixed representation path integral of the fourth statistical moment. Beginning from the "exact" Feynman path integral, we show that heuristic arguments are involved in producing the two-scale zero-order approximation, and that the formulation is not asymptotic. Additionally, the mixed representation that we introduced provides a final link in the unification of all the heuristic approximations previously used: the extended Huygens-Fresnel principle,[5] and the incoming[6] and the outgoing plane wave formulations[7] that are discussed in detail in the second paper[21] in our three-part series. In Section 3, we provide a systematic way to compute path integrals. This method entails orthogonal expansion of the trajectories, previously discussed in the first paper[22] of our series. The preliminary computations were unsatisfactory; hence we suggest in Section 4 a future direction of research.

1.1 Governing equation

The fourth-order coherence function (or the fourth moment) is defined by $\Gamma_4 = \langle U_1\, U_2\, U_3^*\, U_4^* \rangle$. The wave function $U_i = U(\boldsymbol{\rho}_i, z)$ is the envelope of a quasi-monochromatic wave with wave number k. The transverse position vectors are denoted by $\boldsymbol{\rho}_i = (x_i, y_i)$, where $i = 1, 2, 3$, or 4, and z is the forward-propagation range. The symbols $\langle\ \rangle$ and * denote an ensemble average and a complex conjugation, respectively.

The partial differential equation that describes the propagation of the fourth moment is[23]

$$\frac{\partial}{\partial z}\Gamma_4 = \frac{i}{k}\nabla_{\mathbf{r}_1}\cdot\nabla_{\mathbf{r}_2}\Gamma_4 + \frac{i}{k}\nabla_{\mathbf{R}}\cdot\nabla_{\mathbf{r}}\Gamma_4 - \frac{\pi k^2}{4}F_4(\mathbf{r}_1, \mathbf{r}_2, \mathbf{r}, z)\,\Gamma_4 \qquad (1.1)$$

with the initial condition $\Gamma_4^0(\mathbf{r}_1, \mathbf{r}_2, \mathbf{r}, \mathbf{R}) \equiv \Gamma_4(\mathbf{r}_1, \mathbf{r}_2, \mathbf{r}, \mathbf{R}, z = 0)$. The transverse variables $\mathbf{r}_1$, $\mathbf{r}_2$, $\mathbf{r}$, or $\mathbf{R}$ are defined by a linear transformation (1.2a) of $\boldsymbol{\rho}_i$, where $\boldsymbol{\rho}_i$, $i = 1, 2, 3$, or 4, are the position vectors of four observation points at the same propagation range z. This transformation is

$$
\text{(1.2a)}\quad
\begin{cases}
\mathbf{R} = \frac{1}{4}(\boldsymbol{\rho}_1 + \boldsymbol{\rho}_2 + \boldsymbol{\rho}_3 + \boldsymbol{\rho}_4)\,, \\
\mathbf{r}_1 = \frac{1}{2}(\boldsymbol{\rho}_1 - \boldsymbol{\rho}_2 + \boldsymbol{\rho}_3 - \boldsymbol{\rho}_4)\,, \\
\mathbf{r}_2 = \frac{1}{2}(\boldsymbol{\rho}_1 - \boldsymbol{\rho}_2 - \boldsymbol{\rho}_3 + \boldsymbol{\rho}_4)\,, \\
\mathbf{r} = \boldsymbol{\rho}_1 + \boldsymbol{\rho}_2 - \boldsymbol{\rho}_3 - \boldsymbol{\rho}_4.
\end{cases}
\qquad
\begin{cases}
\boldsymbol{\rho}_1 = \mathbf{R} + \frac{\mathbf{r}_1 + \mathbf{r}_2}{2} + \frac{1}{4}\mathbf{r}, \\
\boldsymbol{\rho}_2 = \mathbf{R} - \frac{\mathbf{r}_1 + \mathbf{r}_2}{2} + \frac{1}{4}\mathbf{r}, \\
\boldsymbol{\rho}_3 = \mathbf{R} + \frac{\mathbf{r}_1 - \mathbf{r}_2}{2} - \frac{1}{4}\mathbf{r}, \\
\boldsymbol{\rho}_4 = \mathbf{R} - \frac{\mathbf{r}_1 - \mathbf{r}_2}{2} - \frac{1}{4}\mathbf{r}.
\end{cases}
\quad\text{(1.2b)}
$$

In (1.1), the operator $\nabla_{\boldsymbol{\rho}} = (\partial/\partial\rho_x \;,\; \partial/\rho_y)$ is a two-dimensional gradient, where $\boldsymbol{\rho}$ stands for $\mathbf{R}$, $\mathbf{r}_1$, $\mathbf{r}_2$, or $\mathbf{r}$.

The scattering function, F_4, of the fourth moment is a nonnegative function. It is defined as

$$
F_4(\mathbf{r}_1, \mathbf{r}_2, \mathbf{r}, z) = H(\mathbf{r}_1 + \mathbf{r}/2, z) + H(\mathbf{r}_1 - \mathbf{r}/2, z) + H(\mathbf{r}_2 + \mathbf{r}/2, z) + H(\mathbf{r}_2 - \mathbf{r}/2, z) - H(\mathbf{r}_1 + \mathbf{r}_2, z) - H(\mathbf{r}_1 - \mathbf{r}_2, z)\,. \qquad (1.3)
$$

Here

$$
H(\mathbf{r}, z) = \frac{1}{\pi}\,[A(0, z) - A(\mathbf{r}, z)]\,,
$$

$$
A(\mathbf{r}, z) = \int_{-\infty}^{\infty} \langle\, \tilde{\epsilon}(\,\mathbf{0}, z\,)\, \tilde{\epsilon}(\,\mathbf{r}, z + \varsigma\,)\,\rangle\, d\varsigma\,; \qquad (1.4)
$$

$\tilde{\epsilon}\,(\mathbf{r}, z)$ is the random deviation of the medium dielectric permittivity from its mean $\langle\epsilon\rangle$.

1.2 Salient properties of the fourth moment

We review here properties of the fourth moment related to our discussion.[23] The fourth moment with $\mathbf{r} = \mathbf{0}$ contains information about fluctuations of the instantaneous intensity $I(\boldsymbol{\rho}, z) \equiv U(\boldsymbol{\rho}, z)\, U^*(\boldsymbol{\rho}, z)$. The covariance of intensities measured at points $\boldsymbol{\rho}_a$ and $\boldsymbol{\rho}_b$ in the same propagation range z is

$$
\begin{aligned}
B_I[\boldsymbol{\rho}_a - \boldsymbol{\rho}_b, (\boldsymbol{\rho}_a + \boldsymbol{\rho}_b)/2, z] &\equiv \langle I(\boldsymbol{\rho}_a, z)\, I(\boldsymbol{\rho}_b, z)\rangle - \langle I(\boldsymbol{\rho}_a, z)\rangle\, \langle I(\boldsymbol{\rho}_b, z)\rangle \\
&\equiv \Gamma_4[\mathbf{r}_1 = \boldsymbol{\rho}_a - \boldsymbol{\rho}_b, \mathbf{r}_2 = \mathbf{0}, \mathbf{r} = \mathbf{0}, \mathbf{R} = (\boldsymbol{\rho}_a + \boldsymbol{\rho}_b)/2, z] \\
&\quad - \Gamma_2(\mathbf{r}_2 = \mathbf{0}, \mathbf{R} + \mathbf{r}_1/2 = \boldsymbol{\rho}_a, z)\, \Gamma_2(\mathbf{r}_2 = \mathbf{0}, \mathbf{R} - \mathbf{r}_1/2 = \boldsymbol{\rho}_b, z)\,, \qquad (1.5a)
\end{aligned}
$$

where

$$
\Gamma_2[\mathbf{r}_2, \mathbf{R} \pm \mathbf{r}_1/2, z] \equiv \langle U(\mathbf{R} \pm (\mathbf{r}_1 + \mathbf{r}_2)/2, z)\, U^*(\mathbf{R} \pm (\mathbf{r}_1 - \mathbf{r}_2)/2, z)\rangle \qquad (1.5b)
$$

is the second moment, also called the mutual coherence function (MCF). To get the transformed positions in (1.5a) and (1.5b) we use (1.2a) and (1.2b). The normalized intensity variance at the center of the beam, $\sigma_I^2(z)$ [also denoted by $\beta^2(z)$] is called the scintillation index. It is defined as

$$\sigma_I^2(z) \equiv \frac{\langle I(\mathbf{0}, z)^2\rangle}{\langle I(\mathbf{0}, z)\rangle^2} - 1 = \frac{\Gamma_4(\mathbf{r}_1 = \mathbf{0}, \mathbf{r}_2 = \mathbf{0}, \mathbf{r} = \mathbf{0}, \mathbf{R} = \mathbf{0}, z)}{\Gamma_2^2(\mathbf{0}, \mathbf{0}, z)} - 1 . \quad (1.6)$$

The full fourth moment contains amplitude information and phase information, which are important for imaging and speckle interferometry applications. For irradiance fluctuations, $\mathbf{r} = \mathbf{0}$ is needed. However, note that to get the covariance function, one has to solve (1.1) for the full fourth moment and set $\mathbf{r}_2 = \mathbf{r} = \mathbf{0}$ only in the observation plane.

Using the definition of the fourth moment and its governing equation, it is easy to infer salient properties of the fourth moment. Symmetry of the fourth moment with respect to $\mathbf{r}_2 \longleftrightarrow \mathbf{r}_1$ is an intrinsic property, which follows from its definition. Thus

$$\Gamma_4(\mathbf{r}_1, \mathbf{r}_2, \mathbf{r}, \mathbf{R}, z) = \Gamma_4(\mathbf{r}_2, \mathbf{r}_1, \mathbf{r}, \mathbf{R}, z) . \quad (1.7)$$

When there is large separation of two pairs of points, the fourth moment decorrelates to

$$\lim_{|\mathbf{r}_2| \to \infty} \Gamma_4(\mathbf{r}_1, \mathbf{r}_2, \mathbf{r}, \mathbf{R}, z) = \Gamma_2(\mathbf{r}_1 + \mathbf{r}/2, \mathbf{R}, z)\, \Gamma_2^*(\mathbf{r}_1 - \mathbf{r}/2, \mathbf{R}, z)$$

$$\xrightarrow[|\mathbf{r}_1| >> |\mathbf{r}|]{} |\Gamma_2(\mathbf{R}, \mathbf{r}_1, z)|^2 . \quad (1.8)$$

Finally, the *energy conservation* law for intensity covariances follows:[23]

$$\int_{-\infty}^{\infty} d^2R\, d^2r_1\, B_I(\mathbf{r}_1, \mathbf{R}, z) = 0 . \quad (1.9)$$

This means that the medium fluctuations only redistribute the energy of the beam, such that (1.9) holds. Equation (1.9) applies to a finite beam wave; the plane wave energy conservation law is given in Section 1.3.

1.3 Model problem

The case of an initially plane wave, for which $\Gamma_4^0(\mathbf{r}_1, \mathbf{r}_2, \mathbf{r}, \mathbf{R}) \equiv 1$, simplifies (1.1) considerably. The problem becomes invariant with respect to the average position vector $\mathbf{R}$. Formally this means that $\nabla_{\mathbf{R}} \Gamma_4 \equiv \mathbf{0}$ and that $\mathbf{r}$ appears only in $F_4(\mathbf{r}_1, \mathbf{r}_2, \mathbf{r}, z)$. Now, (1.1) reads[22]

$$\frac{\partial}{\partial z}\Gamma = \frac{i}{k}\nabla_{\mathbf{r}_1} \cdot \nabla_{\mathbf{r}_2} \Gamma - \frac{\pi k^2}{4} F(\mathbf{r}_1, \mathbf{r}_2, z)\, \Gamma , \quad (1.10)$$

where $\Gamma(\mathbf{r}_1, \mathbf{r}_2, z) \equiv \Gamma_4(\mathbf{r}_1, \mathbf{r}_2, \mathbf{r}, z)$, and $F(\mathbf{r}_1, \mathbf{r}_2, z) \equiv F_4(\mathbf{r}_1, \mathbf{r}_2, \mathbf{r}, z)$. For irradiance fluctuations $\mathbf{r}$ may be set to $\mathbf{0}$ at the onset, because $\mathbf{r}$ is merely a parameter. The plane wave model equation is a valuable model for instructive purposes. Additionally it can be shown that this model also applies to a spherical wave propagating in random media, which is physically a very common realistic configuration.[24] It is shown elsewhere[23,25] that this model equation can also be applied to other physical situations that are $\mathbf{R}$-invariant. For this, the initial condition must comply with $\Gamma_4(\mathbf{r}_1, \mathbf{r}_2, \mathbf{r}, \mathbf{R}, z = 0) \equiv \Gamma_4^{\,0}(\mathbf{r}_1, \mathbf{r}_2)$. Even for the finite beam case, which has an obvious dependency on $\mathbf{R}$, the distance from the center of the beam, it can be easily shown by using the δ-function functional[26] that (1.10) conveys the elements necessary to explain our formulation. Thus, subsequently we consider (1.10) with $\Gamma(\mathbf{r}_1, \mathbf{r}_2, z = 0) = \Gamma_0(\mathbf{r}_1, \mathbf{r}_2) = \Gamma_0(\mathbf{r}_2, \mathbf{r}_1)$ to be our model problem. Note that symmetry $\mathbf{r}_2 \longleftrightarrow \mathbf{r}_1$ and decorrelation are kept, and the energy conservation law for the plane wave model assumes the form[23]

$$\int_{-\infty}^{\infty} d^2r_1\, B_I(\mathbf{r}_1, z) = 0 \,, \tag{1.11}$$

where $B_I(\mathbf{r}_1, z)$ is the covariance of intensities for the plane wave case [see (1.5a)].

1.4 Formal solution of Eq. (1.10)

To construct a formal solution of (1.10), one can use a path integral for the random fields, and average the appropriate product according to the moment definitions.[26] However, it is instructive to derive the formal path integral solution directly for (1.10) in a simple way. We slice the extended medium $[0, Z)$ into N slices $\{[z_{n-1}, z_n)|n = 1,\ldots, N\}$, with $z_0 = 0$, and $z_N = Z$. In each subinterval $[z_{n-1}, z_n)$ we find $\Gamma(\cdot, z_{n+1})$, given the distribution $\Gamma(\cdot, z_n)$ by the following two consecutive steps:

(*a*) Apply to $\Gamma(\cdot, z_n)$ the Fresnel diffraction operator for moments over free space of length $\Delta z \equiv z_{n+1} - z_n$:

$$\mathcal{F}_{n+1,n}\Gamma \equiv \left(\frac{k}{2\pi\Delta z}\right)^2 \iint_{-\infty}^{\infty} d^2r_1(z_n)\, d^2r_2(z_n)$$

$$\times \exp\left\{ \frac{ik}{\Delta z} [\mathbf{r}_1(z_n) - \mathbf{r}_1(z_{n-1})] \cdot [\mathbf{r}_2(z_n) - \mathbf{r}_2(z_{n-1})]\right\}\Gamma(\cdot, z_n) \,. \tag{1.12a}$$

(*b*) Multiply (1.12a) by the effective averaged thin phase-screen modulation over the length Δz, lumped at z_{n+1}:

$$\theta_{n+1} \equiv \exp\left\{ - \frac{\pi k^2 \Delta z}{4} F[\mathbf{r}_1(z_{n+1}), \mathbf{r}_2(z_{n+1}), z_{n+1}]\right\}. \tag{1.12b}$$

A block diagram of the combined result is shown in Fig. 1.

$$\boxed{\Gamma(\cdot, z_{n+1})} \leftarrow \boxed{\theta_{n+1}} \times \boxed{\mathcal{F}_{n+1,n}} \times \boxed{\Gamma(\cdot, z_n)}$$

Fig. 1. Transfer between $\Gamma(\cdot, z_n)$ and $\Gamma(\cdot, z_{n+1})$.

This process is repeated recursively, beginning with the initial condition $\Gamma_0[\mathbf{r}_1(0), \mathbf{r}_2(0)]$ with $n = 0$. As $N \to \infty$, and equal subintervals $\Delta z \equiv Z/N \to 0$ are adopted, the combined result assumes the form

$$\Gamma(\mathbf{r}_1, \mathbf{r}_2, Z) = \lim_{N \to \infty} \int \frac{d^2r_1(z_{N-1})\, d^2r_2(z_{N-1})}{(2\pi\Delta z/k)^2} \cdots \int \frac{d^2r_1(z_0)\, d^2r_2(z_0)}{(2\pi\Delta z/k)^2}$$

$$\times \exp\left\{ \frac{ik}{\Delta z} \sum_{n=0}^{N-1} [\mathbf{r}_1(z_{n+1}) - \mathbf{r}_1(z_n)] \cdot [\mathbf{r}_2(z_{n+1}) - \mathbf{r}_2(z_n)] \right\}$$

$$\times \exp\left\{ -\Delta z \frac{\pi k^2}{4} \sum_{n=0}^{N-1} F[\mathbf{r}_1(z_{n+1}), \mathbf{r}_2(z_{n+1}), z_{n+1}] \right\} \Gamma_0[\mathbf{r}_1(0), \mathbf{r}_2(0)] , \tag{1.13}$$

where $z_n \equiv n\, Z/N$, $z_0 = 0$, $z_N = Z$, and we force $\mathbf{r}_1(Z) = \mathbf{r}_1$, $\mathbf{r}_2(Z) = \mathbf{r}_2$.

As $N \to \infty$ and $\Delta z = Z/N \to 0$, a continuous form may be written symbolically as

$$\Gamma(\mathbf{r}_1, \mathbf{r}_2, z) = \int \mathcal{D}^2 r_1(z) \int \mathcal{D}^2 r_2(z)\, \Gamma_0[\mathbf{r}_1(0), \mathbf{r}_2(0)]\; \delta[\mathbf{r}_1(Z) - \mathbf{r}_1]\, \delta[\mathbf{r}_1(Z) - \mathbf{r}_2]$$

$$\times \exp\left\{ ik \int_0^Z dz\, \dot{\mathbf{r}}_1(z) \cdot \dot{\mathbf{r}}_2(z) - \frac{\pi k^2}{4} \int_0^Z dz\, F[\mathbf{r}_1(z), \mathbf{r}_2(z), z] \right\}, \tag{1.14}$$

where $\dot{\mathbf{r}}_j(z) \equiv d\mathbf{r}_j(z)/dz$ with $j = 1$ or 2, and the ends of the trajectories are fixed by the δ-functions. Equation (1.14) is defined in terms of the discrete form of (1.13), and is useful for both shorthand notation and physical interpretation. The differential of the functional integral can be deduced, by comparing (1.13) and (1.14), to be[4]

$$\mathcal{D}^2 r_j = \lim_{N \to \infty} \prod_{n=0}^{N-1} \frac{d^2 r_j(z_n)}{2\pi Z/(Nk)}, \quad j = 1,2 . \tag{1.15}$$

1.5 Green's function

The fourth moment Green's function is the solution of (1.10) for four point sources on the initial plane, $z = Z_0$. For our model problem they lie on the vertices of a parallelogram; i.e.,

$$\Gamma_0[\mathbf{r}_1(0), \mathbf{r}_2(0)] = \delta[\mathbf{r}_1(0) - \mathbf{r}_{1,0}]\, \delta[\mathbf{r}_2(0) - \mathbf{r}_{2,0}] \,. \tag{1.16}$$

Inserting (1.16) into (1.14), we find the expression of Green's function to be

$$G(\mathbf{r}_1, \mathbf{r}_2, Z; \mathbf{r}_{1,0}, \mathbf{r}_{2,0}) \equiv \int \mathcal{D}^2 r_1(z) \int \mathcal{D}^2 r_2(z)$$

$$\times\, \delta[\mathbf{r}_1(0) - \mathbf{r}_{1,0}]\, \delta[\mathbf{r}_1(Z) - \mathbf{r}_1]\, \delta[\mathbf{r}_2(0) - \mathbf{r}_{2,0}]\, \delta[\mathbf{r}_1(Z) - \mathbf{r}_2]$$

$$\times \exp\left\{ ik \int_0^Z dz\, \dot{\mathbf{r}}_1(z) \cdot \dot{\mathbf{r}}_2(z) \;-\; \frac{\pi k^2}{4} \int_0^Z dz\, F[\mathbf{r}_1(z), \mathbf{r}_2(z), z] \right\} \tag{1.17a}$$

with the condition

$$G(\mathbf{r}_1, \mathbf{r}_2, Z = 0; \mathbf{r}_{1,0}, \mathbf{r}_{2,0}) = \delta(\mathbf{r}_1 - \mathbf{r}_{1,0})\, \delta(\mathbf{r}_2 - \mathbf{r}_{2,0}) \,. \tag{1.17b}$$

Because our source is always placed in $z = Z_0 \equiv 0$, our notation $G(\mathbf{r}_1, \mathbf{r}_2, Z; \mathbf{r}_{1,0}, \mathbf{r}_{2,0})$ means in fact $G(\mathbf{r}_1, \mathbf{r}_2, Z; \mathbf{r}_{1,0}, \mathbf{r}_{2,0}, Z_0 = 0)$. Note that $G(\mathbf{r}_1, \mathbf{r}_2, z = Z; \mathbf{r}_{1,0}, \mathbf{r}_{2,0}, Z_0 = 0) = G(\mathbf{r}_{1,0}, \mathbf{r}_{2,0}, Z_0 = 0; \mathbf{r}_1, \mathbf{r}_2, z = Z)$; this property is called *reciprocity*. The normalization of the differential of the functional integral in (1.17a) is most conveniently defined by demanding that for free space, i.e., $F(\cdot) \equiv 0$, the free space Green's function,

$$G_0(\mathbf{r}_1, \mathbf{r}_2, Z; \mathbf{r}_{1,0}, \mathbf{r}_{2,0}) = \left(\frac{k}{2\pi Z}\right)^2 \exp\left[\frac{ik}{Z} (\mathbf{r}_1 - \mathbf{r}_{1,0}) \cdot (\mathbf{r}_2 - \mathbf{r}_{2,0})\right], \tag{1.18}$$

be obtained.

The use of Green's function is useful because it isolates the propagation from the initial condition. The solution is now found by a superposition integral:

$$\Gamma(\mathbf{r}_1, \mathbf{r}_2, Z) = \int d^2 r_{1,0} \int d^2 r_{2,0}\, \Gamma_0(\mathbf{r}_{1,0}, \mathbf{r}_{2,0})\, G(\mathbf{r}_1, \mathbf{r}_2, Z; \mathbf{r}_{1,0}, \mathbf{r}_{2,0}) \,. \tag{1.19}$$

Equation (1.17a), is called the Feynman path integral, or the configuration-space path integral. It was derived by Feynman in 1948 for a study of problems in quantum mechanics.[1] Sometimes it is interpreted as the summation of the contributions of the trajectories $\mathbf{r}_1(z)$ and $\mathbf{r}_2(z)$ that begin at source points $\mathbf{r}_{1,0}$ and $\mathbf{r}_{2,0}$ and end at the observation points $\mathbf{r}_1$ and $\mathbf{r}_2$.

2. Mixed representation

This section derives an interesting variant of the path integral. The configuration-space path integral (1.17a) has fixed initial and end trajectory values. We represent here the two trajectories in the path integral in two ways. The first trajectory $\mathbf{r}_1(z)$ has a fixed initial value and a free end value, and the second trajectory has a free initial value and a fixed end value. Ultimately the free values will be fixed using δ-functions to conform with the meaning of a Green's function. To this end we find the velocity representation of the trajectories extremely useful.

To introduce the mixed representation we represent the trajectories in terms of velocities $\mathbf{v}_1(z)$ and $\mathbf{v}_2(z)$ as

$$\mathbf{r}_1(z) = \bar{\mathbf{r}}_1(z) + \int_0^z d\varsigma\, \mathbf{v}_1(\varsigma)\,, \qquad \mathbf{r}_2(z) = \bar{\mathbf{r}}_2(z) - \int_z^Z d\varsigma\, \mathbf{v}_2(\varsigma) \tag{2.1a}$$

where $\bar{\mathbf{r}}_1(z)$ and $\bar{\mathbf{r}}_2(z)$ are straight lines connecting source and receiver:

$$\bar{\mathbf{r}}_1(z) = \mathbf{r}_{1,0}\left(1 - \frac{z}{Z}\right) + \mathbf{r}_1 \frac{z}{Z}, \qquad \bar{\mathbf{r}}_2(z) = \mathbf{r}_{2,0}\left(1 - \frac{z}{Z}\right) + \mathbf{r}_2 \frac{z}{Z}\,. \tag{2.1b}$$

Note that the trajectories $\mathbf{r}_1(z)$ and $\mathbf{r}_2(z)$, are piecewise linear continuous functions, and the velocities $\mathbf{v}_1(z) \equiv \dot{\mathbf{r}}_1$ and $\mathbf{v}_1(z) \equiv \dot{\mathbf{r}}_1$ are piecewise constant discontinuous functions.

The functional integral becomes

$$G(\mathbf{r}_1, \mathbf{r}_2, Z; \mathbf{r}_{1,0}, \mathbf{r}_{2,0}) = \exp\left[\frac{ik}{Z}(\mathbf{r}_1 - \mathbf{r}_{1,0})\cdot(\mathbf{r}_2 - \mathbf{r}_{2,0})\right]\int D^2\mathrm{v}_1(z)\int D^2\mathrm{v}_2(z)$$

$$\times\, \delta\left[\int_0^Z dz\, \mathbf{v}_1(z)\right]\delta\left[\int_0^Z dz\, \mathbf{v}_2(z)\right]\exp\left[ik\int_0^Z dz\, \mathbf{v}_1(z)\cdot\mathbf{v}_2(z)\right]$$

$$\times\, \exp\left\{-\frac{\pi k^2}{4}\int_0^Z dz\, F\left[\bar{\mathbf{r}}_1(z) + \int_0^z d\varsigma\, \mathbf{v}_1(\varsigma),\, \bar{\mathbf{r}}_2(z) - \int_z^Z d\varsigma\, \mathbf{v}_2(\varsigma),\, z\right]\right\}. \tag{2.2}$$

The normalization of the differential of the functional integral (2.2) differs from that of (1.17a). It is found by demanding that in free space, i.e., $F(\cdot) \equiv 0$, the free-space Green's function (1.18) be obtained.

The boundary conditions are rendered implicit by use of a δ-function,

$$\delta\left[\int_0^Z d\varsigma\ \mathbf{v}_i(\varsigma)\right] = \frac{1}{(2\pi)^2}\int d^2K_j\ \exp\left\{i\left[\frac{k}{Z}(\mathbf{r}_j - \mathbf{r}_{j,0}) - \mathbf{K}_j\right]\cdot\int_0^Z d\varsigma\ \mathbf{v}_i(\varsigma)\right\}, \tag{2.3}$$

where $\{i,j = 1,2 \mid j \neq i\}$. Inserting (2.3) into (2.2) and changing variables to $\mathbf{v}_j \rightarrow \mathbf{v}_j - \mathbf{K}_j/k + (\mathbf{r}_j - \mathbf{r}_{j,0})/Z$, where $j = 1$ or 2, we get

$$G(\mathbf{r}_1, \mathbf{r}_2, Z; \mathbf{r}_{1,0}, \mathbf{r}_{2,0}) = \frac{1}{(2\pi)^4}\int d^2K_1 \int d^2K_2$$

$$\times \exp\left[-i\frac{Z}{k}\mathbf{K}_1\cdot\mathbf{K}_2 + i\mathbf{K}_1\cdot(\mathbf{r}_2 - \mathbf{r}_{2,0}) + i\mathbf{K}_2\cdot(\mathbf{r}_1 - \mathbf{r}_{1,0})\right]$$

$$\times \int D^2\mathrm{v}_1(z)\int D^2\mathrm{v}_2(z)\ \exp\left[ik\int_0^Z d\varsigma\ \mathbf{v}_1(\varsigma)\cdot\mathbf{v}_2(\varsigma)\right]$$

$$\times \exp\left\{-\frac{\pi k^2}{4}\int_0^Z d\varsigma\ F\left[\bar{\bar{\mathbf{r}}}_1(\varsigma) + \int_0^z d\varsigma\ \mathbf{v}_1(\varsigma), \bar{\bar{\mathbf{r}}}_2(\varsigma) - \int_z^Z d\varsigma\ \mathbf{v}_2(\varsigma), z\right]\right\}. \tag{2.4}$$

Now the straight lines assume the form

$$\bar{\bar{\mathbf{r}}}_1(z) = \mathbf{r}_{1,0} + z\,\mathbf{K}_1/k, \quad \bar{\bar{\mathbf{r}}}_2(z) = \mathbf{r}_2 - (Z - z)\mathbf{K}_2/k\ . \tag{2.5}$$

Each one of them has a free end value, which will be fixed finally by the implicit δ-functions, when all the integrations indicated in (2.4) are completed.

The final result is obtained using the superposition integral (1.19). The final expression for the fourth moment is simplified if we represent the initial and the final expressions of the fourth moment in partial Fourier transforms:

$$\hat{\Gamma}(\mathbf{K}_2, \mathbf{r}_2, Z) = \frac{1}{(2\pi)^2}\int d^2r_1\ \Gamma(\mathbf{r}_1, \mathbf{r}_2, Z)\exp(-i\mathbf{r}_1\cdot\mathbf{K}_2)\ , \tag{2.6a}$$

$$\hat{\Gamma}_0(\mathbf{r}_{1,0}, \mathbf{K}_1) = \frac{1}{(2\pi)^2}\int d^2r_{2,0}\ \Gamma_0(\mathbf{r}_{1,0}, \mathbf{r}_{2,0})\exp(-i\mathbf{r}_{2,0}\cdot\mathbf{K}_1)\ . \tag{2.6b}$$

This leads to the final form called the *mixed representation*:

$$\hat{\Gamma}(\mathbf{K}_2, \mathbf{r}_2, Z) = \frac{1}{(2\pi)^2} \int d^2 r_{1,0} \int d^2 K_1 \; \hat{\Gamma}_0(\mathbf{r}_{1,0}, \mathbf{K}_1)$$

$$\times \exp\left(-i \frac{Z}{k} \mathbf{K}_1 \cdot \mathbf{K}_2 + i\mathbf{K}_1 \cdot \mathbf{r}_2 - i\mathbf{K}_2 \cdot \mathbf{r}_{1,0}\right)$$

$$\times \int D^2 \mathrm{v}_1(z) \int D^2 \mathrm{v}_2(z) \; \exp\left[ik \int_0^Z dz \; \mathbf{v}_1(z) \cdot \mathbf{v}_2(z)\right]$$

$$\times \exp\left\{-\frac{\pi k^2}{4} \int_0^Z dz \; F\left[\bar{\bar{\mathbf{r}}}_1(z) + \int_0^z d\zeta \; \mathbf{v}_1(\zeta), \bar{\bar{\mathbf{r}}}_2(z) - \int_z^Z d\zeta \; \mathbf{v}_2(\zeta), z\right]\right\}. \tag{2.7}$$

Here $\bar{\bar{\mathbf{r}}}_1$ and $\bar{\bar{\mathbf{r}}}_2$ are defined in (2.5). Note that (2.7) expresses the fourth moment dependency in terms of a superposition integral in a mixed spatial-spectral representation:

$$\hat{\Gamma}(\mathbf{K}_2, \mathbf{r}_2, Z) = \frac{1}{(2\pi)^2} \int d^2 r_{1,0} \int d^2 K_1 \; \hat{G}(\mathbf{K}_2, \mathbf{r}_2, Z; \mathbf{r}_{1,0}, \mathbf{K}_1) \, \hat{\Gamma}_0(\mathbf{r}_{1,0}, \mathbf{K}_1) \; . \tag{2.8}$$

The mixed transfer function between source and receiver $\hat{G}(\cdot)$ can be easily determined from (2.7) and (2.8). A final note on the notation: the subscripts of $\mathbf{K}_1$ and $\mathbf{K}_2$ are arbitrary, and can be switched.

2.1 Heuristic approximation

If in $F(\cdot)$ of (2.7) we ignore $\mathbf{v}_1(z)$, $\mathbf{v}_2(z)$, or both, the functional integral can be computed using the δ-function functional. As an example, let us ignore the $\mathbf{v}_1(z)$ in $F(\cdot)$. Now $\mathbf{v}_1(z)$ appears only outside of $F(\cdot)$. Therefore we can perform the functional integration over $\mathbf{v}_1(z)$, using the δ-function functional definition[21]

$$\int D^2 \mathrm{v}_1(z) \; \exp\left\{ik \int_0^Z dz \; \mathbf{v}_1(z) \cdot \left[\; \mathbf{v}_2(z) - \mathbf{v}_0(z) \;\right]\right\} \equiv \delta_\infty[\mathbf{v}_2(z) - \mathbf{v}_0(z)] \; ; \tag{2.9a}$$

that is,

$$\int D^2 \mathrm{v}_2(z) \; \delta_\infty[\mathbf{v}_2(z) - \mathbf{v}_0(z)] \; \Phi[\mathbf{v}_2] \equiv \Phi[\mathbf{v}_0] \; , \tag{2.9b}$$

where $\Phi[\cdot]$ is a continuous functional. Following these steps leads to Uscinski's fundamental solution[8] or to the two-scale zero-order approximation:[9,10]

$$\hat{\Gamma}_{ts}(\mathbf{K}_2, \mathbf{r}_2, Z) = \frac{1}{(2\pi)^2} \int d^2r_{1,0} \int d^2K_1 \; \hat{\Gamma}_0(\mathbf{r}_{1,0}, \mathbf{K}_1)$$

$$\times \exp\left[-i\frac{Z}{k}\mathbf{K}_1 \cdot \mathbf{K}_2 + i\mathbf{K}_1 \cdot \mathbf{r}_2 - i\mathbf{K}_2 \cdot \mathbf{r}_{1,0}\right]$$

$$\times \exp\left\{-\frac{\pi k^2}{4} \int_0^Z dz \, F[\bar{\bar{\mathbf{r}}}_1(z), \bar{\bar{\mathbf{r}}}_2(z), z]\right\}, \tag{2.10}$$

where $\bar{\bar{\mathbf{r}}}_1(z)$ and $\bar{\bar{\mathbf{r}}}_2(z)$ are defined in (2.5).

2.2 Connection to other straight-line approximations

There is a strong link between this result, (2.10), and other straight-line approximations. Reference 21 discusses these types of heuristic approximations in detail. It appears that though these approximations never produced a high degree of confidence, they were widely used in many practical applications. Note that there was no asymptotic argument in neglecting the trajectories, which leaves the straight lines as their representatives. If the manifold of trajectories is tightly confined over the straight line, this approximation will be effective. Continuing with this line of argument, we may infer that in long-range propagation in the weak fluctuation regime, this approximation is useful. A propagation through a combination of a thin screen and free space is another situation that may be applicable. Otherwise we could not find a justification for this approximation, except for the need to obtain a tractable expression. Keeping in mind that this approximation is heuristic, we already know of many cases in which it provided qualitatively good results. The next section reviews the properties of the zero-order two-scale expansion to enable the reader to decide when it is possible to use the two-scale zero-order approximation for rough estimates.

2.3 Properties of the two-scale solution

This section compares the properties of the two-scale zero-order approximation with the exact properties mentioned in Section 1.2. The most notable property of the zero-order two-scale expansion is the loss of symmetry, $\mathbf{r}_1 \longleftrightarrow \mathbf{r}_2$, of

$$\Gamma_{ts}(\mathbf{r}_1, \mathbf{r}_2, Z) = \int d^2r_1 \, \hat{\Gamma}_{ts}(\mathbf{K}_2, \mathbf{r}_2, Z) \exp(i\mathbf{r}_1 \cdot \mathbf{K}_2) \,. \tag{2.11}$$

It can be shown that the fourth moment decorrelates as either $|\mathbf{r}_1| \to \infty$ or $|\mathbf{r}_2| \to \infty$, but not at the same rate. The profile $\Gamma_{ts}(\mathbf{r}_1, \mathbf{0}, Z)$ is a better approximation than $\Gamma_{ts}(\mathbf{0}, \mathbf{r}_2, Z)$ for correlation purposes. Also the first profile follows the *energy conservation* law (1.11) whereas the latter does not. Because energy conservation is an integral property, this does not say whether the shape of the correlation function is reasonable. However, a reasonable shape, i.e., two-scaled structure of intensity fluctuations in the regime of saturated fluctuations, was shown in numerous publications. For a plane wave, Gozani[27] found quantitatively that the intensity correlation scale (the small scale) is correct, whereas the large scale starts to deviate from an "exact" numerical solution and from asymptotic expressions, at a certain range larger than the position of the scintillation peak. It was also found that at least in the cases studied (see Ref. 20 for an extensive list), the scintillation index, which is the center point of the correlation function, gives all salient effects: it begins with geometrical optics when the inner scale is present, follows Rytov variance in the weak fluctuation regime, exhibits a peak of scintillation approximately in the correct place, and converges to saturation with a correct power but with an incorrect coefficient. The largest error occurs in the value and location of the scintillation peak, whereas in the perturbation regime the match is exact. The caveat, though, is that this is only true for a constant level of turbulence, C_n^2 = constant. When dealing with Green's function, one may note that the two-scale zero-order approximation violates the *reciprocity* principle. This implies that the backscattering formulation based on the two-scale zero-order approximation is also inaccurate.

The above deficiencies limit the applicability of the zero-order two-scale approximation and urge us to find a correction procedure. Note that the two-scale zero-order approximation offers the same complexity as, but more fidelity than, the other members of the straight-line heuristic legion. We hope that the simple exposition we adopted beginning with the "exact" Feynman path integral, using simple algebraic manipulation, and finally applying a crude (but useful) approximation underlines that the success of this approximation depends much on the problem at hand.

3. Orthogonal expansion

In this section we develop a consistent method to compute functional integrals. The method uses the straight approximations mentioned as heuristic approximations for an initial guess. We show the procedure using the Feynman path integral in the configuration space, (1.17a). Because trajectories begin and end at source and observation points, respectively, we expand the trajectories in a complete orthonormal set about the straight line connecting source and receiver; i.e.,[1,22,28]

$$\mathbf{r}_1(z) = \bar{\mathbf{r}}_1 + \sum_{n=1}^{\infty} \mathbf{a}_n \, \phi_n(z) \,, \qquad \mathbf{r}_2(z) = \bar{\mathbf{r}}_2 + \sum_{n=1}^{\infty} \mathbf{b}_n \, \phi_n(z) \,, \tag{3.1}$$

where $\bar{\mathbf{r}}_1 \equiv \mathbf{r}_{1,0} + (\mathbf{r}_1 - \mathbf{r}_{1,0})z/Z$, $\bar{\mathbf{r}}_2 \equiv \mathbf{r}_{2,0} + (\mathbf{r}_2 - \mathbf{r}_{2,0})z/Z$ are first defined in (2.1b). Thus the mixed term in (1.17a) becomes

$$ik \int_0^Z dz\, \bar{\mathbf{r}}_1(z) \cdot \bar{\mathbf{r}}_2(z) = i\frac{k}{Z}(\mathbf{r}_1 - \mathbf{r}_{1,0}) \cdot (\mathbf{r}_2 - \mathbf{r}_{2,0})$$

$$+ ik \sum_{m=1}^{\infty} \sum_{n=1}^{\infty} \mathbf{a}_m \cdot \mathbf{b}_n \int_0^Z dz\, \dot{\phi}_m(z)\, \dot{\phi}_n(z) \,. \tag{3.2}$$

To diagonalize the summation, we need $\{\dot{\phi}_n(z)\}$ to be a complete orthonormal set; i.e.,

$$\int_0^Z dz\, \dot{\phi}_m(z)\, \dot{\phi}_n(z) = \delta_{m,n}\,, \qquad \sum_{n=1}^{\infty} \dot{\phi}_n(z')\, \dot{\phi}_n(z'') = \delta(z' - z'') \,. \tag{3.3}$$

Inserting the identities (3.3) into (3.2) leads to

$$ik \sum_{m=1}^{\infty} \sum_{n=1}^{\infty} \mathbf{a}_m \cdot \mathbf{b}_n \int_0^Z dz\, \dot{\phi}_m(z)\, \dot{\phi}_n(z) = ik \sum_{n=1}^{\infty} \mathbf{a}_n \cdot \mathbf{b}_n \,. \tag{3.4}$$

We must account for the Jacobian. Because (3.1) is a linear transformation whose Jacobian is constant, taking into account the Jacobian is most conveniently done by demanding that in free space, i.e., when $F(\cdot) \equiv 0$, the free-space Green's function (1.18) be obtained. The final result is

$$G(\mathbf{r}_1, \mathbf{r}_2, Z; \mathbf{r}_{1,0}, \mathbf{r}_{2,0}) = G_0(\mathbf{r}_1, \mathbf{r}_2, Z; \mathbf{r}_{1,0}, \mathbf{r}_{2,0}) \prod_{n=1}^{\infty} \left[\int \frac{d^2a_n\, d^2b_n}{(2\pi/k)^2}\, e^{ik\, \mathbf{a}_n \cdot \mathbf{b}_n} \right.$$

$$\left. \times \exp\left\{ -\frac{\pi k^2}{4} \int_0^Z dz\, F\left[\bar{\mathbf{r}}_1(z) + \sum_{n=1}^{\infty} \mathbf{a}_n\, \phi_n(z),\, \bar{\mathbf{r}}_2(z) + \sum_{n=1}^{\infty} \mathbf{b}_n\, \phi_n(z),\, z \right] \right\} \right]. \tag{3.5}$$

Note that the dimension of $\mathbf{a}_n$, $\mathbf{b}_n$, and ϕ_n is $(\text{length})^{1/2}$.

3.1 Revisiting straight-line approximations

We already used a straight-line approximation in the case of the mixed representation. When we begin with the configuration space path integral, a distinguished approximation, the Huygens-Fresnel principle (phase) approximation[5]

$$G_s(\mathbf{r}_1, \mathbf{r}_2, Z; \mathbf{r}_{1,0}, \mathbf{r}_{2,0}) = G_0(\mathbf{r}_1, \mathbf{r}_2, Z; \mathbf{r}_{1,0}, \mathbf{r}_{2,0})$$

$$\times \exp\left\{ -\frac{\pi k^2}{4} \int_0^Z dz\, F[\bar{\mathbf{r}}_1(z), \bar{\mathbf{r}}_2(z), z] \right\}, \tag{3.6}$$

emerges. This heuristic approximation keeps the $\mathbf{r}_1 \longleftrightarrow \mathbf{r}_2$ symmetry but fails to follow energy conservation and reciprocity. The solution is known to fail for intensity fluctuations of a point source, i.e., when

$$\{\mathbf{r}_1 = \mathbf{0} \quad \text{or} \quad \mathbf{r}_2 = \mathbf{0}\} \text{ and } \Gamma_0(\mathbf{r}_{1,0}, \mathbf{r}_{2,0}) = \delta(\mathbf{r}_{1,0}, \mathbf{r}_{2,0}) . \tag{3.7}$$

It is easy to reproduce this fact by inserting (3.7) in the superposition integral, (1.19). In so doing, we get $\Gamma(\mathbf{0}, \mathbf{0}, Z) = 1$, or $\sigma_I^2(Z) = 0$; i.e., it erroneously predicts no fluctuations!

The plane wave source, $\Gamma_0(\mathbf{r}_{1,0}, \mathbf{r}_{2,0}) \equiv 1$, according to the Huygens-Fresnel principle (phase) approximation,

$$\Gamma_s(\mathbf{r}_1, \mathbf{r}_2, Z) = \int d^2r_{1,0} \int d^2r_{2,0}\, G_0(\mathbf{r}_1, \mathbf{r}_2, Z; \mathbf{r}_{1,0}, \mathbf{r}_{2,0})$$

$$\times \exp\left\{ -\frac{\pi k^2}{4} \int_0^Z dz\, F[\bar{\mathbf{r}}_1(z), \bar{\mathbf{r}}_2(z), z] \right\}, \tag{3.8}$$

is more forgiving. The solution using this approximation exhibits the expected qualitative effects: weak fluctuations, focusing, and saturation, but with incorrect values. In the asymptotic regimes of weak or strong fluctuation, the scintillation index changes with correct power of Rytov variance $\equiv \beta_0^2$, but with incorrect coefficients.

Analogous to the Huygens-Fresnel principle approximation, the incoming wave and outgoing wave approximations and the two-scale expansion are obtained by ignoring trajectories and considering only a straight line connecting source and receiver.[21] It is noteworthy that they do not exhibit such poor results for scintillation due to a point source; however, in general they do not comply with energy conservation and reciprocity principles. Depending on the physical case, few of them are qualitatively better than the others, but no one knows in advance when to apply them. To asses their validity, extensive comparison with other independent techniques were made in the past. Additionally, Aksenov and Mironov[6] and Gozani[19] have shown that they can be used as a first guess in an iterative process. The iteration is computer intensive, however, and needs to be simplified for computations. Similarly, the four approximations are natural for use as a zero-order approximation in our orthogonal expansion scheme. This is the main direction of our future research, where our goal is to find an expansion that will reduce the complexity of the calculation.

3.2 Computation

We begin our study using sinusoidal basis functions $\{\phi_n\}$ where

$$\phi_n(z) = \frac{\sqrt{2Z}}{\pi n} \sin\left(\frac{n\pi z}{Z}\right), \quad n = 1,2,\ldots,N,\ldots, \quad . \tag{3.9}$$

The coefficient's value stems from the normalization in (3.3). In practice, we limit ourselves to a finite N. This produces $4N$ integrations to perform in (3.5). Our goal is to study the convergence rate with this approximation and to compare computational results with analytic expressions that are available in the weak and strong fluctuation regimes (see Table 1).

Table 1. Asymptotic expressions in weak and strong fluctuation regimes.[24]

Regime	Plane Wave	Spherical Wave
Weak, $\beta_0^2 \le 0.25$	$\sigma_I^2 = \beta_0^2$	$\sigma_I^2 \simeq 0.4\,\beta_0^2$
Strong, $\beta_0^2 >> 1$	$\sigma_I^2 \simeq 1 + 0.86\,\beta_0^{-4/5}$	$\sigma_I^2 \simeq 1 + 2.74\,\beta_0^{-4/5}$

The rationale behind our tactics is the following: The scintillation index at the peak is difficult to compute, in contrast to the weak and strong fluctuation regimes where an analytic asymptotic expression is readily available for a medium with a Kolomogorov 5/3 power law.[28] If poor agreement is found in the weak and strong fluctuation regimes, then there is no reason to believe that the approximation will be acceptable near the scintillation peak. In the asymptotic regimes, the analytic approximations using a particular number of sinusoids is obtained[28] by the same procedure[4,24] used to derive the expressions in Table 1. The latter expressions are approached as $N \to \infty$. Thus the resulting expressions are asymptotic in β_0^2 but not in N.

Figures 2*a* and 2*b* show the scintillation index dependence on the number of orthogonal functions for a plane wave and for a spherical wave, respectively. In Fig. 2*a* the asymptotic analytic result is denoted by ∞. The valid range of the computation is approximately when $\sigma_I^2 \le 0.25$, or when $1 \le \sigma_I^2 \le 1.1$. We extended the graphs beyond these regions to enable us to make a clear distinction between the lines. The case of a plane wave with no sinusoids, $N = 0$, relates to (3.8). The asymptotic behaviors in both perturbation and saturation regimes are with a correct power of β_0^2, but the coefficient is incorrect.

In Fig. 2*b*, the case of a spherical wave is shown. The case with no sinusoids, $N = 0$, coincides with the abscissa (i.e., $\sigma_I^2 = 0$). A single sinusoid produces a curve comparable with the plane wave case with $N = 0$. For both plane and spherical waves, the peak value for the two-scale expansion and the simulation by Martin and Flatté are given for comparison and show that additional terms are needed at the scintillation peak.

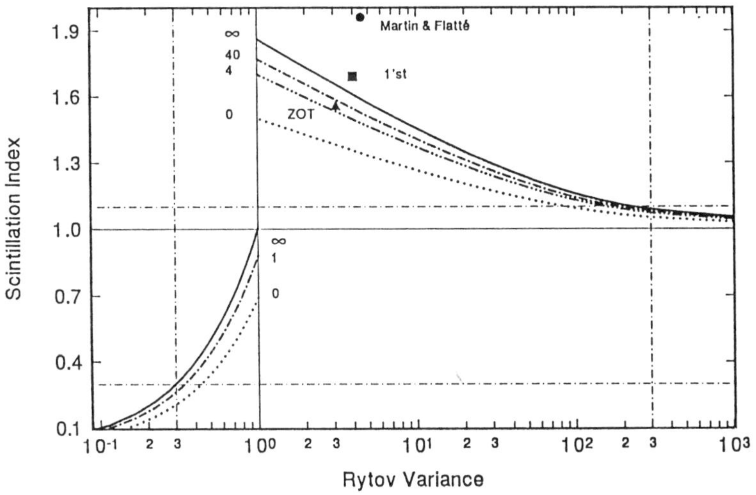

Fig. 2*a* Plane wave: Scintillation index dependence on the number of sinusoids; $N \rightarrow \{0, 1, \infty\}$, $N \rightarrow \{0, 4, 40, \infty\}$ in weak and strong fluctuation regimes, respectively. Triangle - peak by the zero-order two scale approximation, (ZOT);[12] square - peak by including the first-order two-scale approximation, (1'st);[16] circle - peak by Martin and Flatté.[14]

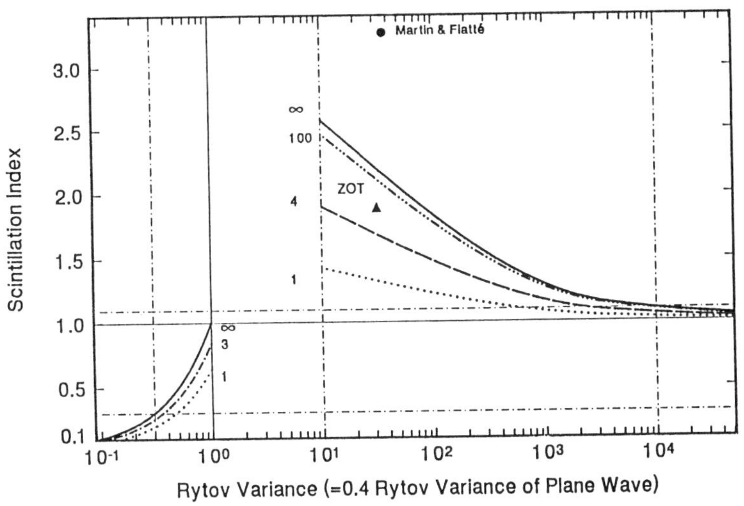

Fig. 2*b*. Spherical wave: Scintillation index dependence on the number of sinusoids; $N \rightarrow \{1, 3, \infty\}$, $N \rightarrow \{1, 4, 100, \infty\}$ in weak and strong fluctuations, respectively. Triangle - peak by the zero-order two-scale approximation (ZOT);[18] circle - peak by Martin and Flatté.[17]

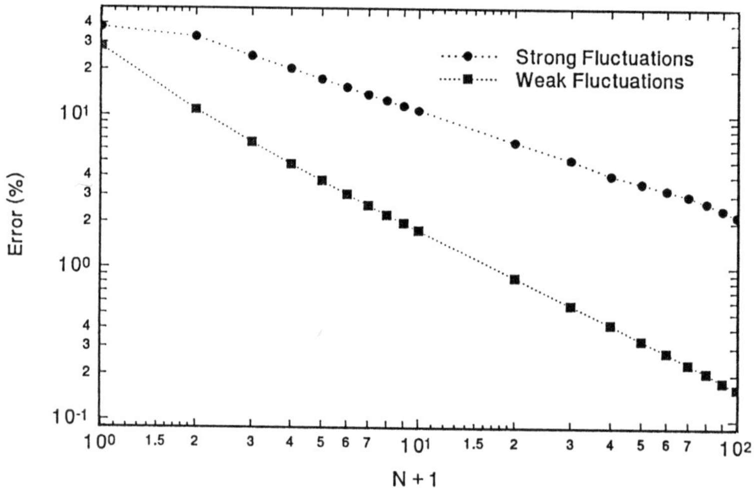

Fig. 3*a*. Plane wave: Scintillation index error vs. number of orthogonal functions for strong and weak fluctuations.

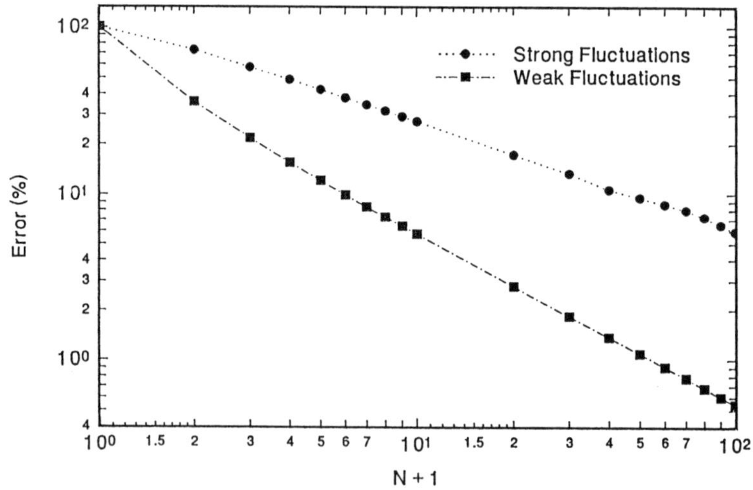

Fig. 3*b*. Spherical wave: Scintillation index error vs. number of orthogonal functions for strong and weak fluctuations.

Figures 3*a* and 3*b* show the error in the scintillation index coefficients in weak, {1, 0.4}, and in strong, {0.86, 2.74}, fluctuation regimes. Note that the abscissa is shifted to $N + 1$ because the scale is logarithmic. For clarity, we give an example showing how to interpret the graph. Let us use the strong fluctuation asymptotic expression in Table 1 for a spherical wave, $1 + 2.74\, \beta_0^{-4/5}$. Here the coefficient is 2.74. For $N = 1$, we have from Fig. 3b 72% error, which means that the actual coefficient is 0.77 = (1. - 0.72) × 2.74; thus the asymptote is $1 + 0.77\, \beta_0^{-4/5}$. For an error of 10%, say, the plane wave needs $N = 2$ or $N = 20$ sinusoids in the weak and strong fluctuation regimes, respectively. The spherical wave needs about $N = 5$ or $N = 50$ sinusoids in the weak and strong fluctuation regimes , respectively. We observe that while the plane wave exhibits a faster convergence than does the spherical wave, both convergence rates are disappointingly slow. We conclude that the sinusoidal basis function converges too slowly, and another orthonormal set should be sought.

4. Conclusions

The first part of this paper introduced the mixed representation. By neglecting the trajectories and leaving only the source-receiver straight line the functional integral can be performed. The expression coincides with Uscinski's fundamental solution. The resulting expression is not asymptotic, and we did not use the smallness of some parameter to get the result.

This approximation was categorized as heuristic like other straight-line approximations: the extended Huygens-Fresnel principle and the incoming-wave and the outgoing-wave phase approximations. The unification shows that all the straight-line approximations are descendants of the Feynman path integral derived by a moderate number of algebraic manipulations.

The goal of the second part was to provide a rigorous and systematic procedure to compute a path integral or to correct the first guess, which can be taken from the collection of the heuristic approximations. For the spherical wave expansion, we used a sinusoidal basis function to expand the velocities $\mathbf{v}_1 \equiv \dot{\mathbf{r}}_1$ and $\mathbf{v}_2 \equiv \dot{\mathbf{r}}_2$, and we realized that the convergence rate was poor. We intend to examine other bases. A better convergence, which means fewer basis functions, will reduce the number of integrals we have to compute. Because of the discontinuous nature of the velocities we intend to use Walsh functions for our future research.

We recommend that researchers use the path integral extensively as a starting point for new applications. This "exact" solution, equipped with heuristic and asymptotic approximations, may lead to better results. Even if a heuristic expression is sometimes unavoidable, it can be used judiciously with the next term in the expansion to assess the error.

The heuristic approximations share the same idea: neglecting the trajectories about a straight line connecting source and receiver. This cautions the user that there is no a priori rule for their success. Sometimes heuristic approaches provide excellent approximations for practical uses, especially in light of comparable experimental errors, but often they are crude. Our experience suggests that the collections of approximations offered by the heuristic approximations[4,21] should be tried and the best of them used.

References

1. R. P. Feynman and A.R. Hibbs, *Quantum Mechanics and Path Integrals*, McGraw-Hill, New York (1965).

2. V. I. Klyatskin, *Statistical Description of Dynamical Systems with Fluctuating Parameters*, Nauka, Moscow (1975) (in Russian).

3. R. Dashen, "Path integrals for waves in random media," *J. Math. Phys.* **20**, 894-920 (1979).

4. V. I. Tatarskii and V. U. Zavorotnyi, "Strong fluctuations in light propagation in a randomly inhomogeneous medium," *Progress in Optics*, **18**, ed. E. Wolf, North-Holland Publishing Co., Amsterdam (1980) 207-255.

5. Z. I. Feyzulin and Yu. A. Kravtsov, "Broadening of a laser beam in a turbulent atmosphere," *Radiophys. Quantum Electron.* **10**, 33-35 (1967).

6. V. P. Aksenov and V. L. Mironov, "Spectral expansion method in problems of laser-beam propagation in a turbulent atmosphere," *Opt. Lett.* **3**, 184-186 (1978).

7. K. S. Gochelashvili, "Saturation of the fluctuations of focused radiation in a turbulent medium," *Radiophys. Quantum Electron.* **14**, 470-473 (1971).

8. B. J. Uscinski, "Intensity fluctuations in a multiple scattering medium: Solution of the fourth moment equation," *Proc. R. Soc. London Ser. A* **380**, 137-169 (1982).

9. C. Macaskill, "An improved solution to the fourth moment equation for intensity fluctuations," *Proc. R. Soc. London Ser. A* **386**, 461-474 (1983).

10. S. Frankenthal, M. J. Beran, and A. M. Whitman, "Caustic corrections using coherence theory," *J. Acoust. Soc. Am.* **71**, 348-358 (1982).

11. S. Frankenthal, A. M. Whitman and M. J. Beran, "Two-scale solutions for intensity fluctuations in strong scattering," *J. Opt. Soc. Am. A* **1**, 585-597 (1984).

12. A. M. Whitman and M. J. Beran, "Two-scale solution for atmospheric scintillation," *J. Opt. Soc. Am. A* 2, 2133-2143 (1985).

13. J. Gozani, "Numerical solution of the fourth order coherence function of a plane wave propagating in a two-dimensional Kolmogorovian medium," *J. Opt. Soc. Am. A* **12**, 2144-2151 (1985).

14. J. M. Martin and S. M. Flatté, "Intensity images and statistics from numerical simulation of wave propagation in 3-D random media," *Appl. Opt.* **27**, 2111-2126 (1988).

15. W. A. Coles and R. G. Frehlich, "Simultaneous measurements of angular

intensity scintillation in the atmosphere," *J. Opt. Soc. Am.* **72**, 1042-1048 (1982).

16. A. M. Whitman and M. J. Beran, "First-order correction for the scintillation index and correlation of the intensity function," *J. Opt. Soc. Am. A* **9**, 974-977 (1992).

17. J. M. Martin and S. M. Flatté, "Simulation of point-source scintillation through three-dimensional random media," *J. Opt. Soc. Am. A* **7**, 838-847 (1990).

18. A. M. Whitman and M. J. Beran, "Two-scale solution for atmospheric scintillation from a point source," *J. Opt. Soc. Am. A* **5**, 735-737 (1988).

19. J. Gozani, "Improvement of the two-scale solution for a wave propagating in a random medium," *Opt. Lett.* **12**, 1-3 (1987).

20. M. J. Beran, S. Frankenthal, R. Mazar, and A. M. Whitman, "Applications of the two-scale embedding technique" (this issue).

21. V. U. Zavorotny, M. I. Charnotskii, J. Gozani and V. I. Tatarskii, "Path integral approach to wave propagation in random media, Part II, Exact formulations and heuristic approximations" (this issue).

22. V. I. Tatarskii, M. I. Charnotskii, J. Gozani and V. U. Zavorotny, "Path integral approach to wave propagation in random media, Part I, Derivations and various formulations" (this issue).

23. V. I. Tatarskii, *The Effect of the Turbulent Atmosphere on Wave Propagation*, National Technical Information Service, Springfield, VA (1971).

24. A. M. Prokhorov, F. V. Bunkin, K. S. Gochelashvili, and V. I. Shishov, "Laser irradiance propagation in turbulent media," *Proc. IEEE* **63**, 1669-1692 (1975).

25. M. Tur and M. J. Beran, "Propagation of a finite beam through a random medium," *Opt. Lett.* **5**, 306-308 (1980).

26. A. E. Novikov, "The solutions of some variational differential equations," *Usp. Mat. Nauk SSSR* **16**, 135-141 (1964) (in Russian).

27. J. Gozani, "Applicability of the two-scale solution of a wave propagating in a random medium," *J. Opt. Soc. Am. B* **3**, 721-727 (1987).

28. V. I. Tatarskii and V. U. Zavorotnyi, "On the connection between the extended Huygens-Fresnel principle and the path integral approximate computation based on orthogonal expansions," *Proc. SPIE* **642**, 276-281 (1986).

THE IMBEDDING METHOD IN STATISTICAL BOUNDARY-VALUE WAVE PROBLEMS

V. I. Klyatskin
Institute of Atmospheric Physics
Russian Academy of Science
3, Pyzhevskii per., Moscow 109017, Russia

Abstract

The statement of the basic boundary-value problem for waves propagating through layered random media with multidimensional inhomogeneities is reviewed. The solution to this problem obtained using the imbedding method is presented. Formulation of statistical problems based on the delta-correlated and diffusion approximations for the parabolic equation are discussed. The exact statistical solution for a point source in layered random media is also considered.

Introduction

At present, the propagation of various kinds of waves in natural media is intensively studied (ocean acoustics, atmospheric optics, ionosphere and plasma physics, and so on). In general, parameters of these media undergo large fluctuations in space and time. Such problems are described mathematically by *the boundary-value problem* for the wave equation with a nonstationary-in-time and nonuniform-in-space refraction factor. The statistical variability of the media parameters produces additional difficulties for the boundary-value problem. In general, such statistical problem solutions are based on one of three approximate methods:

1. The use of *a phenomenological linear theory of a radiative transfer*;

2. The neglect of backward scattering effects and use of *the small angle approximation (the parabolic equation)*;

3. The reformulation of the problem in terms of an integral equation. In analyzing statistical characteristics of this equation, one uses *the Bourret approximation for the Dyson equation and the ladder approximation for the Bethe-Salpeter equation.*

These approximate methods can be substantiated only for problems having the property of *dynamical causality*; that is, the solution is defined only by *previous* (in time or space) parameter values and does not depend on subsequent ones. Boundary-value problems are not of this type. It is desirable in such cases to convert the problem to an evolutionary type with initial conditions. This conversion is indispensable for statistical problems and may be useful for the numerical analysis of deterministic ones. This conversion provides the basis of *the imbedding method.*[1] Many formulations based on this method currently exist for a large number of stationary and nonstationary, linear and nonlinear boundary-value wave problems of different dimensions.[1-3] These equations are, in general, very complicated (for example, they may be nonlinear integro-differential equations in both finite and infinite dimensional spaces), and are difficult if not impossible to solve. An exception to this is the problem of *a plane wave in layered media*, which can be reduced to *a one-dimensional boundary-value problem*, for which a solution can be obtained. Note that such natural media as the ocean and atmosphere can be regarded as layered in the first approximation. Therefore, an analysis of the plane-wave problem, for which a tractable solution exists, can be used to describe the main characteristics of the solution.

We shall examine below the formulation of the *main problems* relating to the propagation of waves in inhomogeneous media, and the technique for their solution based on the imbedding method. In particular, we will consider the fundamental characteristics of the solution for plane waves in layered random media.

1. Formulation of boundary wave problems

Let the layer of inhomogeneity occupy the region $L_0 \leq x \leq L$. A point source is placed at (x_0, ρ_0), where ρ_0 describes a coordinate in the plane perpendicular to the propagation axis x. The wave field inside the layer is then described by the following equation for Green's function:

$$\left\{\frac{\partial^2}{\partial x^2} + \Delta_\rho + k^2\left[1 + \varepsilon(x,\rho)\right]\right\} G\left(x,\rho; x_0,\rho_0\right) = \delta(x - x_0)\,\delta(\rho - \rho_0), \tag{1.1a}$$

where k is the wavenumber and $\varepsilon(x,\rho)$ is the deviation of the refractive index from unity. Let $\varepsilon(x,\rho) = 0$ outside the layer. Then the wave field is described outside the layer by the *Helmholtz equation*,

$$\left\{\frac{\partial^2}{\partial x^2} + \Delta_\rho + k^2\right\} G\left(x,\rho; x_0,\rho_0\right) = 0, \tag{1.1b}$$

At the layer boundary, continuity conditions for the functions G and $\partial G/\partial x$ must be satisfied. In addition, Eq. (1.1b) must satisfy radiation conditions at $x \to \pm\infty$.

The solution of Eq. (1.1b) with radiation conditions can be expressed in the following form:

$$G\left(x,\rho;\, x_0,\rho_0\right) = \begin{cases} \int dq\; T_1(q)\exp\left[-i\,(k^2-q^2)^{1/2}\,(x-L_0)+iq\cdot\rho\right] & \left(x \le L_0\right), \\ \int dq\; T_2(q)\exp\left[\;i\,(k^2-q^2)^{1/2}\,(x-L)\;+\;iq\cdot\rho\right] & \left(x \ge L\right). \end{cases} \tag{1.2}$$

At the layer boundary $x = L_0$, the solution to Eq. (1.1a) is connected with the solution to Eq. (1.1b) by the formula,

$$G\left(L_0,\rho;\, x_0,\rho_0\right) = \int dq\; T_1(q)\exp(iq\cdot\rho)\,. \tag{1.3a}$$

In an analogous way, we obtain for the derivative $\partial G/\partial x|_{x=L_0}$;

$$\begin{aligned} \frac{\partial}{\partial x} G\left(x,\rho;\, x_0,\rho_0\right)|_{x=L_0} &= -\,i\int dq\left(k^2-q^2\right)^{1/2} T_1(q)\exp(iq\cdot\rho) \\ &= -\,i\left(k^2+\Delta_\rho\right)^{1/2}\int dq\; T_1(q)\exp(iq\cdot\rho)\,. \end{aligned} \tag{1.3b}$$

Hence, the boundary condition for Eq. (1.1a) at $x = L_0$ has the form

$$\left\{\frac{\partial}{\partial x} + i\left(k^2+\Delta_\rho\right)^{1/2}\right\} G\left(x,\rho;\, x_0,\rho_0\right)|_{x=L_0} = 0\,. \tag{1.4a}$$

Similarly, we have the boundary condition at $x = L$:

$$\left\{\frac{\partial}{\partial x} - i\left(k^2+\Delta_\rho\right)^{1/2}\right\} G\left(x,\rho;\, x_0,\rho_0\right)|_{x=L} = 0\,. \tag{1.4b}$$

Inside the layer, the field G is continuous at all points, but the quantity $\partial G/\partial x$ has a discontinuity at the source point $x = x_0$:

$$\frac{\partial}{\partial x} G\left(x,\rho;\, x_0,\rho_0\right)|_{x=x_0+0} - \frac{\partial}{\partial x} G\left(x,\rho;\, x_0,\rho_0\right)|_{x=x_0-0} = \delta\left(\rho-\rho_0\right). \tag{1.5}$$

When inhomogeneities are absent $(\varepsilon = 0)$, Green's function in free space,

$$G\left(x,\rho;\, x_0,\rho_0\right)|_{\varepsilon=0} = g\left(x-x_0;\rho-\rho_0\right) = -\frac{1}{4\pi\left|\boldsymbol{r}-\boldsymbol{r}_0\right|}\, e^{ik|\boldsymbol{r}-\boldsymbol{r}_0|},\quad \left(\boldsymbol{r} = \{x,\rho\}\right),$$

is described by the integral form

$$g(x,\rho) = \int dq\, \hat{g}(q) \exp\left[i(k^2 - q^2)^{1/2}|x| + iq\cdot\rho\right],$$
$$\hat{g}(q) = \frac{1}{8i\pi^2}\frac{1}{(k^2 - q^2)^{1/2}}. \qquad (1.6)$$

When the observation point "x" and the source point "x_0" are fixed (for definiteness, we shall assume that $x < x_0$), the function $g(x - x_0, \rho - \rho_0)$, being the function of the parameter x_0, satisfies the following first order equations,

$$\frac{\partial}{\partial x_0} g(x - x_0; \rho - \rho_0) = i\left(k^2 + \Delta_\rho\right)^{1/2} g(x - x_0; \rho - \rho_0),$$
$$\frac{\partial}{\partial x_0} g(x - x_0; \rho - \rho_0) = i\left(k^2 + \Delta_{\rho_0}\right)^{1/2} g(x - x_0; \rho - \rho_0), \qquad (1.7a)$$

with the initial condition

$$g(0; \rho - \rho_0) = g_0(\rho - \rho_0) = -\frac{1}{4\pi|\rho - \rho_0|} \exp\left(ik|\rho - \rho_0|\right). \qquad (1.7b)$$

The equality

$$2i\left(k^2 + \Delta_\rho\right)^{1/2} g_0(\rho) = \delta(\rho) \qquad (1.8a)$$

is a consequence of the integral representation (1.6), and we have from (1.8a)

$$\left(k^2 + \Delta_\rho\right)^{-1/2} \delta(\rho) = 2ig_0(\rho),$$
$$\left(k^2 + \Delta_\rho\right)^{1/2} \delta(\rho) = 2i\left(k^2 + \Delta_\rho\right) g_0(\rho). \qquad (1.8b)$$

The effect of the operator $(k^2 + \Delta_\rho)^{1/2}$ on an arbitrary function $F(\rho)$ can be written in the form of an integral operator,

$$\left(k^2 + \Delta_\rho\right)^{1/2} F(\rho) = \int d\rho' \left(k^2 + \Delta_\rho\right)^{1/2} \delta(\rho - \rho')\, F(\rho')$$
$$\equiv \int d\rho'\, K(\rho - \rho')\, F(\rho'). \qquad (1.9a)$$

Its kernel is defined by the second equality in (1.8b);

$$K(\rho) = \left(k^2 + \Delta_\rho\right)^{1/2} \delta(\rho) = 2i\left(k^2 + \Delta_\rho\right) g_0(\rho) . \tag{1.9b}$$

The first equation of (1.8b) defines the kernel of the inverse operator.

If the point source is placed at the layer boundary $x_0 = L$, the wave field inside the layer $L_0 < x < L$ is described by the equation

$$\left\{ \frac{\partial^2}{\partial x^2} + \Delta_\rho + k^2 \left[1 + \varepsilon(x,\rho)\right] \right\} G\left(x,\rho; L,\rho_0\right) = 0 . \tag{1.10}$$

The condition (1.4a) (at $x_0 = L$) remains the boundary condition at $x = L_0$. For the boundary $x = L$, we have the equality according to (1.5),

$$\frac{\partial}{\partial x} G\left(x,\rho; x_0,\rho_0\right) \big|_{x=L-0} = \frac{\partial}{\partial x} G\left(x,\rho; L-0,\rho_0\right) \big|_{x=L} - \delta\left(\rho - \rho_0\right),$$

which can be rewritten, taking into account (1.4b), in a form

$$\left\{ \frac{\partial}{\partial x} - i\left(k^2 + \Delta_\rho\right)^{1/2} \right\} G\left(x,\rho; L,\rho_0\right) \big|_{x=L} = -\delta\left(\rho - \rho_0\right) . \tag{1.11}$$

The equality (1.11) defines the boundary condition for Eq. (1.10) at $x = L$.

If a wave $\Psi_0(x,\rho)$ is incident from the region $x > L$ on the layer, it creates at the boundary $x = L$ a distribution of sources $f(\rho_0)$ such that

$$\Psi_0(x,\rho) = \int d\rho_0 \; g(x-L; \rho - \rho_0) \; f(\rho_0) . \tag{1.12a}$$

According to (1.8a), the source distribution $f(\rho_0)$ will be described by the expression

$$f(\rho) = 2i\left(k^2 + \Delta_\rho\right)^{1/2} \Psi_0(L,\rho) . \tag{1.12b}$$

In this case, the wave field $\Psi(x,\rho)$ inside the layer is connected with the solution of Eq. (1.10) by the equality

$$\Psi(x,\rho) = \int d\rho_0 \; G\left(x,\rho; L,\rho_0\right) \; f(\rho_0) , \tag{1.13}$$

and, consequently, satisfies the boundary problem

$$\left\{ \frac{\partial^2}{\partial x^2} + \Delta_\rho + k^2 \left[1 + \varepsilon(x,\rho)\right] \right\} \Psi(x,\rho) = 0,$$

$$\left\{ \frac{\partial}{\partial x} + i\left(k^2 + \Delta_\rho\right)^{1/2} \right\} \Psi(x,\rho) \Big|_{x=L_0} = 0, \tag{1.14}$$

$$\left\{ \frac{\partial}{\partial x} - i\left(k^2 + \Delta_\rho\right)^{1/2} \right\} \Psi(x,\rho) \Big|_{x=L} = -2i\left(k^2 + \Delta_\rho\right)^{1/2} \Psi_0(L,\rho).$$

2. Imbedding equations

First let us study Green's function with the source at the point (L,ρ_0) for $x < L$, which is described by the boundary problem (1.10) and (1.11). Expressing the solution of this problem as a function of the parameter L, we obtain[2-5]

$$\left\{ \frac{\partial}{\partial x} - i\left(k^2 + \Delta_{\rho_0}\right)^{1/2} \right\} G(x,\rho;L,\rho_0) = -k^2 \int d\rho_1\, G(x,\rho;L,\rho_1)\, \varepsilon(L,\rho_1)\, H(L;\rho_1;\rho_0). \tag{2.1a}$$

Here $H(L;\rho_1;\rho_0) = G(L,\rho_1;L,\rho_0)$ denotes the field in the plane $x = L$, i.e., the *backscattered wave.* One can consider the equality (2.1a) as an integro-differential equation for G if one adds to it the initial condition at $L \rightarrow x$,

$$G(x,\rho;x,\rho_0)\,|_{L=x} = H(x;\rho;\rho_0). \tag{2.1b}$$

The function $H(L;\rho,\rho_0)$ obeys the closed nonlinear integro-differential equation,

$$\left\{ \frac{\partial}{\partial L} - i\left(k^2 + \Delta_\rho\right)^{1/2} - i\left(k^2 + \Delta_{\rho_0}\right)^{1/2} \right\} H(L;\rho;\rho_0) = \tag{2.2a}$$
$$-\delta(\rho - \rho_0) - k^2 \int d\rho_1\, H(L;\rho;\rho_1)\, \varepsilon(L,\rho_1)\, H(L;\rho_1;\rho_0).$$

The initial condition for H at $L = L_0$ becomes

$$H(L_0;\rho;\rho_0) = g_0(\rho - \rho_0), \tag{2.2b}$$

which corresponds to the source in the free space.

Analogously, taking Green's function $G(x,\rho;x_0,\rho_0)$ as a function of the parameter L, we obtain the equation

$$\frac{\partial}{\partial L} G(x,\rho;x_0,\rho_0;L) = -k^2 \int d\rho_1\, G(x,\rho;L,\rho_1)\, G(x_0,\rho_0;L,\rho_1)\, \varepsilon(L,\rho_1) \tag{2.3a}$$

with the initial condition

$$G(x,\rho;x_0,\rho_0;L)\big|_{L=\max\{x,x_0\}} = \begin{cases} G(x,\rho;\,x_0,\rho_0), & x_0 \geq x; \\ G(x_0,\rho_0;\,x,\rho), & x_0 \leq x, \end{cases} \tag{2.3b}$$

which reflects the continuity condition with respect to the parameter L.

Thus, Eqs. (2.1a), (2.2a), and (2.3a) form the closed system of imbedding equations for the given problem. The limit $L_0 \to -\infty$, $L \to +\infty$ corresponds to the solution for a point source located in the inhomogeneous medium, which occupies the whole space. Equation (2.3a) can be integrated, and we see the field of the point source, placed inside the medium, to be connected in a simple way (by quadrature) to the field when the source is located at the layer boundary.

Let us consider the case of the layered medium for which $\varepsilon(L,\rho) = \varepsilon(L)$. Then all functions G are functions of $\rho - \rho_0$. Using the Fourier transform, it is possible to pass from a system of integro-differential equations to a set of ordinary differential equations:

$$\frac{\partial}{\partial L} G_q(x;x_0;L) = -(2\pi k)^2 \varepsilon(L)\, G_q(x;L)\, G_q(x_0;L)\,, \tag{2.4a}$$

$$\left[\frac{\partial}{\partial L} - i\left(k^2 - q^2\right)^{1/2}\right] G_q(x;L) = -(2\pi k)^2 \varepsilon(L)\, G_q(x;L)\, H_q(L)\,, \tag{2.4b}$$

$$\left[\frac{\partial}{\partial L} - 2i\left(k^2 - q^2\right)^{1/2}\right] H_q(L) = -\frac{1}{4\pi^2} - (2\pi k)^2 \varepsilon(L)\, H_q^2(L)\,,$$

$$H_q(L_0) = g_0(q)\,,\quad G_q(x;L)\big|_{L=x} = H_q(x)\,,\quad G_q = \frac{1}{(2\pi)^2}\int d\rho\, G(\rho)\exp\left[-iq\cdot\rho\right]. \tag{2.4c}$$

Equation (2.4b) describes the propagation of a plane wave incident at some angle to the boundary $x = L$ with an amplitude $g(\boldsymbol{q}) = 1/8i\pi^2(k^2-q^2)^{-1/2}$. After normalizing the incident wave by $\Psi_0(x,\rho) = \exp[-ik(x-L) + i\boldsymbol{q}\cdot\rho]$, we have the following system of equations

$$\frac{\partial}{\partial L}\tilde{G}_q(x;x_0;L) = \frac{ik^2}{2(k^2-q^2)^{1/2}}\varepsilon(L)\tilde{G}_q(x;L)\tilde{G}_q(x_0;L),$$

$$\left[\frac{\partial}{\partial L} - i(k^2-q^2)^{1/2}\right]\tilde{G}_q(x,L) = \frac{ik^2}{2(k^2-q^2)^{1/2}}\varepsilon(L)\tilde{G}_q(x;L)\tilde{H}_q(L),$$

$$\left[\frac{\partial}{\partial L} - 2i(k^2-q^2)^{1/2}\right]\tilde{H}_q(L) = -2i(k^2-q^2)^{1/2} + \frac{ik^2}{2(k^2-q^2)^{1/2}}\varepsilon(L)\tilde{H}_q^2(L),$$

$$\tilde{H}_q(L_0) = 1. \tag{2.5}$$

The case of the normal incidence of the wave on the boundary corresponds to $q = 0$.

Note that the differential equation (2.1) can be rewritten as an integral equation,

$$G(x,\rho;L,\rho_0) = \tilde{g}(x,\rho;L,\rho_0) - k^2\int_x^L dx_1\int d\rho_1\, G(x,\rho;x_1,\rho_1)\,\varepsilon(x_1,\rho_1)\,\tilde{g}(x_1,\rho_1;L,\rho_0),$$

which gives

$$\Psi(x,\rho;L) = \Psi_0(x,\rho;L) - k^2\int_x^L dx_1\int d\rho_1\,\tilde{g}(x,\rho;x_1,\rho_1)\,\varepsilon(x_1,\rho_1)\,\Psi(x_1,\rho_1;L), \tag{2.6a}$$

where

$$\tilde{g}(x,\rho;L,\rho_0) = \exp\left\{i\left(k^2+\Delta_{\rho_0}\right)^{1/2}(L-x)\right\}H(x;\rho;\rho_0),$$

$$\Psi_0(x,\rho;L) = \int d\rho_0\;\tilde{g}(x,\rho;x_0,\rho_0)\;f(\rho_0),$$

and $f(\rho_0)$ is the distribution of sources at the boundary $x = L$.

Equation (2.2) describes the field on the boundary $x = L$ (i.e., the backscattered field). The backscattering effect is essentially nonlinear, and is described by the last term in (2.2). If we neglect this term, the solution to the remaining equation has the form $H(L;\rho;\rho_0) = g(\rho-\rho_0)$, which corresponds to the presence of the incident wave only in the plane $x = L$. In this case, $\tilde{g}(x,\rho;L,\rho_0) = g(x-L;\rho-\rho_0)$ and Eq. (2.6) takes the form of the causality integral equation

$$\Psi(x,\rho) = \Psi_0(x,\rho) - k^2 \int_x^L dx_1 \int d\rho_1\, g(x-x_1;\rho-\rho_1)\, \varepsilon(x_1,\rho_1)\, \Psi(x_1,\rho_1), \tag{2.6b}$$

This result includes scattering of the incident wave through large angles (i.e., up to but not exceeding $\pi/2$). It can be rewritten in the form of an operator equation. Differentiating (2.6b) with respect to x and using (1.8a), we obtain

$$\begin{gathered} \left[\frac{\partial}{\partial x} + i\left(k^2 + \Delta_\rho\right)^{1/2}\right]\Psi(x,\rho) = \\ -i\frac{k^2}{2}\left(k^2+\Delta_\rho\right)^{-1/2}\{\varepsilon(x,\rho)\,\Psi(x,\rho)\},\ \Psi(L,\rho) = \Psi_0(L,\rho). \end{gathered} \tag{2.7}$$

For the case of narrow-angle scattering ($\Delta_\rho << k^2$), we obtain the parabolic equation

$$\frac{\partial}{\partial x}\Psi(x,\rho) = -ik\Psi - \frac{i}{2k}\Delta_\rho\Psi - i\frac{k}{2}\varepsilon(x,\rho)\,\Psi(x,\rho),\ \Psi(L,\rho) = \Psi_0(\rho),$$

which, for the function $u(x,\rho)$ defined by the equality

$$\Psi(x,\rho) = u(x,\rho)\exp\left[-ik(x-L)\right],$$

takes the well-known form

$$\frac{\partial}{\partial x}u(x,\rho) = \frac{i}{2k}\Delta_\rho u + i\frac{k}{2}\varepsilon(x,\rho)\,u(x,\rho),\ u(L,\rho) = u_0(\rho). \tag{2.8}$$

Let us introduce Green's function $G(x,\rho;x',\rho')$ for Eq. (2.8), which satisfies Eq. (2.8) with the initial condition

$$G(x,\rho;x',\rho')\big|_{x=x'} = \delta(\rho-\rho').$$

If inhomogeneities are absent, i.e., $\varepsilon = 0$, we have

$$G(x,\rho;x',\rho')\big|_{\varepsilon=0} = g(x-x';\rho-\rho') = \frac{k}{2i\pi(x-x')}\exp\left[\frac{ik(\rho-\rho')^2}{2(x-x')}\right]. \tag{2.9}$$

The wave field $u(x,\rho)$ is a functional of the field $\varepsilon(x,\rho)$. The variational derivative of the field $u(x,\rho)$ with respect to ε is expressed in terms of the function G using the equality

$$\frac{\delta}{\delta\varepsilon(x',\rho')} u(x,\rho) = i\frac{k}{2} G(x,\rho;x',\rho') u(x',\rho'). \tag{2.10}$$

Fields in different planes are connected by the equality

$$u(x,\rho) = \int d\rho' \, G(x,\rho;x',\rho') \, u(x',\rho'), \quad x \geq x'. \tag{2.11}$$

Green's function satisfies the orthogonality condition[1,3]

$$\int d\rho \, G(x,\rho;x',\rho') \, G^*(x,\rho;x'',\rho'') = \delta(\rho' - \rho''),$$

and hence, we obtain from (2.11)

$$u(x',\rho') = \int d\rho \quad G^*(x,\rho;x',\rho') \, u(x,\rho), \quad x' \leq x. \tag{2.12}$$

Using (2.12) it is possible to rewrite (2.10) in the form

$$\frac{\delta}{\delta\varepsilon(x',\rho')} u(x,\rho) = i\frac{k}{2} G(x,\rho;x',\rho') \int d\rho'' \, G^*(x,\rho'';x',\rho') \, u(x,\rho''), \tag{2.13}$$

which expresses the variational derivative $u(x,\rho)$ in terms of the field $u(x,\rho')$ in the same plane.

3. Statistical averaging

Let us now discuss ordinary methods of the statistical analysis. In the general case, the boundary problems (1.1), (1.4a), and (1.4b) have no dynamical causality property because the solution of the problem $G(x,\rho;x_0,\rho_0)$ for fixed values of x and x_0 are described by fluctuations $\varepsilon(x,\rho)$ in the whole layer $L_0 < x,x_0 < L$. Equations for $H(L;\rho;\rho_0)$, $G(x,\rho;L,\rho_0)$, and $G(x,\rho;x_0,\rho_0;L)$, however, possess the dynamical causality with respect to the additional parameter L, because by using the imbedding method we constructed the auxiliary initial-value problem out of the original boundary-value problem. These equations have not been studied in the general case. The analysis of Eq. (2.2) for the backscattered field[6-9] is an exception. Typically, investigators have restricted themselves to the analysis of the parabolic equation (2.8) (i.e., see Refs. 1,10,11), which has a large enough field of applicability in a number of cases.

The main method of statistical analysis of Eq. (2.8) is the approximation of the delta-correlated field $\varepsilon(x,\rho)$. It means that the correlation function $B_\varepsilon(x,\rho;x',\rho') = \langle\varepsilon(x,\rho)\ \varepsilon(x',\rho')\rangle$ can be approximated by an *effective correlation function*

$$B_\varepsilon^{\mathrm{eff}}(x,\rho;x',\rho') = \delta(x-x')\ A(x;\rho,\rho'), \tag{3.1a}$$

where

$$A(x;\rho,\rho') = \int_{-\infty}^{\infty} dx'\ B_\varepsilon(x,\rho;x',\rho').$$

For homogeneous fluctuations $\varepsilon(x,\rho)$,

$$A(x;\rho,\rho') = A(\rho-\rho'). \tag{3.1b}$$

We assume $\varepsilon(x,\rho)$ to be a Gaussian random field with zero mean value.

It is also assumed in this approximation that the longitudinal correlation radius is small in comparison with all parameters arising in the propagation process. The method for deriving the equations for statistical characteristics of the wave field is well known (i.e., see Refs. 1,3,11), and we shall not dwell on them here.

As an example, let us discuss the mean field. Averaging Eq. (2.8) over an ensemble of realizations of the function $\varepsilon(x,\rho)$, we have

$$\left(2ik\frac{\partial}{\partial x} + \Delta_\rho\right)\langle u(x,\rho)\rangle + k^2\left\langle\varepsilon(x,\rho)\ u(x,\rho)\right\rangle = 0. \tag{3.2}$$

To obtain an expression for the last term in (3.2), let us use the Furutsu-Novikov formula

$$\left\langle\varepsilon(\boldsymbol{r})\ R[\varepsilon]\right\rangle = \int d\boldsymbol{r}'\ \langle\varepsilon(\boldsymbol{r})\ \varepsilon(\boldsymbol{r}')\rangle\left\langle\frac{\delta R[\varepsilon]}{\delta\varepsilon(\boldsymbol{r}')}\right\rangle, \quad \left(\boldsymbol{r} = \{x,\rho\}\right),$$

which is valid for a Gaussian field $\varepsilon(\boldsymbol{r})$ and its arbitrary functional R. Then using (2.10), we obtain

$$\left(2ik\frac{\partial}{\partial x}+\Delta_\rho\right)\langle u(x,\rho)\rangle + i\frac{k^2}{2}\int_0^x dx'\int d\rho'\,B_\varepsilon(x,\rho;x',\rho')\,\langle G(x,\rho;x',\rho')\,u(x',\rho')\rangle = 0. \tag{3.3}$$

Substituting (2.12) into (3.3), it is possible to rewrite Eq. (3.3) in the form

$$\left(2ik\frac{\partial}{\partial x}+\Delta_\rho\right)\langle u(x,\rho)\rangle = -i\frac{k^3}{2}\int_0^x dx'\int d\rho'\,B_\varepsilon(x,\rho;x',\rho')\int d\rho''\langle G(x,\rho;x',\rho')\,G^*(x,\rho'';x',\rho')\,u(x,\rho'')\rangle. \tag{3.4}$$

For the model of the delta-correlated field $\varepsilon(x,\rho)$ [(3.1a), (3.1b)], we obtain the following closed equation for the mean field,

$$\left(2ik\frac{\partial}{\partial x}+\Delta_\rho\right)\langle u(x,\rho)\rangle + i\frac{k^3}{4}A(0)\langle u(x,\rho)\rangle = 0. \tag{3.5}$$

It is possible to take into account the finiteness of the longitudinal correlation radius $l_\parallel$ of the field $\varepsilon(x,\rho)$. Suppose that medium fluctuations do not significantly effect the wave field on the distances of order $l_\parallel$. In this case, it is possible to substitute G in (3.4) by g from (2.9). As a result, this operation gives a closed integro-differential equation, which corresponds to the *diffusion approximation* (i.e., see Ref. 12),

$$\left(2ik\frac{\partial}{\partial x}+\Delta_\rho\right)\langle u(x,\rho)\rangle = \int d\rho''\,D(x,\rho;\rho'')\,\langle u(x,\rho'')\rangle, \tag{3.6}$$

where

$$D(x,\rho;\rho'') = -i\frac{k^3}{2}\int_0^x dx'\int d\rho'\,B_\varepsilon(x,\rho;x',\rho')\,g(x-x';\rho-\rho')\,g^*(x-x';\rho'-\rho'').$$

This equation has a larger field of applicability than the approximation of the delta-correlated random field.

For spatially homogeneous fluctuations of ε and an incident plane wave, we can express the solution of Eq. (3.6) in the following form:

$$\langle u(x,\rho)\rangle = \exp\left\{\frac{k^2}{4}\, x \int_0^\infty d\xi \int d\rho'\, B_\varepsilon(\xi,\rho')\, g(\xi,\rho')\right\}, \quad x \gg l_\parallel . \tag{3.7}$$

The condition of applicability of the diffusion approximation for homogeneous fluctuations $\varepsilon(x,\rho)$ is then

$$\frac{k^2}{4}\, l_\parallel \left|\int_0^\infty d\xi \int d\rho\, B_\varepsilon(\xi,\rho)\, g(\xi,\rho)\right| \ll 1 .$$

Note that neither the delta-correlated random field $\varepsilon(x,\rho)$ nor the diffusion approximation are applicable when $\varepsilon(x,\rho) = \varepsilon(\rho)$ does not depend on x (what takes place, for example, in the layered medium with respect to z). In this case, the random field ε has an infinite correlation radius along the x axis. The next part of the work is dedicated to the discussion of this problem.

4. Point source in a layered random media

Let us now examine the wave field of a point source in a multidimensional randomly inhomogeneous layered medium,[13] for which the Green function satisfies the equation

$$\left\{\frac{\partial^2}{\partial z^2} + \Delta_\perp + k^2\left[1 + \varepsilon(z)\right]\right\} G(\rho, z; z_0) = \delta(z - z_0)\, \delta(\rho) , \tag{4.1}$$

where $\rho = (x,y)$, $\Delta_\perp = \partial^2/\partial x^2 + \partial^2/\partial y^2$.

We have the following integral representations of Green's function in a one-, two-, and three-dimensional space, respectively,

$$G^{(1)}(z; z_0) = \frac{1}{2ik} \int_0^\infty dt\, \exp(ikt/2)\, \Psi(t; z; z_0) ,$$

$$G^{(2)}(x,z; z_0) = \frac{1}{2ik}\left(\frac{k}{2\pi i}\right)^{1/2} \int_0^\infty dt\, t^{-1/2} \exp\left[ik(x^2 + t^2)/2t\right] \Psi(t; z; z_0) ,$$

$$G^{(3)}(\rho,z; z_0) = -\frac{1}{4\pi} \int_0^\infty dt\, t^{-1} \exp\left[ik(\rho^2 + t^2)/2t\right]\, \Psi(t; z; z_0) , \tag{4.2}$$

where $\Psi(t,z;z_0)$ is defined by the equation

$$\frac{\partial}{\partial t}\Psi(t,z;z_0) = \frac{i}{2k}\left[\frac{\partial^2}{\partial z^2} + k^2\varepsilon(z)\right]\Psi(t,z;z_0),$$
$$\Psi(0,z;z_0) = \delta(z-z_0). \tag{4.3}$$

This formula expresses the *factorization property* of the Helmholts equation in a layered medium.

The evolutionary problem (4.3) is to be supplemented with boundary conditions along z. We consider the following boundary problems:

(a) the source is located in the unbounded space with a radiation condition at $|z| \to \infty$;

(b) the source is located on the reflective boundary, for which $\partial\Psi/\partial z|_{z=z_0-0} = 0$, and there is a radiation condition at $z \to \infty$;

(c) the source is located on the boundary of a homogeneous half-space with a radiation condition at $|z| \to \infty$.

When $x \to \infty$ or $|\rho| \to \infty$, we have, respectively, the asymptotics

$$G^{(2)}(x,z;z_0) \cong \frac{1}{2ik}\exp\left(ik|x|\right)\Psi\left(|x|,z;z_0\right),$$
$$G^{(3)}(\rho,z;z_0) \cong \frac{1}{4\pi}\left(\frac{2\pi i}{k|\rho|}\right)^{1/2}\exp\left(ik|\rho|\right)|\rho|^{-1/2}\,\Psi(|\rho|,z;z_0), \tag{4.4}$$

if the function $\Psi(t,z;z_0)$ does not have exponential behavior with respect to t. Formulae (4.4) correspond to *narrow-angle scattering (the parabolic approximation).* In order to analyze the contribution of scattering at wide angles, the exact representation (4.2) of Green's function should be used.

One writes $\Psi(t,z;z_0)$ via Fourier transformation

$$\Psi_\omega(z;z_0) = \int_{-\infty}^{\infty} dt\, e^{-i\omega t}\,\Psi(t,z;z_0),$$

and we see that $\Psi_\omega(z;z_0)$ satisfies the equation

$$\left[\frac{\partial^2}{\partial z^2} - 2k\omega + k^2\varepsilon(z)\right]\Psi_\omega(z;z_0) = 2ik\,\delta(z-z_0). \tag{4.5}$$

The solution $\Psi_{\omega<0}(z;z_0)$ corresponds to *propagating waves* when $\omega < 0$ and to *evanescent waves* when $\omega > 0$.

We shall be interested in the asymptotic behavior of Green's functions at $x,\rho \to \infty$. On this limit, we have

$$G^{(1)}(z;z_0) = \frac{1}{2ik}\Psi_{-k/2-i0}(z;z_0),$$

$$G^{(2)}(x,z;z_0) = \frac{1}{4i\pi k}\int_{-\infty}^{\infty}\frac{d\omega}{(1+2\omega/k)^{1/2}}\Psi_{\omega-i0}(z;z_0)\exp\left[ikx\,(1+2\omega/k)^{1/2}\right],$$

$$G^{(3)}(\rho,z;z_0) = -\frac{1}{8\pi^2}\left(\frac{2\pi i}{k\rho}\right)^{1/2}\int_{-\infty}^{\infty}\frac{d\omega}{(1+2\omega/k)^{1/4}}\Psi_{\omega-i0}(z;z_0)\exp\left[ikx(1+2\omega/k)^{1/2}\right]. \tag{4.6}$$

Formulae (4.6) reduce to (4.4) if the condition $2\,\omega/k \ll 1$ is satisfied.

Let us first discuss the statistical behavior of Green's functions in the parabolic approximation. The representation of the function $\Psi(t,z;z_0)$ can then be written in the form

$$\Psi(t,z;z_0) = \frac{1}{2\pi}\int_0^{\infty}d\omega\left(\frac{k}{2\omega}\right)^{1/2}e^{-i\omega t}\,\tilde{\Psi}_\omega(z;z_0) + \frac{1}{2\pi}\int_0^{\infty}d\omega\, e^{i\omega t}\,\Psi_\omega(z;z_0), \tag{4.7}$$

where $\tilde{\Psi}_\omega(z;z_0)$ is described by the equation

$$\left[\frac{\partial^2}{\partial z^2} + 2k\omega + k^2\varepsilon(z)\right]\tilde{\Psi}_\omega(z;z_0) = 2ik\,(2k\omega)^{1/2}\,\delta(z-z_0). \tag{4.8}$$

For the evanescent waves, the influence of fluctuations $\varepsilon(z)$ is not essential and $\Psi_{\omega<0}(z,z_0) = -i\,(k/2\omega)^{1/2}\exp\{-(2k\omega)^{1/2}|z-z_0|\}$, which corresponds to a free space field. The unknown function $\tilde{\Psi}_\omega(z;z_0)$ is a solution of Eq. (4.8), the statistical analysis of which has been carried out elsewhere.[14,15]

The most important feature of a one-dimensional statistical problem is that any small but finite absorption $\gamma \ll 1$ has to be taken into account. We introduce it via the function $\varepsilon(z) = \varepsilon_1(z) + i\gamma$, where $\varepsilon_1(z)$ is a real random function. The quantity of interest is the average intensity $\langle I(t,z;z_0)\rangle = \langle \Psi(t,z;z_0)\, \Psi^*(t,z;z_0)\rangle$ at sufficiently large values of t. We can express this quantity as

$$\langle I(t,z;z_0)\rangle = I_{\text{fluc}}(t,z;z_0) + I_1(t,z;z_0) + I_2(t,z;z_0), \tag{4.9}$$

where the functions $I_{\text{fluc}}(t,z;z_0)$ and $I_2(t,z;z_0)$ describe the contributions of the first and second terms on the right side of (4.7) into the average intensity, and $I_1(t,z;z_0)$ takes into account the cross term, $\langle \tilde{\Psi}_{\omega_1}(z;z_0)\, \Psi^*_{\omega_2}(z;z_0)\rangle$.

First let us consider

$$I_{\text{fluc}}(t,z;z_0) = \frac{k}{2(2\pi)^2} \int_0^\infty d\Omega \int_{-2\Omega}^{2\Omega} \frac{d\omega}{(\Omega^2 - \omega^2/4)^{1/2}}\, e^{-i\omega t} \langle I_{\Omega,\omega}(z;z_0)\rangle,$$

where $\langle I_{\Omega,\omega}(z;z_0)\rangle = \langle \tilde{\Psi}_{\Omega+\omega/2}(z;z_0)\, \tilde{\Psi}^*_{\Omega-\omega/2}(z;z_0)\rangle$ is the two-frequency correlation of the corresponding boundary problem solution. When $t \to \infty$, the values of the integrand have to be evaluated at $\omega \to 0$.

The statistical characteristics of the solution of Eq. (4.8) are defined by the statistics of the reflection coefficient $R_{z_0}(\omega)$ of the plane wave incident on the half-space $z_0 > z$ from a homogeneous space. The function $R_{z_0}(\omega)$ satisfies the Riccati equation,

$$\frac{d}{dz_0} R_{z_0}(\omega) = \left[2i\,(2k\omega)^{1/2} - k\,(k/2\omega)^{1/2}\,\gamma\right] R_{z_0}(\omega) + i\,\frac{k}{2}\left(\frac{k}{2\omega}\right)^{1/2} \varepsilon_1(z_0)\left[1 + R_{z_0}(\omega)\right]^2.$$

This yields the equation for the two-frequency function, $W_{z_0}(\Omega,\omega) = R_{z_0}(\Omega+\omega/2)\, R^*_{z_0}(\Omega-\omega/2)$ at $\omega \to 0$:

$$\frac{d}{dz_0} W_{z_0}(\Omega,\omega) = -2\left(\frac{k}{2\Omega}\right)^{1/2}(k\gamma - i\omega)\, W_{z_0}(\Omega,\omega) - i\,\frac{k}{2}\left(\frac{k}{2\Omega}\right)^{1/2} \varepsilon_1(z_0)\left[R_{z_0}(\Omega) - R^*_{z_0}(\Omega)\right]\left[1 - W_{z_0}(\Omega,\omega)\right].$$

(We neglect terms proportional to $\omega\gamma, \omega\varepsilon_1$.) For simplicity, we suppose $\varepsilon_1(z)$ to be the Gaussian *white noise* with the characteristics

$$\langle\varepsilon_1(z)\rangle = 0, \ \langle\varepsilon_1(z)\ \varepsilon_1(z')\rangle = 2\sigma_\varepsilon^2\, l_0\, \delta(z-z').$$

Then the one-frequency statistical characteristics are expressed by the parameters

$$\beta = \frac{k\gamma}{D}\left(\frac{k}{2\Omega}\right)^{1/2}, \quad D = D_0\frac{k}{2\Omega}, \quad D_0 = k^2\sigma_\varepsilon^2\, l_0/2\,,$$

with the functional dependence on β being defined by the boundary conditions. In this case, there is probability distribution $P(u)$ of $u = (1+W)/(1-W)$ which does not depend on z_0 (half-space)

$$P(u) = e^{-\beta(u-1)}. \tag{4.10}$$

Determination of the two-frequency function $W_{z_0}(\Omega,\omega)$ at small ω is reduced to the analytical continuation of the corresponding one-frequency solution with respect to β,

$$\beta \Rightarrow \left(\frac{k}{2\Omega}\right)^{1/2}\frac{k\gamma - i\omega}{D} = -i\left(\frac{2\Omega}{k}\right)^{1/2}\frac{\omega + ik\gamma}{D_0}. \tag{4.11}$$

We investigate some problems for different boundary conditions.

a. Source location in the unbounded space

Consider the average intensity at $z = z_0$. The one-frequency characteristic is $\langle I_{\Omega,0}(z_0;z_0)\rangle = \langle\tilde{\Psi}_\Omega(z_0;z_0)\ \tilde{\Psi}^*_\Omega(z_0;z_0)\rangle = 1 + 1/\beta$. Performing the analytical continuation with respect to β, one has

$$I_{\text{fluc}}(t,z_0;z_0) = \frac{kD_0}{2i(2\pi)^2}\int_0^\infty d\Omega\int_{-2\Omega}^{2\Omega}\frac{d\omega}{(\Omega^2-\omega^2/4)^{1/2}}\, e^{-i\omega t}\left(\frac{k}{2\Omega}\right)^{1/2}\frac{1}{\omega + i\gamma}. \tag{4.12}$$

Expression (4.12) for $I_{\text{fluc}}(t,z_0;z_0)$ exists for a point source only in the case of a finite absorption γ. If $\gamma \to 0$, we have

$$I_{\text{fluc}}(t,z_0;z_0) \propto D_0 k/\gamma^{1/2}.$$

Hence, the presence of a small but finite absorption is important for the description of the wave field characteristics (the result is identical to the one-dimensional problem).

b. Source location on the reflective boundary

For this case, we find that all formulated conclusions for the source in the unbounded space are also valid for the source on the reflective boundary.

c. Source location on the half-space boundary

If the source is placed on the boundary $z = z_0$ of the medium, the one-frequency characteristic has the form

$$\left\langle I_{\Omega,0}(z_0;z_0)\right\rangle = 1 + \left\langle W_{\Omega,0}\right\rangle = 1 + \left\langle |R_{z_0}(\Omega)|^2\right\rangle ,$$

where the mean value is calculated with the help of the statistical probability distribution (4.10). Then the average intensity is then given by

$$\left\langle I(t,z_0;z_0)\right\rangle = I_{\text{free}}(t,z_0;z_0) + \tilde{I}_{\text{fluc}}(t,z_0;z_0) ,$$

$$\tilde{I}_{\text{fluc}}(t,z_0;z_0) = \frac{k}{2(2\pi)^2}\int_0^\infty d\Omega \int_{-2\Omega}^{2\Omega} \frac{d\omega \exp(-i\omega t)\,\beta(\Omega,\omega)}{(\Omega^2-\omega^2/4)^{1/2}} \int_0^\infty \frac{u\,du}{u+2} \exp[-\beta(\Omega,\omega)] ,$$

and the parameter $\beta(\Omega,\omega)$ is given by the formula (4.11). Integration with respect to ω, u yields the expression ($\gamma \to 0$)

$$\tilde{I}_{\text{fluc}}(t,z_0;z_0) = \frac{k^{3/2}D_0}{2(2\pi)^{1/2}} \int_0^\infty d\Omega\, \Omega^{-3/2} \left[2 + D_0\left(\frac{k}{2\Omega}\right)^{1/2} t\right]^{-2} .$$

In the limit $t \to \infty$, we have

$$\tilde{I}_{\text{fluc}}(t,z_0;z_0) = I_{\text{free}}(t,z_0;z_0) = k/2\pi t ,$$

and the average intensity of the wave field is doubled, as in the one-dimensional problem.

Let us discuss now an exact description of the problem of point source. One considers the function $G^{(2)}(x,z;z_0)$ (4.6). We divide the integration region into $(-\infty, -k/2)$, $(-k/2, 0)$, and $(0, +\infty)$. The contribution of the first region to Green's function is approximately $\Psi_{-k/2}(z_0;z_0)/kx$ because of a decreasing exponent at $x \to +\infty$. It gives rise to the term

$$\left\langle I_2^{(2)}(x)\right\rangle \propto D_0/(kx)^2 k\gamma$$

for the average intensity. In the third region, the field $\Psi_\omega(z_0; z_0)$ corresponds to a wave propagating in free space and generates the contribution $<I_3^{(2)}(x)> \propto 1/kx$ to the average intensity. To evaluate the contribution of the region $(-k/2, 0)$, we act by analogy with the parabolic equation. The corresponding term in the average intensity is then

$$\left\langle I_1^{(2)}(x) \right\rangle \propto D_0/k\gamma^{1/2} \tag{4.13}$$

for sufficiently large values of x, but $k\gamma x << 1$. It is noted that there are cross terms which are produced by fields from different regions. There are, however, no powers of γ in denominators of the corresponding asymptotical expressions. The combination of these terms yields the condition $\gamma^{3/4} << k\gamma x << 1$, which defines the leading order of $<I_2^{(2)}(x)>$ in the average intensity.

We analyzed the three-dimensional case by analogy with the two-dimensional one:

$$\left\langle I_1^{(3)}(\rho) \right\rangle \propto D_0\, k/\gamma\,(k\rho)^4, \quad \left\langle I_2^{(3)}(\rho) \right\rangle \propto D_0/\rho\gamma^{1/2}, \quad \left\langle I_3^{(3)}(\rho) \right\rangle \propto 1/(k\rho)^2,$$

for $\rho \to \infty$, and $k\gamma\rho << 1$. Under the condition $k\gamma\rho >> \gamma^{5/6}$, the function $<I_2^{(3)}(\rho)>$ gives the main contribution to the average intensity of a point source at $z = z_0$. The result corresponds to a cylindrical wave and is determined by the absorption γ. Analogous results are true for a point source on the reflective boundary.

If the source is located on the boundary of a randomly inhomogeneous half-space, the average intensity is

$$\left\langle I(x,z_0;z_0) \right\rangle = I_{\text{free}}(x,z_0;z_0) \left(1 + 2D_0x \int_0^1 \frac{ds}{\left[D_0x + 2s\,(1-s^2)^{1/2}\right]^2} \right).$$

We are interested in the two limits: $D_0x << 1$ (but $kx >> 1$), and $D_0x >> 1$. In the first case,

$$\left\langle I(x,z_0;z_0) \right\rangle = 2\,I_{\text{free}}(x,z_0;z_0).$$

This result is analogous to that obtained using the parabolic equation method and indicates that scattering at wide angles does not influence the statistics. When $D_0x >> 1$, the average intensity is

$$\left\langle I(x,z_0;z_0) \right\rangle = I_{\text{free}}(x,z_0;z_0)\,(1 + 2/D_0x),$$

and scattering at wide angles is essential in the formation of the statistics, since it appears to be an additional attenuating factor of the intensity inside the empty space. We have the same result in the three-dimensional case.

We have shown that a significant feature of the one-dimensional problem related to plane waves in layered random media is that a small but finite absorption has to be taken into account (a parameter γ). The wave field statistics are caused by interference of multiple reflected waves in the medium giving rise to the dependence of the average intensity $<I>$ on γ (e.g., $<I> \propto 1/\gamma$ for a point source in an unbounded space). In multidimensional layered problems, spatial diffraction appears to play the role of an "absorption." It allows calculation of the statistical characteristics by means of an analytical continuation with respect to γ. We hoped that the diffraction effects would remove the dependence of the wave field statistics on γ, and that a finite limit at $\gamma \to 0$ would exist in multidimensional problems. This, however, was not the case. Diffraction effects were shown to diminish the power of γ, e.g., $<I> \propto 1/\gamma^{1/2}$ for a point source in an unbounded space, but absorption remained an important feature for the multidimensional problem as in the one-dimensional case.

References

1. V. I. Klyatskin, *Stochastic Equations and Waves in Randomly Inhomogeneous Media*, Nauka, Moscow (in Russian) (1980).

2. V. I. Klyatskin, *The Imbedding Method in Wave Propagation Theory*, Nauka, Moscow (in Russian) (1986).

3. V. I. Klyatskin, *Ondes et Équations Stochastiques dans les Milieux Aléatoirement non Homogènes*, Editions de Physique, Besançon-Cedex (in French) (1985).

4. G. I. Babkin, V. I. Klyatskin, and L. Ya. Lubavin, "Theory of invariant imbedding and waves in statistically nonuniform media," *Sov. Phys. Dokl.* **25**(2), 98-100 (1980).

5. G. I. Babkin and V. I. Klyatskin, "Invariant imbedding method for wave problems," *Wave Motion* **4**(3), 195-207 (1982).

6. B. M. Shevtsov, "Three-dimensional problem of inverse scattering in stratified randomly inhomogeneous media," *Radiophys. Quantum Electron.* **25**(9), 742-749 (1982).

7. B. M. Shevtsov, "Statistical theory of backscattering in randomly nonhomogeneous media," *Radiophys. Quantum Electron.* **24**(11), 913-917 (1981).

8. B. M. Shevtsov, "The probe of backscattering in three-dimensional, randomly nonhomogeneous media," *Radiophys. Quantum Electron.* **26**(4), 326-331 (1983).

9. Yu. N. Barabanenkov, D. I. Kryukov, "Functional Fokker-Planck formalism for wave propagation in random media under delocalized and weak localized regimes," *Waves in Random Media* **2**(1), 1-6 (1992).

10. V. I. Tatarskii, *The Effects of the Turbulent Atmosphere on Wave Propagation*, National Technical Information Service, Springfield, VA (1971).

11. S. M. Rytov, Yu. A. Kravtsov, and V. I. Tatarskii, *Principles of Statistical Radiophysics*, vol. 1-4, Springer Verlag, Berlin (1987-1989).

12. V. I. Klyatskin, "Approximations by delta-correlated random processes and diffusive approximation in stochastic problem," *Lectures in Appl. Math.* **27**, 447-476 (1991).

13. M. A. Guzev, V. I. Klyatskin, "Approximation of the parabolic equation and the wavefield of a point source in a layered random medium," *Waves in Random Media* **1**(4), 275-286 (1991).

14. V. I. Klyatskin, "Statistical theory of radiation transport in stratified random media," *Izvestiya, Atmosph. Ocean. Phys.* **27**(1), 31-44 (1991).

15. V. I. Klyatskin, "A statistical theory of radiative transfer in layered random media," in *First International Conference on Mathematical and Numerical Aspects of Wave Propagation Phenomena*, SIAM, Philadelphia (1991), 595-608.

Simulation of Wave Propagation in Random Media: Theory and Applications

Jan Martin
Jet Propulsion Laboratory
California Institute of Technology
Pasadena, CA 91109, USA

1. Introduction

Wherever waves travel through any medium, fluctuations in the composition of that medium induce fluctuations in the wave fronts. Nearly every real medium has fluctuations in its refractive index — if they are great enough, significant variation in the wave phase and amplitude will develop, and can be observed and interpreted to gain information about the wave source or the medium (mean structure or fluctuations). Such fluctuations have been observed in many situations of scientific interest, such as electromagnetic waves through atmospheres, ionospheres, and interplanetary and interstellar media, acoustic waves in the ocean and atmosphere, seismic waves in the earth.

The effects of the random medium are well characterized by Rytov theory[1] if the intensity fluctuations are less than about 0.1 times the mean intensity. If the scattering is sufficiently strong that the intensity fluctuations are near the mean in size, asymptotic theory can be applied to obtain accurate estimates of the wave statistics[1,2]. However, many interesting phenomena and experimental conditions lie in the very large region between weak and strong scattering, and simulations are currently the only tool that provides accurate information about this "strong focusing" parameter regime. Simulations are also effective in investigations of scattering when boundary conditions or large-scale medium variation have complex geometry and interact strongly with the waves, e.g. the ocean sound channel[3]. Simulations are also useful by providing benchmarks to establish the accuracy or validity limits of more easily computed, but also more approximate theories of scattering.

The parabolic approximation to the wave equation and the split-step solution have been used extensively in a variety of contexts to simulate propagation phenomena, initially to model acoustic waves propagating in both deterministic[4,5] and random[6] ocean channels. The single thin phase screen has been used as a theoretical model for the ionosphere effect on radio signals[7,8], and more complex spatial spectra of its random variations have been used in simulations of transionospheric radio propagation[9,10]and radio propagation through the solar wind[11]. A natural extension of this concept is the multiple phase screen model of propagation through extended random media[12,13], which is the same as the split-step PE method of Tappert. With the availability of more powerful computing resouces, such simulations have been extended to a variety of problems, some of which we will discuss in this review. It should also be noted that there is an extensive body of applications of this technique to deterministic media as well, especially the ocean acoustic channel[14–16], but also to optical fibers [17] and to solving the Schrödinger equation [18].

This paper has two main sections: the first will describe the basic methods of simulating wave propagation through a random medium and discuss the limits of applicability and accuracy; the second section will discuss several applications of the simulation techniques, such as plane wave propagation in idealized media, laser propagation in the atmosphere, radio waves in the ionosphere, and seismic waves in the earth. We will show results of simulations compared with theory and experiment, and discuss the enhancements to the basic techniques required for dealing with these various media.

2. Basic Theory and Methods

We begin with the reduced form of the scalar wave equation with time harmonic dependence removed: the result is the scalar Helmholtz equation

$$\nabla^2 U + k_0^2 n^2(\mathbf{R}) U = 0 \tag{1}$$

where U is the pressure, component of EM field, etc., $k_0 = 2\pi/\lambda$ is the free-space wavenumber, and $n(\mathbf{R})$ is the spatially varying refractive index. If we for the moment consider propagation in rectangular coordinates, and a wave propagating generally in the $+\hat{z}$ direction, we can substitute $U = ue^{ik_0 z}$ and we obtain

$$\frac{\partial^2 u}{\partial z^2} + \partial_\perp^2 u + 2ik_0 \frac{\partial u}{\partial z} + k_0^2(n^2 - 1)u = 0 \tag{2}$$

which can be factored[17,19] formally as

$$\left(\frac{\partial}{\partial z} + ik_0 + ik_0 Q\right)\left(\frac{\partial}{\partial z} + ik_0 - ik_0 Q\right) u + ik_0 \left[Q, \frac{\partial}{\partial z}\right] u = 0 \tag{3}$$

where

$$Q = \left(1 + \frac{1}{k_0^2}\nabla_\perp^2 + (n^2 - 1)\right)^{1/2} \tag{4}$$

and $\nabla_\perp^2 = \frac{\partial^2}{\partial x^2} + \frac{\partial^2}{\partial y^2}$. If the range dependence of n is weak, the commutator can be neglected, and if only the forward-propagating term of (3) is retained, we have the "generalized parabolic equation"[19]

$$\frac{\partial u}{\partial z} = ik_0(Q - 1)u \tag{5}$$

Since (5) is parabolic in z, it allows a marching-type attack, where the solution $u(z+\delta z)$ may be found from the solution $u(z)$. The operator Q, which is the square-root of an operator, can be implemented in many different ways, yielding various parabolic equation (PE) methods, incorporating varying degrees of approximation. In cylindrical and spherical coordinates, the same equation (5) for outgoing waves is derived, except the propagation direction is $\hat{r}$ and the definitions of U and $\nabla_\perp^2$ differ.

2.1 Split-step methods

In this section we will discuss in some detail the approximations, errors and implementation of three methods to solve (5). By split-step here, we mean that the terms dealing with field derivatives and terms dealing with the medium refractive index are additively separable. The simplest approximation to Q is a Taylor expansion

$$Q_1 = 1 + \frac{1}{2k_0^2}\nabla_\perp^2 + \frac{1}{2}(n^2 - 1) \tag{6}$$

which leads to the standard parabolic equation (SPE) first proposed by Leontovich and Fock[20], then later popularized and applied to the ocean acoustic channel by Tappert[4,5]. Another approximation to Q may be derived by assuming only $n^2 - 1$ is small, and doing a similar expansion [17,19,21]

$$Q_2 = (1 + \frac{1}{k_0^2}\nabla_\perp^2)^{1/2} + (n - 1) \tag{7}$$

where $1/2(n^2 - 1)$ has also been replaced by $n - 1$. That this is a significant change is indicated by the results of a simple error analysis[19] where $Q_1^2 - Q^2$ and $Q_2^2 - Q^2$ were calculated assuming constant strength of fluctuations and wave propagation at a given angle. The results showed Q_2 was significantly better than Q_1 or Q_2 with $1/2(n^2 - 1)$ for all angles and for $\delta n \approx 0.1$. We therefore use this approximation even for the SPE, which becomes

$$Q_1 = 1 + \frac{1}{2k_0^2}\nabla_\perp^2 + (n - 1) \tag{8}$$

and the error measure of the new Q_1 and Q_2 reveals relatively small difference in performance for propagation angles less than 20°–25° if $\delta n < 0.1$, which is nearly always true.

Yet another approach to deriving a PE-type equation is to approximate Q in such as way as to more closely follow the rays of the original Helmholtz equation[22,23]. The result of this effort is the modified log PE

$$Q_3 = 1 + \frac{1}{2}\log[\cos^2\left(\frac{i}{k_0}\frac{\partial}{\partial x}\right)] + \log(n) \tag{9}$$

appropriate for a 2-d propagation problem. The ray equation for the Helmholtz equation is

$$\frac{d^2x(z)}{dz^2} = \left[1 + \left(\frac{dx}{dz}\right)^2\right]\left(\frac{\partial \log(n)}{\partial x} - \frac{\partial x}{\partial z}\frac{\partial \log(n)}{\partial z}\right) \tag{10}$$

while the ray equation for (5) with (10) substituted for Q is

$$\frac{d^2x(z)}{dz^2} = \left[1 + \left(\frac{dx}{dz}\right)^2\right]\frac{\partial \log(n)}{\partial x} \tag{11}$$

Thus, if the variation of n in the propagation direction is weak, the final term of (10) will be small, and the rays of (11) will be very similar to the rays of the Helmholtz equation. This is important in media such as the ocean, where the ray structure generated by the action of the sound channel are a crucial element of the propagation problem. A comparison of simulations using (9) and (7) shows somewhat better performance for the modified log PE if the propagation angles are large or the propagation distances are large[23].

In each of the approximation described above, the resulting form for the propagation equation is

$$\frac{\partial u}{\partial z} = ik_0(Q-1)u = ik_0(D+N)u \tag{12}$$

where D is a diffraction operator and N reflects the effect of the medium refractive index. For the SPE,

$$D = \frac{1}{2k_0^2}\nabla_\perp^2 \qquad \text{and} \qquad N = n-1 \tag{13}$$

Formally then, we can write a solution for a small step in the propagation direction as

$$u(\mathbf{r}, z+\delta z) \approx \exp\left[ik_0 \int_z^{z+\delta z} D\,dz\right] \exp\left[ik_0 \int_z^{z+\delta z} N\,dz\right] u(\mathbf{r}, z) \tag{14}$$

thus separating the diffraction from the medium effect.

The two operators are applied sequentially and so may be solved separately. The medium effect is the solution to (12) with D removed, and the result is a simple spatially varying phase change applied to $u(\mathbf{r}, z)$. The solution for the D operator is just the solution of (12) with N removed, which may be easily solved by Fourier transform. The method of u propagation is therefore

$$\tilde{u}(\boldsymbol{\kappa}, z) = \mathcal{F}[e^{i\theta(\mathbf{r})}u(\mathbf{r}, z)] \tag{15a}$$

$$\tilde{u}(\boldsymbol{\kappa}, z+\delta z) = \exp(ik_0\tilde{D}\delta z)\tilde{u}(\mathbf{r}, z) \tag{15b}$$

$$u(\mathbf{r}, z+\delta z) = \mathcal{F}^{-1}[\tilde{u}(\boldsymbol{\kappa}, z+\delta z)] \tag{15c}$$

where

$$\theta(\mathbf{r}) = \int_z^{z+\delta z} N(\mathbf{r}, z')\,dz'; \qquad \tilde{D} = \mathcal{F}[D] \tag{16}$$

so for each of the methods discussed above, we have

$$\begin{aligned} \tilde{D}_1(\kappa) &= -\frac{\kappa^2}{2k_0^2} \\ \tilde{D}_2(\kappa) &= \left(1 - \frac{\kappa^2}{2k_0^2}\right)^{1/2} - 1 \\ \tilde{D}_3(\kappa) &= \frac{1}{2}\log[\cos^2\left(\frac{\kappa}{k_0}\right)] \end{aligned} \tag{17}$$

Physically, this technique corresponds to dividing the medium into slabs of length δz, each of which introduces a spatially varying contribution to the wave phase defined by geometrical optics in the slab, but essentially no amplitude variation. Each slab is effectively compressed to a phase screen located at the slab center, and the wave is then propagated through a uniform background medium from screen to screen. Amplitude fluctuations build up by diffraction over many δz. It has been observed that this algorithm is the discrete analog to the path integral method of wave scattering analysis and therefore is an exact solution to the parabolic wave equation in the limit of small δz and adequate grid sampling, even when scattering is strong[3,24].

Generation of the phase fluctuations produced by the variations of $n(\mathbf{r}, z)$ within each slab is straightforward. The deterministic part of θ can be calculated directly from the deterministic part of n through (16). The random part of θ may be generated from the power spectrum of θ

$$\Phi_\theta(\boldsymbol{\kappa}) = 2\pi k_0^2 \delta z \Phi_n(\boldsymbol{\kappa}, \kappa_z = 0) \tag{18}$$

via the standard method of filtering complex Gaussian pseudorandom field by (18) and inverse transforming to obtain a spatial field with the desired second-order statistics[12,25].

If the separation of screens or the width of the slabs is greater than the correlation length of the medium, each phase screen can be generated independently, greatly reducing storage and computational requirements. This is the standard simulation procedure in modeling WPRM. While it is relatively easy currently to generate correlated screens in 2-d scattering simulations[25,26], full 3-d correlated multiple screen generation with arbitrary correlation functions are not presently feasible, requiring very large 3-d FFTs, and somewhat defeat the purpose of having a marching algorithm in the first place. However, simulations are ideally suited to modeling media with slowly range-dependent statistical properties, a common situation in practice which is not easy to deal with analytically[12].

One exception occurs if the correlation function in the $\hat{z}$ direction can be modeled with an exponential $A\exp(-\alpha z)$. In this case

$$\theta^{l+1} = e^{-\alpha\delta z}\theta^l + (1 - e^{-2\alpha\delta z})^{1/2}\theta' \tag{19}$$

where θ^l is the phase screen at position $l\delta z$ and θ' is a new uncorrelated phase field generated with the desired transverse spatial spectrum and variance σ_θ^2. The variance of every screen thus generated will also be σ_θ^2 and the covariance is

$$C_\theta(l\delta z) = \langle \theta^k \theta^{k+l} \rangle = e^{-\alpha l \delta z}\sigma_\theta^2 \tag{20}$$

as desired. The same result could also be obtained by generating correlated white Gaussian fields using (19), then filtering as in the standard method.

2.2 Point Source Modeling

Since source structure has an influence on the wave statistics after propagation through a random medium, modeling of the source structure is an important element in any accurate simulation. While plane waves are often a useful approximation, in many circumstances a more point like source with significant angular spread is required. This effect may modeled in several ways: (1) construction of a small source area with appropriate phase modulation and rectangular medium coordinates; (2) scaling of medium coordinates in the source region followed by rescaling in the propagation region, but rectangular coordinates still used for the medium; (3) propagation in cylindrical or spherical coordinates. Which method is preferable depends on the degree of accuracy required for the source modeling and the ease of representation of the deterministic or random parts of the medium structure.

Medium generation and propagation procedures are the same as plane waves if the medium coordinates remain rectangular and the point source is modeled as a Gaussian source extending over a small area [27]. This source structure produces a diffraction-broadened Gaussian beam, but constraints on the allowable propagation distance (see section 2.3) and sampling grid size do not always allow a sufficiently broad central area for the observing region. It was found that the introduction of an amplitude modulation on the beam

$$u(\mathbf{r},0) = \exp\left(\frac{-r^2}{2\sigma^2}\right)\cos\left(\frac{r^2}{2\rho^2}\right) \tag{21}$$

where σ and ρ are constants of the order of δx, the transverse grid spacing, resulted in a more satisfactory beam shape with wider, flatter central region at the cost of small (1%) intensity variations.

The only modification of the propagation procedure is the introduction of an absorbing boundary region at the screen edges to remove the waves which encounter the screen edge and which, due to Fourier wrap-around, would in the absence of such a region propagate into the region of interest in the beam interior and cause spurious interference. This is implemented by adding a deterministic imaginary component to the refractive index in the absorbing region[6,27].

One limitation of the first method is that it does not allow very accurate source modeling due to the small number of points in the source representation—typically the source size is about the same as the grid spacing. Going to much large grids is an expensive option, but the second method of source construction allows more precise representation of source structure by setting the sampling interval to be small near the source, then expanding the grid with increasing propagation distance. One way this can be accomplished is with a Talanov lens transformation[13,28].

Consider a general pointlike source $U(\mathbf{r},0)$: a portion of the parabolic phase front can be extracted leaving $u'(\mathbf{r},0)$ a smoother and more easily modeled function

$$u(\mathbf{r},0) = \exp\left(-ik\frac{r^2}{2f_0}\right)u'(\mathbf{r},0) \tag{22}$$

where $-f_0$ is the focal length of the lens transformation which determines how much phase curvature is removed from $u(\mathbf{r},0)$. Then it can be shown[13] that

$$u(\mathbf{r}F,\delta z) = \frac{1}{F}u'(\mathbf{r},\frac{\delta z}{F})\exp\left(-ik_0\frac{r^2F^2}{(\delta z+f_0)}\right) \tag{23}$$

where $F = 1+\delta z/f_0$ and u' has been propagated by the standard Fresnel propagator (15) a distance $\delta z/F$ instead of δz. The transverse sampling interval has increased by the factor F, and the propagation process may be continued by the phase multiplication step (remembering that the change in sampling interval will change any absolute scales in the medium spectrum, such as inner or outer scales). At this point the final factor in (23) may be included and propagation continued as in the plane wave case, or the final phase correction omitted and another Talanov step calculated, this time with $f_1 = \delta z + f_0$.

Finally, the pointlike source problem may be handled by stating the problem in spherical or cylindrical coordinates. The SPE in spherical coordinates can be written as[29]

$$2ik_0\frac{\partial u}{\partial r} + \frac{1}{r^2}\left(\frac{\partial^2 u}{\partial \phi^2} + \frac{\partial^2 u}{\partial \nu^2}\right) + 2k_0(n-1)u = 0 \tag{24}$$

where ϕ and ν are angular variables similar to longitude and latitude, respectively. This expression is approximate, but will be accurate if $\phi, \nu \ll 1$ and $1/\delta\nu \gg \nu$, where $\delta\nu$ is the angular sampling interval. Consider the medium divided into spherical shells at boundaries r_j and define $\delta r_{j+1/2} = r_{j+1} - r_j$ and $r_{j+1/2} = (r_j + r_{j+1})/2$. Then we can write the equivalents to (18) and (15b) in this parabolic approximation to a spherical coordinate system as

$$\Phi_\theta(\kappa) = 2\pi k_0^2 \frac{\delta r_{j+1/2}}{r_{j+1/2}^2}\Phi_n(0, \frac{\kappa_\phi}{r_{j+1/2}}, \frac{\kappa_\nu}{r_{j+1/2}}) \tag{25}$$

and

$$\tilde{u}(r_{j+1}, \kappa_\phi, \kappa_\nu) = \exp\left(i\frac{\kappa_\nu^2 + \kappa_\phi^2}{2k_0}\frac{\delta r_{j+1/2}}{r_j r_{j+1}}\right)\tilde{u}(r_j, \kappa_\phi, \kappa_\nu) \tag{26}$$

The Fourier transform steps are carried out as before. This method allows high resolution near the source and decreasing resolution with increasing sampling interval as the propagation distance increases. Note that this method, as well as the closely related lens transformation methods, requires knowledge of absolute position of each screen with respect to the source position, while the plane wave method is independent of position.

2.3 Errors and Limits of Application

Several approximations have been made in representing solutions of the original Helmholtz equation (1) by the split-step solutions of equations (15abc). In this section we will identify those approximations, discuss their parametric dependence, and conclude with some practical requirements and limits of application of these simulation techniques to real experiments.

The first approximation is the use of the parabolic approximation to the the Helmholtz equation. The conditions for accuracy of this approximation are well known[1,30]. Briefly, they are: (1) backscatter must be negligible; (2) the cone of wave propagation angles must be in the near-forward direction, i.e. $\lambda/l_0 \ll 1$ where l_0 is the scale of medium fluctuations; and (3) the medium must be weakly fluctuating

$$k_0 l_0 \sqrt{\sigma_n^2} \ll 1 \tag{27}$$

where σ_n^2 is the variance of refractive index. We have already discussed the errors incurred by approximating the Helmholtz operator in section 2.1.

A second approximation is made when expressing the solution of (1) by a marching solution of (5). Formally,

$$u(z+\delta z) \approx \exp\left[\int_z^{z+\delta z} ik_0(Q-1)\,dz\right]u(z) \tag{28}$$

It can be shown[26] that the error in this expression is approximately

$$\epsilon = \left[Q-1, \frac{\partial}{\partial z}(n^2-1)\right](\delta z)^3 \tag{29}$$

and no solution method based on (28) can have any smaller error or depend only on a higher power of δz.

A third source of error is the splitting of the Q operator into diffraction and phase addition steps implemented in sequence. For each propagation step, the errors incurred by this approximation are $\mathcal{O}(\delta z^2)$[13,26], but increase as $\partial u/\partial z$ increases with increasing z, for fixed δz. Similarly, Spivack and Uscinskii[26] have found the error produced in the second moment over a single step to be smaller, $\mathcal{O}(\delta z^4)$. Generally, however, we are more interested in the error accumulated over many propagation steps: the errors in the field and second moment have been shown to be[26] $\mathcal{O}(\delta z)$ and $\mathcal{O}(\delta z^2)$, respectively. This analysis has also been extended to higher moments[31] with the results the same as for the second moment. Note that in all cases the accuracy of the higher moments is greater than that of the field itself. This expected dependence has been verified by 2d simulations using correlated screens, and the error contribution has been shown to be very small if $\langle \theta^2 \rangle < 0.25$ when $\delta z \geq l_0$, i.e. accurate simulations are obtained with step sizes greater than the medium correlation length (allowing independent generation of screens) if (27) is satisfied[26,31].

There are also several practical considerations which must be satisfied for accurate simulations of wave propagation in an extended, continuous random medium[27,32]. Clearly the phase spectrum must be contained[12]within the spatial frequency band determined by the grid spacing δ and the number of grid points N_g. and edge effects must be removed if wave propagation back to the observing region will result in interference, as discussed above for point-source simulations. To insure that the split-step method is accurate, the scattering must be weak over the screen-to-screen propagation distance δz: $m_R^2(\delta z) < 0.1$ where m_R^2 is the Rytov intensity variance; and the geometrical optics must be valid within within the slab: $\delta z < k_0 l_0^2$. To model a continuous medium, $m_R^2(\delta z) < 0.1 m_R^2(L)$, which ensures that even under relatively weak scattering case, the whole medium is not collapsed to a single screen. Finally, the most important limit on the applicability of a simulation is the finite spatial bandwidth set by N_g. Schematically, the spatial spectrum of intensity fluctuations is shown in Fig. 1, and it must be contained within the spatial bandwidth of the simulations

$$\frac{\kappa_{\max}}{\kappa_{\min}} \leq N_g \qquad \text{and} \qquad \kappa_{\max} < \frac{\pi}{\delta} \tag{30}$$

Since $\kappa_{\max}/\kappa_{\min}$ is approximately proportional[27]to the Rytov intensity variance, a given grid size can only accomodate a limited strength of scattering—this limit for $N_g = 512$ is about $m_R^2 = 15$. Larger values of m_R^2 produce serious aliasing errors.

2.4 Finite Difference Methods

The split-step solutions to the various parabolic-type equations previously discussed very efficiently provide accurate results if the angles of scattering are relatively small and boundary interactions are not important. To loosen these restrictions various wider-angle versions of the basic parabolic equation have been devised, which more accurately approximate the dispersion relation of the Helmholtz equation over large angles[33]. If the diffractive and refractive-index parts of Q are

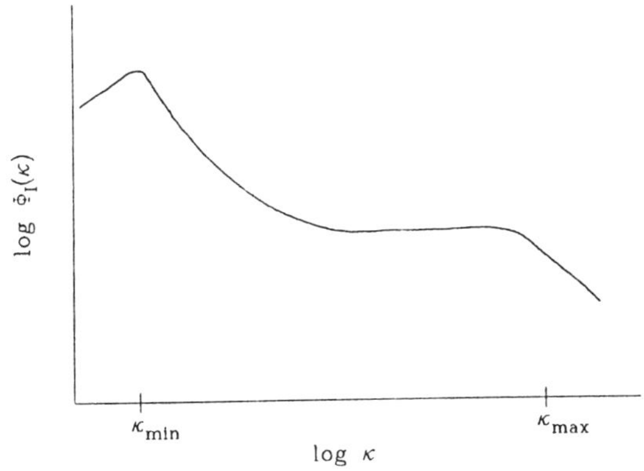

Fig. 1. Schematic form of intensity spectrum after passage through a random medium.

defined as

$$D = \frac{1}{k_0^2}\nabla_\perp^2 \qquad \text{and} \qquad N = n^2 - 1 \tag{31}$$

then approximations to (4) can be expressed as functions of D and N. One way to approximate the square root accurately is to use a rational function or Padé approximant [33,34]

$$Q_3 = 1 + \sum_{j=1}^{M} \frac{a_j(D+N)}{1 + b_j(D+N)} \tag{32}$$

which has the advantage of being first-order in $(D + N)$, and thus simpler to solve. The square root can also be approximated by a second-order polynomial (the Lee-Saad-Schultz form)[35]

$$Q_4 = 1 + \frac{1}{2}(D+N) - \frac{1}{8}(D+N)^2 \tag{33}$$

which can also be implemented using ratios of first-order polynomials. Ratios of higher order polynomials[36]may also be used, but result in more complex methods of solution. Note that in (32) and (33) the functions of D and N are not additively separable, so the Fourier split-step method of (14) can not be applied. Finite difference methods, however, such as Crank-Nicholson, pure implicit, or alternating-direction implicit methods are effective[16,37]. Using these methods, it is clear why first-order expressions, such as Q_3 and Q_4 are preferable—one must only solve a tridiagonal matrix to step the solution, for which very efficient algorithms exist.

These methods are much more accurate than SPE or any of the wide-angle split-step methods shown in section 2.1 when large-angle propagation is encountered[21,38] and can accurately include range variations of boundary position and absorptivity, refractive index and medium density (for acoustic propagation)[5,39–41]. Time-domain versions have been produced for pulse-propagation analysis[34], and these methods have been extended to three dimensions by including azimuthal coupling[35,42] and approximate backscattering[43]

Finite-difference solutions can be very efficient modeling deterministic channels[44], but finite-difference solutions of the wide-angle PEs discussed above have not been widely applied to random media due to their sampling requirements.

While the split-step multiple phase screen methods need only realize one phase screen to represent a medium slab many correlation lengths thick in the propagation direction (20 screens are sufficient to represent the atmosphere over kilometer paths[32]), the finite difference methods must sample the medium in every direction with sample spacing much less than the medium correlation scale[45]. Finite difference methods may be modified to use space phase screens as the medium representation[25], but that removes the very wide-angle handling capability for which these methods were introduced.

3. Applications

In this section we will present some of the recent results of simulations in various fields ranging from general propagation experiments in idealized media as a comparison to analytic approximation to more complex media, such as ionospheric reflection, atmospheric propagation, and elastic wave propagation. In some cases, the techniques discussed in section 1 have been extended to handle the new problems, and we will present these additional techniques in the relevant sections.

3.1 Plane Wave Propagation

Plane wave propagation through an idealized random medium has been simulated with specialization to the case of the ionosphere[12]and for Gaussian and exponentially correlated media[25], both using 2d media. Knepp demonstrated the technique's usefulness by showing qualitative agreement with experimental data. By expanding the pulse into many fourier components, propagating each separately, then recombining at the observing plane, Knepp was able to show the temporal distortion of a pulse through a disturbed ionosphere and obtained qualitative agreement with satellite-source radio data propagated through a barium-release cloud.

Macaskill and Ewart[25]compared theoretical spatial spectra and correlations of intensity with split-step simulations through 2d extended random media with Gaussian and modified exponential correlated media. The results they obtained showed agreement with the theoretical spectra[46,47] to about 10%, within the expected error for the number of realizations in the simulation. The agreement was not as good for the intensity variance.

Simulations using the split-step technique have also been done for 3d wave propagation through a variety of simulated random media: pure power law, power law with inner scale, power law with outer scale, and Gaussian spectra[32]. Some results from this study are shown in Fig. 2, where intensity variance is plotted as a function of m_R^2 (called U in the Figure) for various values of the inner scale parameter $\beta = R_f/3$, R_f, and $3R_f$ where $R_f = L/k_0$ is the Fresnel length and β defines the high-frequency cutoff of the spatial spectrum of refractive index irregularities

$$\Phi_n = C_n^2 \kappa^{-\alpha-2} \exp\left(\frac{-\kappa^2\beta^2}{4}\right). \tag{34}$$

Three regions can be defined by this figure: (1) the weak-scattering region defined by $m_R^2 < 0.1$, (2) the asymptotic strong scattering region defined by the return

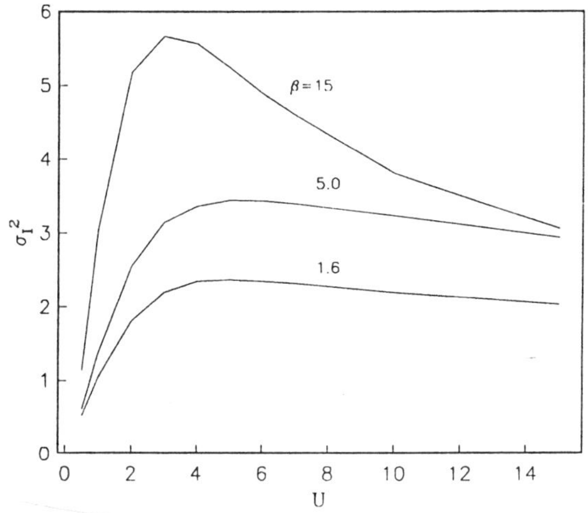

Fig. 2. Simulation results for intensity variance as a function of $U = m_R^2$ for $\beta = R_f/3, R_f, 3R_f$ ($R_f = 5.0$)and power law $\alpha = 1.7$.

of σ_I^2 to unity for very large m_R^2, and (3) the strong-focusing region defined by the peak in σ_I^2. Note that for $\beta = 3R_f$, σ_I^2 can become as large as 5–6, and even for $\beta = R_f$, the maximum intensity variance is about 3, significantly above unity. Although laser propagation data have shown intensity variance values of 4-5, the medium inner scales implied by $\beta = 3R_f$ are too large to be physically plausible.

A comparison of 3d plane wave simulation to a two-scale theory [48] showed differences of about 23% for no inner scale, and significantly larger differences in the strong-focussing region when the medium had significant inner scale[32], suggesting more than first-order terms in the theory were needed. Recently a new reorganization of the conventional asymptotic series has yielded an asymptotic form that converges more rapidly[49,50], thus significantly expanding its region of validity. Fig. 3 shows the comparison of numerical simulation results for $\langle I^2 \rangle$ for U ($= m_R^2$) values from 0 to 50 to the reorganized and traditional series for 1 and 3 terms. The 3-term reorganized series shows clear agreement within expected errors of the simulation for $m_R^2 \geq 10$, a result obtained for the third and fourth moments of intensity as well.

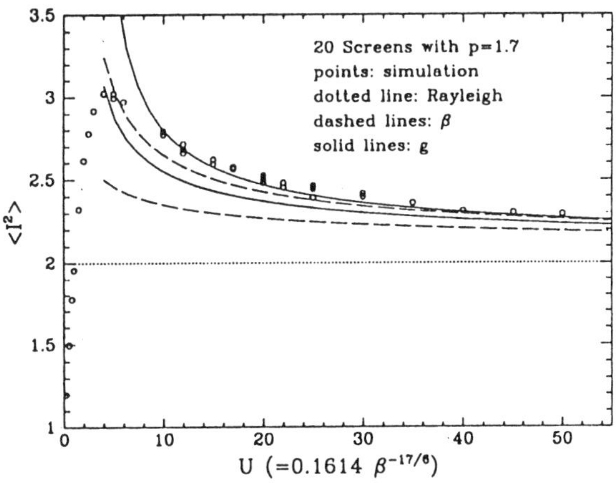

Fig. 3. Comparison of simulation intensity second moment (points) with traditional (dashed) and reorganized (solid) asympototic series, both 1- and 3-term forms.

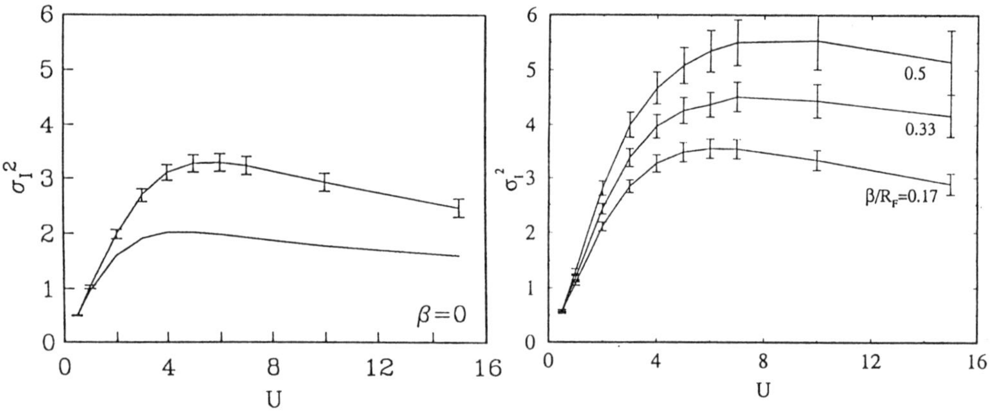

Fig. 4. Intensity variance versus m_R^2 (U) for point source, power law medium with $\alpha = 1.7$, and $\beta = 0$, $0.17R_f$, $0.33R_f$, and $0.5R_f$. Error bars show 1σ uncertainties. Plane wave result shown for camparison with no error bars, as they are negligible.

An area of continuing interest in WPRM research has been the determination of the correct form for the intensity probability density function (PDF) for various scattering regimes[30,51–53]. The generalized gamma distribution has been investigated as a model for the intensity PDF[54,55]by comparison with 2d simulations through a medium with a power law spectrum with exponent 3. These results show that by fitting the three parameters of the distribution, very accurate (95% confidence level) fits can be obtained. Three-dimensional simulations with both null and finite inner scales have also been compared with several distribution families[56]. These results show the distribution is log-normal for weak scattering as expected, goes through a transition in the strong-focusing region, then asymptotically goes to a K distribution when inner scale is significant, but goes to a convolution of log-normal and exponential distribution[52]for pure power law medium. For very strong scatter, the distribution becomes exponential, independent of inner scale.

3.2 Point Source and Laser Propagation in the Atmosphere

Source structure can have an important influence on the intensity and phase statistics of wave propagated in random media, so that structure must be modeled in order to simulate the wave propagation accurately. We have already discussed methods for simulating point-source propagation; the first results presented in this section are from Martin and Flatté[27], and are obtained using the first method: small source, rectangular coordinates. Intensity variance as a function of m_R^2 is shown in Fig. 4 for β values from 0 to $0.5R_f$. Note the peak values are consistently much higher for the point-source data than for the corresponding plane wave data. If β is increased to $R_f/3$, the peak value of σ_I^2 increases 40% for the point source, but only 10% for the plane wave, and σ_I^2 of 5.5 is obtained with a β only 1/6 the corresponding plane wave value. This suggests that the point source structure with its smaller near-source scales allows more efficient focusing by fluctuations of smaller inner scales than the plane wave source. This hypothesis is also supported

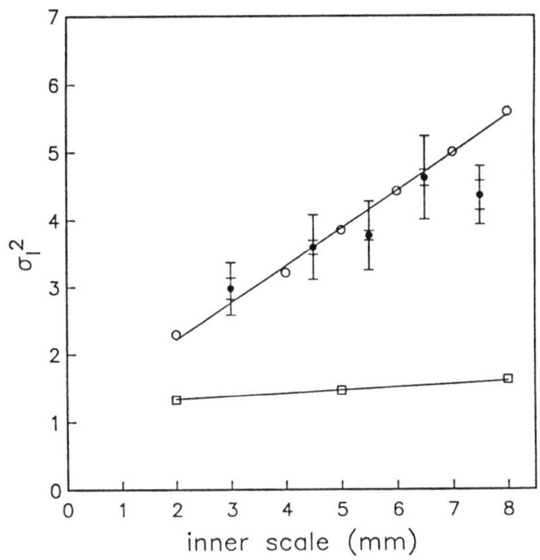

Fig. 5. Intensity variance as a function of inner scale for laser data of Consortini *et al.*(•) compared with simulation results using atmospheric spectrum model for point (○) and plane wave (□) sources.

by examination of individual point-source and plane-wave intensity image fields: they are similar in general appearance, but the intensity peaks are much higher for the point source[27].

The first experimental point-source laser scintillation measurements in the atmosphere[57] indicated large peak values of σ_I^2 of 5-6, which are consistent with the values obtained for point-source simulations with simple spectral cutoff at 0.7 to 1.1 cm [27]. The experimentally measured inner scales were somewhat smaller, 0.35 cm, but were not measured at the same time as the optical experiment. A new, more completely characterized optical point-source propagation experiment has been completed by Consortini *et al.*[58] in which measurements of the strength and inner scale of the refractive index fluctuations were measured[59] for paths near the scintillation path. These results have been compared to simulations[60] in the region beyond the peak in intensity variance, which may be characterized approximately by a linear dependence on $\beta_0 = (0.497C_n^2k^{7/6}L^{11/6})^{1/2}$. The slope of this dependence is used to translate all the measurements for a given subrange of inner scales to a common equivalent β_0, then the results were averaged to clearly display the dependence on inner scale, and the results are plotted as the filled circles in Fig. 5. The simulation results using the atmospheric spectrum model[61] are shown as circles (point source) and squares (plane wave). The agreement with the point-source simulations is good, and even the error bar size may be accounted for by errors in the inner scale estimates.

3.3 Electromagnetic Waves in the Atmosphere and Ionosphere

Complex phenomena are often encountered in the study of HF and microwave propagation in the atmosphere, such as ducting, random refraction, and surface boundary interactions, and simulations allow estimation of mean signal strength and statistics[62]. The propagation components of the EM wave can be cast in

SPE-like form

$$2ik\frac{\partial u}{\partial z} + \frac{\partial^2 u}{\partial x^2} + k^2\left[n^2 - 1 + \frac{2x}{R_e}\right]u = 0 \tag{35}$$

where $\hat{z}$ is the propagation direction approximately parallel to the earth's surface, $\hat{x}$ is the transverse, vertical coordinate, and R_e is the earth radius. This equation can be solved by the split-step Fourier method with simple boundary conditions, and the results can reflect the effects of random and deterministic variations in the refractive index.

A more realistic boundary condition at the earth surface is a finite conductivity condition

$$\left(\frac{\partial u}{\partial x} + \alpha u\right)_{x=0} = 0 \tag{36}$$

where α is a complex function of the complex permittivity of the earth surface[62]. Kuttler and Dockery[63]have shown that this boundary condition can be automatically satisfied using a mixed Fourier transform pair:

$$\tilde{u}(\kappa, z) = \int_0^\infty u(x,z)[\alpha\sin(\kappa x) - \kappa\cos(\kappa x)]\,dz \tag{37a}$$

$$u(x,z) = \frac{2}{\pi}\int_0^\infty \tilde{u}(\kappa, z)\frac{\alpha\sin\kappa x - \kappa\cos\kappa x}{\alpha^2 + \kappa^2}\,d\kappa + K(z)e^{-\alpha x} \tag{37b}$$

where

$$K(z) = K(z_0)\exp\left[i(n^2 - 1 + \frac{2x}{R_e} + \alpha^2)\frac{z}{2k}\right] \tag{38}$$

instead of the standard Fourier transform in (15), with no loss in efficiency. Comparisons of simulations by this method with a (much more computationally expensive) mode calculation for a surface-based duct showed very good agreement in signal intensity.Although use of this algorithm has been in the characterization of a deterministic channel, random variations in refractive index can easily be included.

Studies of transionospheric radio propagation have employed the standard techniques aready described, and the vertically propagating ionospheric reflection problem can be solved similarly, with the addition of a special phase screen and the reflection height[64]. However, simulations of the full, oblique problem in result in some modifications to the standard technique[65]. If the problem is limited to the x-z plane and the medium properties are functions of z only ($n = n_d(z) + n_1(x,z)$), then the field can be expressed as

$$U = U_0(z)u(x,z)e^{ik_0\sin\theta_0 x} \tag{39}$$

where θ_0 is the initial propagation angle and $U_0(z)$ is the solution to the deterministic part of the refractive index profile

$$\frac{\partial^2 U_0}{\partial z^2} + k_0^2(n_d^2(z) - \sin^2\theta_0)U_0 = 0 \tag{40}$$

Thus the solution u(x,z) is an expansion around the WKB ray solution $U_0(z)$. If the medium is divided into slabs, foreach slab $U_0(z)$ defines a local propagation direction θ_j and a local wavenumber k_j.

Each propagation step consists of a phase addition and a Fresnel propagator implemented in the Fourier domain. The phase screen step not only adds phase, but also shifts the field horizontally

$$u(x, z+) = u(x - \delta z \tan\theta_j, z)e^{i\psi} \tag{41}$$

where the phase field $\psi(x)$ is a Gaussian random process with spectrum proportional to $\Phi_n(\kappa/\cos\theta_j)$ due to path obliquity. The Fresnel propagator (15b) becomes

$$\tilde{u}(\kappa, z_{j+1}) = \tilde{u}(\kappa, z_j)\exp\left[-i\frac{\delta z}{\cos\theta_j}\left(\frac{\kappa^2}{2k_j\cos^2\theta_j} + \kappa\sin\theta_j\right)\right] \tag{42}$$

At the turning point level, the up- and down-going waves are highly correlated, but the effect of propagation in this region can also be modeled as a phase screen and Fresnel propagation operation[66]. These simulations have been employed to evaluate the phase and amplitude spectra, and intensity variance of signals reflected from the ionosphere[65], phase and angle-of-arrival statistics[66], and the two-frequency mutual coherence function[67]in model ionospheres. Such oblique propagation in layered media is often encountered in propagation through geophsyical media such as atmospheres and oceans, often with the added complication of an anisotropic irregularity spectrum.

Finally we consider the generalization associated with the propagation of vector waves in anisotropic random media. A vector simulation to model propagation in a magnetized plasma has been devised by Brent *et al.*[68]. which includes depolarization, an effect impossible to model with a scalar simulation. The essence of the problem is to construct a parabolic approximation to the vector Helmholtz equation. The first-order variation in the propagation direction is removed, and the parabolic approximation is then applied to the resulting propagation equations which gives

$$\left[2ik_0\frac{\partial}{\partial z} + \nabla_\perp^2 + ik_0^2(n(\mathbf{R}) - 1) + ik_0^2\gamma(\mathbf{R})\begin{pmatrix} 0 & -1 \\ 1 & 0 \end{pmatrix}\right]\begin{pmatrix} u_1 \\ u_2 \end{pmatrix} = 0 \tag{43}$$

where u_1 and u_2 are the vector components along and transverse to the propagation direction, $n(\mathbf{R})$ is the medium refractive index, and γ is a function which produces the coupling between the field components due to refractive index fluctuations.

The Fresnel propagation step is identical to (15b), since the refractive index terms are ignored for that step. The phase addition step is the solution to (43) ignoring the diffraction term $\nabla_\perp^2$, which can be written as

$$\mathbf{u}(x, z+) = \exp(\mathbf{G}\delta z)\mathbf{u}(x, z). \tag{44}$$

where

$$\mathbf{G} = -i\frac{k_0}{2}\left[(n-1)\mathbf{I} + i\gamma\begin{pmatrix} 0 & -1 \\ 1 & 0 \end{pmatrix}\right] \tag{45}$$

and the exponential is calculated using eigenvalues and eigenvectors in the standard way. Brent *et al.*show that for the EM case, the matrix breaks up into a phase screen factor and a depolarization matrix $\mathbf{M}$. In general, the eigenvalues and eigenvectors would have to be found for each step, but in many cases of practical interest, the structure of the problem allows them to be calculated analytically, retaining the full efficiency of the FFT split-step technique.

3.4 Acoustic Waves in the Ocean

Simulation has been important in modeling propagation of acoustic waves in the ocean sound channel since Tappert's early work[5], but it has largely been applied only to deterministic channels at lower frequencies where bottom interactions, wide angle propagation, and density fluctuations are the dominant considerations and where finite difference methods are most useful. Random fluctuations in sound speed are produced by internal waves, which generally can be modeled by the Garrett-Munk spectrum. Flatté *et al.*[6,69]have developed simulations of long-path high-frequency pulse propagation using measurements of mean ocean sound-speed profiles and Garrett-Munk internal wave spectra. The results of the simulations are consistent with theoretical expectations and show effects observed in experiments, such as the pulse dispersion in time, depth, and angle-of-arrival observed in the Slice89 experiment[70]. More details of the simulations and the experiment comparison can be found the review by Flatté contained in these proceedings[71].

3.5 Elastic Waves

Propagation of seismic waves in an earth model with random irregularities in wave speed has been studied using finite difference methods[45,72]and analytic techniques[73], but with limited applicability due to constraints on grid size and scattering strength. A phase-screen method has been implemented by Fisk and McCartor[74], which can be more efficient than the standard finite difference code and can handle strong-scattering conditions, as well as the special conditions of seismic propagation. First, seismic waves are generally of low fequency and long wavelength (up to a few km for 1 Hz waves) and the irregularity spectrum often contains components near this scale, thus producing scattering at larger angles than usually accomodated by SPE. Second, the irregularities can produce depolarization of transverse elastic waves, and mode conversion between transverse and longitudinal waves, which must be included in a simulation.

Only an outline of the method will be presented; a detailed presentation is available in Fisk and McCartor. The starting point for the derivation is again the Helmholtz equation, which will be expressed in terms of the wave function in the Fourier domain

$$\mathbf{u}(\mathbf{R}) = \frac{1}{(2\pi)^3} \int \tilde{u}(\mathbf{k}) e^{i\phi(\mathbf{R},\mathbf{k})} e^{i\mathbf{k}\cdot\mathbf{R}}\, d^3k \tag{46}$$

which is unusual in that the phase ϕ is a function of wavenumber as well as position. If the medium is again divided into slabs of thickness δz, then the phase function at $N\delta z$ is

$$\phi(\rho, N\delta z; \mathbf{k}) = \int_{(N-1)\delta z}^{N\delta z} (k^2 n^2(\rho, z') - \kappa^2)^{1/2} - (k^2 - \kappa^2)^{1/2}\, dz' + \phi(\rho, (N-1)\delta z; \mathbf{k}) \tag{47}$$

where $\boldsymbol{\kappa} = (k_x, k_y)$ is the transverse component of the wavenumber and the dispersion relation $k^2 = k_z^2 + \kappa^2$ has been assumed. The field at $N\delta z$ can be expressed in terms of the Fourier components of the field at $(N-1)\delta z$

$$\mathbf{u}_N(\boldsymbol{\rho}) = \int \tilde{\mathbf{u}}_{N-1} \exp[i\Delta_N(\boldsymbol{\rho},\boldsymbol{\kappa})] \exp(i\boldsymbol{\kappa}\cdot\boldsymbol{\rho}) d^2\kappa \tag{48}$$

where

$$\Delta_N(\boldsymbol{\rho};\boldsymbol{\kappa}) = \int_{(N-1)\delta z}^{N\delta z} (k^2 n^2(z') - \kappa^2)^{1/2}\, dz' \tag{49}$$

It can be shown that if $k \gg \kappa$, (48) reduces to the standard split-step method. In the derivation, the Fourier transform $\tilde{\mathbf{u}}_N$ of the displacement $\mathbf{u}_N$ can be decomposed into P- and S-wave components, each component is propagated individually, combined into a new total displacement, then decomposed into P and S components again. This defines a procedure for propagating the P- and S-wave Fourier coefficients which includes mode conversion and depolarization.

The propagation is carried out in the Fourier domain by a matrix multiplication. If the Δ-phase multiplier is expanded

$$\exp[i\Delta_N(\boldsymbol{\rho},\boldsymbol{\kappa})] = \frac{1}{(2\pi)^2} \int D_N(\boldsymbol{\kappa},\boldsymbol{\kappa}') \exp(i\boldsymbol{\kappa}'\cdot\boldsymbol{\rho})\, d^2\kappa' \tag{50}$$

it can be substituted into the Fourier transform of (48) and the result is

$$\tilde{u}^{(N)} = \int \tilde{u}^{(N-1)}(\boldsymbol{\kappa}') D_N(\kappa',\kappa-\kappa')\, d^2\kappa' \tag{51}$$

The actual method is the discrete analog of (51) with careful inclusion of all the proper additional factors related to the propagation velocity of each mode and the subsequent decomposition[74]. To force conservation of energy, the propagation matrix is factorized by singular value decomposition and the eigenvalues set to unity. This technique is only efficient when the number of frequency components is much smaller than the number of spatial components in the grid. This condition is satisfied in the seismic case, due to the very low frequency, and the cutoff this produces in the propagating frequency components. The optical or radio propagation simulations previously discussed typically use as many spatial frequency components as spatial grid points (indeed, this limit on the available spatial bandwidth is an important factor limiting application of simulations to strong scattering), so this technique would not be efficient under those conditions.

In comparison to a theoretical result for propagation of elastic waves in a medium with sharp variation of n in the transverse direction, but no variation in z, the new phase screen simulations were much more accurate than the SPE method. Inclusion of the κ dependence in the Δ-phase function produces that accuracy, but at the cost of computing the integral (48), which need be done only once for a z-invariant medium, but which must be done at every step for a random medium. This technique has also been compared to a standard finite difference technique[75] simulating propagation in a 2D random medium. In this case the phase dependence on the transverse wavenumber was ignored, and Δ was just the standard phase screen. The comparison showed agreement when backscatter was insignificant and scattering angles small, such as in the first-arrival part of the seismic pulse. At later times when larger angle scattering becomes important, there was significant disagreement, but possible error sources in the finite difference simulations could account for a portion of that difference. Under small-angle scattering conditions where its use is appropriate, the phase screen method produced accurate results with increase in speed of 2 orders of magnitude over the finite difference algorithm, and allowed 3D simulations to be run which were impractical using finite difference.

3. Summary

Simulation of wave propagation through random media using the split-step Fourier algorithm and its wider-angle variants has proved an efficient and easily implemented technique for modeling various propagation phenomena. As long as the requirements of small-angle scattering, no backscatter, weakly flutuating media, and sufficient spatial bandwidth are met, the results are accurate and agree with analytic approximations when those approximations are expected to be valid. Simulations in 2 and 3 dimensions modeling plane, line, and point sources have shown that the dimensionality and the source structure, in addition to the medium statistics, are important determinants of wave amplitude and phase variations.

Simulations are currently the only method capable of theoretically exploring the strong-focusing regime between the weak-scattering ($m_R^2 < 0.1$) and the asymptotic ($m_R^2 > 15$) regimes. In this region of parameter space simulations provide benchmark results to check the accuracy of analytic approximations and allow controlled numerical experiments with a variety of spatially or spectrally complicated media. Fields are obtained at every propagation step, allowing detailed investigation of amplitude and phase fluctuation development along the path, and computation of any desired field statistic. Propagation of multiple-frequency waves also allows study of pulse statistics by Fourier synthesis.

Scalar simulations have been used to model laser propagation in the atmosphere, radio propagation in the atmosphere, ionosphere, and interplanetary medium, and acoustic propagation in the ocean. The technique has been enhanced and extended to model various source structures, and to include vector EM and elastic waves with depolarization and mode conversion, and oblique propagation through layered random media. Many more investigations can make use of simulations in the future, such as investigations of pulse broadening and wander statistics, more detailed investigation into the strong-focusing regime and the role of source structure and inner scale, and the effects of spatial variation of scattering strength and intermittency of turbulence on scattering statistics, for example. With the growth in available computing power, it seems clear that the use of simulations to aid in understanding and predicting complex scattering phenomena will continue to expand.

Acknowledgements

I would like to express my gratitude to Len Tyler for introducing me to the WPRM business, and to Stanley Flatté for opening my eyes to the possibilities offered by simulation and giving me the opportunity to do scattering experiments on the computer. Thanks also to my colleagues at Stanford and UC Santa Cruz for many provocative and enlightening discussions.

References

1. V. I. Tatarskii, *The Effects of the Turbulent Atmosphere on Wave Propagation*, National Technical Information Service, 1971, Gov't ordering no. TT-68-50464.

2. J. L. Codona, D. B. Creamer, S. M. Flatté, R. G. Frehlich & F. S. Henyey, "Solution for the Fourth Moment of Waves Propagating in Random Media," *Radio Sci.* **21**(4), 929–935 (1986).

3. S. M. Flatté, R. Dashen, W. H. Munk & K. M. Zachariasen, *Sound Transmission Through a Fluctuating Ocean*, Cambridge Univ. Press, New York, NY, 1979.

4. F. D. Tappert & R. H. Hardin, "Computer simulation of long-range ocean acoustic propagation using the parabolic equation method," in *Proceedings of the Eighth International Congress on Acoustics* **#2**, Goldcrest, London, 1974, 452.

5. F. D. Tappert, "The Parabolic Approximation Method," in *Wave Propagation and Underwater Acoustics*, J. B. Keller & J. S. Papadakis, eds., Lect. Notes in Phys. **#70**, Springer-Verlag, New York–Heidelberg–Berlin, 1977.

6. Stanley M. Flatté & Frederick D. Tappert, "Calculation of the effect of internal waves on oceanic sound transmission," *J. Acoust. Soc. Amer.* **58**(6), 1151–1159 (1975).

7. J. A. Ratcliffe, "Some aspects of diffraction theory and their application to the ionosphere," *Rep. Progr. Phys.* **19**, 188–268 (1956).

8. E. N. Bramley, "The accuracy of computing ionospheric radio-wave scintillation by the thin phase-screen approximation," *J. Atmos. Terr. Phys.* **39**(3), 367–373 (1977).

9. R. Buckley, "Diffraction by a random phase-changing screen: A numerical experiment," *J. Atmos. Terr. Phys.* **37**(11), 1431–1446 (1975).

10. C. L. Rino & J. Owen, "Numerical simulations of intensity scintillation using the power law phase screen model," *Radio Sci.* **19**(3), 891–908 (1984).

11. J. P. Filice, "Studies of the Microscale Density Fluctuations in the Solar Wind Using Interplanetary Scintillation," University of California, San Diego, PhD Thesis, 1984.

12. Dennis L. Knepp, "Multiple Phase-Screen Calculation of the Temporal Behavior of Stochastic Waves," *IEEE Proc.* **71**(6), 722–737 (1983).

13. J. A. Fleck, Jr., J. R. Morris & M. D. Feit, "Time-Dependent Propagation of High Energy Laser Beams through the Atmosphere," *Appl. Phys.* **10**(2), 129–160 (1976).

14. D. Lee, R. L. Sternberg & M. H. Schultz, eds., *Computational Acoustics: Wave Propagation*, North-Holland, Amsterdam, 1988.

15. D. Lee, A. Cakmak & R. Vichnevetsky, eds., *Computational Acoustics: Ocean-Acoustic Models and Supercomputing*, North-Holland, Amsterdam, 1990.

16. D. Lee & S. T. McDaniel, *Ocean Acoustic Propagation by Finite Difference Methods*, Pergamon, Oxford, 1988.

17. M. D. Feit & J. A. Fleck, Jr., "Light propagation in graded-index optical fibers," *Appl. Opt.* **17**(24), 3990–3998 (1978).

18. M. D. Feit, J. A. Fleck, Jr. & A. Steiger, "Solution of the Schrödinger Equation by a Spectral Method," *J. Comput. Phys.* **47**(3), 412–433 (1982).

19. D. J. Thomson & N. R. Chapman, "A wide-angle split-step algorithm for the parabolic equation," *J. Acoust. Soc. Amer.* **74**(6), 1848–1854 (1983).

20. M. A. Leontovich & V. A. Fock, "Solution of the problem of propagation of electromagnetic waves along the Earth's surface by method of parabolic equations," *J. Phys. USSR* **10**(1), 13–23 (1946).

21. David J. Thomson, "Wide-angle parabolic equation solutions to two range-dependent benchmark problems," *J. Acoust. Soc. Amer.* **87**(4), 1514–1520 (1990).

22. A. Tolstoy, D. H. Berman & E. R. Franchi, "Ray theory versus the parabolic equation in a long-range ducted environment," *J. Acoust. Soc. Amer.* **78**(1), 176–189 (1985).

23. David H. Berman, Evan B. Wright & Ralph N. Baer, "An optimal PE-type wave equation," *J. Acoust. Soc. Amer.* **86**(1), 228–233 (1989).

24. R. Dashen, "Path Integrals for Waves in Random Media," *J. Math. Phys.* **20**(5), 894 (1979).

25. C. Macaskill & T. E. Ewart, "Computer Simulation of Two-dimensional Random Wave Propagation," *IMA J. Appl. Math.* **33**(1), 1–15 (1984).

26. M. Spivack & B. J. Uscinski, "The split-step solution in random wave propagation," *J. Computat. Appl. Math.* **27**(3), 349–361 (1989).

27. J. M. Martin & Stanley M. Flatté, "Simulation of point-source scintillation through three-dimensional random media," *J. Opt. Soc. Amer. A* **7**(5), 838–847 (1990).

28. V. I. Talanov, "Focusing of light in cubic media," *JETP Lett.* **11**(6), 199–201 (1970).

29. L. C. Lee, "Theory of Thin-Screen Scintillations for a Spherical Wave," *Astrophys. J.* **218**(2), 468–476 (1977).

30. B. J. Uscinski, *The Elements of Wave Propagation in Random Media*, McGraw-Hill, New York, NY, 1983.

31. M. Spivack, "Accuracy of the moments from simulation of waves in random media," *J. Opt. Soc. Amer. A* **7**(4), 790–793 (1990).

32. J. M. Martin & Stanley M. Flatté, "Intensity images and statistics from numerical simulation of wave propagation in 3-D random media," *Appl. Opt.* **27**(11), 2111–2126 (1988).

33. Jon F. Claerbout, *Fundamentals of Geophysical Data Processing*, McGraw-Hill, New York, NY, 1989.

34. Michael D. Collins, "Applications and time-domain solution of higher-order parabolic equations in underwater acoustics," *J. Acoust. Soc. Amer.* **86**(3), 1097–1102 (1989).

35. Ding Lee, George Botseas & William L. Siegmann, "Examination of three-dimensional effects using a propagation model with azimuth-coupling capability (FOR3D)," *J. Acoust. Soc. Amer.* **91**(6), 3192–3202 (1992).

36. George H. Knightly, Ding Lee & Donald F. St. Mary, "A higher-order parabolic wave equation," *J. Acoust. Soc. Amer.* **82**(2), 580–587 (1987).

37. William H. Press, Brian P. Flannery, Saul A. Teukolsky & William T. Vetterling, *Numerical Recipes*, Cambridge Univ. Press, New York, NY, 1986.

38. R. J. Hill, "Wider-angle parabolic wave equation," *J. Acoust. Soc. Amer.* **79**(5), 1406–1409 (1986).

39. Frederick D. Tappert & Ding Lee, "A range refraction parabolic equation," *J. Acoust. Soc. Amer.* **76**(6), 1797–1803 (1984).

40. Michael D. Collins & Evan K. Westwood, "A higher-order energy-conserving parabolic equation for range-dependent ocean depth, sound speed, and density," *J. Acoust. Soc. Amer.* **89**(3), 1068–1075 (1991).

41. I. W. Schurman, W. L. Siegmann & M. J. Jacobson, "An energy-conserving parabolic equation incorporating range refraction," *J. Acoust. Soc. Amer.* **89**(1), 134–144 (1991).

42. William L. Siegmann, Gregory A. Kriegsmann & Ding Lee, "A wide-angle three-dimensional parabolic wave equation," *J. Acoust. Soc. Amer.* **78**(2), 659–664 (1985).

43. Michael D. Collins & Richard B. Evans, "A two-way parabolic equation for acoustic backscattering in the ocean," *J. Acoust. Soc. Amer.* **91**(3), 1357–1368 (1992).

44. R. Scarmozzino & R. M. Osgood, Jr., "Comparison of finite-difference and Fourier-transform solutions of the parabolic wave equation with emphasis on integrated-optics applications," *J. Opt. Soc. Amer. A* **8**(5), 724–731 (1991).

45. A. Frankel & R. W. Clayton, "A finite difference simulation of wave propagation in two-dimensional random media," *J. Geophys. Res.* **74**, 2167–2186 (1984).

46. B. J. Uscinski, "Intensity fluctuations in a multiple scattering medium. Solution of the fourth moment equation," *Proc. Roy. Soc. London Ser. A* **380**, 137–169 (1982).

47. C. Macaskill, "An improved solution to the fourth moment equation for intensity fluctuations," *Proc. Roy. Soc. London Ser. A* **386**, 461–474 (1983).

48. A. M. Whitman & M. J. Beran, "Two-scale solution for atmospheric scintillation," *J. Opt. Soc. Amer. A* **2**(12), 2133 (1985).

49. Guang-Yu Wang & Roger Dashen, "Asymptotic Scheme for Waves in Random Media," *Opt. Lett.* **17**(2), 91–93 (1992).

50. Roger Dashen, Guang-Yu Wang, Stanley M. Flatté & Charles Bracher, "Moments of Intensity and Log-Intensity: New Asymptotic Results for Waves in Power-Law Media," *J. Opt. Soc. Amer. A, in press* (1993).

51. L. C. Andrews & R. L. Phillips, "I-K distribution as a universal propagation model of laser beams in atmospheric turbulence," *J. Opt. Soc. Amer. A* **2**(2), 160–163 (1985).

52. J. H. Churnside & R. J. Hill, "Probability density of irradiance scintillations for strong path-integrated refractive turbulence," *J. Opt. Soc. Amer. A* **4**(4), 727–733 (1987).

53. R. Dashen, "Distribution of intensity in a multiply scattering medium," *Opt. Lett.* **9**(4), 110–112 (1984).

54. T. E. Ewart & D. B. Percival, "Forward scattered waves in random media—The probability distribution of intensity," *J. Acoust. Soc. Amer.* **80**(6), 1745–1753 (1986).

55. T. E. Ewart, "A model of the intensity probability distribution for wave propagation in random media," *J. Acoust. Soc. Amer.* **86**(4), 1490–1498 (1989).

56. Stanley M. Flatté & Charles Bracher, "Probability Density Distributions of Intensity for Waves in Three-Dimensional Random Media by Numerical Simulation," *submitted to Phys. Rev. Lett.* (1992).

57. W. A. Coles & R.G. Frehlich, "Simultaneous measurements of angular scattering and intensity scintillation in the atmosphere," *J. Opt. Soc. Amer.* **72**(8), 1042–1048 (1982).

58. A. Consortini, F. Cochetti, J. H. Churnside & R. J. Hill, "Inner-scale effect on intensity variance measured for weak, intermediate, and strong atmospheric scintillation," *J. Opt. Soc. Amer. A* (1993).

59. Reginald J. Hill, "Comparison of scintillation methods for measuring the inner scale of turbulence," *Appl. Opt.* **27**(11), 2187–2193 (1988).

60. Stanley M. Flatté, Guang-Yu Wang & Jan Martin, "Intensity Variance of Optical Waves through Atmospheric Turbulence by Numerical Simulation and Comparison with Experiment," *submitted to J. Opt. Soc. Amer. A* (1993).

61. R. J. Hill, "Spectra of fluctuations in refractivity, temperature, humidity, and the temperature-humidity cospectrum in the inertial and dissipation ranges," *Radio Sci.* **13**(6), 953–961 (1978).

62. G. Daniel Dockery, "Modeling Electromagnetic Wave Propagation in the Troposphere Using the Parabolic Equation," *IEEE Trans. Antennas Propag.* **36**(10), 1464–1470 (1988).

63. James R. Kuttler & G. Daniel Dockery, "Theoretical description of the parabolic approximation/Fourier split-step method of representing electromagnetic propagation in the troposphere," *Radio Sci.* **26**(2), 381–393 (1991).

64. Yean-Woei Kiang & C. H. Liu, "Multiple phase-screen simulation of HF wave propagation in the turbulent stratified ionosphere," *Radio Sci.* **20**(3), 652–668 (1985).

65. J.-F. Wagen & K. C. Yeh, "A numerical study of waves reflected from a turbulent ionosphere," *Radio Sci.* **21**(4), 583–604 (1986).

66. J.-F. Wagen & K. C. Yeh, "Simulation of HF propagation and angle of arrival in a turbulent ionosphere," *Radio Sci.* **24**(2), 196–208 (1989).

67. J.-F. Wagen & K. C. Yeh, "Numerical investigations of two-frequency mutual coherence functions of an ionospheric reflection channel," *Radio Sci.* **24**(2), 209–223 (1989).

68. R. I. Brent, W. L. Siegmann, M. J. Jacobson & G. M. Jacyna, "Anisotropic electromagnetic wave propagation modeling using parabolic approximations," *Radio Sci.* **25**(6), 1121–1136 (1990).

69. J. A. Colosi, S. M. Flatté & C. Bracher, "Numerical Simulation of Pulse Propagation through a Random Field of Internal Waves," *International Meeting on Wave Propagation in Random Media* (Aug., 1992).

70. T. F. Duda, S. M. Flatté, J. A. Colosi, B. D. Cornuelle, J. A. Hildebrand, W. S. Hodgkiss, Jr., P. F. Worcester, B. M. Howe, J. A. Mercer & R. C. Spindel, "Measured wavefront fluctuations in 1000-km pulse propagation in the Pacific Ocean," *J. Acoust. Soc. Amer.* **92**, 939–955 (1992).

71. S. M. Flatté, "Random-Media Effects in Ocean Acoustics: An Introduction," *International Meeting on Wave Propagation in Random Media, Proceedings* (Aug., 1992).

72. A. Frankel & R. W. Clayton, "Finite difference simulations of seismic scattering: Implications for the propagation of short-period seismic waves in the crust and modelas of crustal heterogeneity," *J. Geophys. Res.* **91**, 6465–6489 (1986).

73. S. M. Flatté & R. S. Wu, "Small-scale structure in the lithosphere and asthenosphere deduced from arrival time and amplitude fluctuations at NORSAR," *J. Geophys. Res.* **93**(B6), 6601–6614 (1988).

74. M. D. Fisk & G. D. McCartor, "The Phase Screen Method for Vector Elastic Waves," *J. Geophys. Res.* **96**(B4), 5985–6010 (1991).

75. M. D. Fisk, E. E. Charrette & G. D. McCartor, "A Comparison of Phase Screen and Finite Difference Calculations for Elastic Waves in Random Media," *J. Geophys. Res.* **97**(B9), 409–423 (1992).

Author Index